Chemical Process Technology

Chemical Process Technology

SECOND EDITION

JACOB A. MOULIJN
MICHIEL MAKKEE
ANNELIES E. VAN DIEPEN

Catalysis Engineering, Department of Chemical Engineering, Delft University of Technology,
The Netherlands

A John Wiley & Sons, Ltd., Publication

This edition first published 2013
© 2013 John Wiley & Sons Ltd

Registered office
John Wiley & Sons Ltd, The Atrium, Southern Gate, Chichester, West Sussex, PO19 8SQ, United Kingdom

For details of our global editorial offices, for customer services and for information about how to apply for permission to reuse the copyright material in this book please see our website at www.wiley.com.

Library of Congress Cataloging-in-Publication Data

Moulijn, Jacob A.
  Chemical process technology / Jacob A. Moulijn, Dr. Michiel Makkee, Dr. Annelies van Diepen. – Second edition.
    1 online resource.
  Includes index.
  Description based on print version record and CIP data provided by publisher; resource not viewed.
  ISBN 978-1-118-57073-9 – ISBN 978-1-118-57074-6 – ISBN 978-1-118-57075-3 (ePub) – ISBN 978-1-4443-2024-4 (cloth)
1. Chemical processes.   I. Makkee, Michiel.   II. Diepen, Annelies van.   III. Title.
  TP155.7
  660'.281–dc23

                                                                                                    2013014661

A catalogue record for this book is available from the British Library.

Cloth ISBN: 9781444320244
Paper ISBN: 9781444320251

Typeset in 10/12pt Times by Aptara Inc., New Delhi, India

Contents

Preface

This book is largely the result of courses we have given. Its main purposes are to bring alive the concepts forming the basis of the Chemical Process Industry and to give a solid background for innovative process development. We do not treat Chemical Process Technology starting from unifying disciplines like chemical kinetics, physical transport phenomena, and reactor design. Rather, we discuss actual industrial processes that all present fascinating challenges chemical engineers had to face and deal with during the development of these processes. Often these processes still exhibit open challenges. Our goal is to help students and professionals in developing a vision on chemical processes taking into account the microscale ((bio)chemistry, physics), the mesoscale (reactor, separation units), and the macroscale (the process).

Chemical process technology is not exclusively the domain of chemical engineers; chemists, biologists, and physicists largely contribute to its development. We have attempted to provide students and professionals involved in chemical process technology with a fresh, innovative background, and to stimulate them to think "out of the box" and to be open to cooperation with scientists and engineers from other disciplines. Let us think in "conceptual process designs" and invent and develop novel unit operations and processes!

We have been pragmatic in the clustering of the selected processes. For instance, the production of syngas and processes in which syngas is the feedstock are treated in two sequential chapters. Processes based on homogeneous catalysis using transition metal complexes share similar concepts and are treated in their own chapter. Although in the first part of the book many solid-catalyzed processes are discussed, for the sake of "symmetry" a chapter is also devoted to heterogeneous catalysis. This gave us the opportunity to emphasize the concepts of this crucial topic that can be the inspiration for many new innovations. In practice, a large distance often exists between those chemical reaction engineers active in homogeneous catalysis and those in heterogeneous catalysis. For a scientist these sectors often are worlds apart, one dealing with coordination chemistry and the other with nanomaterials. However, for a chemical reaction engineer the kinetics is similar but the core difference is in the separation. When, by using a smart two-phase system or a membrane, the homogeneous catalyst (or the biocatalyst) is kept in one part of the plant without a separation step, the difference between homogeneous and heterogeneous catalysis vanishes. Thus, the gap between scientists working in these two areas can be bridged by taking into account a higher level of aggregation.

From the wealth of chemical processes a selection had to be made. Knowledge of key processes is essential for the understanding of the culture of the chemical engineering discipline. The first chapters deal with processes related to the conversion of fossil fuels. Examples are the major processes in an oil refinery, the production of light alkenes, and the production of base chemicals from synthesis gas. In this second edition we have added biomass as an alternative to feedstocks based on fossil fuels. Analogously to the oil refinery, the (future) biorefinery is discussed. Biomass conversion processes nicely show the benefit of having insight into the chemistry, being so different from that for processes based on the conversion of the conventional feedstocks. It is fair to state that chemical engineers have been tremendously successful in the bulk chemicals industry. In the past, in some other important sectors, this was not the case, but today also in these fields chemical engineers are becoming more and more important. Major examples are the production of fine chemicals and biotechnological processes. These subjects are treated in separate chapters. Recently, the emphasis in chemical engineering has shifted to Sustainable Technology and, related to that, Process Intensification. In this edition we have added a chapter devoted to this topic.

In all chapters the processes treated are represented by simplified flow schemes. For clarity these generally do not include process control systems, and valves and pumps are only shown when essential for the understanding of the process concept.

This book can be used in different ways. We have written it as a consistent textbook, but in order to give flexibility we have not attempted to avoid repetition in all cases. Dependent on the local profile and the personal taste of the lecturer or reader, a selection can be made, as most chapters are structured in such a way that they can be read separately. At the Delft University of Technology, a set of selected chapters is the basis for a compulsory course for third-year students. Chapter 3 is the basis for an optional course "Petroleum Conversion". In addition, this book forms a basis for a compulsory design project.

It is not trivial how much detail should be incorporated in the text of a book like the present one. In principle, the selected processes are not treated in much detail, except when this is useful to explain concepts. For instance, we decided to treat fluid catalytic cracking (FCC) in some detail because it is such a nice case of process development, where over time catalyst improvements enabled improvements in chemical engineering and vice versa. In addition, its concepts are used in several new processes that at first sight do not have any relationship with FCC. We also decided to treat ammonia synthesis in some detail with respect to reactors, separation, and energy integration. If desired this process can be the start of a course on Process Integration and Design. The production of polyethene was chosen in order to give an example of the tremendously important polymerization industry and this specific polymer was chosen because of the unusual process conditions and the remarkable development in novel processes. The production of fine chemicals and biotechnology are treated in more detail than analogous chapters in order to expose (bio)chemistry students to reactor selection coupled to practice they will be interested in.

To stimulate students in their conceptual thinking a lot of questions appear throughout the text. These questions are of very different levels. Many have as their major function to "keep the students awake", others are meant to force them in sharpening their insights and to show them that inventing new processes is an option, even for processes generally considered to be "mature". In chemical engineering practice often there is not just a single answer to a question. This also applies to most questions in this book. Therefore, we decided not to provide the answers: the journey is the reward, not the answer as such!

Most chapters in the book include a number of "boxes", which are side paths from the main text. They contain case studies that illustrate the concepts discussed. Often they give details that are both "nice to know" and which add a deeper insight. While a box can be an eye opener, readers and lecturers can choose to skip it.

We are grateful for the many comments from chemists and chemical engineers working in the chemical industry. These comments have helped us in shaping the second edition. For instance, we added a section on the production of chlorine to the chapter on inorganic bulk chemicals. This gives insight in electrochemical processing and gives a basis for considering this technology for a chemical conversion process.

We hope that the text will help to give chemical engineers sufficient feeling for chemistry and chemists for chemical engineering. It is needless to say that we would again greatly appreciate any comments from the users of this book.

Jacob A. Moulijn
Michiel Makkee
Annelies E. van Diepen
Delft, The Netherlands, October 2012

1

Introduction

Chemical process technology has had a long, branched road of development. Processes such as distillation, dyeing, and the manufacture of soap, wine, and glass have long been practiced in small-scale units. The development of these processes was based on chance discoveries and empiricism rather than thorough guidelines, theory, and chemical engineering principles. Therefore, it is not surprising that improvements were very slow. This situation persisted until the seventeenth and eighteenth centuries.[1] Only then were mystical interpretations replaced by scientific theories.

It was not until the 1910s and 1920s, when continuous processes became more common, that disciplines such as thermodynamics, material and energy balances, heat transfer, and fluid dynamics, as well as chemical kinetics and catalysis became (and still are) the foundations on which process technology rests. Allied with these are the unit operations including distillation, extraction, and so on. In chemical process technology various disciplines are integrated. These can be divided according to their scale (Table 1.1).

Of course, this scheme is not complete. Other disciplines, such as applied materials science, information science, process control, and cost engineering, also play a role. In addition, safety is such an important aspect that it may evolve as a separate discipline.

In the development stage of a process or product all necessary disciplines are integrated. The role and position of the various disciplines perhaps can be better understood from Figure 1.1, in which they are arranged according to their level of integration. In process development, in principle the x-axis also roughly represents the time progress in the development of a process. The initial phase depends on thermodynamics and other scale-independent principles. As time passes, other disciplines become important, for example, kinetics and catalysis on a micro/nanolevel, reactor technology, unit operations and scale-up on the mesolevel, and process technology, process control, and so on on the macrolevel.

Of course, there should be intense interaction between the various disciplines. To be able to quickly implement new insights and results, these disciplines are preferably applied more or less in parallel rather than in series, as can also be seen from Figure 1.1. Figure 1.2 represents the relationship between the different levels of development in another way. The plant is the macrolevel. When focusing on the chemical

---

[1] This remark is not completely fair. Already in the sixteenth century Agricola published his book "De Re Metallica" containing impressive descriptions of theory and practice of mining and metallurgy, with relevance to chemical engineering.

Chemical Process Technology, Second Edition. Jacob A. Moulijn, Michiel Makkee, and Annelies E. van Diepen.
© 2013 John Wiley & Sons, Ltd. Published 2013 by John Wiley & Sons, Ltd.

Table 1.1 *Chemical process technology disciplines*

Scale	Discipline
Scale independent	Chemistry, biology, physics, mathematics
	Thermodynamics
	Physical transport phenomena
Micro/nanolevel	Kinetics
	Catalysis on a molecular level
	Interface chemistry
	Microbiology
	Particle technology
Mesolevel	Reactor technology
	Unit operations
Macrolevel	Process technology and process development
	Process integration and design
	Process control and operation

conversion, the reactor would be the level of interest. When the interest goes down to the molecules converted, the micro-and nanolevels are reached.

An enlightening way of placing the discipline Chemical Engineering in a broader framework has been put forward by Villermeaux [personal communication], who made a plot of length and time scales

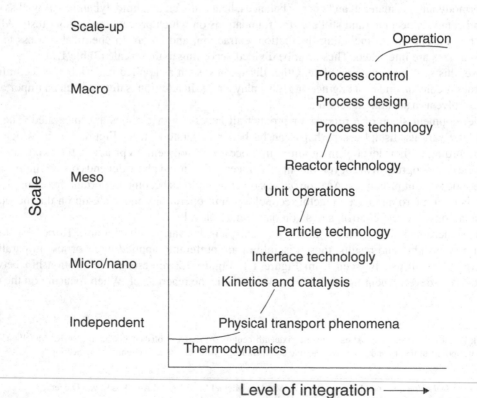

Figure 1.1 *Disciplines in process development organized according to level of integration.*

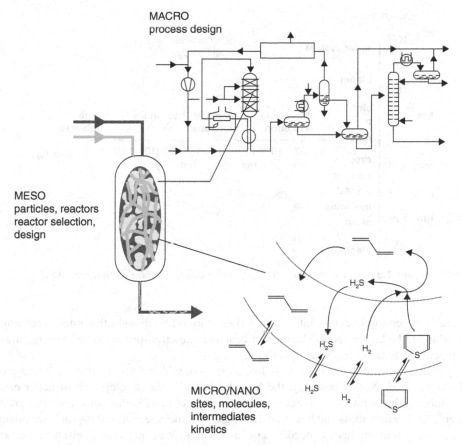

MACRO
process design

MESO
particles, reactors
reactor selection,
design

MICRO/NANO
sites, molecules,
intermediates
kinetics

Figure 1.2 *Relationship between different levels of development.*

(Figure 1.3). From this figure it can be appreciated that chemical engineering is a broad integrating discipline. On the one hand, molecules, having dimensions in the nanometer range and a vibration time on the nanosecond scale, are considered. On the other hand, chemical plants may have a size of half a kilometer, while the life expectancy of a new plant is 10–20 years. Every division runs the danger of oversimplification. For instance, the atmosphere of our planet could be envisaged as a chemical reactor and chemical engineers can contribute to predictions about temperature changes and so on by modeling studies analogous to those concerning "normal" chemical reactors. The dimensions and the life expectancy of our planet are fortunately orders of magnitude larger than those of industrial plants.

Rates of chemical reactions vary over several orders of magnitude. Processes in oil reservoirs might take place on a time scale of a million years, processes in nature are often slow (but not always), and reactions in the Chemical Process Industry usually proceed at a rate that reactor sizes are reasonable, say smaller than 100 m^3. Figure 1.4 indicates the very different productivity of three important classes of processes.

It might seem surprising that despite the very large number of commercially attractive catalytic reactions, the commonly encountered reactivity is within a rather narrow range; reaction rates that are relevant in practice are rarely less than one and seldom more than ten mol m_R^3 s^{-1} for oil refinery processes and processes in the chemical industry. The lower limit is set by economic expectations: the reaction should take place in a reasonable amount of (space) time and in a reasonably sized reactor. What is reasonable is determined by

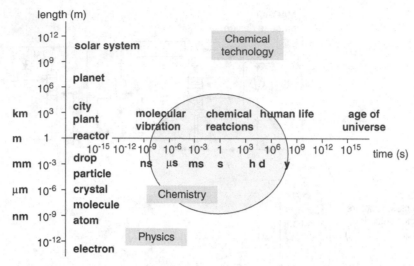

Figure 1.3   *Space and time scales [J. Villermaux, personal communication].*

physical (space) and economic constraints. At first sight it might be thought that rates exceeding the upper limit are something to be happy about. The rates of heat and mass transport become limiting, however, when the intrinsic reaction rate far exceeds the upper limit.

A relatively recent concept is that of Process Intensification, which aims at a drastic decrease of the sizes of chemical plants [2, 3]. Not surprisingly, the first step often is the development of better catalysts, that is, catalysts exhibiting higher activity (reactor volume is reduced) and higher selectivity (separation section reduced in size). As a result, mass and heat transfer might become rate determining and equipment allowing higher heat and mass transfer rates is needed. For instance, a lot of attention is given to the development of compact heat exchangers that allow high heat transfer rates on a volume basis. Novel reactors are also promising in this respect, for instance monolithic reactors and microreactors. A good example of the former is the multiphase monolithic reactor, which allows unusually high rates and selectivities [4].

In the laboratory, transport limitations may lead to under- or overestimation of the local conditions (temperature, concentrations) in the catalyst particle, and hence to an incorrect estimation of the intrinsic reaction rate. When neglected, the practical consequence is an overdesigned, or worse, underdesigned reactor. Transport limitations also may interfere with the selectivity and, as a consequence, upstream and downstream processing units, such as the separation train, may be poorly designed.

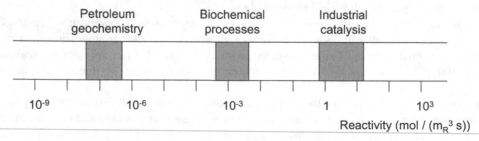

Figure 1.4   *Windows on reality for useful chemical reactivity [1].*

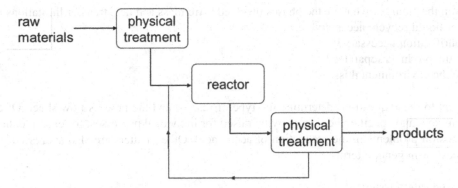

Figure 1.5 *Typical chemical process structure.*

QUESTIONS

What would have been the consequence of much lower and of much higher reactivity of petroleum geochemistry for humanity?

Which factors determine the lower and upper limits of the window for biochemical processes?

Given a production rate of between 10^4 and 10^6 t/a of large volume chemicals (bulk chemicals), estimate required reactor volumes. Do the same for the production of petroleum products (10^6–10^8 t/a).

A → B → C kinetics in which B is the desired product is often encountered. Explain why the particle size of the catalyst influences the observed selectivity to B.

How would you define the "intrinsic" reaction rate?

Every industrial chemical process is designed to produce economically a desired product or range of products from a variety of raw materials (or feeds, feedstocks). Figure 1.5 shows a typical structure of a chemical process.

The feed usually has to be pretreated. It may undergo a number of physical treatment steps, for example, coal has to be pulverized, liquid feedstocks may have to be vaporized, water is removed from benzene by distillation prior to its conversion to ethylbenzene, and so on. Often, impurities in the feed have to be removed by chemical reaction, for example, desulfurization of the naphtha feed to a catalytic reformer, making raw synthesis gas suitable for use in the ammonia converter, and so on. Following the actual chemical conversion, the reaction products need to be separated and purified. Distillation is still the most common separation method, but extraction, crystallization, membrane separation, and so on can also be used.

In this book, emphasis is placed on the reaction section, since the reactor is the heart of any process, but feed pretreatment and product separation will also be given attention. In the discussion of each process, typically the following questions will be answered:

- Which reactions are involved?
- What are the thermodynamics of the reactions, and what operating temperature and pressure should be applied?
- What are the kinetics, and what are the optimal conditions in that sense?
- Is a catalyst used and, if so, is it heterogeneous or homogeneous? Is the catalyst stable? If not, what is the deactivation time scale? What are the consequences for process design? Are conditions feasible where deactivation is minimized? Is regeneration required?

- Apart from the catalyst, what are the phases involved? Are mass and heat transfer limitations important?
- Is a gas or liquid recycle necessary?
- Is feed purification necessary?
- How are the products separated?
- What are the environmental issues?

The answers to these questions determine the type of reactor and the process flow sheet. Of course, the list is not complete and specific questions may be raised for individual processes, for example, how to solve possible corrosion problems in the production of acetic acid. Other matters are also addressed, either for a specific process or in general terms:

- What are the safety issues?
- Can different functions be integrated in one piece of equipment?
- What are the economics (comparison between processes)?
- Can the sustainability of the technology be improved?

References

[1] Weisz, P.B. (1982) The science of the possible. *CHEMTECH*, **12**, 424–425.
[2] Stankiewicz, A. and Drinkenburg, A. (2003) Process intensification, in *Re-Engineering the Chemical Processing Plant* (eds. A. Stankiewicz and J.A. Moulijn), CRC Press, pp. 1–32.
[3] Stankiewicz, A.I. and Moulijn, J.A. (2000) Process intensification: transforming chemical engineering. *Chemical Engineering Progress*, **96**, 22–33.
[4] Kapteijn, F., Heiszwolf, J., Nijhuis, T.A., and Moulijn, J.A. (1999) Monoliths in multiphase processes - aspects and prospects. *CATTECH*, **3**, 24–40.

General Literature

Douglas, J.M. (1988) *Conceptual Design of Chemical Processes*. McGraw-Hill, New York.
Kirk Othmer Encyclopedia of Chemical Technology (1999–2011) Online edition. John Wiley & Sons, Inc., Hoboken. doi: 10.1002/0471238961
Levenspiel, O. (1999) *Chemical Reaction Engineering*, 3rd edn, John Wiley & Sons, Inc. New York.
Seider, W.D., Seader, J.D. and Lewin, D.R. (2008) *Product and Process Design Principles. Synthesis, Analysis, and Evaluation*, 3rd edn, John Wiley & Sons, Inc. Hoboken.
Sinnot, R.K. (2005) *Coulson and Richardson's Chemical Engineering*, vol. **6**, 4th edn, Elsevier Butterworth-Heinemann, Oxford.
Ullman's Encyclopedia of Industrial Chemistry (1999–2011) Online edition, John Wiley & Sons, Inc. Hoboken. doi: 10.1002/14356007
Westerterp, K.R., van Swaaij, W.P.M. and Beenackers, A.A.C.M (1984) *Chemical Reactor Design and Operation*, 2nd edn. John Wiley & Sons, Inc., New York.

2

The Chemical Industry

2.1 A Brief History

Chemical processes like dyeing, leather tanning, and brewing beer were already known in antiquity, but it was not until around 1800 that the modern chemical industry began in the United Kingdom. It was triggered by the industrial revolution, which began in Europe with the mechanization of the textile industry, the development of iron making techniques, and the increased use of refined coal, and rapidly spread all over the world. One of the central characteristics of the chemical industry is that it has experienced a continuous stream of process and product innovations, thereby acquiring a very diverse range of products. Table 2.1 shows a number of selected milestones in the history of the chemical industry.

2.1.1 Inorganic Chemicals

Sulfuric acid and sodium carbonate were among the first industrial chemicals. "Oil of vitriol", as the former was known, was an essential chemical for dyers, bleachers, and alkali[1] manufacturers. In 1746, John Roebuck managed to greatly increase the scale of sulfuric acid manufacture by replacing the relatively expensive and small glass vessels used with larger, less expensive chambers made of lead, the lead chamber process. Sulfuric acid is still the largest volume chemical produced.

QUESTIONS:

> *What is special about lead? What materials are used in modern sulfuric acid plants?*
> *Why are these materials preferred?*

The demand for sodium carbonate in the glass, soap, and textile industries rapidly increased. This led the French Academy of Sciences, at the end of the eighteenth century, to establish a contest for the invention of a method for manufacturing inexpensive sodium carbonate (Na_2CO_3). It took Nicholas Leblanc five years to stumble upon the idea of reacting sodium chloride with sulfuric acid and then converting the sodium sulfate

---

[1] Alkali was an all-encompassing term for all basic compounds such as sodium and potassium carbonates and hydroxides (Na_2CO_3, K_2CO_3, NaOH, KOH).

Chemical Process Technology, Second Edition. Jacob A. Moulijn, Michiel Makkee, and Annelies E. van Diepen.
© 2013 John Wiley & Sons, Ltd. Published 2013 by John Wiley & Sons, Ltd.

Table 2.1   Selected events in the history of the chemical industry.

Year	Event
1746	John Roebuck starts producing moderately concentrated sulfuric acid in the lead chamber process on an industrial scale.
1789	Nicholas LeBlanc develops a process for converting sodium chloride into sodium carbonate. In many ways, this process began the modern chemical industry. From its adoption in 1810, it was continually improved over the next 80 years, until it was replaced by the Solvay process.
1831	Peregrine Phillips patents the contact process for manufacturing concentrated sulfuric acid, the first mention of heterogeneous catalysis for a large-scale process. For various reasons, the process only became a success at the end of the nineteenth century.
1850	The first oil refinery, consisting of a one-barrel still, is built in Pittsburgh, Pennsylvania, USA, by Samuel Kier.
1856	Seeking to make quinine, William Henry Perkin, at the age of 18, synthesizes the first synthetic aniline dye, mauveine, from coal tar. This discovery is the foundation of the dye synthesis industry, one of the earliest successful chemical industries.
1863	Ernest Solvay perfects his method for producing sodium carbonate. This process started to replace Leblanc's process in 1873.
1864	The British government passes the "Alkali Works Act" in an effort to control environmental emissions; the first example of environmental regulation.
1874	Henry Deacon develops the Deacon process for converting hydrochloric acid into chlorine.
~1900	With the coming of large-scale electrical power generation, the chlor-alkali industry is born.
1905	Fritz Haber and Carl Bosch develop the Haber process (sometimes referred to as the Haber–Bosch process) for producing ammonia from its elements, a milestone in industrial chemistry. The process was first commercialized in 1910.
1907	Wilhelm Normann introduces the hydrogenation of fats (fat hardening).
1909	Leo Baekeland patents Bakelite, the first commercially important plastic, which was commercialized shortly after.
1920	Standard Oil Company begins large-scale industrial production of isopropanol from oil, the first large-scale process using oil as feedstock.
1923	Matthias Pier of BASF develops a high-pressure process to produce methanol. This marks the emergence of the synthesis of large-volume organic chemicals.
	Franz Fischer and Hans Tropsch develop the Fischer–Tropsch process, a method for producing synthetic liquid fuels from coal gas. The process was used widely by Germany during World War II for the production of aviation fuel.
1926	Fritz Winkler introduces a process for commercial fluidized-bed coal gasification at a BASF plant in Leuna, Germany.
1930	First commercial steam reforming plant is constructed by the Standard Oil Company.
	First commercial manufacture of polystyrene by IG Farben.
	Wallace Carrothers discovers nylon, the most famous synthetic fiber. Production by DuPont began in 1938; output was immediately diverted to parachutes for the duration of World War II.
1931	Development of ethene epoxidation process for the production of ethene oxide.
1933	Polyethene[2] discovered by accident at ICI by applying extremely high pressure to a mixture of ethene and benzaldehyde.
1934	First American car tire produced from a synthetic rubber, neoprene.
1936	Eugène Houdry develops a method of industrial scale catalytic cracking of oil fractions, leading to the development of the first modern oil refinery. The most important modification was the introduction of Fluid Catalytic Cracking in 1941.

[2] The IUPAC naming terminology has been used but some traditional names for polymers are retained in common usage, for example, polyethylene (polyethene) and polypropylene (polypropene).

Table 2.1   Selected events in the history of the chemical industry (Continued).

Year	Event
1938	Commercialization of the alkylation process for the production of high-octane alkylate.
	Otto Roelen discovers the hydroformylation reaction for the formation of aldehydes from alkenes.
1939	Start of large-scale low-density polyethene (LDPE) production at ICI.
	Start of large-scale poly(vinyl chloride) (PVC) production in Germany and USA.
1940	Standard Oil Company develops catalytic reforming to produce higher octane gasoline.
1953	Karl Ziegler introduces the Ziegler catalyst for the production of high-density polyethene (HDPE).
1954	Introduction of chromium-based catalysts for the production of HDPE by Phillips petroleum. This process became the world's largest source of polyethene in 1956.
1955	Start of large-scale production of poly(ethene terephthalate) (PET).
1957	First commercial production of isotactic polypropene, based on research by Giulio Natta. Made possible by the development of the Ziegler catalyst.
1960	Commercialization of two industrially important processes: ethene to acetaldehyde (Wacker), and acrylonitrile production (SOHIO).
1963	BASF develops the first commercial methanol carbonylation process for the production of acetic acid based on a homogeneous cobalt catalyst. In 1970, Monsanto builds the first methanol-carbonylation plant based on a homogeneous rhodium catalyst.
1964	Many new and improved catalysts and processes are developed, for example, a zeolite catalyst for catalytic cracking (Mobil Oil) and the metathesis of alkenes.
1966	First low-pressure methanol synthesis commercialized by ICI.
1970s	Continued invention of new and improved processes and catalysts.
	Birth of environmental catalysis. Development of a catalytic converter for Otto engine exhaust gas (1974).
	Energy crises (1973 and 1979).
	Start of large-scale bioethanol production in USA and Brazil.
1980s	Several new catalytic processes are introduced. One of the most important is the selective catalytic reduction (SCR) for controlling NO_x emissions. A revolutionary development in polymerization is the development of a process for the production of linear low-density polyethene (LLDPE) by Union Carbide and Shell.
	Biotechnology emerges.
1990s	Sumio Iijima discovers a type of cylindrical fullerene known as a carbon nanotube (1991). This material is an important component in the field of nanotechnology.
	Improved NO_x abatement in exhaust gases by NO_x trap (Toyota).
2000s	Introduction of soot abatement for diesel engines by Peugeot (2000).
	Ultra-efficient production of bulk chemicals.
	Green chemistry and sustainability, including biomass conversion, become hot topics.
	New paradigms and concepts, Product Technology, Process Intensification

formed into sodium carbonate. Although Leblanc never received his prize, Leblanc's process is commonly associated with the birth of the modern chemical industry.

QUESTIONS:

> *What reactions take place in Leblanc's process?*
> *What were the emissions controlled by the United Kingdom's "Alkali Works Act"?*

The Haber process for the production of ammonia, which was first commercialized in 1910, can be considered the most important chemical process of all times. It required a major technological breakthrough – that of being able to carry out a chemical reaction at very high pressure and temperature. It also required a systematic investigation to develop efficient catalysts for the process, a scientific revolution. Originally developed to provide Europe with a fertilizer, most of the ammonia at that time ended up in nitrogen-based explosives in World War I.

2.1.2 Organic Chemicals

In 1856, the English chemist William Henry Perkin was the first chemist to synthesize an organic chemical for commercial use, the aniline dye mauveine (Box 2.1). Until then dyes had been obtained from natural sources. The development of synthetic dyes and, subsequently, of other synthetic organic chemicals in the following decades, mainly by German chemists, sparked a demand for aromatics, which were mostly obtained from coal tar, a waste product of the production of town gas from coal.

QUESTION:

> *Many chemical concepts have evolved from the coal industry. Companies producing coal-based chemicals often were major producers of fertilizers. Explain.*

Box 2.1 Mauveine

The serendipitous discovery of mauveine by William Henry Perkin was the birth of the modern organic chemical industry. In 1856, Perkin, at the age of 18, was trying to make the antimalarial drug quinine, as a challenge from his professor. At that time, only the molecular formula of quinine was known and there was no idea of how atoms are connected. By simply balancing the masses of an equation, Perkin thought that two molecules of N-allyl toluidine upon oxidation with three oxygen atoms (from potassium dichromate) would give one molecule of quinine and one molecule of water (Scheme B2.1.1).

$$2\ C_{10}H_{13}N + 3\ O \longrightarrow C_{20}H_{24}N_2O_2 + H_2O$$

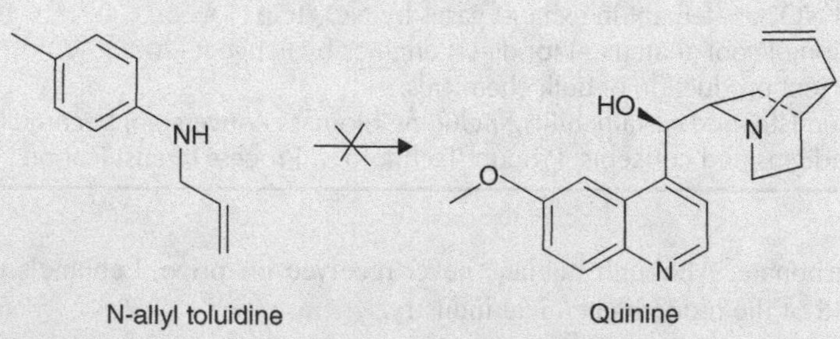

N-allyl toluidine   Quinine

Scheme B2.1.1   *Perkin's first attempt at the synthesis of quinine.*

The experiment failed to produce quinine, but produced a solid brown mess instead, which is not at all surprising knowing the structures of N-allyl toluidine and quinine. In fact, the complete synthesis of quinine was only achieved 88 years later [1].

To simplify the experiment, Perkin oxidized aniline sulfate, which was contaminated with toluidines, with potassium dichromate, again producing a very unpromising mixture, this time a black sludge. While trying to clean out his flask, Perkin discovered that some component of the mixture dissolved in ethanol to yield a striking purple solution, which deposited purple crystals upon cooling. This was the first ever synthetic dye, aniline purple, better known as mauveine. Perkin successfully commercialized the dye, replacing the natural dye, Tyrian purple, which was enormously expensive.

The actual molecular structure of mauveine was only elucidated in 1994 by Meth-Cohn and Smith [2]. They found that the major constituents are mauveines A and B (Figure B2.1.1). In 2008, Sousa *et al.* [3] found that mauveine contains at least ten other components in smaller amounts.

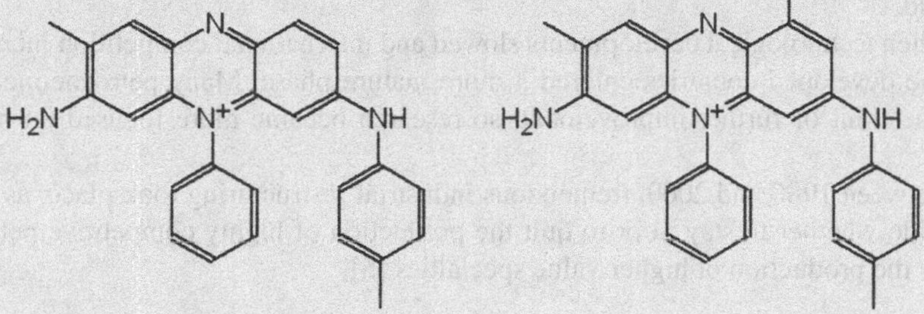

Mauveine A ($C_{26}H_{23}N_4$)   Mauveine B ($C_{27}H_{25}N_4$)

Figure B2.1.1 *Mauveine.*

A modern laboratory procedure for the synthesis of mauveine consists of dissolving a 1 : 2 : 1 mixture of aniline, *o*-toluidine, and *p*-toluidine (Figure B2.1.2) in sulfuric acid and water followed by addition of potassium dichromate [4].

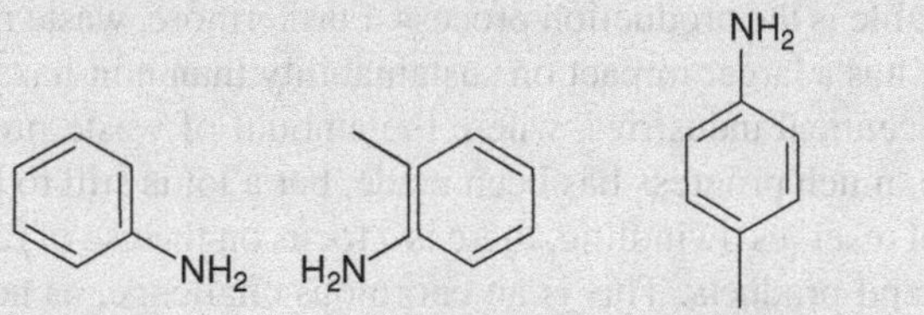

Aniline   *o*-Toluidine   *p*-Toluidine

Figure B2.1.2 *Reactants in the synthesis of mauveine.*

2.1.3 The Oil Era

The use of crude oil (and associated natural gas) as a raw material for the manufacture of organic chemicals started in the 1930s in the United States, where oil-derived hydrocarbons were recognized as superior feedstocks for the chemical industry relatively early. This so-called petrochemical industry got a further boost in World War II, when North American companies built plants for the production of aromatics for high-octane aviation fuel. In Europe, the shift from coal to crude oil only took place at the end of World War II and in Japan the shift took place in the 1950s.

The invention of the automobile shifted the demand to gasoline and diesel, which remain the primary refinery products today. Catalytic cracking of oil fractions, a process developed by Houdry in 1936, resulted in much higher gasoline yields. This process is one of the most important chemical processes ever developed.

The use of crude oil as a substitute for coal provided the chemical industry with an abundant, cheap raw material that was easy to transport. The period from the 1930s to the 1960s was one of many innovations in the chemical industry, with an appreciable number of scientific breakthroughs. Many new processes were developed, often based on new developments in catalysis. Production units multiplied in the United States, Europe, and Japan. Synthetic polymers were the major growth area during this period. With an increase in understanding of the structure of these materials, there was a rapid development in polymer technology.

During the late 1960s and early 1970s, the world started to become aware of the environmental impact of the chemical industry and the discipline of environmental catalysis was born. The most notable example is the catalytic cleaning of car exhaust gases. The exhaust gas catalyst system is now the most common catalytic reactor in the world.

In the 1980s, when technological developments slowed and international competition increased, the chemical industry in the developed countries entered a more mature phase. Many petrochemical processes had started to reach the limit of further improvement, so research became more focused on high-value-added chemicals.

In the period between 1980 and 2000, tremendous industrial restructuring took place, as traditional companies had to decide whether to stay in or to quit the production of highly competitive petrochemicals and whether to shift to the production of higher value specialties [5].

2.1.4   The Age of Sustainability

During the first decade of the twenty-first century, the concept of sustainability has become a major trend in the chemical industry. It has become clear that for chemical companies not only economic aspects (investment costs, raw material costs, etc.) are important, but also environmental issues (greenhouse gas emissions, wastes, etc.) and social matters (number of employees, number of work accidents, R&D expenses, etc.).

A company needs to use economic resources at least as efficiently as its peers. However, this is not sufficient. The environmental impact of chemical production processes needs to be as small as possible; the more waste is produced the less sustainable is the production process. Furthermore, waste may be more or less hazardous. Evidently, hazardous waste has a larger impact on sustainability than non-hazardous waste. Especially in the fine chemicals and pharmaceutical industries, where the amount of waste produced per amount of product has traditionally been large, much progress has been made, but a lot is still to be gained in this respect.

Currently, with fossil fuel reserves dwindling, there is a focus on the use of renewable feedstocks (biomass) and recycling of materials and products. This is an enormous challenge, as new synthesis routes have to be developed that must be competitive from an economic viewpoint as well. These new routes must be more sustainable than existing routes and production processes. Incentives are the reduction of the amount of (hazardous) waste produced and better energy efficiency, which leads to reduced CO_2 emissions.

QUESTIONS:

Greenhouse gases are generally considered to be an environmental burden. How can their emissions be minimized? What is your opinion on the recent increase of the use of coal in this respect? Explain. What is the position of nuclear energy?

How would you assess the number of employees of a company? From a social perspective, is a large number positive? Is R&D primarily a cost item in view of sustainability or not?

What is the relation between large-scale power generation and the chlor-alkali industry? What does the acronym BASF stand for?

2.2   Structure of the Chemical Industry

In the chemical industry, raw materials are converted into products for other industries and consumers. The range of products is enormous, but the vast majority of these chemicals, about 85%, are produced from a very limited number of simple chemicals called *base chemicals*, which in turn are produced from only about ten *raw materials*. These raw materials can be divided into inorganic and organic materials. Inorganic raw materials include air, water, and minerals. Oil, coal, natural gas – together termed the fossil fuels – and biomass belong to the class of organic raw materials. Conversion of base chemicals can produce about 300 different *intermediates*, which are still relatively simple molecules. Both the base chemicals and the intermediates can be classified as *bulk chemicals*. A wide variety of *advanced chemicals, industrial specialty chemicals,* and *consumer products* can be obtained by further reaction steps. Figure 2.1 shows this tree-shaped structure of the chemical industry.

Figure 2.2 presents a survey of the petrochemical industry. Crude oil and natural gas are the primary raw materials for the production of most bulk organic chemicals. The first stage in the petrochemical industry is conversion of these raw materials into base chemicals:

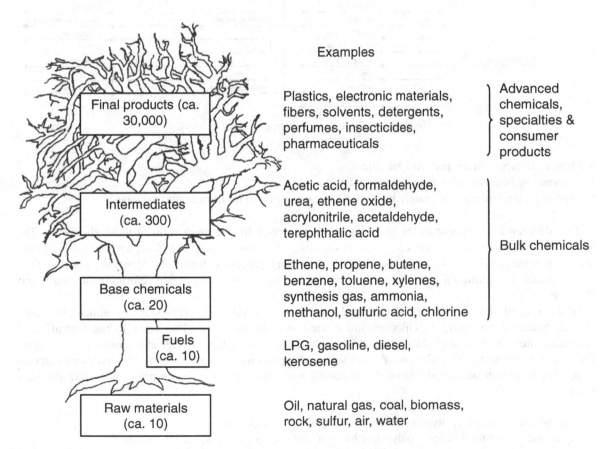

Figure 2.1   *Structure of the chemical industry.*

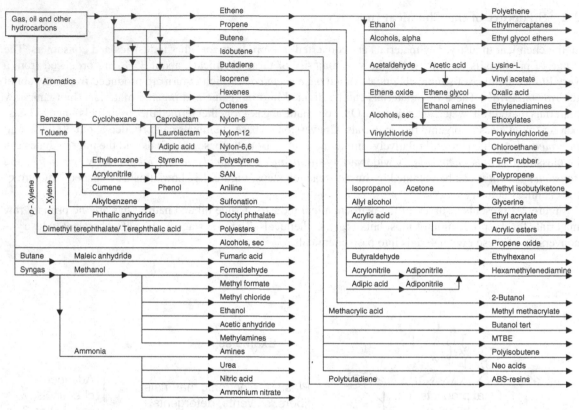

Figure 2.2   *Survey of the petrochemical industry.*

- lower alkenes: ethene, propene, butadiene;
- aromatics: benzene, toluene, xylenes ("BTX");
- synthesis gas (mixture of mainly hydrogen and carbon monoxide), ammonia, methanol.

This division also represents the most important processes for the production of these chemicals. The lower alkenes are mainly produced by steam cracking of ethane or naphtha (Chapter 4), while aromatics are predominantly produced in the catalytic reforming process (Chapter 3). Synthesis gas, which is the feedstock for ammonia and methanol, is predominantly produced by steam reforming of natural gas (Chapter 5).

In the second stage, a variety of chemical operations is conducted, often with the aim to introduce various hetero-atoms (oxygen, chlorine, sulfur, etc.) into the molecule. This leads to the formation of chemical intermediates, such as acetic acid, formaldehyde, and ethene oxide, and monomers like acrylonitrile, terephthalic acid, and so on. A final series of operations – often consisting of a number of steps – is needed to obtain advanced chemicals, industrial specialties, and consumer products. These products include:

- plastics: for example, poly(vinylchloride) (PVC), polyacrylonitrile;
- synthetic fibers: for example, polyesters like poly(ethene terephthalate) (PET), nylon-6;

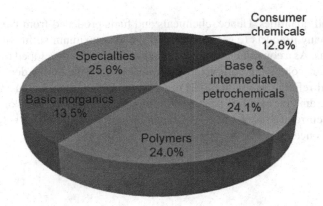

Figure 2.3 *EU chemical market excluding pharmaceuticals (2010). Total sales: € 491 billion (US$ 685 billion) [6].*

- elastomers: for example, polybutadiene;
- paints and coatings;
- herbicides, insecticides, and fungicides (agrochemicals);
- fertilizers: for example, ammonium nitrate;
- vitamins;
- flavors and fragrances;
- soaps, detergents, and cosmetics (consumer chemicals);
- pharmaceuticals.

In general, with each stage, the complexity of the molecules becomes larger and the added value of the chemicals becomes higher. Of course, this three-stage classification shows exceptions. For instance, ethene, a base chemical formed from hydrocarbons present in oil, is a monomer for the direct formation of plastics like polyethene, so the second stage does not occur. Another example is acetic acid. Depending on its use, it can be classified as an intermediate or as a consumer product.

The output of the chemical industry can be conveniently divided into five broad sectors, for which Figure 2.3 shows the 2010 European Union (EU) sales figures.

Petrochemicals have a relatively low added value but, due to their large production volumes, they still hold a large part of the market. This is not surprising since they are feedstocks for the other sectors except for most of the inorganic chemicals. However, even some inorganic chemicals are produced from petrochemicals: ammonia, which is a large-volume inorganic compound, is produced mainly from oil or natural gas. An interesting case is sulfur (Box 2.2).

Box 2.2 Sulfur

The consumption of sulfur falls into two main sectors; approximately 65% is used to produce fertilizers via sulfuric acid and approximately 35% is used in the chemical industry. Sulfur compounds are present in fossil fuels in small quantities. Hence, chemicals and fuels produced from these raw materials also contain sulfur. For many products (e.g., gasoline and diesel) maximum sulfur limits exist, mainly for environmental reasons. As a consequence, many processes have been developed to meet these standards by removing the sulfur-containing compounds. The majority of the sulfur demand is now produced as a by-product of oil refining and natural gas processing. In fact, with increasingly stricter demands on sulfur content, the amount of sulfur produced in refineries already exceeds or soon will exceed the demand! Therefore, currently several new uses, such as sulfur concrete and sulfur-enhanced asphalt modifier, are being pioneered.

2.3   Raw Materials and Energy

2.3.1   Fossil Fuel Consumption and Reserves

From the previous section, it is clear that feedstocks for the production of chemicals and energy are closely related. Indeed, the main raw materials for the chemical industry are the fossil fuels. These are also the most important sources of energy, as is clear from Figure 2.4. Until 1973 energy consumption was increasing

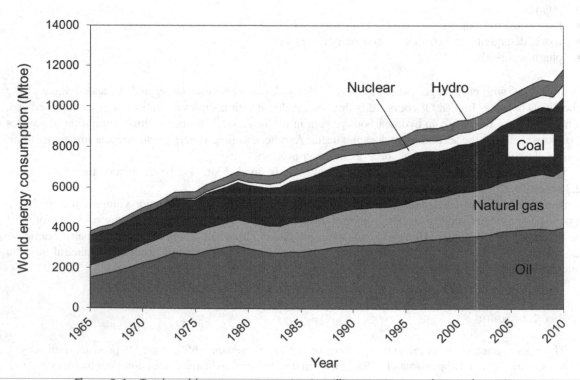

Figure 2.4   *Total world energy consumption in million metric tons oil equivalent [7].*

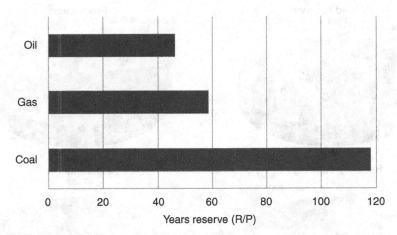

Figure 2.5 *Fossil fuel reserves, 2010; R/P = reserves-to-production ratio. If the total proved reserves at the end of a year are divided by the production in the same year, the result is the length of time that the reserves would last if production were to continue at that rate. Shale gas reserves are not included in the natural gas reserves [7].*

exponentially, rising faster than the world population. Then, until the mid-1980s, consumption was fairly stable. Since then it has been growing, but at a lower rate than in the years prior to 1973, despite continuing economic growth. The main reason for this is the more efficient use of energy. In the first decade of the twenty-first century, however, the average annual energy consumption is increasing faster again. The use of other energy sources, such as nuclear energy and hydroelectricity, has also increased but the overall picture has not changed: fossil fuels are still the main energy carriers.

QUESTIONS:

Why are fossil fuels still the major source of energy?
What are advantages and disadvantages of fossil fuels compared to other energy sources?
What are advantages and disadvantages of the individual fossil fuels?

The major source of energy is still oil, currently accounting for 34% of total energy consumption. The share of natural gas has increased from about 16% in 1965 to 24% in 2010. Coal accounts for 30% of total energy consumption. The reserves of fossil fuels do not match the current consumption pattern, as illustrated in Figure 2.5.

The coal reserves are by far the largest. Although far less abundant, natural gas reserves exceed the oil reserves. In fact, the proved natural gas reserves have more than doubled since 1980 and despite high rates of increase in natural gas consumption, most regional reserves-to-production ratios have remained high. About 65% of the natural gas reserves are located in a relatively small number (310) of giant gas fields ($>10^{11}$ m^3), while the remainder is present in a large number ($\sim$25 000) of smaller fields [8].

The natural gas reserves used for the calculation of the R/P ratio in Figure 2.5 do not include so-called shale gas (Box 2.3). Current estimates of recoverable shale gas reserves at least equal those of estimates for conventional natural gas [7, 9].

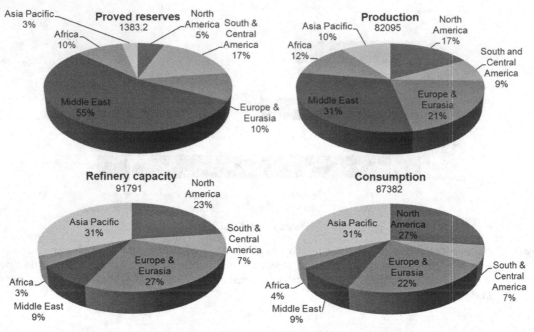

Figure 2.6   *Global distribution of proved crude oil reserves, production, refinery capacity and consumption. Numbers are in thousand million barrels at the end of 2010 for the proved reserves and in thousand barrels daily in 2010 for the others (1 barrel = 0.159 m^3 ≈ 0.136 metric ton) [7].*

QUESTIONS:

In 1973, at a share of 48%, the contribution of oil to the world energy consumption was at its peak. What happened in 1973? From 2000 on, the share of coal has been increasing again, after a steady decline from a share of 39% in 1965 (similar to that of oil) to 25% in 1999. What would be the explanation for this?

Find out where the mega giant (>10^{13} m^3) and super giant (>10^{12} m^3) gas fields are located.

Figure 2.6 shows the reserves, production, refinery capacity, and consumption of crude oil. Crude oil reserves are distributed over the world very unevenly: more than half the reserves are located in the Middle East, the reserves in the rest of the world are much smaller. The production of crude oil shows a different picture. Although the reserves are limited, the production in the Western world is relatively high.

Refinery capacity shows a similar picture to the crude oil consumption, which is very different from that of the reserves.

Box 2.3   Shale Gas

Conventional natural gas formations are generally found in some type of porous and permeable rock, such as sandstone, and can be extracted through a simple vertical well.

In contrast, shale gas is natural gas that is trapped in non-porous rock (shale) formations that tightly bind the gas, so its extraction requires a different approach. Recently, improved (horizontal) drilling and hydraulic fracturing technologies have enormously increased the exploration and production capability of shale gas. Hydraulic fracturing, or "fracking", uses high-pressure water mixed with sand and chemicals to break gas-rich shale rocks apart and extract the gas.

Extracting the reserves of unconventional natural gas (also including gas from coal seams) was long thought to be uneconomical, but rising energy prices have made exploration worthwhile. Shale gas has thus become an increasingly important source of natural gas in the United States over the past decade, and interest has spread to potential gas shales in the rest of the world. Shale gas as a percentage of the total North American gas production has increased from virtually nothing in 2000 to over 25% in 2012!

Recently, however, criticism has been growing louder too. There are increasing concerns about air and groundwater pollution and the risk of earthquakes. Another fear is that the exploitation of shale gas as a relatively clean fuel may slow the development of renewable energy sources.

In the near future, energy consumption patterns are likely to show major changes. Conversion of coal into gas (coal gasification) and production of liquid fuels from both coal gas and natural gas is technically possible and has been demonstrated on a large scale. Undoubtedly, these processes will gain importance. Another reason for natural gas becoming more important is that it is a convenient source of energy and it is well-distributed over the regions where the consumers live. In fact, the trend away from oil and towards natural gas and coal is already apparent.

Renewables will also play a more important role in the future; renewables are based on biomass, but hydroelectricity, wind and solar energy also can be considered as part of this class. In the more distant future, solar energy might well become a major source for electricity production. Although the contribution of renewables other than hydroelectricity to the energy pool is still minor (<1%), its percentage growth is appreciable. For example, the production of biofuels, that is bio-ethanol and biodiesel, has increased at an average of about 20% per year during the first decade of this century (from circa 9000 metric tons oil equivalent (toe) in 2000 to circa 60 000 toe in 2010).

Only a small fraction of the crude oil demand is used as a raw material for the petrochemical industry, while most of the remainder is used for fuel production (Figure 2.7). This explains why, despite the relative scarcity of oil, the driving force for finding alternative raw materials for the chemical industry is smaller than is often believed. Nevertheless, as in the production of fuels, a trend can be observed towards the production of chemicals from natural gas and coal, mostly through synthesis gas (CO/H_2), the so-called C_1-chemistry. In addition, biomass conversion is increasingly considered for the production of chemicals.

2.3.2   Biomass as an Alternative for Fossil Fuels

From the point of view of desired sustainability there is an increasing global urgency to reduce the dependence on oil and other fossil fuels. Biomass holds the promise of allowing a really sustainable development for the production of energy, fuels, and chemicals. The first question to be answered is: "Is there sufficient biomass available?"

Biomass is the product of photosynthesis. The energy flow of the sun reaching the earth is roughly 1 kW/m². This number translates into the huge amount of four million EJ/a (exajoules/a). In the order of 0.1% of these photons is captured in photosynthesis, resulting in an energy production of 4000 EJ/a. Is this a high number? In Table 2.2, this number is compared with the present energy consumption (440 kJ/a) and the consumption data for fossil fuels and the present usage of biomass.

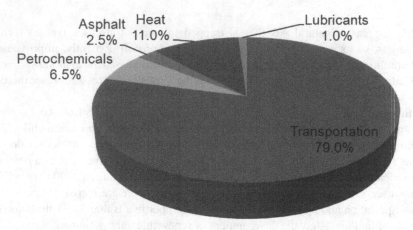

Figure 2.7   *Petrochemical share of total world oil demand.*

QUESTIONS:

Find the heats of combustion for different fuels. From the value for biomass calculate the production in kg/m². Is this a reasonable number?

Thus, the present worldwide energy consumption corresponds to about 10% of the biomass produced. In the year 2050 this number is expected to increase to 40%. At first sight these numbers seem slightly alarming. Several research groups have evaluated the potential for an increase in biomass production [10]. These studies show that waste streams are sizeable but not sufficient and that the agricultural methodology for the production of biomass has to be critically evaluated. Present agriculture cannot provide the amounts of biomass needed without affecting the food and feed industries. Surplus land, which is not used for forestry production, nature reserves, or animal grazing, has to be involved in biomass production; the good news is that this land is available in the world. In the studies, algae were not taken into account. Undoubtedly, these can provide significant additional amounts of energy.

Table 2.2   *Comparison of energy production by photosynthesis with current and future consumption. Source: IEA World Energy Outlook 2008.*

	Energy flow			
	TW	EJ/a	Remarks	Years reserve (R/P)
Radiation reaching earth	125 000	4 000 000	1 kW/m²	sustainable
Natural photosynthesis	130	4000		sustainable
Energy consumption (2005)	14	440		
Biomass	1.5	50	mainly cooking, heating	sustainable
Coal	4.4	140		190
Oil	5.2	170		42
Natural gas	3.6	110		63
Energy consumption (2050 estimate)	49	1550	3.5 times that of 2005	

Biomass is not only a sustainable basis for energy production. Similar to fossil fuels it can be used as feedstock for the chemical industry. Assuming that the global chemical industry is responsible for not more than 10% of energy consumption (which is a reasonable assumption, compare Figure 2.7), it is clear that there is plenty of biomass available to serve as a feedstock for this industry.

The source crops for biomass production can be grown renewably and in most climates around the world, which makes biomass a good alternative for the production of fuels and chemicals.

2.3.3 Energy and the Chemical Industry

The chemical industry uses a lot of energy. The amount of hydrocarbons used to provide the required energy is of the same order as the quantity of hydrocarbons used as feedstock. Fuel is used in direct heaters and furnaces for heating process streams, and for the generation of steam and electricity, the most important utilities.

2.3.3.1 *Fuels for Direct Heaters and Furnaces*

The fuel used in process furnaces is often the same as the feedstock used for the process. For instance, in steam reforming of natural gas (Chapter 5), natural gas is used both as feedstock and as fuel in the reformer furnace. Fuel oil, a product of crude oil distillation, which is less valuable than crude oil itself, is often used in refineries; for example, to preheat the feed to the crude oil fractionator.

2.3.3.2 *Steam*

The steam system is the most important utility system in most chemical plants. Steam has various applications; for example, for heating process streams, as a reaction medium, and as a distillation aid. Steam is generated and used saturated, wet, or superheated. Saturated steam contains no moisture or superheat, wet steam contains moisture, and superheated steam contains no moisture and is above its saturation temperature. Steam is usually generated in water-tube boilers (Figure 2.8) using the most economic fuel available.

The boiler consists of a steam drum and a so-called mud drum located at a lower level. The steam drum stores the steam generated in the riser tubes and acts as a phase separator for the steam/water mixture. The saturated steam that is drawn off may re-enter the furnace for superheating. Circulation of the water/steam mixture usually takes place by natural convection. Occasionally, a pump is used. The saturated water at the bottom of the steam drum flows down through the downcomer tubes into the mud drum, which collects the solid material that precipitates out of the boiler feed water due to the high pressure and temperature conditions of the boiler.

The flue gases discharged during the combustion of the fuel serve to heat the boiler feed water to produce steam. In steam reforming of natural gas, the natural gas is usually also used for heating, and steam is generated in a so-called waste heat boiler by heat exchange with both the furnace off gases and the synthesis gas produced.

Steam is generally used at three different pressure levels. Table 2.3 indicates the pressure levels and their corresponding saturation temperatures. The exact levels depend on the particular plant. In the flow sheets presented in this book, the steam pressures are generally not shown. When an indication is given (HP, MP or LP), the values shown in Table 2.3 can be assumed to be valid. The thermodynamic properties of saturated and superheated steam have been compiled in so-called steam tables, which can be found in numerous texts [11]. Figure 2.9 shows the saturation pressure as a function of temperature.

Table 2.3 Typical steam pressure levels.

	Operating conditions		
	Pressure (bar)	Temperature (K)	Saturation temperature (K)
HP (high pressure) steam	40	683	523
MP (medium pressure) steam	10	493	453
LP (low pressure) steam	3	463	407

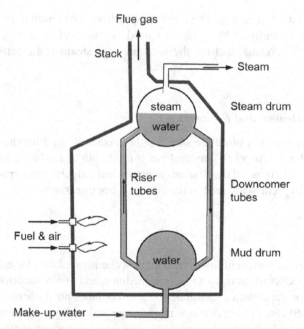

Figure 2.8 Water-tube boiler based on natural convection. Only one riser and one downcomer tube of the multiple tubes are shown.

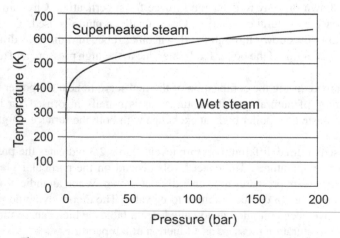

Figure 2.9 Steam saturation pressure versus temperature.

2.3.3.3   *Electricity*

Electricity can either be generated on site in steam turbines or be purchased from the local supply company. On large sites, reduction of energy costs is possible if the required electrical power is generated on site in steam turbines and the exhaust steam from the turbines used for process heating. It is often economical to drive large compressors, which demand much power, with steam turbines. The steam produced can be used for local process heating. A recent development is the building of so-called cogeneration plants, in which heat and electricity are generated simultaneously, usually as joint ventures between industry and public organizations. Examples are central utility boilers that provide steam for electricity generation and supply to a local heating system, and combined cycle power plants that combine coal gasification with electricity generation in gas turbines and steam turbines (Chapter 5).

2.3.4   Composition of Fossil Fuels and Biomass

All three fossil fuels (oil, coal, and natural gas) have in common that they mainly consist of carbon and hydrogen, while also small amounts of hetero-atoms like nitrogen, oxygen, sulfur and metals are present. However, the ratio of these elements is very different, which manifests itself in the very different molecular composition (size, type, etc.) and physical properties. In addition to carbon and hydrogen, biomass contains a relatively large amount of oxygen. The C/H ratio is a characteristic feature of hydrocarbons. Figure 2.10 shows the C/H ratio for the major fossil fuels, biomass, and some other hydrocarbons. Clearly, the relative amount of carbon in coal is much larger than in crude oil. Methane (CH_4) obviously has the lowest C/H ratio of all hydrocarbons. The C/H ratio of natural gas is very similar to that of methane, because methane is the major constituent of natural gas.

QUESTION:

> CO_2 *is the most important greenhouse gas. Do the various hydrocarbons differ in their contribution to the greenhouse effect?*

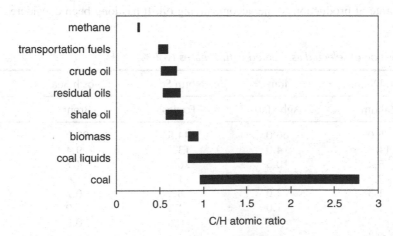

Figure 2.10   *C/H atomic ratios of hydrocarbon sources and some products.*

Table 2.4   *Composition of selected non-associated natural gases (vol.%) [12, 13].*

Area	Algeria	France	Holland	New Zealand	North Sea	N. Mexico	Texas	Texas	Canada
Field	Hasi-R'Mel	Lacq	Gron.	Kapuni	West Sole	Rio Arriba	Terrell	Cliffside	Olds
CH_4	83.5	69.3	81.3	46.2	94.4	96.9	45.7	65.8	52.4
C_2H_6	7.0	3.1	2.9	5.2	3.1	1.3	0.2	3.8	0.4
C_3H_8	2.0	1.1	0.4	2.0	0.5	0.2	—	1.7	0.1
C_4H_{10}	0.8	0.6	0.1	0.6	0.2	0.1	—	0.8	0.2
$C_5{}^+$	0.4	0.7	0.1	0.1	0.2	—	—	0.5	0.4
N_2	6.1	0.4	14.3	1.0	1.1	0.7	0.2	26.4	2.5
CO_2	0.2	9.6	0.9	44.9	0.5	0.8	53.9	—	8.2
H_2S	—	15.2	trace	—	—	—	—	—	35.8

2.3.4.1   *Natural Gas*

Natural gas is a mixture of hydrocarbons with methane as the main constituent. It can be found in porous reservoirs, either associated with crude oil ("associated gas") or in reservoirs in which no oil is present ("non-associated gas"). Natural gas is of great importance not only as a source of energy, but also as a raw material for the petrochemical industry. Besides hydrocarbons, natural gas usually contains small (or sometimes large) amounts of non-hydrocarbon gases such as carbon dioxide, nitrogen, and hydrogen sulfide. Tables 2.4 and 2.5 show the composition of a wide variety of natural gases.

Natural gas is classified as "dry" or "wet". The term "wet" refers not to water but to the fact that wet natural gas contains substantial amounts of ethane, propane, butane, and C_5 and higher hydrocarbons, which condense on compression at ambient temperature forming "natural gas liquids". Dry natural gas contains only small quantities of condensable hydrocarbons. Associated gas is invariably wet, whereas non-associated gas is usually dry. The terms "sweet" and "sour" denote the absence or presence of hydrogen sulfide and carbon dioxide.

Non-associated gas can only be produced as and when a suitable local or export market is available.

Associated natural gas, on the other hand, is a coproduct of crude oil and, therefore, its production is determined by the rate of production of the accompanying oil. It has long been considered a waste product

Table 2.5   *Composition of selected associated natural gases (vol.%) [12]*

Area	Abu Dhabi	Iran	North Sea	North Sea	N. Mexico
Field	Zakum	Agha Jari	Forties	Brent	San Juan County
CH_4	76.0	66.0	44.5	82.0	77.3
C_2H_6	11.4	14.0	13.3	9.4	11.2
C_3H_8	5.4	10.5	20.8	4.7	5.8
C_4H_{10}	2.2	5.0	11.1	1.6	2.3
$C_5{}^+$	1.3	2.0	8.4	0.7	1.2
N_2	1.1	1.0	1.3	0.9	1.4
CO_2	2.3	1.5	0.6	0.7	0.8
H_2S	0.3	—	—	—	—

Table 2.6   *Composition of selected shale gases (vol.%) [14–16].*

Shale	Barnett		Marcellus		Antrim	
Well[a]	1	4	1	4	1	4
CH_4	80.3	93.7	79.4	95.5	27.5	85.6
C_2H_6	8.1	2.6	16.1	3.0	3.5	4.3
C_3H_8	2.3	0.0	4.0	1.0	1.0	0.4
N_2	7.9	1.0	0.4	0.2	65.0	0.7
CO_2	1.4	2.7	0.1	0.3	3.0	9.0

[a]For each gas shale data were reported for four different wells (1–4). The gas compositions of wells 2 and 3 are in between those of wells 1 and 4.

and most of it was flared (for safety reasons). With the present energy situation, however, associated gas represents a feedstock and energy source of great potential value. Moreover, utilizing, instead of flaring, is in better agreement with environment protection measures. An increasing number of schemes are being developed to utilize such gas.

QUESTION:

> *Why would associated natural gas always be "wet"?*

Table 2.6 shows the composition of shale gas from two wells for three gas shales in the United States.

QUESTION:

> *No data for C_4 or C_5^+ hydrocarbons or H_2S have been reported for shale gases. Would these components be present?*
> *Where would exploration companies preferentially develop new wells to drill for shale gas, in regions containing dry or wet gas?*

Natural gas from the well is treated depending on the compounds present in the gas. A dry gas needs little or no treatment except for H_2S and possibly CO_2 removal if the amounts are appreciable (sour gas). Condensable hydrocarbons are removed from a wet gas and part can be sold as liquefied petroleum gas (LPG, propane, butane) or be used for the production of chemicals (ethane). The other part, C_5^+ hydrocarbons, can be blended with gasoline. So, once natural gas has been purified and separated, a large part of the gas is a single chemical compound, methane.

2.3.4.2 Crude Oil

The composition of crude oil is much more complex than that of natural gas. Crude oil is not a uniform material with a simple molecular formula. It is a complex mixture of gaseous, liquid, and solid hydrocarbon compounds, occurring in sedimentary rock deposits throughout the world. The composition of the mixture depends on its location. Two adjacent wells may produce quite different crudes and even within a well the composition may vary significantly with depth. Nevertheless, the elemental composition of crude oil varies over a rather narrow range, as shown in Table 2.7.

Although at first sight these variations seem small, the various crude oils are extremely different. The high proportion of carbon and hydrogen suggests that crude oil consists largely of hydrocarbons, which indeed

Table 2.7   *Elemental composition of crude oil.*

Element	Percentage range (wt%)
C	83−87
H	10−14
N	0.2−3
O	0.05−1.5
S	0.05−6
Trace metals	<0.1

is the case. Detailed analysis shows that crude oil contains alkanes, cycloalkanes (naphthenes), aromatics, polycyclic aromatics, and nitrogen-, oxygen-, sulfur- and metal-containing compounds.

QUESTION:

> *Does the elemental composition agree with the presence of these compounds?*

The larger part of crude oil consists of alkanes, cycloalkanes, and aromatics (Figure 2.11). Both linear and branched alkanes are present. In gasoline applications the linear alkanes are much less valuable than the branched alkanes, whereas in diesel fuel the linear alkanes are desirable. One of the aims of catalytic reforming is to shift the ratio of branched/linear alkanes in the desired direction. Aromatics have favorable properties for the gasoline pool. However, currently their potential dangerous health effects are receiving increasing attention. The most important binuclear aromatic is naphthalene.

QUESTION:

> *Crude oil does not contain alkenes. Is this surprising? Why or why not?*

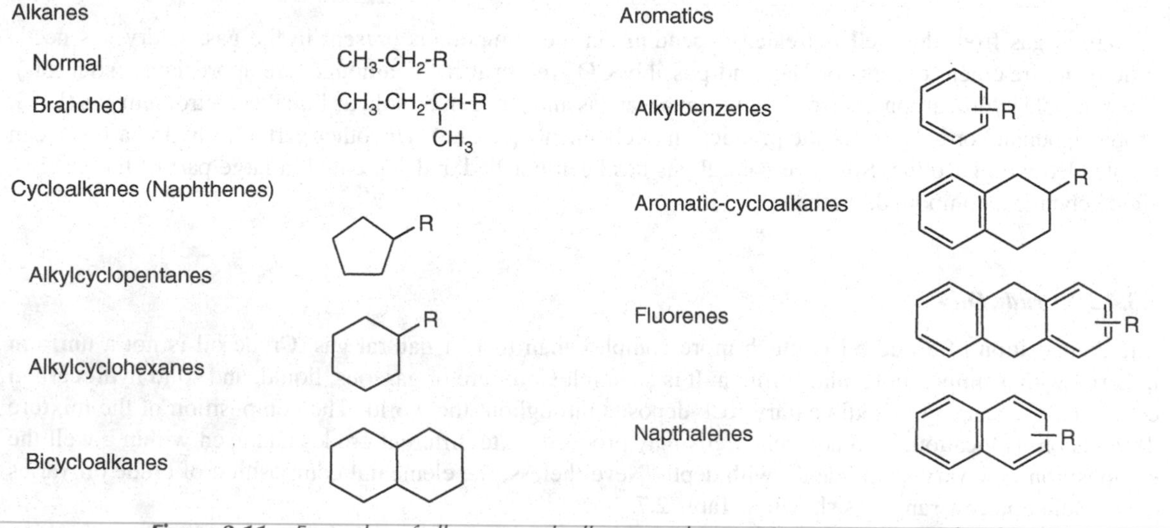

Figure 2.11   *Examples of alkanes, cycloalkanes and aromatics present in crude oil.*

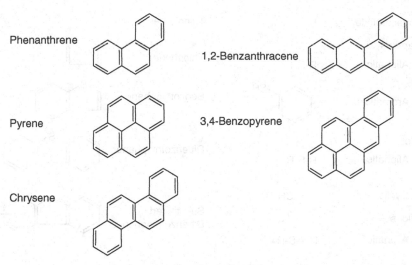

Phenanthrene

1,2-Benzanthracene

Pyrene

3,4-Benzopyrene

Chrysene

Figure 2.12 Examples of polycyclic, polynuclear aromatics in crude oil.

The heavier the crude the more polycyclic aromatic compounds it will contain. Generally, heavier crudes yield a lower proportion of useful products. A further disadvantage of the heavier crudes is that they contain more polynuclear aromatics (PNAs), which tend to lead to carbonaceous deposits ("coke") during processing. The implications of coke formation play an important role. Figure 2.12 shows examples of polycyclic polynuclear aromatics.

Crudes do not consist exclusively of carbon and hydrogen; minor amounts of so-called hetero-atoms are also present, the major ones being sulfur, nitrogen, and oxygen. Of these, sulfur is highly undesirable because it leads to corrosion, poisons catalysts, and is environmentally harmful. Therefore, it must be removed. Chapter 3 discusses the technology for sulfur removal from oil fractions.

Figure 2.13 shows examples of sulfur-containing compounds. They are ranked according to reactivity: sulfur in mercaptans is relatively easy to remove in various chemical reactions, whereas sulfur in thiophene and benzothiophenes has an aromatic character, resulting in high stability. Aromatic sulfur compounds are present in heavy crudes in particular.

The nitrogen content of crude oil is lower than the sulfur content. Nevertheless, nitrogen compounds (Figure 2.14) deserve attention because they disturb major catalytic processes, such as catalytic cracking and hydrocracking. Basic nitrogen compounds react with the acid sites of the cracking catalyst and thus destroy its acidic character. Nitrogen compounds, as are aromatic sulfur compounds, are present in the higher boiling hydrocarbon fractions in particular.

Although the oxygen content of crude oil usually is low, oxygen occurs in many different compounds (Figure 2.15). A distinction can be made between acidic and non-acidic compounds. Organic acids and phenols are counted as part of the class of acidic compounds.

Metals are present in crude oil only in small amounts. Even so, their occurrence is of considerable interest, because they deposit on and thus deactivate catalysts for upgrading and converting oil products. Part of the metals is present in the water phase of crude oil emulsions and may be removed by physical techniques. The other part is present in oil-soluble organometallic compounds and can only be removed by catalytic processes. Figure 2.16 shows examples of metal containing hydrocarbons.

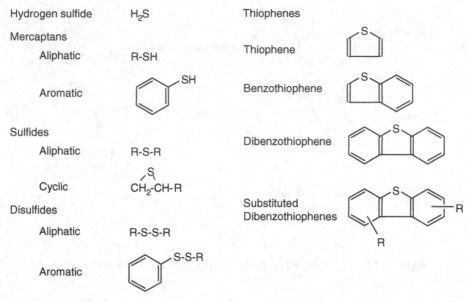

Figure 2.13   *The most important sulfur-containing compounds in crude oil.*

Figure 2.17 shows the metals contents of various crudes. Most of the metal-containing compounds are present in the heavy residue of the crude oil. Clearly, the metal contents vary widely. The most abundant metals are nickel, iron, and vanadium.

2.3.4.3   *Coal*

In contrast to crude oil and natural gas, the elemental composition of coal varies over a wide range (Table 2.8). The composition range is based on only the organic component of coal. In addition, coal contains an appreciable amount of inorganic material (minerals), which forms ash during combustion and gasification

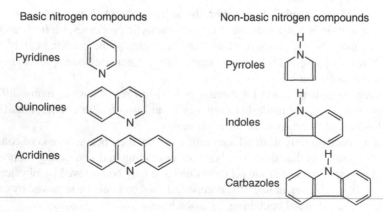

Figure 2.14   *The most important nitrogen-containing compounds present in crude oil.*

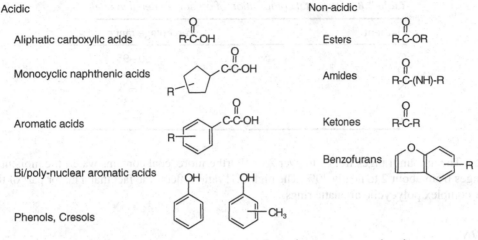

Figure 2.15 *Oxygen-containing compounds present in crude oil.*

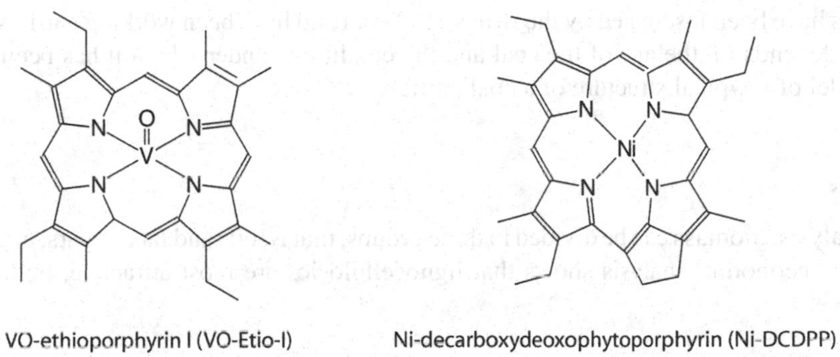

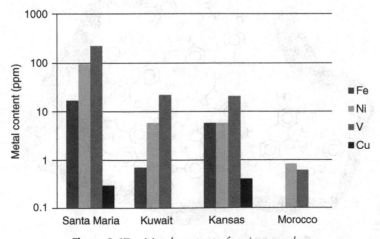

VO-ethioporphyrin I (VO-Etio-I)   Ni-decarboxydeoxophytoporphyrin (Ni-DCDPP)

Figure 2.16 *Examples of metal-containing compounds in crude oil.*

Figure 2.17 *Metal content of various crudes.*

Table 2.8   *Elemental composition of organic material in coal.*

Element	Percentage range (wt%)
C	60–95
H	2–6
N	0.1–2
O	2–30
S	0.3–13

(Chapter 5). The amount ranges from 1 to over 25%. Furthermore, coal contains water; the moisture content of coal ranges from about 2 to nearly 70%. The high C/H ratio reflects the fact that a major part of the coal is built up of complex polycyclic aromatic rings.

QUESTION:

> *Show that most coals must be aromatic.*

Many scientists have been fascinated by the structure of coal and have been working hard trying to elucidate it. This structure depends on the age of the coal and the conditions under which it has been formed. Figure 2.18 shows a model of a typical structure of a coal particle.

2.3.4.4   *Biomass*

In a simplified analysis, biomass can be divided in three groups, that is, oils and fats, sugars, and lignocellulosic biomass. A simple economic analysis shows that lignocellulosics are most attractive, both with respect to resources and price.

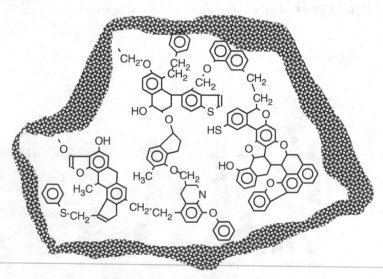

Figure 2.18   *Model of a typical coal structure.*

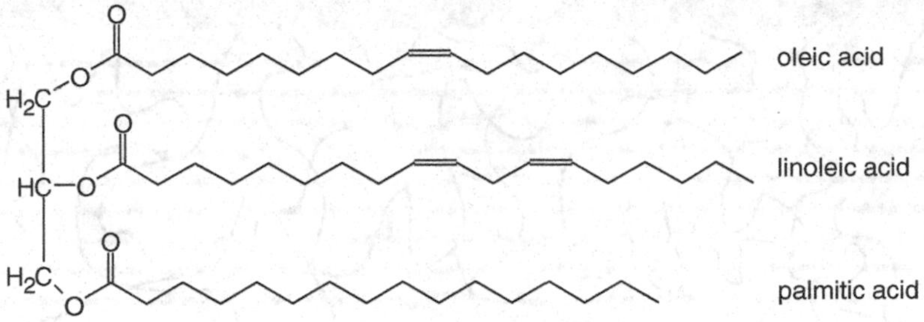

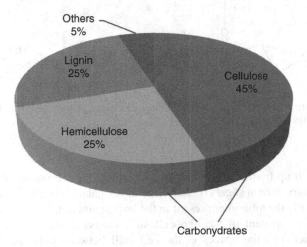

oleic acid

linoleic acid

palmitic acid

Figure 2.19 *Typical vegetable oil triglyceride.*

Animal fats and vegetable oils are primarily composed of triglycerides, esters of fatty acids with glycerol. Figure 2.19 shows an example of a typical vegetable oil triglyceride.

Lignocellulosic biomass consists mainly of three components: cellulose (35–50 wt%), hemicellulose (15–25 wt%), and lignin (15–30 wt%). Plant oils, proteins, different extractives, and ashes make up the rest of the lignocellulosic biomass structure. Figure 2.20 shows an average composition. The structure of lignocellulosics is complex. A schematic representation is given in Figure 2.21. Although the material is very important, there is no full agreement on the structure (yet). Recent research suggests that the lignocellosic biomass mainly consists of cell wall components built up from cellulose and hemicellulose held together by lignin. The hemicellulose in turn is a matrix containing the cellulose fibers.

Cellulose is the most abundant organic polymer on earth and its chemical structure, which is largely crystalline, is remarkably simple (Figure 2.22). It consists of linear polymers of cellobiose, a dimer of glucose. The multiple hydroxyl groups of the glucose molecule form hydrogen bonds with neighbor cellulose chains making cellulose microfibrils of high strength and crystallinity.

Hemicellulose is chemically related to cellulose in that it comprises a carbohydrate backbone. However, due to its random and branched structure hemicellulose is amorphous. It also has a more complex composition than cellulose. Figure 2.23 shows the structure of xylan, a representative polymer of hemicellulose. Whereas

Others
5%

Lignin
25%

Cellulose
45%

Hemicellulose
25%

Carbonydrates

Figure 2.20 *Average composition of lignocellulosic biomass.*

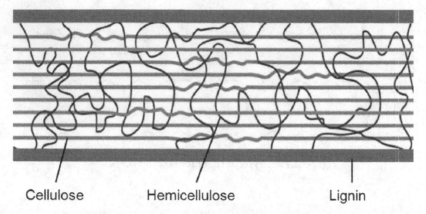

Cellulose   Hemicellulose   Lignin

Figure 2.21   *Structure of lignocellulosic biomass.*

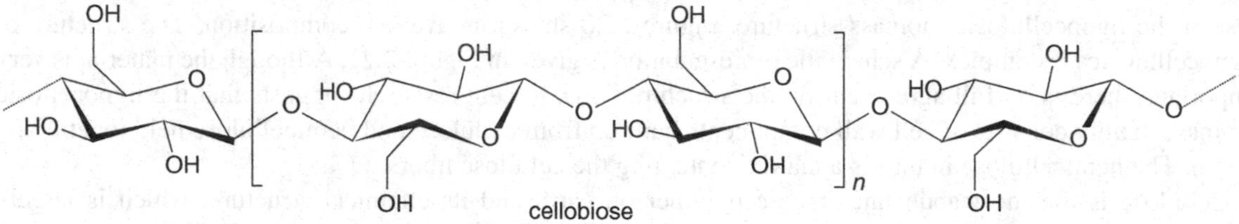

Figure 2.22   *Chemical composition of cellulose; n = 2500–5000.*

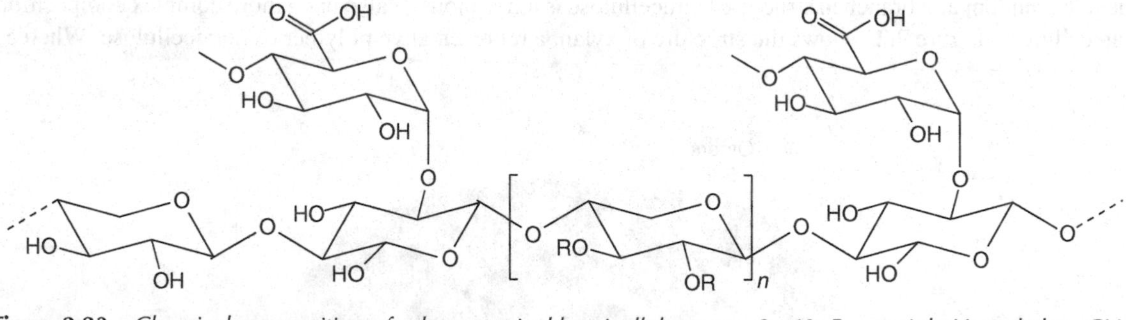

cellobiose

Figure 2.23   *Chemical composition of xylan, a typical hemicellulose; n = 3–40; R = mainly H, and also –CH$_3$, –(CO)CH$_3$ (acetyl) and C$_5$ or C$_6$ sugars.*

cellulose is completely built up from glucose monomers, hemicellulose consists of a mixture of five-carbon sugars (xylose, arabinose), six-carbon sugars (glucose, mannose, galactose), and uronic acids (e.g., glucoronic acid) (Figure 2.24). Xylose is the monomer present in the largest amount.

Lignin, the third main component of the lignocellosic biomass (Figure 2.25), is an amorphous three-dimensional polymer which fills the spaces in the cell wall between cellulose and hemicellulose. It is aromatic and hydrophobic in comparison with cellulose and hemicellulose. The complexity and variability

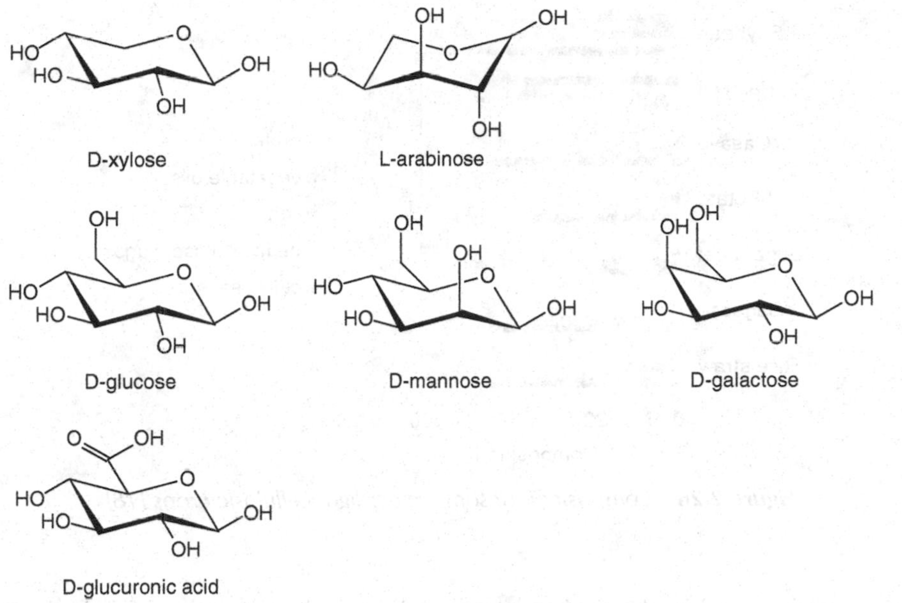

Figure 2.24　*Monomers present in hemicellulose.*

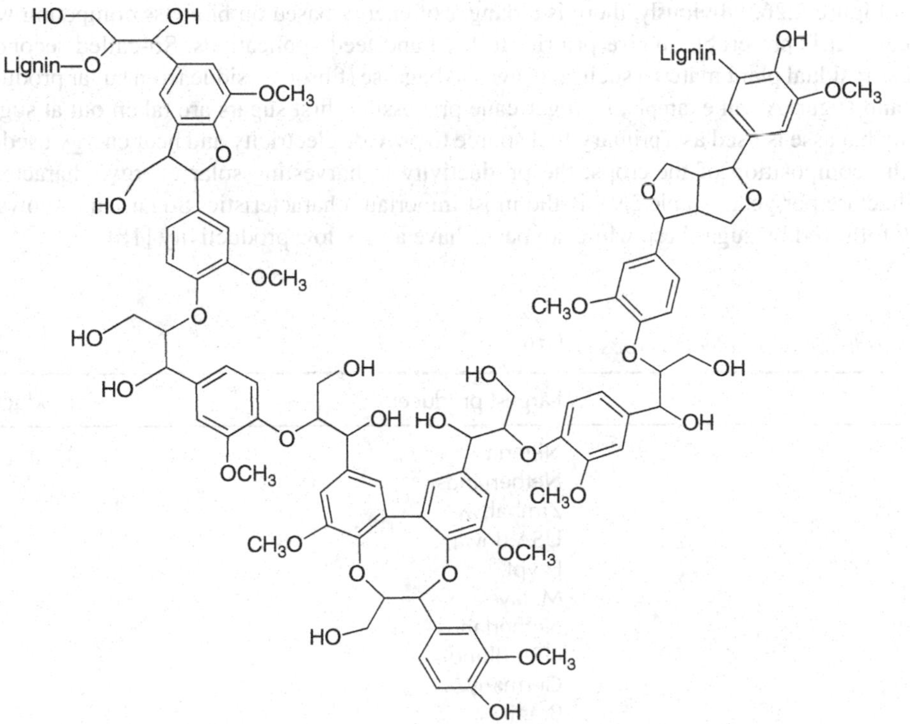

Figure 2.25　*Chemical composition of a small piece of a lignin polymer.*

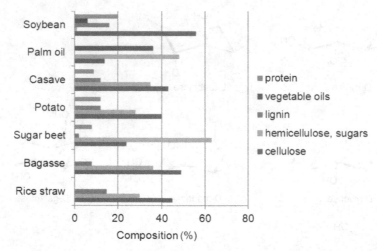

Figure 2.26   *Composition of some major lignocellulosic crops [18].*

of the lignin composition and its chemical resistance make its utilization quite difficult. A comprehensive review on the catalytic valorization of lignin has been published [17].

Not surprisingly, the composition of different crops is very different; some typical representatives are compared in Figure 2.26. Obviously, there is a danger of energy based on biomass competing with the food chain. Ethically, it is preferable to give priority to food and feed applications. So-called second-generation plants process residual plant material such as straw and bagasse (fibrous residue from sugar production) rather than starch and sugar. As an example, in sugar cane processing, first sugars are taken out at sugar mills and the remaining bagasse is used as a primary fuel source to provide electricity and heat energy used in the mills.

Besides the composition of the crops, the productivity in harvesting solar energy, characterized as dry weight per hectare per year, (Table 2.9), is the most important characteristic. Sugar cane shows the highest productivity followed by sugar beet, while soybeans have a very low productivity [18].

Table 2.9   *Productivity of selected crops [18, 19].*

Crop	Largest producer	Production[a] (t/ha/a)
Cassava	Nigeria	50
Grass	Netherlands	14
Jatropha seeds	Zimbabwe	8
Maize	USA (Iowa)	14
Rice straw	Egypt	11
Palm oil	Malaysia	25
Potato	Netherlands	65
Soybean	USA (Illinois)	4
Sugar beet	Germany	100
Sugarcane	Brazil	125
Sunflower	France	6

[a] Best practice technology.

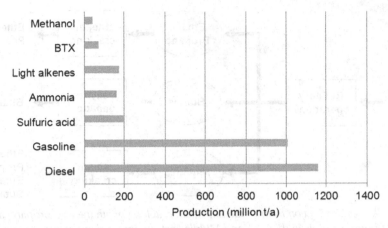

Figure 2.27 *Production of some base chemicals and fuels (2010). Based on data from various sources.*

2.4 Base Chemicals

The vast majority of the products from the chemical industry are manufactured from a very limited number of simple chemicals, called *base chemicals*. The most important organic base chemicals are the lower alkenes (ethene, propene, and butadiene), the aromatics (benzene, toluene, and xylene, also referred to as "BTX"), and methanol. Synthesis gas, a mixture of hydrogen and carbon monoxide in varying ratio, can also be considered an organic base chemical. Sulfuric acid and ammonia are the two most important inorganic base chemicals. Most chemicals are produced directly or indirectly from these compounds, which can be considered real building blocks. Figure 2.27 shows the world production of base chemicals in comparison with two major energy carriers, gasoline and diesel. Although base chemicals account for most of the chemicals production, gasoline and diesel production individually are larger than the combined production of these base chemicals, which illustrates the huge consumption of these fuels.

The feedstock chosen for the production of a base chemical will depend on the production unit, the local availability, and the price of the raw materials. For the production of light alkenes, in broad terms, there is a difference between the USA and the Middle East and the rest of the world, as illustrated by Figures 2.28 and 2.29.

One of the reasons for these differences lies in the consumer market: in the USA, the production of gasoline is even more dominant than in the rest of the developed world. As a consequence, lower alkenes are primarily produced from sources that do not yield gasoline. This development became possible by the discovery of natural gas fields with high contents of hydrocarbons other than methane (Table 2.6). Natural gas from these fields is very suitable for the production of lower alkenes. The availability of low-cost gas in the Middle East has resulted in a large growth of the production of lower alkenes in this region.

Historically, the production of the simple aromatics has been coal based: coal tar, a by-product of coking ovens, was the main source of benzene, toluene, and so on. At present, nearly 95% of the aromatic base chemicals are oil based. Figure 2.30 shows the main processes involved in aromatics production.

The main sources of aromatics, that is, catalytic reforming and steam cracking, are also producers of gasoline. Therefore, production of aromatics is closely related to fuel production; aromatics and gasoline have to compete for the same raw materials. The demand for benzene is greater than the demand for the other aromatics. Hence, part of the toluene produced is converted to benzene by hydrodealkylation (and by disproportionation to benzene and p-xylene).

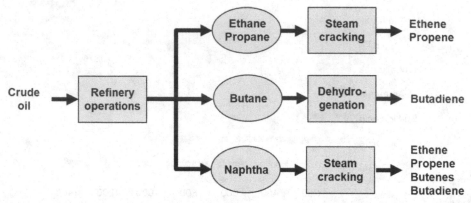

Figure 2.28   *Light alkenes production from crude oil; steam cracking of ethane and propane and dehydrogenation of butane mainly in the USA and Middle East, steam cracking of naphtha mainly in other parts of the world.*

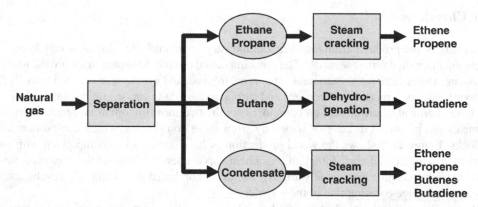

Figure 2.29   *Light alkenes production from natural gas; mainly in the USA and Middle East.*

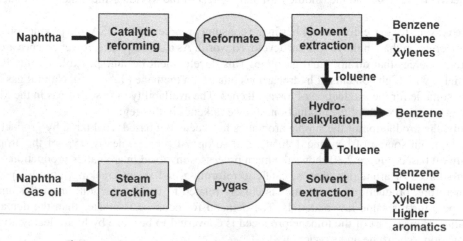

Figure 2.30   *Aromatics production (Pygas = pyrolysis gasoline).*

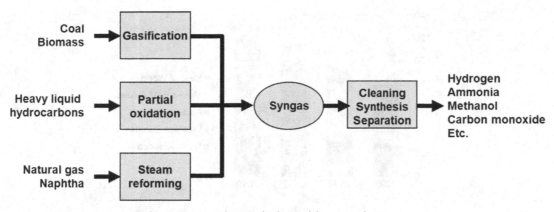

Figure 2.31 *Chemicals derived from synthesis gas.*

Ammonia and methanol are two other important base chemicals that are derived from organic raw materials. Their production generally involves the conversion of synthesis gas, a mixture of mainly hydrogen and carbon monoxide, which can be produced from any hydrocarbon source. Figure 2.31 shows the processes used for the conversion of different raw materials.

Most of the processes referred to in Figures 2.28 to 2.31 will be the topics of subsequent chapters.

2.5   Global Trends in the Chemical Industry

As mentioned earlier, organic chemicals are produced mainly from a few base chemicals. The production of these base chemicals is thus a useful indicator of the growth of the petrochemical industry. In the 1950s through to the 1970s, essentially exponential growth of the petrochemical industry took place in North America and Western Europe, as illustrated in Figure 2.32. In the 1980s and 1990s, the chemical market

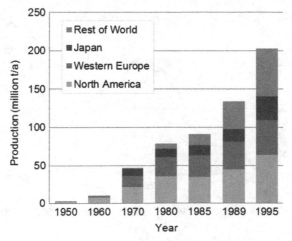

Figure 2.32 *Growth of the petrochemical industry: production of six major base chemicals (ethene, propene, butadiene, benzene, toluene, and xylenes) [20, 21].*

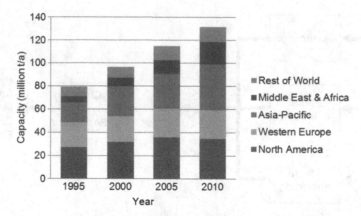

Figure 2.33   *Growth of the ethene production capacity. Data from various sources.*

growth was only modest in these regions, among other things as a result of a steep rise in competition from producers in Japan and other parts of the world, referred to as "Rest of World" in Figure 2.32. The latter category includes countries in the Asia-Pacific region (e.g., Singapore, Taiwan, Korea, China, and India) and the Middle East. It is not surprising that growth is taking place in these regions in particular.

Figure 2.33 clearly shows that in the last few decades the growth in the production capacity of ethene, the largest volume organic base chemical, has mainly been the result of added capacity in the Asia-Pacific region and the Middle East.

A more general indicator of a nation's industrial strength is its sulfuric acid production. Principal uses of sulfuric acid include fertilizer production, ore processing, oil refining, and chemical synthesis, but there are many more. In fact, the uses of sulfuric acid are so varied that the volume of its production provides an approximate index of general industrial activity. Figure 2.34 compares sulfuric acid production in the USA and China and also shows the total world production.

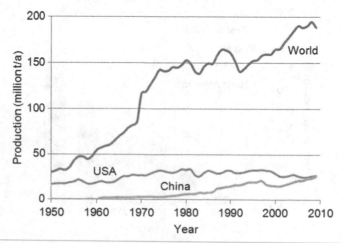

Figure 2.34   *Sulfuric acid production in the USA, China and worldwide. Data from [22] and various other sources.*

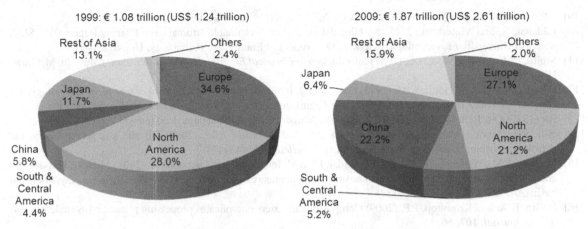

Figure 2.35 *World chemical sales by region [6].*

Not only has the production of chemicals been shifting from West to East, but so too has consumption, as can be seen from the sales figures in Figure 2.35. In the Western world, the chemical industry in many respects is mature and investments tend to be stable or even decreasing for various reasons. Firstly, partly due to good maintenance the lifetime of the plants is much longer than anticipated. Secondly, capacities of existing plants have been increased by good engineering ("debottlenecking"). For instance, new packings in distillation units, new catalysts, and new control systems can rejuvenate a plant. Thirdly, notably due to their feedstock advantages, countries in the Middle East attract very high investments in petrochemicals.

Meanwhile, new drivers for radically novel plants have been emerging. For instance, environmental concerns call for new technology. It is generally accepted that a sustainable technology is desired. Elements are renewable feedstocks and energy, less polluting production routes, and more efficient processes. Many new concepts fall under the umbrella of "Process Intensification".

Although in the past decade much progress has been made in the conversion of biomass, a lot of work still needs to be done before biomass can replace fossil fuels as a primary source of energy and/or chemicals.

References

[1] Woodward, R.B. and Doering, W.E. (1944) The total synthesis of quinine. *Journal of the American Chemical Society*, **66**, 849.

[2] Meth-Cohn, O. and Smith, M. (1994) What did W.H. Perkin actually make when he oxidised aniline to obtain mauveine? *Journal of the Chemical Society, Perkin Transactions 1*, 5–7.

[3] Sousa, M.M., Melo, M.J., Parola, A.J., Morris, P.J.T., Rzepa, H.S., and de Melo, J.S.S. (2008) A study in mauve: unveiling Perkin's dye in historic samples. *Chemistry – A European Journal*, **14**, 8507–8513.

[4] Scaccia, R.L., Coughlin, D., and Ball, D.W. (1998) A microscale synthesis of mauve. *Journal of Chemical Education*, **75**, 769.

[5] Spitz, P.H. (2003) *The Chemical Industry at the Millennium: Maturity, Restructuring and Globalization*. Chemical Heritage Press, Philadelphia.

[6] Cefic (2011) Facts and Figures 2011, The European Chemical Industry Council (Cefic), Brussels, pp. 1–51.

[7] BP (2011) BP Statistical Review of World Energy June 2011, BP, London, www.bp.com/statisticalreview (last accessed 17 December 2012).

[8] Chabrelie, M.-F. (2006) Current status of the world's gas giants. International Gas Union. 23rd World Gas Conference, 6.

[9]   KPMG International (2011) Shale Gas – A Global Perspective, KPMG International, pp. 1–28.

[10] Ladanai, S. and Vinterbäck, J. (2009) Global Potential of Sustainable Biomass for Energy, Report 013, SLU, Swedish University of Agricultural Sciences, Department of Energy and Technology, Uppsala, 1–32.

[11] Smith, J.M. and Van Ness, H.C. (1987) *Introduction to Chemical Engineering Thermodynamics*, 4th edn. McGraw-Hill, New York.

[12] Woodcock, K.E. and Gottlieb, M. (1994) Gas, natural. In: *Kirk Othmer Encyclopedia of Chemical Technology*, 4th edn, vol. **12** (eds J.I. Kroschwitz and M. Howe-Grant). John Wiley & Sons, Inc., New York, pp. 318–340.

[13] BP (1977) Our Industry; Petroleum, 5th edn. The British Petroleum Company, London.

[14] Hill, R.J., Jarvie, D.M., Zumberge, J., Henry, M., and Pollastro, R.M. (2007) Oil and gas geochemistry and petroleum systems of the Fort Worth Basin, *AAPG Bulletin*, **91**, 445–473.

[15] Martini, A.M., Walter, L.M., Ku, T.C.W., Budai, J.M., McIntosh, J.C., and Schoell, M. (2003) Microbial production and modification of gases in sedimentary basins: a geochemical case study from a Devonian shale gas play, Michigan basin. *AAPG Bulletin*, **87**, 1355–1375.

[16] Bullin, K.A. and Krouskop, P.E. (2009) Compositional variety complicates processing plans for US shale gas. *Oil & Gas Journal*, **107**, 50–55.

[17] Zakzeski, J., Bruijnincx, P.C.A., Jongerius, A.L., and Weckhuysen, B.M. (2010) The catalytic valorization of lignin for the production of renewable chemicals. *Chemical Reviews*, **110**, 3552–3599.

[18] Brehmer, B., Boom, R.M., and Sanders, J. (2009) Maximum fossil fuel feedstock replacement potential of petrochemicals via biorefineries. *Chemical Engineering Research and Design*, **87**, 1103–1119.

[19] Moulijn, J.A. and Babich, I.V. (2011) The potential of biomass in the production of clean transportation fuels and base chemicals, in *Production and Purification of Ultraclean Transportation Fuels*, 1088, American Chemical Society, pp. 65–77.

[20] Shell (1990) Chemicals Information Handbook 1990, Shell International Chemical Company Ltd.

[21] BP (1998) BP Statistical Review of World Energy, The British Petroleum Company, London.

[22] Kelly, T.D. and Matos, G.R. (2011) Historical Statistics for Mineral and Material Commodities in the United States, USGS, http://minerals.usgs.gov/ds/2005/140/ (last accessed 17 December 2012).

General Literature

Gary, J.H. and Handwerk, G.E. (1994) *Petroleum Refining, Technology and Economics*, 3rd edn. Marcel Dekker, Inc., New York

McKetta, J.J. and Cunningham, W.A. (eds) (1976) *Encyclopedia of Chemical Processing and Design*, Marcel Dekker, Inc. New York.

Schobert, H.H. (1990) *The Chemistry of Hydrocarbon Fuels*, Buttersworth, London.

Van Krevelen, D.W. (1993) *Coal: Typology – Physics – Chemistry – Constitution*, 3rd edn. Elsevier, Amsterdam.

3

Processes in the Oil Refinery

Refining of crude oil is a key activity in the process industry. About 700 refineries are in operation worldwide with a total annual capacity of over 4000 million metric tons and individual capacities of about one to over 30 Mt/a [1]. The goal of oil refining is threefold: (i) production of fuels for transportation, power generation, and heating purposes; (ii) production of specialties such as solvents and lube oils; and (iii) production of intermediates, especially for the chemical industry.

3.1 The Oil Refinery — An Overview

Oil refineries have become increasingly complex over the years. They constitute a relatively mature and highly integrated industrial sector. Besides physical processes such as distillation and extraction, a large number of chemical processes are applied, many of which are catalytic. Figure 3.1 shows a flow scheme of a typical complex modern refinery.

After desalting and dehydration (not shown in Figure 3.1), the crude oil is separated into fractions by distillation. The distilled fractions cannot be used directly in the market. Many different processes are carried out in order to produce the required products. In this chapter, most of the processes in Figure 3.1 are discussed. They can be divided in physical and chemical processes, as shown in Table 3.1. Oil refineries show large differences and, besides the processes mentioned here, many others are applied. For example, Figure 3.1 and Table 3.1 do not show processes for treating refinery off-gases and sulfur recovery (discussed in Chapter 8) and facilities for wastewater treatment, and so on.

The reason that such a complex set of processes is needed is the difference between the properties of the crude oil delivered and the requirements of the market. Especially for fuels in the transport sector, extensive processing is required in order to obtain products with satisfactory performance. Another reason for the complexity of an oil refinery lies in environmental considerations. Legislation calls for increasingly cleaner processes and products. In fact, at present legislation related to minimizing the environmental impact is the major drive for process improvement and the development of novel processes.

Chemical Process Technology, Second Edition. Jacob A. Moulijn, Michiel Makkee, and Annelies E. van Diepen.
© 2013 John Wiley & Sons, Ltd. Published 2013 by John Wiley & Sons, Ltd.

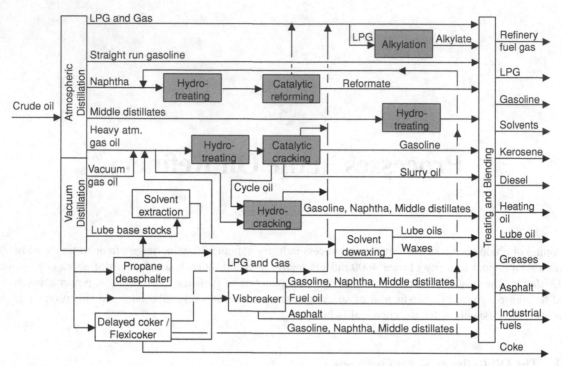

Figure 3.1   *Flow scheme of a typical complex modern oil refinery; catalytic processes are indicated by gray blocks.*

3.2   Physical Processes

3.2.1   Desalting and Dehydration

Crude oil often contains water, inorganic salts, suspended solids, and water-soluble trace metals. These contaminants can cause corrosion, plugging, and fouling of equipment and poisoning of catalysts in catalytic

Table 3.1   *Typical physical and chemical processes in a complex oil refinery.*

Physical processes	Chemical processes	
	Thermal	Catalytic
Desalting	Visbreaking	Hydrotreating
Dehydration	Delayed coking	Catalytic reforming
Distillation	Flexicoking	Catalytic cracking
Solvent extraction[a]		Hydrocracking
Propane deasphalting		Catalytic dewaxing[a]
Solvent dewaxing[a]		Alkylation
Blending[a]		Polymerization
		Isomerization

[a] Not discussed in this book; see, for example, References [2, 3].

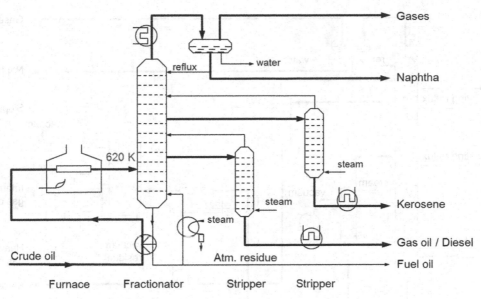

Figure 3.2 *Simple atmospheric crude distillation.*

processes. Therefore, the first step in the refining process is desalting and dehydration. This can be accomplished by adding hot water with added surfactant to extract the contaminants from the oil. Upon heating, the salts and other impurities dissolve into the water or attach to it. The oil and water phases are then separated in a tank, where the water phase settles out. Wastewater and contaminants are discharged from the bottom of the settling tank and the desalted crude is drawn from the top and sent to the crude distillation tower.

3.2.2 Crude Distillation

The desalted crude has to be separated into products, which each have specific uses. Crude oil consists of thousands of different compounds and, as a consequence, it is impossible and undesirable to separate it into chemically pure fractions. The central separation step in every oil refinery is the distillation of the crude, which separates the various fractions according to their *volatility*. Figure 3.2 shows a schematic of a simple atmospheric distillation unit.

After heat exchange with the residue the crude oil feed is heated further to about 620 K in a furnace. It is discharged to the distillation tower as a foaming stream. The vapors flow upward and are fractionated to yield the products given in Figure 3.2. A distillation column typically is 4 m in diameter and 20–30 m in height and contains 15–30 trays. The stripping columns, which may be 3 m high with a diameter of 1 m, serve to remove the more volatile components from the side streams.

QUESTION:

Assuming a distillation capacity of 10 Mt/a of crude oil, estimate the flow rate in the distillation tower (in m³/h).

With this simple distillation unit a satisfactory set of products is not obtained. The market calls for clean products (no sulfur, nitrogen, oxygen, metals, etc.), gasoline (high octane number), more diesel (high cetane number), specific products (aromatics, alkenes, etc.), and less low-value fuel oil. The logical steps to meet

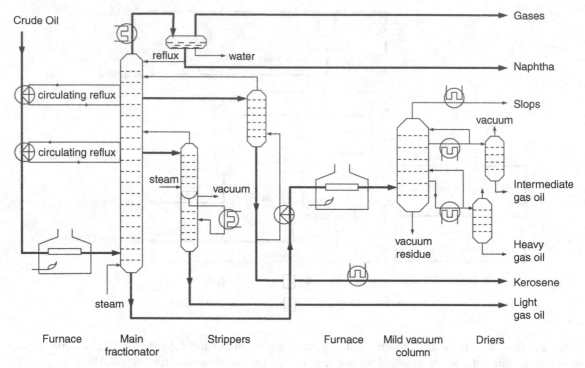

Figure 3.3 *Modern crude distillation unit consisting of atmospheric and vacuum distillation.*

these demands are more sophisticated distillation and other physical separation steps, followed by chemical conversion steps.

The major limitation of atmospheric distillation is the fact that hydrocarbons should not be heated to very high temperature. The reason is that thermal decomposition reactions take place above a temperature of about 630 K. Decomposition reactions ("thermal cracking" or "pyrolysis") are highly undesirable because they result in the deposition of carbonaceous material, also referred to as coke, on the tube walls in heat exchangers and furnaces.

Figure 3.3 shows a typical modern two-column distillation unit, consisting of an atmospheric distillation column and a column operating under reduced pressure (<0.1 bar) for fractionation of the atmospheric residue. This advanced distillation requires a substantial amount of energy and, therefore, optimal heat recovery is essential. Preheating of the feed by heat exchange with the residue, as is done in a simple distillation unit, is not convenient because the residue is further distilled. Therefore, the crude oil passes through several heat exchangers, where it recovers heat from circulating reflux streams. This provides an optimal heat recovery and temperature control system.

The feed enters the first distillation column at 570–620 K, depending on the feedstock and the desired product distribution. As the vapor passes up through the column heat is removed and the vapor is partially condensed by the circulating reflux streams, thus providing a flow of liquid down the column. As a result, the circulating reflux streams also improve the composition of the fractions that are withdrawn as side streams by establishing a new vapor–liquid equilibrium. The light gas oil side stream is stripped with steam to recover more volatile components. The kerosene side stream is heat exchanged with the column bottom product. In this case no steam is used for stripping in order to meet the water content specification for use of the kerosene as aviation fuel.

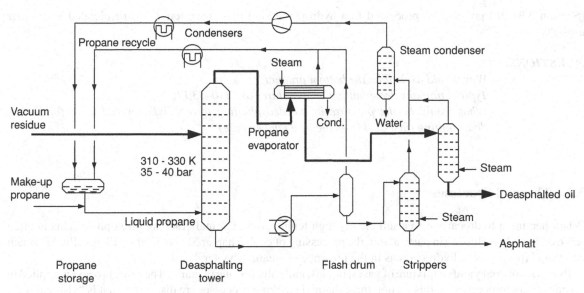

Figure 3.4   *Propane deasphalting process.*

The heavy liquid fraction leaving the bottom of the main fractionator is sent to a mild-vacuum column for further fractionation. Heat is withdrawn from the vacuum column at various levels by circulating reflux streams. The residue from the mild-vacuum column may be further processed in a high-vacuum fractionator to recover additional distillates, which are used as fuel oil or as feedstock for catalytic cracking.

A vacuum column usually has a much larger diameter (up to 15 m) than an atmospheric distillation column because of the large volumetric flow rate due to the low vapor pressures.

QUESTION:

Packed columns allow larger liquid-gas contact areas than tray columns. Two types of packing can be applied, that is, a structured and a random packing. What would be the preferred packing for a vacuum distillation column?

3.2.3   Propane Deasphalting

The coke-forming tendencies of heavier distillation products (mainly atmospheric and vacuum residue) can be reduced by removal of "asphaltenic" materials (large aromatic compounds) by means of extraction. Liquid propane generally appears to be the most suitable solvent, but butane and pentane are also commonly applied. Propane deasphalting is based on the *solubility* of hydrocarbons in propane, that is, on the type of molecule, rather than on molecular weight as in distillation. Figure 3.4 shows a schematic representation of the propane deasphalting process [3, 4].

The heavy feed, usually vacuum residue, is fed to the deasphalting tower and is contacted with liquefied propane countercurrently. Alkanes that are present in the feed dissolve in the propane, whereas the asphaltenic materials, the "coke precursors", do not dissolve. Propane evaporates on depressurization and is condensed and recycled to the deasphalting tower. The remaining deasphalted oil is stripped with steam to remove any residual propane. The asphaltenic material that leaves the deasphalting unit undergoes similar treatment. The asphalt residue is then sent to thermal processes, such as visbreaking, delayed coking, and "Flexicoking"

(Section 3.3). It may also be processed in a hydroprocessing plant (Section 3.4.5) or blended with other asphalts.

QUESTIONS:

Why would asphalt be the bottom product?
Typical propane deasphalting temperatures are 310–330 K.
What would be the pressure in the deasphalting tower? What would it be for butane deasphalting (first decide on the best temperature)?

3.3   Thermal Processes

When heating a hydrocarbon to a sufficiently high temperature, thermal cracking takes place. This is often referred to as "pyrolysis" (in particular in the processing of coal, Chapter 5) and, slightly illogically, as "steam cracking" (pyrolysis of hydrocarbons in the presence of steam, Chapter 4).

Pyrolysis of coal yields a mixture of gases, liquids, and solid residue ("char"). The same process is applied in oil conversion processes. In this section three thermal cracking processes are discussed, that is "visbreaking", delayed coking, and the more recent "Flexicoking" process. Thermal processes are flexible but a disadvantage is that, in principle, large amounts of low-value products are formed. This is the reason for innovations like the Flexicoking process and processes optimized towards the production of certain types of high-quality coke.

3.3.1   Visbreaking

Visbreaking, that is, viscosity reduction or breaking (Figure 3.5), is a relatively mild thermal cracking process in which the viscosity of vacuum residue is reduced. The severity of the visbreaking process depends on temperature (710–760 K) and reaction time (1–8 minutes). Usually, less than 10 wt% of gasoline and lighter products are produced. The main product (about 80 wt%) is the cracked residue, which is of lower viscosity than the vacuum residue.

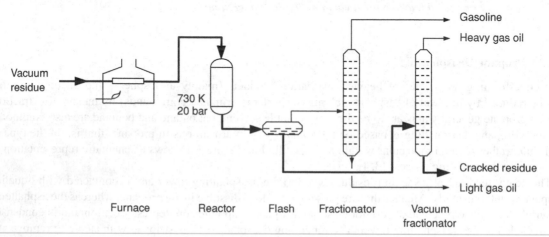

Figure 3.5   *Schematic of the visbreaking process.*

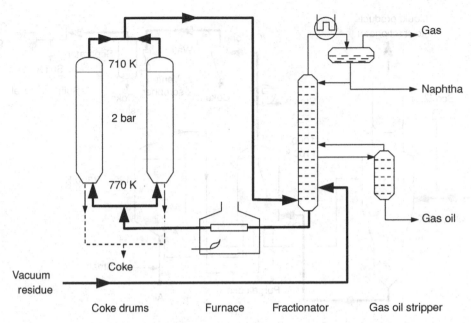

Figure 3.6   *Schematic of the delayed coking process.*

3.3.2   Delayed Coking

Delayed coking [2] is a thermal cracking process with long residence times, making it much more severe than visbreaking. As a consequence, a solid residue, petroleum coke or simply coke, is also formed:

$$\text{heavy feed} \diagDown \begin{array}{l} \text{gas} \\ \text{liquid} \\ \text{petroleum coke} \end{array} \tag{3.1}$$

Coke is similar to highly ordered and refractory coal and is typically used as fuel in power plants. Originally, the goal of delayed coking was minimization of the output of residual oil and other high-molecular-weight material, but over the years it was found profitable to produce certain types of cokes. A good example is the production of coke for electrodes, which in fact is the largest non-fuel end use for petroleum coke.

Figure 3.6 shows a typical delayed coking process. The feed to the delayed coker can be any undesirable heavy stream containing a high metal content. A common feed is vacuum residue. The heated feed is introduced in a coke drum, which typically has a height of about 25 m and a diameter of 4–9 m. Usually, a coke drum is on stream for 16–24 hours, during which it becomes filled with porous coke. The feed is subsequently introduced in the second coke drum, while coke is removed from the first drum with high-pressure water jets.

QUESTIONS:

> *Explain the names "visbreaking" and "delayed coking".*
> *In delayed coking gas is formed. Is this also the case in visbreaking? If so, in what stream in* Figure 3.5 *is it present?*

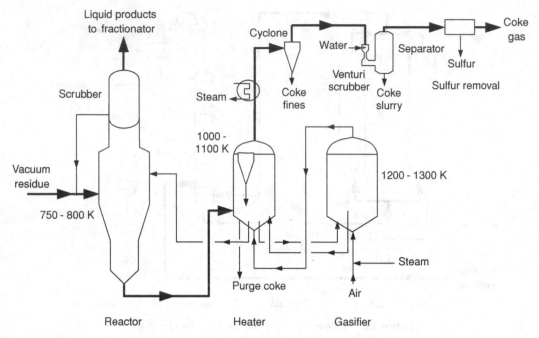

Figure 3.7   *Schematic of the Flexicoking process.*

3.3.3   Flexicoking

The Flexicoking process has been developed by Exxon to minimize coke production [3, 5]. Figure 3.7 shows a flow scheme of this process. A residual oil feed is injected into the reactor and contacted with a hot fluidized bed of coke. Thermal cracking takes place producing gas, liquids, and more coke. Coke constitutes approximately 30 wt% of the cracking products. The coke particles are transported to a heater, which is also a fluidized bed. The primary function of the heater is to transfer heat from the gasifier to the reactor. Part of the hot coke is recirculated to the reactor, while another part is fed to the gasifier, which too is a fluidized bed. Here the coke is gasified by reaction with steam and oxygen (comparable to coal gasification, Chapter 5):

$$C + O_2/H_2O \rightarrow H_2, CO, CO_2, CH_4, \text{pollutants} \qquad (3.2)$$

The gas leaves the top of the gasifier and flows to the heater, where it serves to fluidize the coke particles and provide part of the heat required in the reactor. The gas leaving the heater is cooled by steam generation and then passes through a cyclone and a venturi type gas scrubber for the removal of coke fines. A venturi scrubber is a device in which dust (in this case coke fines) is removed from a gas stream by washing with a liquid, usually water. The turbulence created by the flow of gas and liquid through the narrow part of the venturi tube promotes the contact between the dust particles and liquid droplets. The resulting slurry is separated from the gas in a separator. Finally, hydrogen sulfide is removed from the gas and converted to elemental sulfur (Chapter 8).

The recirculating coke particles serve as nuclei for coke deposition and as heat carriers. Part of the coke is withdrawn to prevent accumulation of metals and so on.

QUESTIONS:

How is the heat required for thermal cracking produced? Guess the composition of the gases produced. The gases produced are of low calorific value. What is the reason?
Why is a cyclone required in the heater but not in the gasifier?
When the Flexicoking units were built as part of the Exxon refinery in Rotterdam, The Netherlands, a hydrogen plant was added also. Why?

3.4 Catalytic Processes

Catalytic processes are the pillars of oil processing. In fact, chemists and chemical engineers from this sector have had tremendous impact on the discipline of catalysis and chemical engineering. The most important catalytic processes in the oil refinery, in terms of throughput, are fluid catalytic cracking (FCC), hydrotreating, hydrocracking, catalytic reforming, and alkylation.

3.4.1 Octane and Cetane Numbers

The major part of the oil fractions is used in the transport sector, especially as gasoline and diesel. Table 3.2 gives a survey of the octane number of various gasoline-range hydrocarbons and their boiling points. Table 3.3 gives typical octane numbers of refinery gasoline streams.

QUESTION:

Some octane numbers are larger than 100. Explain why this is possible. From the difference between RON and MON for the refinery products can you guess the approximate composition (alkanes, alkenes, aromatics)?

The octane number is a measure of the quality of the gasoline. Modern gasoline-powered cars require a minimum octane number of 95 or 98 (RON). Gasoline with a lower octane number can cause irreversible damage to the engine. The amount of gasoline obtained from simple distillation is too low. Therefore, either high-octane number compounds have to be added or additional conversion steps such as catalytic cracking and hydrocracking, which convert heavy hydrocarbons into lighter ones, are required. Moreover, the octane number of "straight-run" gasoline is much lower than required. Therefore, steps like catalytic reforming, which literally re-forms gasoline-range hydrocarbons into hydrocarbons of higher octane number, and alkylation, which combines lower alkenes with isobutane to form high-octane number gasoline components, are commonly carried out. Moreover, ethers like MTBE and currently especially ETBE as a renewable feedstock (both RON 118), are added at a level of a small percentage.

Diesel is another important transportation fuel. The diesel equivalent of the octane number is the cetane number, which in Europe should be at least 49 [6]. The diesel fraction of the hydrocracking process especially contributes to a high cetane number. Table 3.4 shows cetane numbers of hydrocarbons in the diesel fuel range and of diesel refinery fractions.

Cetane numbers are highest for linear alkanes, while naphthalenes, which are very aromatic compounds (Chapter 2), have the lowest cetane numbers. Note that this situation is the opposite for octane numbers (Table 3.2). In general, the more suitable a hydrocarbon is for combustion in a gasoline engine, the less favorable it is for combustion in a diesel engine (Box 3.1).

QUESTION:

ETBE and especially MTBE are currently seen as a burden for the environment. What is the reason for this? (Hint: are these ethers polar or apolar?)

Table 3.2   Octane numbers of various hydrocarbons and typical octane numbers of main refinery gasoline streams.

Hydrocarbon	Octane number[a]	Boiling point (K)
n-pentane	62	309
2-methylbutane	90	301
cyclopentane	85	322
n-hexane	26	342
2-methylpentane	73	333
2,2-dimethylbutane	93	323
1-hexene	63	337
2-hexene	81	341
benzene	>100	353
cyclohexane	77	354
heptane	**0[b]**	**371**
n-octane	<0[b]	399
2-methylheptane	13	391
2,2,3-trimethylpentane	**100[b]**	**372**
1-octene	35	395
2-octene	56	398
3-octene	68	396
xylenes	>100	≈415
ethylbenzene	98	410
1,2-dimethylcyclohexane	79	403
ethylcyclohexane	41	403
methyl tert-butyl ether (MTBE)	118	328
ethyl tert-butyl ether (ETBE)	118	345
tert-amyl methyl ether (TAME)	111	359

[a]Research octane number (RON).
[b]By definition.

Table 3.3   Typical octane numbers of gasoline base stocks.

Gasoline	RON[a]	MON[a]
Light straight-run gasoline	68	67
Isomerate	85	82
FCC light gasoline	93	82
FCC heavy gasoline	95	85
Alkylate	95	92
Reformate	99	88

[a]Research octane number (RON); the motor octane number (MON) is generally lower, depending on the particular compound. The difference is particularly large for alkenes and aromatics.

Table 3.4 Cetane numbers of various hydrocarbon classes and of main refinery gas oil streams (typical ranges).

Hydrocarbon class	Cetane number
n-**hexadecane (cetane)**	**100**[a]
n-alkanes	100–110
iso-alkanes	30–70
alkenes	40–60
cyclo-alkanes	40–70
alkylbenzenes	20–60
naphthalenes	0–20
1-methylnaphthalene	**0**[a]
straight-run gas oil	40–50
FCC cycle oil	0–25
Thermal gas oil	30–50
Hydrocracking gas oil	55–60

[a] By definition.

Box 3.1 Gasoline versus Diesel Engines

The differences between the required properties of gasoline and diesel fuel result from the difference between the ignition principles of gasoline and diesel engines. In gasoline engines the premixed air/fuel mixture is ignited by a spark and combustion should proceed in a progressing flame front. Hence, uncontrolled self-ignition of gasoline during the compression stroke is highly undesired. In contrast, in diesel engines the fuel is injected in hot compressed air present in the cylinder and self-ignition has to occur.

3.4.2 Catalytic Cracking

The incentive for catalytic cracking [3, 7–9] is the need to increase gasoline production. Originally cracking was performed thermally but nowadays cracking in the presence of a catalyst predominates. Feedstocks are heavy oil fractions, typically vacuum gas oil. Cracking is catalyzed by solid acids, which promote the rupture of C–C bonds. The crucial intermediates are carbocations (positively charged hydrocarbon ions). They are formed by the action of the acid sites of the catalyst. The nature of the carbocations is still subject of debate. In the following it will be assumed that they are "classical" carbenium ions and protonated cyclopropane derivatives.

Besides C–C bond cleavage a large number of other reactions occur:

- isomerization;
- protonation, deprotonation;
- alkylation;
- polymerization;
- cyclization, condensation (eventually leading to coke formation).

Catalytic cracking thus comprises a complex network of reactions, both intramolecular and intermolecular. The formation of coke is an essential feature of the cracking process. It will be shown that although coke deactivates the catalyst, its presence has enabled the development of an elegant practical process.

Catalytic cracking is one of the largest applications of catalysis: worldwide cracking capacity exceeds 600 million t/a. Catalytic cracking was the first large-scale application of fluidized beds. This explains the name "Fluid Catalytic Cracking" (FCC). Nowadays, in the actual cracking section entrained-flow reactors are used instead of fluidized beds, but the name "FCC" is still retained.

3.4.2.1 Cracking Mechanism

Catalytic cracking proceeds via a chain reaction mechanism in which organic ions are the key intermediates. The role of the catalyst is to initiate the chain reactions.

Alkenes can abstract a proton from a Brønsted site of the catalyst to form carbenium ions:

$$R-CH=CH_2 \ + \ H^+ \ \longrightarrow \ R-CH_2-\overset{\overset{\displaystyle H}{|}}{\underset{\underset{\displaystyle H}{|}}{C^+}} \ \text{ or } \ R-\overset{\overset{\displaystyle H}{|}}{\underset{\underset{\displaystyle H}{|}}{C^+}}-CH_3 \tag{3.3}$$

Here, the alkene is a terminal alkene and either a primary or a secondary carbenium ion is formed. Branched alkenes can also lead to tertiary carbenium ions. The probability for the existence of these ions is not random, because (i) their stability differs profoundly and (ii) they are interconverted (e.g., primary to secondary). The relative stability of the carbenium ions decreases in the order:

$$\text{tertiary} > \text{secondary} > \text{primary} > \text{ethyl} > \text{methyl}$$

In reality, these ions are not present as such but they form ethoxy species. For the discussion here it suffices to treat them as adsorbed carbenium ions.

QUESTION:

Why are ethyl and methyl carbenium ions less stable than other primary carbenium ions?

Alkanes are very stable and only react at high temperature in the presence of strong acids. They react via carbonium ions with the formation of H_2:

$$R-CH_2-CH_3 \ + \ H^+ \ \longrightarrow \ R-\overset{\overset{\displaystyle H}{|}}{\underset{\underset{\displaystyle H}{|}}{C^+}}-CH_3 \ \longrightarrow \ R-\overset{\overset{\displaystyle H}{|}}{\underset{\underset{\displaystyle }{}}{C^+}}-CH_3 \ + \ H_2 \tag{3.4}$$

The ion formed upon proton addition contains a penta-coordinated carbon atom and is referred to as a carbonium ion. It decomposes easily into a carbenium ion and H_2.

In a medium where carbenium ions are present, transfer of a hydride ion to a carbenium ion can be the predominant route:

$$\tag{3.5}$$

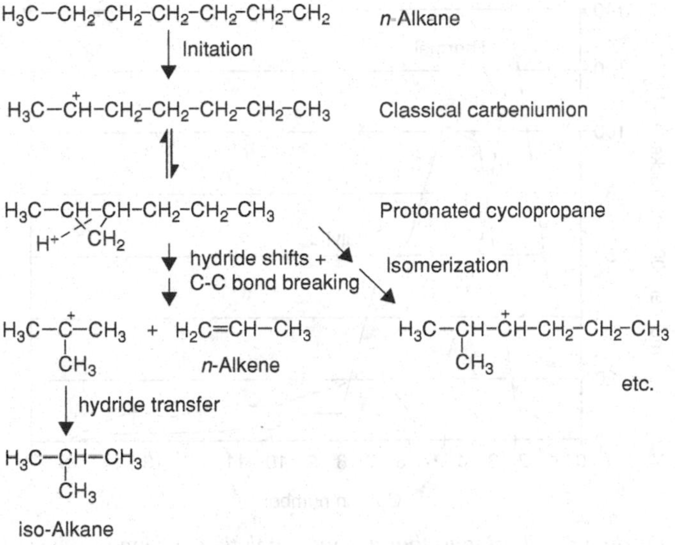

H₃C—CH₂-CH₂-CH₂-CH₂-CH₂-CH₂   *n*-Alkane

$$H_3C-CH_2-CH_2-CH_2-CH_2-CH_2-CH_2$$

| Initation

H₃C—ĊH-CH₂-CH₂-CH₂-CH₂-CH₃   Classical carbeniumion

⇅

H₃C—CH-CH-CH₂-CH₂-CH₃   Protonated cyclopropane
      H⁺⁻CH₂

↓ hydride shifts +   Isomerization
   C-C bond breaking

H₃C—Ċ⁺-CH₃ + H₂C=CH—CH₃   H₃C—CH-ĊH-CH₂-CH₂-CH₃
     CH₃   *n*-Alkene   CH₃
                                                    etc.

↓ hydride transfer

H₃C—CH—CH₃
      CH₃
iso-Alkane

Scheme 3.1   *Mechanism for catalytic cracking of alkanes, including isomerization.*

The presence of carbenium ions gives rise to a variety of reactions. Isomerization takes place via protonated cyclopropane intermediates (Scheme 3.1). The essential reaction of carbenium ions in catalytic cracking is *scission of C–C bonds*. For example:

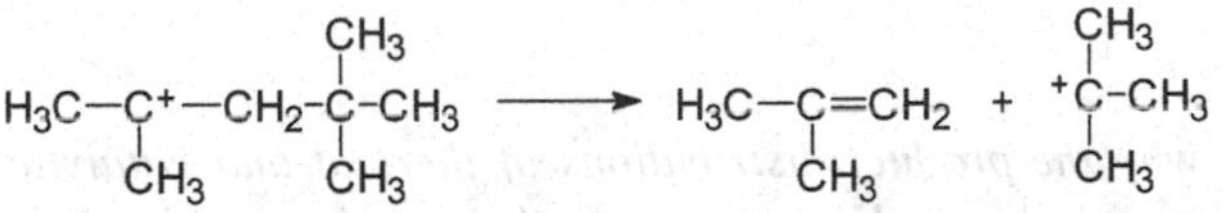

$$H_3C-C^+-CH_2-\overset{CH_3}{\underset{CH_3}{C}}-CH_3 \longrightarrow H_3C-\overset{}{\underset{CH_3}{C}}=CH_2 + {}^+\overset{CH_3}{\underset{CH_3}{C}}-CH_3$$

(3.6)

This reaction generates a smaller carbenium ion and an alkene molecule. The reaction rate depends on the relative stability of the reactant and product carbenium ions. When both are tertiary, as in the above example, then the reaction is fast. However, if a linear carbenium ion undergoes C–C bond breaking, a highly unstable primary carbenium ion would be formed, and the reaction would be very slow. Therefore, the reaction is believed to proceed via a protonated cyclopropane derivative as shown in Scheme 3.1.

A number of hydride shifts and the actual C–C bond breaking result in the formation of a linear alkene and a tertiary carbenium ion. The latter is converted to an iso-alkane by hydride transfer to a neutral molecule. The neutral molecule then becomes a new carbenium ion and the chain continues. Isomerization of linear alkanes into branched alkanes can take place in a similar way.

Of course, cracking of isomerized alkanes and isomerization of sufficiently large cracking fragments is always possible.

The following rules hold for catalytic cracking:

- Cracking occurs on β versus + charge;
- 1-Alkenes will be formed;
- Shorter chains than C_7 are not or hardly cracked.

QUESTION:

Explain these rules from the reaction mechanism shown in Scheme 3.1. Why is it improbable that C–C bond breaking of the "classical" carbenium ion would take place directly?

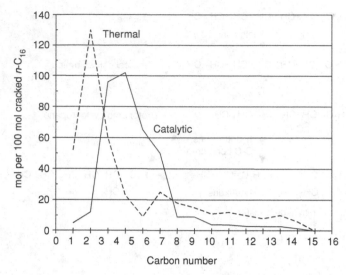

Figure 3.8   Product distribution in thermal and catalytic cracking of n-hexadecane [10].

This cracking mechanism explains why catalytic cracking is preferred over thermal cracking (Section 3.3 and Chapter 4) for the production of gasoline; in thermal cracking, bond rupture is random, while in catalytic cracking it is more ordered and, therefore, selective. Figure 3.8 shows the difference in product distribution resulting from thermal and catalytic cracking of *n*-hexadecane.

QUESTIONS:

> *Explain why the product distributions of thermal and catalytic cracking peak at different carbon numbers. (Hint: compare the reaction mechanisms of thermal (Chapter 4) and catalytic cracking.) Are the units of the y-axis correct in Figure 3.8? Alkylation and polymerization only play a minor role in catalytic cracking. Why?*

There are many pathways leading to coke formation. Important reactions are cyclization of alkynes, di- and poly-alkenes, followed by condensation and aromatization. By these reactions poly-aromatic compounds are formed, which are referred to as coke. The coke is deposited on the catalyst and causes deactivation.

QUESTIONS: *In "hydrocracking" the catalyst performs two catalytic functions, that is, hydrogenation of alkenes into alkanes and cracking reactions are catalyzed simultaneously. Why would the temperature for hydrocracking be lower than for FCC? Why would the reaction be carried out at high hydrogen partial pressure?*

3.4.2.2 Catalysts for Catalytic Cracking

The types of catalysts used in catalytic cracking have changed dramatically over the years. In the past, aluminum trichloride (AlCl$_3$) solutions were used, which resulted in large technical problems such as corrosion and extensive waste streams. Subsequently, solid catalysts were used. From the cracking mechanism described

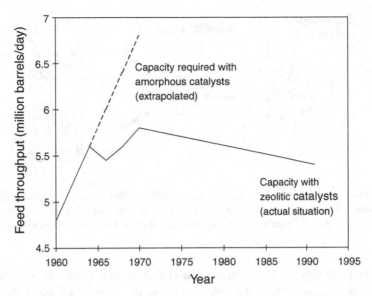

Figure 3.9   *Growth of catalytic cracking capacity in the United States [10].*

previously it is evident that catalytic cracking calls for acid catalysts. The improvement of cracking catalysts has been crucial in the development of the catalytic cracking process. Initially, acid-treated clays were used. Later, it appeared that synthetic materials, namely *amorphous silica-aluminas*, had superior properties due to their higher activity, higher thermal and attrition stability, and better pore structure. A great breakthrough in catalytic cracking was the discovery of *zeolites*, which are even better catalysts: they are more active and more stable (less coke, higher thermal stability).

Figure 3.9 shows that with zeolites the feed throughput can be much lower for the same product yield. With current zeolitic catalysts much less gas oil needs to be processed in an FCC unit than would have been the case with the early amorphous catalysts.

In contrast to amorphous silica-alumina, which contains sites with very different acidic strengths, many zeolites contain very well defined (strong) acid sites. The zeolites used in FCC are three-dimensional structures constructed of SiO_4 and AlO_4 tetrahedra, which are joined by sharing the oxygen ions [11]. The definitive structure depends on the Si/Al ratio, which may vary from 1 to > 12. Figure 3.10 shows examples of zeolitic structures based on so-called sodalite cages (composed of SiO_4 and AlO_4 tetrahedra).

QUESTIONS:

Why would of the three zeolites in Figure 3.10 only the Y zeolite be applied in FCC? Earlier it was stated that carbenium ions do not occur as such but in the form of ethoxy species. Explain this from the structure of the solid acids used.

Zeolites have many interesting characteristics (see also Chapter 10). Important for FCC are their high acidity, their low tendency towards coke formation, and their high thermal stability.

Zeolite crystals are not used as such. They are too active in the catalytic cracking reactions; the reactivity is approximately 10^6 mol/($m_R^3 \cdot$s), which cannot be dealt with in a practical industrial reactor (see Chapter 1). In addition, the pores are too small for a large part of the feed. Moreover, due to the small pore diameters,

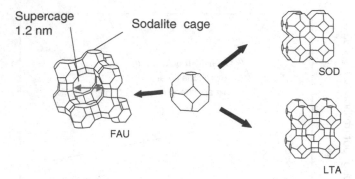

Figure 3.10   *Examples of zeolites based on sodalite cages; left: zeolite Y (Faujasite), used in FCC; middle: part of sodalite cage (truncated octahedron); right: sodalite mineral (top), zeolite A 9 (bottom), used in drying of gas and liquid streams and separation of linear and branched alkanes.*

crystal sizes have to be extremely small to minimize internal mass transfer limitations. Therefore, the zeolite particles are diluted with silica-alumina, a macro/meso-porous matrix material (see Table 3.5).

Although the matrix is only moderately active, its function must not be underestimated: the FCC catalyst matrix provides primary cracking sites, generating intermediate feed molecules for further cracking into desirable products by acidic zeolitic sites. The synthesis of commercial cracking catalysts is described in Box 3.2.

3.4.2.3   *Product Distribution*

The feed to a catalytic cracker is usually an atmospheric or vacuum gas oil, but other heavy feeds can also be processed. Figure 3.11 shows the improvement in product selectivity with time. Gasoline yields with the present catalysts are between 40 and 50 wt%.

Coke formation is a mixed blessing; coke deactivates the catalyst, so regeneration of the catalyst is necessary. Regeneration is carried out by combustion in air at 973 K:

$$C \xrightarrow{O_2} CO/CO_2 \qquad\qquad (3.7)$$

This reaction is highly exothermic. When considering that cracking is an endothermic process, an excellent combination of these processes suggests itself: the two are coupled and the regeneration provides the necessary energy for the endothermic cracking operation!

Table 3.5   *Typical pore sizes.*

Porosity range	Pore diameter (nm)
Micropores	<2
Mesopores	2−50
Macropores	>50

Box 3.2 Synthesis of Commercial Cracking Catalysts

Figure B3.2.1 illustrates the production of commercial cracking catalysts. The zeolite crystals are formed in a stirred-tank reactor. After washing they are ion exchanged with rare earth (RE) ions, introduced as the chloride: $RECl_3$ (zeolites are not acidic in the sodium form, after exchanging sodium ions with rare earth ions they are). In another stirred-tank reactor amorphous silica-alumina is produced. The two slurries are mixed and dried in a spray dryer, producing a fine powder of catalyst particles consisting of a silica-alumina "matrix" with the zeolite crystals dispersed in it.

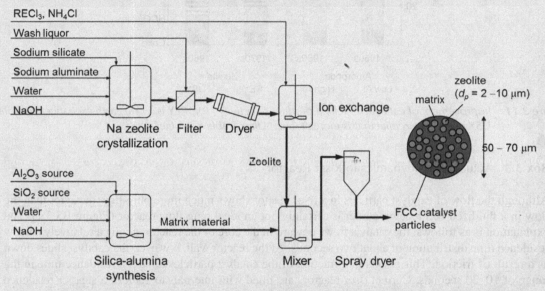

Figure B3.2.1 *Production of commercial cracking catalysts.*

3.4.2.4 Processes

Originally, fixed bed and moving bed reactors were used for the catalytic cracking process, but nowadays the predominant process is Fluid Catalytic Cracking (FCC). The first FCC processes were based on fluidized beds (Figure 3.12).

In the reactor heat is consumed by the cracking reactions, while in the regeneration unit heat is produced. The catalyst is circulated continuously between the two fluidized bed reactors and acts as a vehicle to transport heat from the regenerator to the reactor.

The use of a fluidized bed reactor for catalytic cracking has some disadvantages. Firstly, relatively large differences occur in the residence time of the hydrocarbons, which results in a far from optimal product distribution. Secondly, the fluidized bed reactor acts as a continuously stirred-tank reactor (CSTR) regarding the catalyst particles, causing age distribution. In modern processes, riser reactors are applied for the cracking reaction, because they are closer to plug-flow reactors (Box 3.3).

In the past, riser reactors could not be employed because the catalyst originally was not sufficiently active to apply riser technology. So, the change in technology was driven by a combination of catalyst improvement and chemical reaction engineering insights.

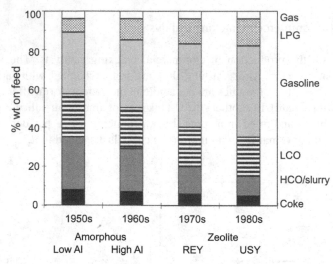

Figure 3.11   *Improvement of catalytic cracking product selectivity [10]; REY is a rare earth exchanged Y-zeolite; USY refers to a zeolite that is more stable, "Ultra Stable Y zeolite".*

Box 3.3   Actual Hydrodynamics in Riser Reactor

Although the flow of catalyst particles in a riser reactor shows much more plug-flow behavior than the flow in a fluidized bed reactor, a riser is certainly not an ideal plug-flow reactor (Figure B3.3.1). The explanation is as follows: the catalyst flows upward in the core of the reactor, giving a relatively narrow residence time distribution of about two seconds. At the reactor wall, however, the catalyst slides down as a result of friction. This results in a fraction of the catalyst particles having a residence time in the range of 10–20 seconds. Current riser reactors are fitted with internals to achieve a sharper residence time distribution.

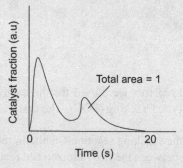

Figure B3.3.1   *Typical residence time distribution in an industrial FCC riser unit [12].*

QUESTION:

>   *Riser reactors are not ideal plug-flow reactors. Nevertheless, practice, FCC riser reactors have been reported to operate very close to plug flow. What would be the reason? (Hint: the activity of fresh catalyst is much higher than that of deactivated catalyst: the activity drops significantly within a second.)*

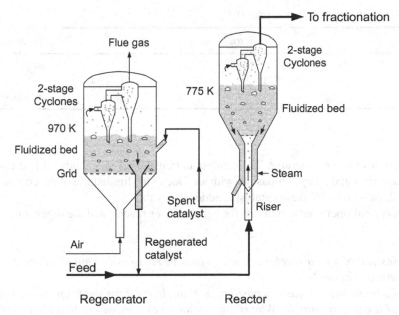

Figure 3.12 *Early FCC process based on combination of two fluidized beds.*

Figure 3.13 shows a flow scheme of a modern fluid catalytic cracking unit employing riser technology. The cracker feed is diluted with steam for better atomization and fed to the riser reactor together with regenerated catalyst. The mixture flows upward and cracking takes place in a few seconds. The spent catalyst is separated from the reaction mixture in a cyclone. Steam is added to the downcomer in order to strip adsorbed heavy hydrocarbons off the catalyst. It also creates a buffer between the reducing environment in the riser and the

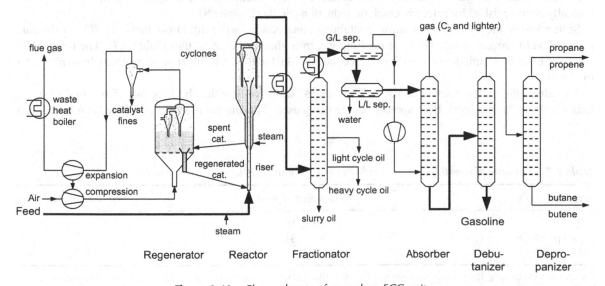

Figure 3.13 *Flow scheme of a modern FCC unit.*

Table 3.6 *Typical operation conditions in riser FCC.*

	Reactor	Regenerator
Temperature (K)	775	973
Pressure (bar)	1	2
Residence time	1–5 s	minutes/half hour

oxidizing environment in the regenerator. The catalyst is transported to a fluidized bed regenerator where coke is removed from the catalyst by combustion with air. Due to the limited life of the commercial catalysts – approximately 30 days – up to 5% fresh catalyst is added every day.

Table 3.6 shows typical operating conditions for both the riser reactor and the regenerator.

QUESTIONS:

> *Instead of a riser reactor a downer could be envisaged. What would be the advantages and disadvantages?*
>
> *Estimate the flue gas composition. Schematically draw the temperature versus time profile of a catalyst particle. Why is regeneration not carried out in a riser reactor? Could you design an improved reactor? Why does the regenerator operate at higher pressure than the riser reactor?*

3.4.2.5 Environment

The FCC unit is one of the most polluting units in the refinery [13]. During catalyst regeneration sulfur oxides, referred to as SO_x (SO_2 and SO_3 with $SO_2/SO_3 > 10$), are formed by oxidation of the sulfur present in the coke. In addition to SO_x, NO_x (NO and NO_2) is formed. Depending on regulations, the SO_x and NO_x emissions in the flue gas might have to be reduced. The reduction of the NO_x content in the flue gas is typically accomplished by selective catalytic reduction (SCR) (Chapter 8).

Sulfur leaves the FCC unit as sulfur-containing compounds in the liquid products, as SO_x in the flue gas, and as hydrogen sulfide (H_2S) in the light gas from the fractionation unit (Table 3.7). The total sulfur emission of an FCC unit depends on the sulfur content of the feed. This sulfur content may vary from about 0.3 to 3 wt%.

To reduce the sulfur content in the FCC products, sulfur must either be kept out of the FCC feed or removed from the products. Both approaches are being used. The amount of sulfur in the feed can be reduced

Table 3.7 *Example of the distribution of sulfur in FCC products [14].*

"Product"	% of sulfur in feed	ton/day* S (SO_2)
H_2S	50 ± 10	83.6
liquid products	43 ± 5	71.9
coke → SO_x	7 ± 3	11.7 (23.3)

* 50 000 barrels/day unit, 500 tons catalyst inventory, 50 000 ton/day catalyst circulation rate, catalyst to oil ratio = 6 kg/kg, feedstock 2 wt% sulfur.

by hydrotreating (Section 3.4.5.1), thus providing lower sulfur amounts in all three product streams. If the feed is not hydrotreated or if the SO_x emissions from an FCC unit are still too high, they can be reduced in two ways. A standard technology is flue gas desulfurization (FGD), which is often applied in power generation [15]. Another method of SO_x control is reserved for the FCC process only (Box 3.4).

Box 3.4   Reduction of SO_x Emissions from FCC Units

Figure B3.4.1 shows the principle of the removal of SO_2 in an FCC unit. It involves the addition of a metal oxide (e.g., MgO, CeO or Al_2O_3), which captures the SO_x in the regenerator and releases it as H_2S in the reactor and stripper. SO_x removal is based on the difference in stability of sulfates and sulfites. An oxide is selected such that the sulfate (formed by reaction with SO_x) is stable under oxidizing conditions (in the regenerator) and unstable under reducing conditions (in the riser).

Oxidation of SO_2 to SO_3 in regenerator and subsequent adsorption on metal oxide (MO):

$$2\,SO_2 + O_2 \longrightarrow 2\,SO_3$$

$$SO_3 + MO \longrightarrow MSO_4 \qquad \text{(stable in regenerator)}$$

Reduction of the metal sulfate in the riser reactor and release of H_2S:

$$MSO_4 + H_2 \longrightarrow MSO_3 + H_2O \qquad \text{(unstable in riser)}$$

$$MSO_3 \xrightarrow{\;H_2\;} \begin{array}{l} MO + H_2S \\ MS + H_2O \end{array}$$

Regeneration to form the metal oxide in the stripper with release of H_2S:

$$MS + H_2O \longrightarrow MO + H_2S$$

Figure B3.4.1   *Removal of SO_2 in an FCC unit.*

The metal oxide can either be incorporated in the FCC catalyst or added as a separate solid phase. The latter option has the advantage that the metal oxide supply is flexible and can be adjusted for feeds with different sulfur contents. The commercially available sulfur traps can remove up to 80% of the sulfur from the regenerator.

The hydrogen sulfide released in the reactor and stripper can be removed by absorption and subsequent conversion in a Claus plant (Chapter 8) together with the normal quantities of hydrogen sulfide formed.

0.51 - 0.55 nm

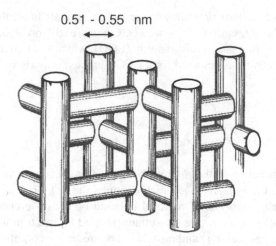

Figure 3.14   Channel system in zeolite ZSM-5.

QUESTIONS:

In principle, hydrotreating can be applied to the feed of an FCC unit or to the products (gasoline, middle distillates) [16,17]. Which option would you choose and why? Does your choice depend on the type of feedstock and/or on regulations (concerning sulfur content in products, flue gas)?

With the demand for gasoline increasing, and the market for heavy fuel oil declining, more and more heavier feedstocks with accompanying higher sulfur contents are being processed by FCC units. For these heavy feeds certainly, hydrotreating is necessary in order to comply with the gasoline and flue gas standards.

3.4.2.6   Production of Lower Alkenes

There is an increasing incentive for the production of larger amounts of C_3 and C_4 alkenes in FCC units, because these are valuable as petrochemical feedstocks (see also Chapter 4). In fact, FCC units in the Middle and Far East have the sole purpose of producing petrochemical feedstocks, in particular propene and aromatics.

   The production of larger amounts of lower alkenes can be achieved by the addition of a small amount of ZSM-5 (Zeolite Socony Mobil 5, Figure 3.14) to a conventional Y zeolite catalyst [18, 19]. ZSM-5 has narrower pores and thus is only accessible to linear or slightly branched alkanes and alkenes and not to the more branched ones and aromatic compounds. Therefore, low-octane hydrocarbons in particular are cracked in ZSM-5, which has the additional advantage of improving gasoline quality due to the formation of aromatics, although at the expense of gasoline yield. This is an example of the advantageous use of the shape selectivity of zeolite catalysts (see also Chapter 10).

QUESTION:

Does the statement that only linear and some monobranched alkanes are converted in the ZSM-5 zeolite added to the catalyst conflict with the theory that branched alkanes are cracked most efficiently during FCC?

3.4.2.7 *Processing of Heavier Feedstocks*

Many modern FCC units are designed to process significant amounts of vacuum residue [16]. Catalytic cracking of this heavy residue has become possible as a result of improved reactors and strippers, improved feed injection and gas/solid separation, and the use of catalyst cooling in the regenerator or a second regeneration zone to remove excess heat. This is necessary because cracking of vacuum residue generates much more coke than cracking of conventional FCC feeds, and excess heat is generated when the extra coke is burned off the catalyst.

Another problem of vacuum residue is that its metal content (mainly nickel and vanadium) can be very high, resulting in destruction of the catalyst, increased coke formation, and reduced product yield. Furthermore, vacuum residue contains more sulfur than conventional feeds, which leads to a high amount of sulfur in products and regenerator off-gas. Therefore, the residue is first hydrotreated before sending it to the FCC unit. Hydrotreating removes the metals (hydrodemetallization (HDM)) and sulfur from the feed (Section 3.4.5).

QUESTION:

> *Why is excess heat production in the regenerator an issue in FCC of heavy feedstocks?*
> *Discuss the pros and cons of catalyst cooling and the use of a second regeneration zone.*

3.4.3 Catalytic Reforming

Catalytic reforming is a key process in the production of gasoline components with a high octane number. It also plays an important role in the production of aromatics for the chemical industry. Furthermore, catalytic reforming is a major source of hydrogen. Feedstocks are straight-run naphtha and other feeds in the gasoline boiling range (about C_6–C_{11}). During catalytic reforming the change in molecular weight of the feed is relatively small, as the process mainly involves internal rearrangement of hydrocarbons.

3.4.3.1 *Reactions and Thermodynamics*

Scheme 3.2 shows examples of major reactions that occur during catalytic reforming.

Reaction	Example		$\Delta_r H_{298}$ (kJ/mol)
Isomerization			-4
Cyclization		$+ H_2$	33
Aromatization		$+ 3 H_2$	205

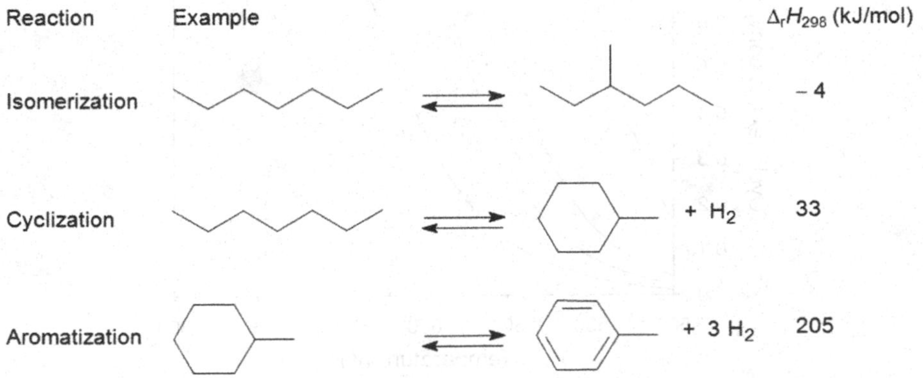

Scheme 3.2 *Reactions occurring during catalytic reforming.*

Table 3.8 *Typical catalytic reforming feedstock and product analysis (vol%) [2].*

Component	Feed	Product
Alkanes	45–55	30–50
Alkenes	0–2	0
Naphthenes	30–40	5–10
Aromatics	5–10	45–60

Except for isomerization, which only involves a minor enthalpy change, all reactions that occur during catalytic reforming are moderately to highly endothermic, particularly the dehydrogenation of naphthenes (aromatization), which is the predominant reaction during reforming (Table 3.8).

A particularly undesirable side reaction that has to be dealt with is the formation of coke as a result of thermal cracking. Additional methane is formed by reaction of coke with hydrogen. Other, undesired, reactions that occur are the formation of light alkanes ($<C_5$) by hydrocracking of alkanes and naphthenes and by hydrodealkylation of aromatics such as toluene and ethylbenzene.

Thermodynamically favorable reaction conditions are high temperature and low pressure. Figure 3.15 illustrates this for the conversion of cyclohexane into benzene. At atmospheric pressure the thermodynamically attainable conversion is essentially 100% at temperatures over 620 K. At higher pressure much higher temperatures are required for high conversions. The equilibrium conversions shown in Figure 3.15 were obtained without addition of hydrogen to the feed. In practice, however, product hydrogen is recycled to limit coke formation. This has an adverse effect on the cyclohexane conversion, as shown in Figure 3.16.

QUESTION:

Explain the difference between the influence of total pressure (large) and the H_2/cyclohexane feed ratio at constant total pressure (small) on the cyclohexane conversion.

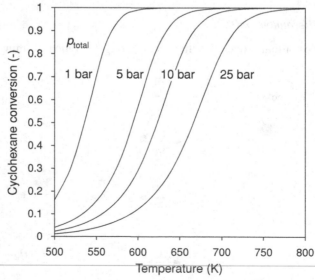

Figure 3.15 *Effect of temperature and pressure on the aromatization of cyclohexane without hydrogen addition.*

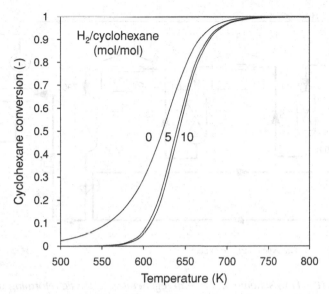

Figure 3.16 *Effect of temperature and H$_2$/cyclohexane feed ratio on the aromatization of cyclohexane at constant total pressure of 10 bar.*

The dehydrogenation and aromatization reactions are catalyzed by metal catalysts. The isomerization and cracking reactions proceed via carbenium ions and, therefore, require an acidic catalyst (Section 3.4.2). In catalytic reforming, the catalysts consist of platinum metal (which explains why the name Platforming is also often used) dispersed on alumina. New generation bimetallic catalysts also contain rhenium ("Rheniforming") or iridium in order to increase catalyst stability. The alumina support has some acidity, which is enhanced by adding chlorine. Figure 3.17 shows a schematic of a catalytic reforming catalyst.

QUESTION:
> *Explain why alumina becomes strongly acidic by chlorination. (Hint: compare CH$_3$COOH and CH$_2$ClCOOH; which is most acidic?)*

3.4.3.2 Feed Pretreatment

Platinum, the main metal of most catalytic reforming catalysts, is sensitive to deactivation by hydrogen sulfide, ammonia and organic nitrogen and sulfur compounds [2]. Therefore, naphtha pretreatment by hydrotreating (Section 3.4.5.1) is required. The organic sulfur and nitrogen compounds are catalytically converted with hydrogen into hydrogen sulfide (H$_2$S) and ammonia (NH$_3$), which are then removed from the hydrocarbon stream in a stripper. Hydrogen needed for the hydrotreater is obtained from the catalytic reformer.

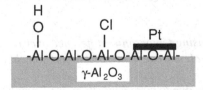

Figure 3.17 *Schematic of catalytic reforming catalyst.*

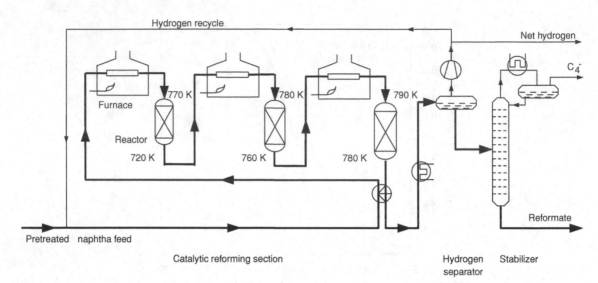

Figure 3.18   *Flow scheme of the semi-regenerative catalytic reforming process.*

3.4.3.3   Reforming Processes

The most important issue in the development of a catalytic reforming process is to find a compromise between optimal octane number and product yield (low hydrogen pressure optimal) and optimal catalyst stability, that is, minimal coke formation (high hydrogen pressure optimal). The two main types of processes applied in catalytic reforming are *semi-regenerative reforming* (SRR) and *continuous-regenerative reforming* (CRR).

3.4.3.3.1   Semi-Regenerative Reforming (SRR)

SRR units consist of several (three to four) adiabatic fixed bed reactors in series (Figure 3.18) with intermediate heating. Early SRR units (~1950) were operated at relatively high pressure (25–35 bar) in order to maximize catalyst life [3]. The development of improved catalysts that had a lower tendency for coke formation enabled operation at lower pressure (15–20 bar), while maintaining the same catalyst life (typically one year).

The feed is preheated and sent to the first reactor. In this reactor the kinetically fastest reactions occur. These are the dehydrogenation reactions of naphthenes to aromatics. To maintain high reaction rates the product gases are reheated in a furnace before being sent to a second reactor, and so on. The temperatures shown in Figure 3.18 are indicative of industrial practice.

The effluent from the last reactor is cooled by heat exchange with the feed. Then hydrogen is separated from the liquid product and partly recycled to the process. The net hydrogen production is used in other parts of the refinery, for instance in the naphtha hydrotreater. The liquid product is fed to a stabilizer in which the light hydrocarbons are separated from the high-octane liquid reformate.

QUESTIONS:

> *Explain the decreasing temperature difference over the reactors.*
> *What can you say about reactor sizes? (Would you choose all reactors of equal size?)*

Some catalyst deactivation, mainly due to coke deposition, is unavoidable. It is normal practice to balance decreased activity by gradually increasing the temperature. This temperature increase, however, results in lower selectivities and higher rates of coke deposition. The catalyst is regenerated after 0.5−1.5 years by carefully burning off the coke deposits in diluted air with the unit taken off-stream. The catalyst is also treated with chlorine (Cl_2) in order to replenish the Cl-content of the catalyst, which is reduced by the coke removal step, and to re-disperse the platinum.

3.4.3.3.2 Fully-Regenerative Reforming

For further reduction of the operating pressure to improve the product octane number and the yield on liquid products and hydrogen, catalyst development is not sufficient. Therefore, process innovations were needed. In the 1960s, *fully-regenerative* or *cyclic regenerative reforming* was developed. This technology still employs fixed bed reactors, but an additional "swing" reactor is present. It is thus possible to take one reactor off line, regenerate the catalyst, and then put the reactor back on line. In this way regeneration is possible without taking the reforming unit off stream and losing production [3], which means that a lower hydrogen partial pressure, and thus operating pressure, can be used.

3.4.3.3.3 Continuous-Regenerative Reforming (CRR)

A real innovation was the catalytic reformer with a moving catalyst bed, which enabled continuous regeneration of the catalyst, that is, the CRR process (Figure 3.19). This process, introduced by UOP and IFP, enabled the use of a much lower operating pressure (3–4 bar), since coke deposits were no longer such an issue, as they were continuously being burned off. Over 95% of new catalytic reformers presently designed are reformers with continuous regeneration [3].

Regenerated catalyst enters the first reactor at the top. The catalyst withdrawn from the bottom enters the next reactor, and so on. The catalyst from the fourth reactor is sent to the regenerator. A continuous flow of catalyst particles is crucial for this system. In practice, spherical particles are used. Downstream processing of the crude reformate is similar as in the SRR process.

QUESTION:
> *What would be the most practical way of arranging the four reactors spatially?*

Summarizing, the different processes mainly differ in the manner and frequency of catalyst regeneration, which in turn determines the operating pressure and temperature, and thus the product quality and yield. Table 3.9 shows typical operating conditions used in practice for the three reformer types.

Catalytic reforming is primarily applied to improve the gasoline properties, but it is also used for the production of aromatics. In this case the operating conditions are somewhat more severe (higher temperature, lower pressure) than those for the production of high-octane gasoline.

Table 3.9 *Operating conditions in catalytic reforming.*

	Semi	Fully	Continuous
Feed H_2/HC (mol/mol)	10	4−8	4−8
Pressure (bar)	15−35	7−15	3−4
Temperature (K)	740−780	740−780	770−800
Catalyst life	0.5 − 1.5 y	days − weeks	days − weeks

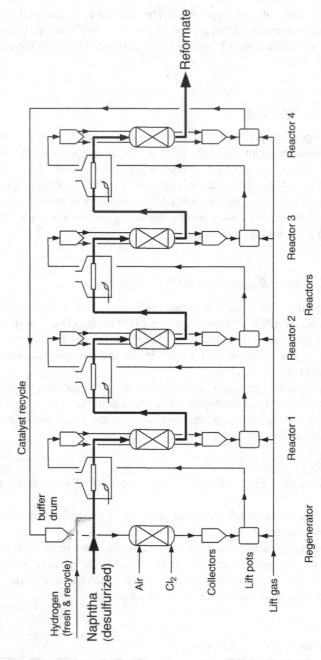

Figure 3.19 Flow scheme of the continuous-regenerative reforming process (IFP).

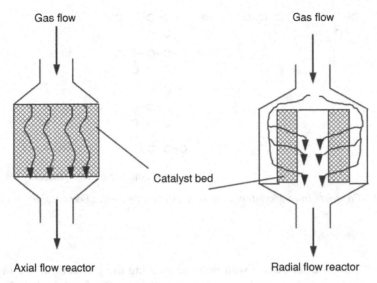

Figure 3.20 *Gas flow in reactors used for catalytic reforming.*

QUESTIONS:

Why, despite the fact that hydrogen is produced during reforming reactions, is hydrogen added to the reforming process?

How much is the octane number increased for a conceptual feedstock represented by cyclohexane, going from the semi- to the continuous-regenerative process? (Assume equilibrium.)

3.4.3.4 Reactors

During reaction the atmosphere is reducing, but during regeneration it is oxidizing. In the semi-regenerative and fully-regenerative processes this calls for materials that can withstand these cyclic conditions.

In all catalytic reforming processes that have been discussed adiabatic fixed or moving bed reactors are employed. Initially, axial flow reactors were used, but later often radial flow reactors were introduced because of their lower pressure drop (Figure 3.20) [20].

QUESTIONS:

Discuss potential material problems (corrosion, reactor walls, attrition of catalyst particles, etc.) for the three reforming technologies.

3.4.4 Alkylation

Alkylation is the reaction of isobutene with alkenes to form higher branched alkanes ("alkylate"). The aim is the production of gasoline components with a high octane number from low molecular weight alkenes (propene, butenes, and pentenes) and isobutane. The advantages of alkylation are that gas-phase molecules are eliminated and a valuable liquid product is formed, that is, gasoline with high octane number (87 up to 98).

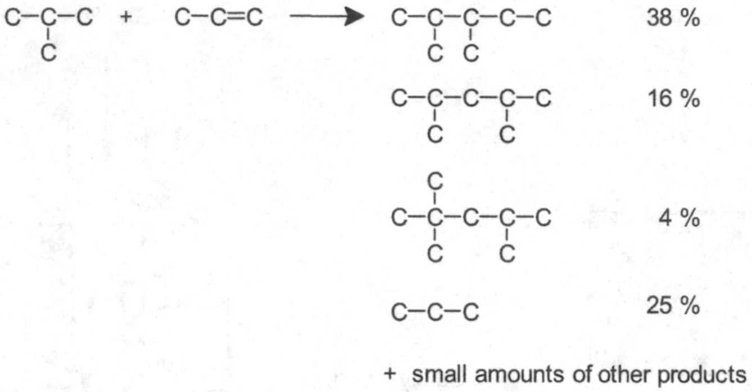

$$
\begin{array}{ccc}
\underset{\overset{|}{C}}{C-C-C} \;+\; C-C{=}C & \longrightarrow & \underset{\overset{|}{C}\;\overset{|}{C}}{C-C-C-C-C} \qquad 38\,\% \\
\end{array}
$$

C–C–C–C–C   16 %
   | |
   C   C

   C
   |
C–C–C–C–C   4 %
   | |
   C   C

C–C–C   25 %

+ small amounts of other products

Scheme 3.3   *Products from alkylation of isobutane with propene. Percentage data from [21].*

3.4.4.1   Reactions

An example is the alkylation of isobutane with propene yielding the products shown in Scheme 3.3. The alkylation reaction occurs via carbenium ions, as shown in Scheme 3.4 for the alkylation of propene with isobutane.

Besides alkylation, undesired side reactions occur, in particular oligomerization/polymerization of the alkene. In that case the cation formed in the initiation reaction may react with an alkene molecule (in this example propene) resulting in oligomers ($2C_3H_6 \rightarrow C_6H_{12} \rightarrow C_9H_{15}$, etc.). This is both a highly undesired and a relatively easy reaction, which may prevail unless isobutane is present in large excess. Other side reactions are hydrogen transfer, disproportionation, and cracking.

QUESTIONS:

Why is isobutane used and not n-butane? Which alkene will produce the highest octane alkylate?
What would be the source(s) of isobutane and alkenes?

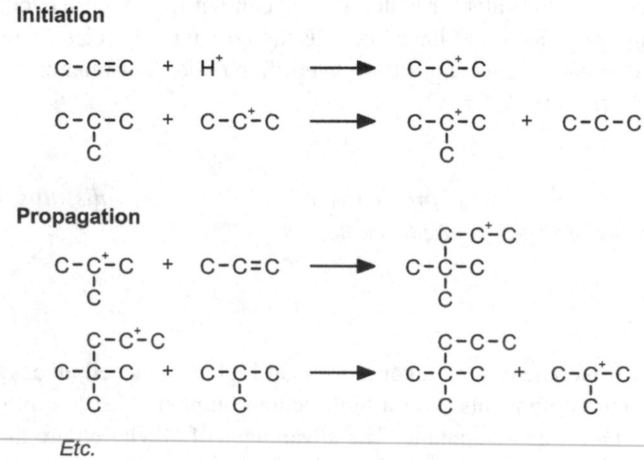

Initiation

C–C=C + H⁺ ⟶ C–C⁺–C

C–C–C + C–C⁺–C ⟶ C–C⁺–C + C–C–C
   | |
   C   C

Propagation

                               C–C⁺–C
                                  |
C–C⁺–C + C–C=C ⟶ C–C–C
   | |
   C   C

   C–C⁺–C   C–C–C
      | |
C–C–C + C–C–C ⟶ C–C–C + C–C⁺–C
   | | | |
   C   C   C   C

Etc.

Scheme 3.4   *Mechanism of alkylation of isobutane with propene.*

Alkylation is an exothermic reaction. The heat evolved when isobutane is alkylated with propene is about 80 kJ/mol alkylate produced. Alkylation is thus favored thermodynamically by low temperature. It is also favored by high pressure, since the number of moles is reduced in the reaction. Unfortunately, the same is true for the undesired side and consecutive reactions. The pressures applied in practice are low.

Excess isobutane is fed to the process to help the alkylation reaction to proceed to the right and to minimize polymerization of propene. Typical isobutane-to-propene feed ratios are in the range of 8 to 12.

Alkylation used to be a thermal process (770 K, 200–300 bar) but nowadays a catalytic process (<340 K, 2–20 bar) is employed using liquid acid catalysts.

QUESTION:
> *Why would the pressure be so high in the thermal process?*

3.4.4.2 Processes Using Liquid Acids as Catalyst

Two types of liquid acid catalysts are used, hydrofluoric acid (HF) and sulfuric acid (H_2SO_4) [3]. The sulfuric acid process is the oldest of the two. In this process, the reactor temperature has to be kept below 293 K to prevent excessive acid consumption as a result of oxidation–reduction reactions, which result in the formation of tars and sulfur dioxide. Therefore, cryogenic cooling is required. There are two main technologies for alkylation with sulfuric acid, the autorefrigeration process licensed by Exxon Research and Engineering and the effluent (or indirect) refrigeration process, licensed by Stratford Engineering Corporation.

Figure 3.21 shows a flow scheme of the autorefrigeration process. A cascade of reaction stages is combined in one reactor vessel. Every compartment is stirred. Isobutane and acid are fed to the first reactor compartment and pass through the reactor. The alkene feed is injected into each of the stages.

The reaction pressure is approximately two bar to keep the temperature at about 278 K. At this low temperature the reaction is rather slow, so relatively long contact times are required (20–30 minutes).

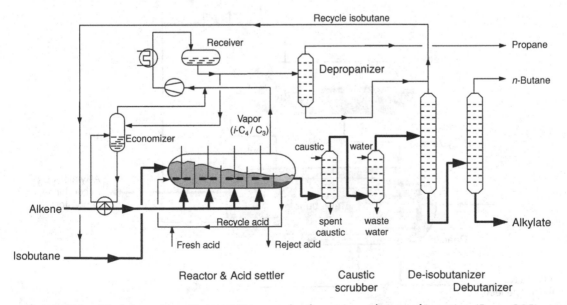

Figure 3.21 *Alkylation with sulfuric acid in cascade of reactors with autorefrigeration (Exxon R&E).*

The hydrocarbon/acid mixture is an emulsion, which is separated in the last compartment of the reactor vessel, the settler. The acid phase is recycled to the reactor. The hydrocarbon phase is neutralized by a caustic wash (e.g., with NaOH) and then fractionated in two distillation columns. Firstly, isobutane is recovered for recycle; then any *n*-butane present is removed to yield alkylate as the bottom product. Fractionation can also be carried out in one tower, with removal of *n*-butane as a side stream.

The heat of reaction is removed by evaporation of hydrocarbons, in particular isobutane and some propane, which may be present in the feed or formed in the reaction. The vapors are compressed and liquefied. A portion of this liquid is vaporized in an economizer, which is an intermediate-pressure flash drum, to cool the alkene feed. The remainder is sent to a depropanizer for removal of propane, which would otherwise accumulate in the system. Isobutane from the bottom of the depropanizer is recycled to the reactor.

QUESTIONS:

> In the past a single stirred-tank reactor was used for alkylation with sulfuric acid. The acid was removed in a separate settler. Give the advantages of the scheme in Figure 3.21 over the old process.
> What is meant by the term "cryogenic cooling"? Why is the need for cryogenic cooling generally a disadvantage of a process?

Figure 3.22 shows the reactor section of the effluent refrigeration process, licensed by Stratford Engineering Corporation.

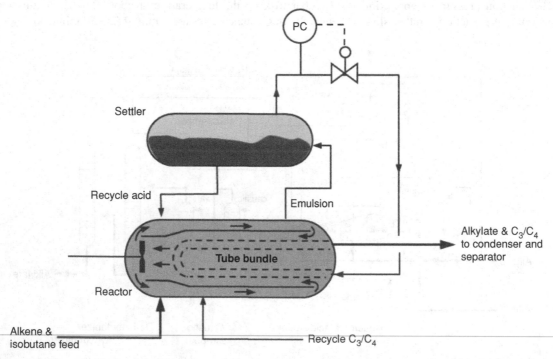

Figure 3.22   Schematic of alkylation with sulfuric acid in a Stratco contactor (Stratford Engineering Corp.).

The pressure in this reactor system is kept relatively high (4 bar) to prevent vaporization of light hydrocarbons in the reactor and settler [22]. The hydrocarbon feed and recycle acid form an emulsion by the action of the impeller. This emulsion circulates through the reactor. Part of the emulsion flows to the settler, where the hydrocarbon phase is separated from the acid phase.

Hydrocarbons from the settler are flashed across a pressure control valve and partially evaporated. The cold two-phase hydrocarbon mixture is passed through a large number of cooling coils contained within the reactor and is further evaporated. The vapors from the cooling coils are routed to the refrigeration system after separation from the alkylate.

QUESTIONS:

> *Explain the names for the two sulfuric acid-catalyzed processes described.*
> *Compare the energy requirements of the processes shown in Figures 3.21 and 3.22. Also compare the two processes with respect to concentration and temperature profiles. Which one is the most favorable in this respect?*

A major problem of the sulfuric acid alkylation process is the consumption of acid (about 100 kg/t of product!). The alternative hydrofluoric acid processes are more favorable in this respect. With these processes, the acid consumption is two orders of magnitude lower. The process scheme for alkylation with hydrofluoric acid is comparable to that of sulfuric acid alkylation. The most important difference is the installation of a regenerator for hydrofluoric acid in the hydrofluoric acid process.

Another advantage of the hydrofluoric acid alkylation process is that reaction can be carried out at higher temperature; because hydrofluoric acid is a non-oxidizing gaseous acid, the reaction can be carried out at 310 K, which eliminates the need for cryogenic cooling. Table 3.10 shows typical process conditions for isobutane/butene alkylation.

Both the hydrofluoric acid and sulfuric acid catalyzed processes suffer disadvantages, such as severe pollution and safety and corrosion problems. Of the two, sulfuric acid is safer because it stays liquid at ambient conditions. A leak of sulfuric acid is much more easily contained than a leak of hydrofluoric acid, which is highly volatile and toxic under ambient conditions. The sulfuric acid and hydrofluoric acid processes hold approximately equal market shares.

QUESTION:

> *Why would the sulfuric acid process produce much more waste than the hydrofluoric acid process?*

Table 3.10 *Typical process conditions for isobutane/butene alkylation.*

Catalyst	H_2SO_4	HF	zeolite
Temperature (K)	277–283	298–313	323–363
Pressure (bar)	2–6	8–20	20
Residence time (min)	20–30	5–20	200–250
Isobutane/butene feed ratio	8–12	10–20	6–15
Exit acid strength (wt%)	89–93	83–92	not applicable
Acid per reaction volume (vol%)	40–60	25–80	20–30
Acid consumption per mass of alkylate (kg/t)	70–100 (liquid)	0.4–1 (gas)	not applicable

3.4.4.3   Processes Using Solid Acids as Catalyst

Much research has been and is being carried out to replace the liquid acid processes by a safer and more environmentally benign solid-catalyzed process. This development is in harmony with the history of fluid catalytic cracking, which also has progressed from a thermal process via homogeneous catalysis to the present state of the art using zeolite catalysts. In fact, one of the furthest developed solid acid alkylation processes also uses riser technology.

The overwhelming success of zeolites in FCC in the 1960s triggered the exploration of the potential of these catalysts for isobutane/alkene alkylation, which can be viewed as the reverse of cracking. Various zeolites and other strong solid acids, among which are the so-called solid super-acids (with an acid strength higher than that of 100% sulfuric acid), have been tested for alkylation and found to be capable, in principle, of catalyzing the reaction with the same alkylate quality as that produced by the traditional liquid acid processes. Extensive research and development has been performed [21, 23, 24].

In the process development research it was found that all solid catalysts lose their activity and selectivity as a result of the formation of carbonaceous deposits (coke) by alkene polymerization reactions. From a chemical engineering viewpoint the deactivation problem has presented a challenge in choosing a suitable reactor configuration; the formation of coke must preferably be limited and at the same time relatively frequent regeneration must be possible.

Currently, four main technologies exist: the Alkylene process (UOP) [23, 24], the AlkyClean process (Lummus/Albemarle) [25], the ExSact process (Exelus) [23, 26, 27], and the Eurofuel process (Lurgi/Süd-Chemie) [28, 29].

Figure 3.23 shows a flow scheme of the Alkylene process developed by UOP during the late 1990s. The alkene feed is first dried and treated to remove impurities such as dialkenes, oxygenates, nitrogen compounds, and sulfur compounds. Then the alkene feed and recycled isobutane are combined and injected into the riser reactor. Regenerated catalyst is mixed with the feed at the bottom of the riser. The liquid-phase reactants

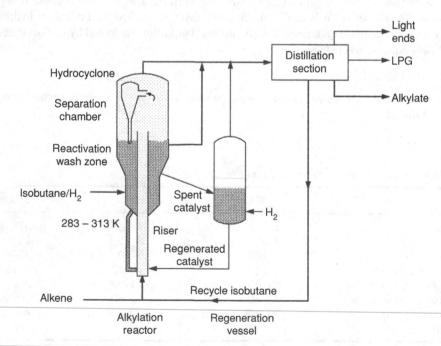

Figure 3.23   *Alkylene solid-catalyzed alkylation process; the catalyst used is AlCl$_3$/Al$_2$O$_3$.*

transport the catalyst up the riser in which the alkylation reaction occurs. Upon exiting the riser, the reaction products, excess isobutane and used catalyst enter into a separation chamber. The low liquid velocity in this chamber allows the catalyst particles to settle downward in the reactivation wash zone, where the catalyst is continuously reactivated by hydrogenation of the carbonaceous deposits with hydrogen dissolved in isobutane. Although the reactivation is nearly complete, some strongly adsorbed material remains on the catalyst surface, requiring a separate regeneration vessel where a hydrogen stream is used for periodic more complete reactivation in the vapor phase.

The hydrocarbons enter into a liquid-solid separator (hydrocyclone), where any entrained catalyst is separated from the hydrocarbons. Fractionation of the alkylate product is similar to that in liquid acid processes.

QUESTION:

> *For the Alkylene process, it is explicitly mentioned that dienes, oxygenates, and nitrogen and sulfur compounds are removed in a pretreatment step. Explain.*

Figure 3.24 shows a simplified diagram of the AlkyClean process. The process utilizes serial reaction stages with distributed (pretreated) alkene feed injection for high internal isobutane/alkene ratios. The reactor is claimed to achieve a high degree of mixing. Multiple reactors are used, which swing between reaction and regeneration. Similar to the Alkylene process, two regeneration phases with different severity are employed. A mild regeneration at reaction temperature and pressure with hydrogen dissolved in isobutane is performed frequently. When necessary, the catalyst is fully regenerated at higher temperature in a stream of hydrogen.

QUESTIONS:

> *The product of alkylation is used in the gasoline pool. Compare alkylate with the other components of the gasoline pool. Rank them with respect to environmental impact.*
>
> *Why is a high degree of mixing advantageous? How is this mixing achieved in the AlkyClean process? (Hint: How can a fixed bed reactor be transformed into a continuous stirred-tank reactor (CSTR) without changing the interior of the reactor?)*
>
> *Another option for catalyst regeneration is combustion of coke, as in the FCC process. Why is this method not suitable for this process?*
>
> *Despite the fact that technology is available for solid-catalyzed alkylation, refiners are reluctant to implement the process. What would be the reason(s)?*

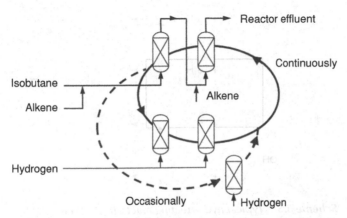

Figure 3.24 *AlkyClean reactor operating scheme; the catalyst is zeolite based.*

3.4.5 Hydroprocessing

Hydroprocessing is a class of conversions involving reaction with hydrogen. Hydroprocessing includes hydrotreating and hydrocracking.

3.4.5.1 Hydrotreating

The term "hydrotreating" is limited to hydrogenation and hydrogenolysis reactions in which removal of hetero-atoms (especially sulfur, nitrogen, and oxygen) and some hydrogenation of double bonds and aromatic rings take place. In contrast to hydrocracking (Section 3.4.5.2), in which size reduction is the main objective, in hydrotreating the molecular size is not drastically altered. Hydrogenation occurs by addition of hydrogen, for example, to double or triple bonds, whereas hydrogenolysis involves breaking of C–S, C–N bonds, and so on. Hydrotreating as defined here is sometimes called hydropurification.

The major objectives of hydrotreating are protection of downstream catalysts from hetero-compounds, which often act as poisons; improvement of gasoline properties such as odor, color, stability, and corrosion; and protection of the environment.

In hydrotreating the following terms are used, referring to the hetero-atom removed, that is, S, N, O and M (metal): hydrodesulfurization (HDS), hydrodenitrogenation (HDN), hydrodeoxygenation (HDO), and hydrodemetallization (HDM).

3.4.5.1.1 Reactions and Thermodynamics

Scheme 3.5 shows typical reactions that occur during hydrotreating. Figure 3.25 gives equilibrium data for selected HDS reactions.

The equilibrium coefficients are large and positive for all reactions considered under reaction conditions used in practice (625–700 K). Note that all hydrotreating reactions are exothermic. Although the equilibrium for all reactions is to the desulfurized products at these temperatures, the rates at which the equilibria are

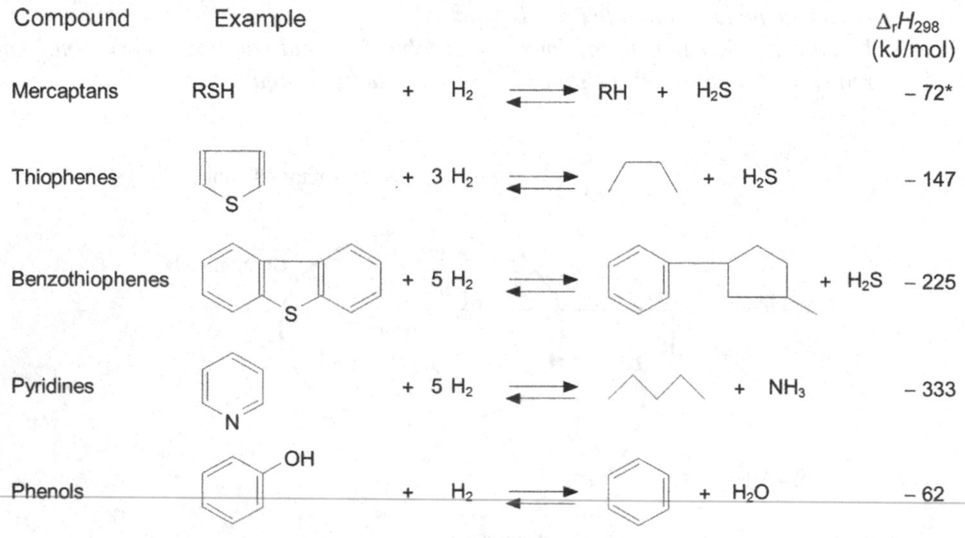

Scheme 3.5   Typical hydrotreating reactions; *with R = CH$_3$.

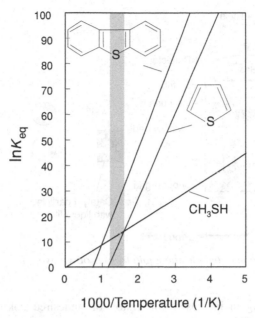

Figure 3.25 *Equilibrium data of selected HDS reactions; shaded area: temperature range of processes used in practice.*

achieved are very different for the various sulfur compounds. Substituted benzothiophenes, which are mostly present in heavy oil fractions (Figure 2.13), in particular show a low reactivity (see also Figure 3.30). Therefore, compared with light oil fractions, hydrotreating of heavy oil fractions requires higher temperature and hydrogen pressure.

QUESTION: *It has been proposed to use adsorption technology instead of hydrotreating. Give the pros and cons.*

3.4.5.1.2 Processes

Hydrotreating is essential as a pretreatment step for catalytic reforming of naphtha (Section 3.4.3) to protect the platinum-containing catalyst against poisoning by sulfur [20]. The sulfur content of the feed to the reforming unit should not exceed 1 ppm. The hydrogen used in naphtha hydrotreating is a by-product of catalytic reforming. On the other hand, when hydrotreating of heavy residues is performed, separate hydrogen production units are often required. In simple naphtha hydrotreating, the feed is evaporated and led through a fixed bed reactor. In hydrotreating of gas oils, evaporation is not possible due to the high boiling point of these feeds, and hence "trickle flow" operation is more suitable (Figure 3.26). In this mode, liquid (the heavy part of the hydrocarbon feed) and gas (hydrogen, the evaporated part of the feed) flow downward cocurrently.

Most hydrotreating of gas oils is performed in trickle bed (or trickle flow) reactors. The major concern is to prevent maldistribution, that is, an uneven distribution of the feed over the cross-section of the bed. In addition, wetting of the catalyst particles is a point of concern: dry regions might exhibit

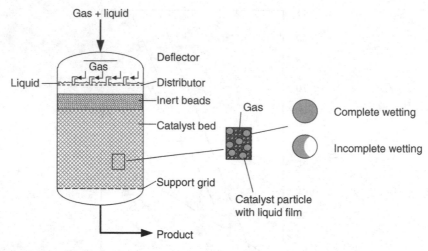

Figure 3.26   *Trickle bed reactor; the liquid trickles down, wetting the catalyst particles.*

relatively large rates, resulting in increased conversion or undesired coking because of the increased temperature.

Figure 3.27 shows a simplified flow scheme of a hydrotreating process involving a trickle bed reactor. The feed is mixed with fresh and recycle hydrogen and preheated by heat exchange with the reactor effluent. This reduces the fuel consumption of the furnace. The reactor usually contains several catalyst beds with intermediate cooling by hydrogen injection (quench). In the reactor, the sulfur and nitrogen compounds present in the feed are converted to hydrogen sulfide and ammonia, respectively. The effluent is cooled and sent to a hot high-pressure (HP) separator, where hydrogen is separated for recycle. The liquid from the HP separator is let down in pressure, sent to the low-pressure (LP) separator, and then to the product stripper.

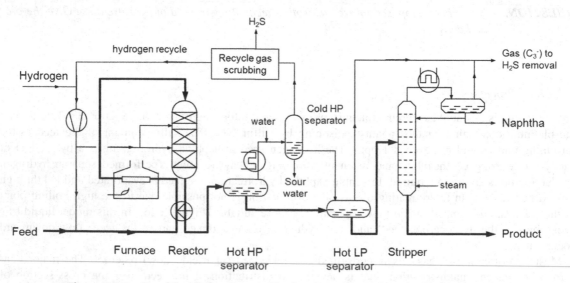

Figure 3.27   *Simplified flow scheme of hydrotreating involving a trickle bed reactor.*

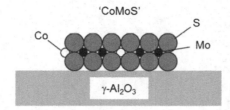

Figure 3.28 *Model of a hydrotreating catalyst.*

The hydrogen-rich vapor from the hot HP separator is cooled and water is injected to absorb ammonia and hydrogen sulfide. The cold HP separator then separates the vapor, liquid water, and liquid naphtha. Further removal of hydrogen sulfide from the hydrogen recycle stream is done by amine scrubbing (see also Chapter 6). In the hot LP separator a stream rich in hydrogen sulfide is removed, which is sent to a gas treating unit. The liquid stream is normally sent to a stripping column, where any remaining gases are removed by stripping with steam.

QUESTIONS:

Is hydrogen present in excess? What is the composition of the gas from the last separator? Estimate typical reactor sizes of trickle bed reactors (see Table 3.11).

Wetting is an issue in hydrotreating based on trickle bed reactor technology. In general, partial wetting is considered to be undesired. Is this true in all respects? (Are there also advantages?)

Do you expect all HDS processes to be gas-liquid processes?

Draw a flow scheme for hydrotreating of naphtha.

Table 3.11 shows typical hydrotreating process conditions. The severity of the processing conditions depends on the feed; for light hydrocarbon fractions it will be milder than for vacuum gas oils, while heavy residues require the highest temperatures and hydrogen pressures [30]. Moreover, it is common practice to compensate for deactivation of the catalyst by increasing the temperature of the reactor.

Typical hydrotreating catalysts are mixed metal sulfides, for example, CoS/MoS_2 on γ-Al_2O_3. Because of the high practical relevance a lot of research has been performed in optimizing the catalysts. It has been shown that the Co–Mo catalyst consists of a so-called CoMoS phase, as shown in Figure 3.28.

3.4.5.1.3 Environment

From the viewpoint of clean technology, hydrotreating is gaining increasing interest. A recent example is diesel fuel. Sulfur compounds in diesel fuel cause serious difficulties in catalytic cleaning of the exhaust gases

Table 3.11 *Typical process conditions in hydrotreating of naphtha and gas oils.*

	Naphtha	Gas oil
Temperature (K)	590–650	600–670
Pressure (bar)	15–40	40–100
H_2/hydrocarbon feed (Nm^3/kg)	0.1–0.3	0.15–0.3
WHSV (kg feed/($m^3_{cat} \cdot h$))	2000–5000	500–3000

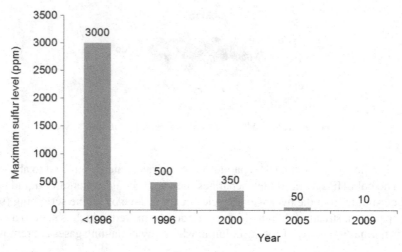

Figure 3.29   Development of maximum sulfur content in automotive diesel in the European Union.

from diesel fueled cars. Moreover, they contribute to the emission of particulate matter. Figure 3.29 shows the development of standards for the maximum sulfur level in automotive diesel fuel in Europe. Currently, the maximum sulfur level in diesel fuels is 10–15 ppm. This low level requires so-called deep desulfurization to produce ultra-low-sulfur diesel (ULSD).

QUESTIONS:

> *How do sulfur compounds contribute to particulate emissions?*
> *The sulfur level of gas oils may be assumed to be about 1.5 wt%. Estimate the required reactor sizes to reduce this level to 500 ppm and to 50 ppm, respectively.*

Figure 3.30 shows the activity of conventional and more advanced hydrotreating catalysts for HDS of a straight-run gas oil that was first hydrotreated to a level of 760 ppm S [31]. Only the most refractory sulfur compounds are still present in this pretreated feed, since the more reactive ones have already been converted in the first step. The figure suggests that with the conventional hydrotreating catalysts (CoMo and NiMo mixed sulfides with a structure like the one shown in Figure 3.28) the low sulfur levels required today cannot be achieved, except possibly with inconveniently large reactors.

As can be seen from Figure 3.30, supported noble metal catalysts like platinum and especially Pt/Pd on amorphous silica-alumina (ASA) are also active and can achieve much better conversions. However, these catalysts are expensive. In fact, a tremendous amount of research has been done aiming at improving the "classical" CoMo and NiMo catalysts. This research has been remarkably successful: the best CoMo and NiMo catalysts on the market are now able to realize sulfur contents down to 1 ppm [32]. It has recently been reported that adding noble metals as promotors to CoMo and NiMo catalysts also improves the HDS activity of classical catalysts [33, 34].

During the last decades, many new engineering concepts have been developed to desulfurize the least reactive sulfur species [34]. Most of these processes consist of two stages with intermediate hydrogen sulfide and ammonia removal. In the first stage mainly HDS and HDN occur over a conventional hydrotreating catalyst, while in the second stage ultra-deep HDS and hydrogenation of aromatics takes place on improved catalysts.

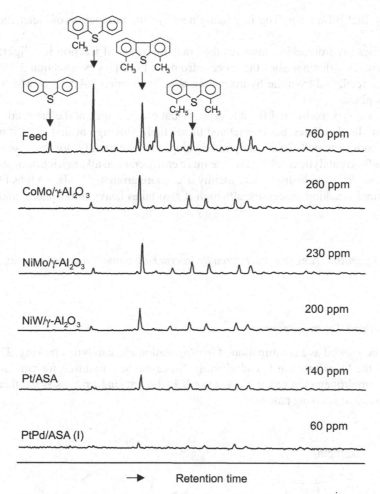

Figure 3.30 *Activity of various catalysts for HDS of gas oil that had undergone HDS up to 760 ppm; 613 K; 50 bar; ASA = amorphous silica-alumina [31].*

QUESTIONS:

Suggest a reactor configuration for deep desulfurization of straight-run gas oil.
Why should aromatics be hydrogenated?
Why is a two-stage process with intermediate hydrogen sulfide and ammonia removal favorable?

3.4.5.2 *Hydrocracking*

Hydrocracking [2, 3, 18, 35, 36] is a catalytic oil refinery process of growing importance. Heavy gas oils and vacuum gas oils are converted into lighter products, that is, naphtha, kerosene, and diesel fuel. Factors contributing to its growing use are the increasing demand for transportation fuels, especially diesel, and the decline in the heavy fuel oil market. The increasing need for the production of clean fuels has also had a significant impact.

As the name implies, hydrocracking involves the cracking of an oil fraction into lighter products in the presence of hydrogen. This distinguishes the process from the FCC process (Section 3.4.2), which does not have hydrogen in the feed, and from the hydrotreating process (Section 3.4.5.1), in which virtually no C–C bond breaking takes place.

Hydrocracking is a very versatile and flexible process that can be aimed at the production of naphtha or at the production of middle distillates, namely jet and diesel fuel. Although at first sight it might be expected that hydrocracking competes with fluid catalytic cracking, this is certainly not the case; the processes are complementary. The fluid catalytic cracker takes the more easily cracked alkane-rich atmospheric and vacuum gas oils as feedstocks, while the hydrocracker mainly uses more aromatic feeds, such as FCC cycle oils and distillates from thermal cracking processes, although it also takes heavy atmospheric and vacuum gas oils and deasphalted oil.

QUESTION:

> *Will gasoline-range products from hydrocracking contain more or less sulfur than those from FCC?*

3.4.5.2.1   Reactions and Thermodynamics

Hydrocracking can be viewed as a combination of hydrogenation and catalytic cracking. The former reaction is exothermic while the latter reaction is endothermic. Since the heat required for cracking is less than the heat released by the hydrogenation reaction the overall hydrocracking process is exothermic. Scheme 3.6 shows examples of reactions taking place.

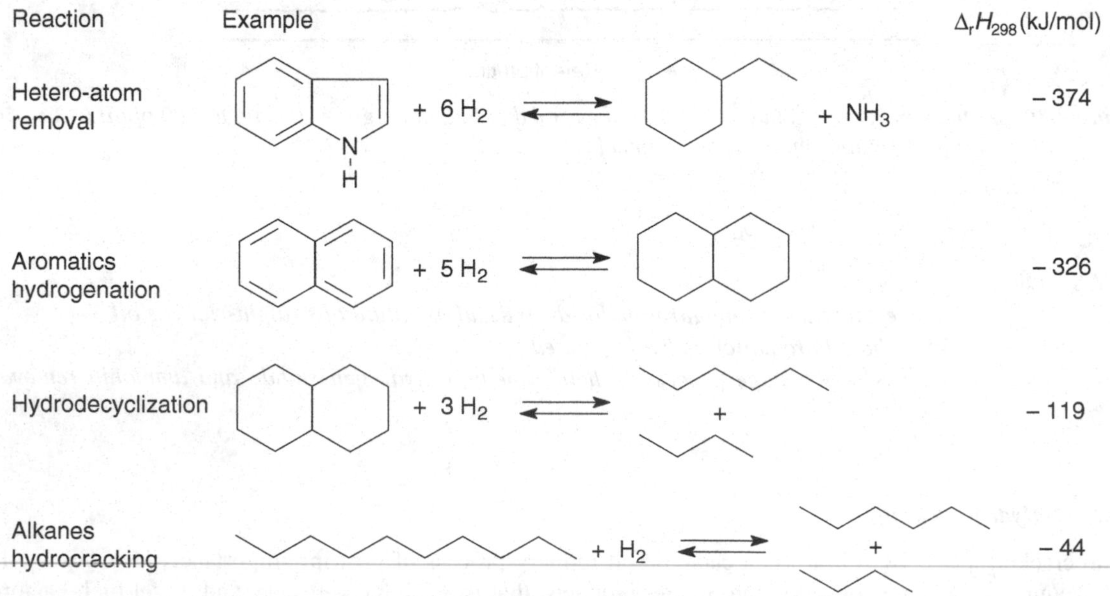

Scheme 3.6   *Examples of reactions during hydrocracking.*

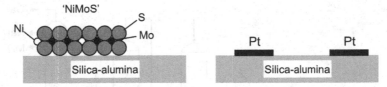

Figure 3.31 *Models of hydrocracking catalysts; left: NiS/MoS$_2$/silica-alumina; right: Pt/amorphous silica-alumina.*

Dehydrogenation, hydrogenation, and hetero-atom removal require a hydrogenation catalyst, while the actual cracking reactions proceed via carbenium ions, and therefore require an acidic catalyst. Various catalysts are in use. As in hydrotreating, mixed metal sulfides are used for the hydrogenation function. Noble metals are also used. The additional cracking function is fulfilled by a more acidic carrier than that in hydrotreating, for instance silica-alumina and, increasingly, zeolites. Figure 3.31 shows a model of a typical hydrocracking catalyst system.

QUESTION:

Explain why aromatic feeds are more easily processed in a hydrocracker than in an FCC unit.

3.4.5.2.2 Processes

Nitrogen compounds, many of them basic, present in the feed play an important role in hydrocracking because they adsorb on the acidic catalyst and thus inhibit the cracking reactions. Therefore, in most processes, hydrodenitrogenation (HDN) is necessary as a first step before the actual hydrocracking.

Various process configurations have been developed, which can be classified as single-stage, two-stage, and series-flow hydrocracking (Figure 3.32). In the design of a hydrocracking process, hydrogenation and cracking reactions have to be balanced carefully.

Single-stage once-through hydrocracking is very similar to hydrotreating (Figure 3.27), except for the different catalyst and more severe process conditions (Table 3.12). It is the simplest configuration of the hydrocracking process with the lowest investment costs. Apart from producing middle distillates, this process produces naphtha, which can be used as, for example, feed to a catalytic reforming unit (Section 3.4.3) or a naphtha cracker for ethene production (Chapter 4).

The conversion of the feedstock is not complete, that is, there is still material present with the same molecular weight range as the feedstock, but this "unconverted" product is highly saturated and free of feed contaminants; it has been hydrotreated and thus is an excellent feedstock for other processes, such as FCC and lube oil production. The process is optimized for hydrogenation rather than for cracking.

A relatively new technology is so-called mild hydrocracking [35]. The advantage of this process is that it can be implemented in existing hydrotreaters by increasing the severity of operation: with a relatively small investment the flexibility of the oil refinery is increased.

QUESTION:

Explain the term "mild hydrocracking". Also explain the term "hydrowax".

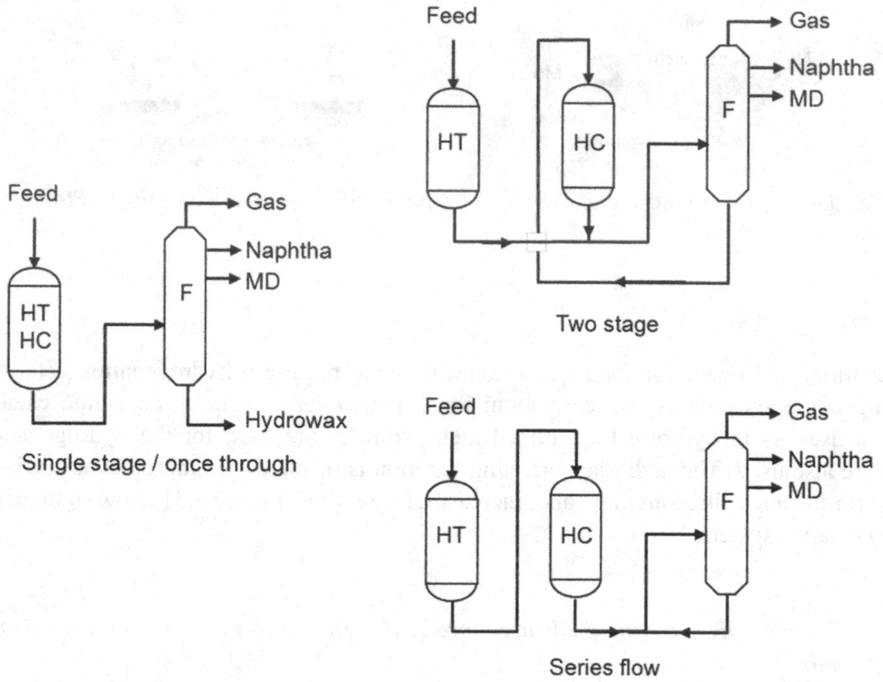

Figure 3.32   *Hydrocracking process configurations; HT = hydrotreating; HC = hydrocracking; MD = middle distillates.*

In two-stage hydrocrackers (Figure 3.33) the conversion of nitrogen and sulfur compounds and the hydrocracking reactions are carried out in two separate reactors with intermediate removal of ammonia and hydrogen sulfide. The effluent from the first reactor, after cooling and removal of hydrogen sulfide and ammonia, is fractionated and the bottom stream of the fractionator is subsequently hydrocracked in the second reactor.

The feed is completely converted into lighter products, in contrast with the single-stage process. The product yield can be tailored towards maximum naphtha production or maximum production of middle distillates (high-quality diesel) by either changing the fractionator operation or using alternative catalysts for the second stage [37].

In the series-flow process, the product from the hydrotreating reactor is directly fed to the hydrocracking reactor, without prior separation of hydrogen sulfide and ammonia. This means that the catalyst in the hydrocracking reactor has to operate under ammonia-rich conditions, resulting in a lower activity due to adsorption of ammonia on the catalyst. As a consequence, development of the series-flow process has only become possible thanks to the development of catalysts that are less sensitive towards ammonia.

QUESTION:

> *Hydrocracking catalysts can serve for two years before they have to be regenerated (coke removal). Explain this extreme difference with catalytic cracking catalysts (which only last for a few seconds).*

Table 3.12 summarizes the processing conditions used in the various hydrocracking processes. Regarding the catalyst, it is also possible to place different catalysts in different beds in one reactor. This somewhat blurs the distinction between the processes.

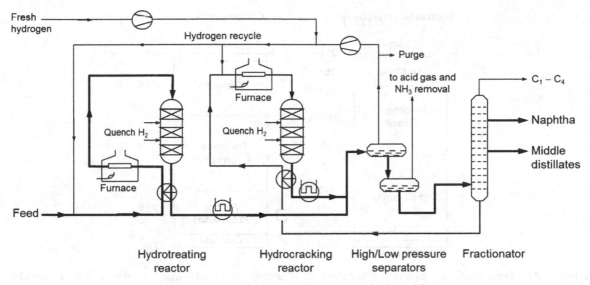

Figure 3.33 *Flow scheme of a two-stage hydrocracker.*

QUESTIONS:

What are advantages and disadvantages of the series-flow hydrocracking process compared to the two-stage process? (Think of investment costs, energy requirements, process flexibility, catalyst stability, etc.)

Compare the FCC process and the hydrocracking process in terms of feedstocks, products, environmental issues, operation costs, and so on.

Under practical conditions slow catalyst deactivation occurs. This deactivation is combatted by increasing the reactor temperature. The undesired consequence is an increase in aromatic contents of the product. Explain. (Hint: consider the data on the aromatization of cyclohexane in Figure 3.15.)

3.4.5.3 Hydroprocessing of Heavy Residues

Even in a complex refinery, including FCC, hydrocracking, and so on, the processing of a light crude oil does not yield a satisfactory product distribution (Figure 3.34). The amounts of fuel oil are too high.

Table 3.12 *Summary of processing conditions of hydrocracking processes.*

	Mild	Single stage/first stage	Fecond stage
Temperature (K)	670–700	610–710	530–650
H_2 pressure (bar)	50–80	80–130	80–130
Total pressure (bar)	70–100	100–150	100–150
Catalyst	Ni/Mo/S/γ-Al$_2$O$_3$-P[a]	Ni/Mo/S/γ-Al$_2$O$_3$-P[a]	Ni/W/S/USY zeolite

[a]P increases activity for nitrogen removal.

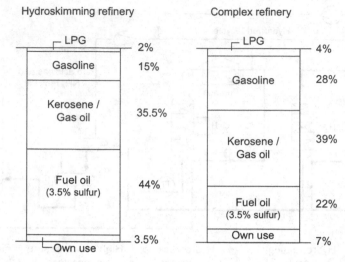

Figure 3.34   *Light crude oil product distribution in a simple ("hydroskimming") refinery and a complex refinery.*

For heavy oil the situation is even worse (Figure 3.35): about 50% fuel oil is produced. It should be noted that fuel oil is worth less than the original crude oil; the value of the products decreases in the order: gasoline > kerosene/gas oil > crude oil > fuel oil.

There are several reasons for an increased incentive to convert fuel oil into lighter products:

• The demand for light products such as gasoline and automotive diesel fuels continues to increase, while the market for heavy fuel oil is declining;

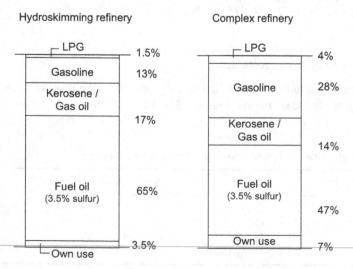

Figure 3.35   *Heavy crude oil product distribution in a simple ('hydroskimming') refinery and a complex refinery.*

- Environmental restrictions become more important. Fuel oil contains a large amount of impurities such as sulfur, nitrogen, and metals, so measures must be taken to lower the emissions;
- The quality of crude oils in general shows a worsening trend. It becomes heavier, with higher amounts of hetero-atoms, so more extensive processing is required to obtain the same amount and quality of products.

QUESTION:

> *Explain the term "hydroskimming" refinery.*

In principle, two solutions are feasible for upgrading residual oils and for obtaining a better product distribution. These are *carbon out* and *hydrogen in* processes.

QUESTION:

> *Discuss the logic of this statement. How do the thermal processes (visbreaking, delayed coking, and Flexicoking (Section 3.3)) fit in this respect?*

Examples of relatively recently developed carbon rejection processes are the Flexicoking process (Section 3.3.3) and the FCC process for the processing of heavy residues (Section 3.4.2).

Catalytic hydrogenation of residues is a "hydrogen in" route. It serves two general purposes: removal of sulfur, nitrogen, and metal compounds, and the production of light products. The reactions taking place are very similar to those occurring during hydrotreating and hydrocracking of lighter oil fractions. However, there are two important differences. Firstly, residues contain much higher amounts of sulfur, nitrogen, and polycyclic aromatic compounds (asphalthenes). Secondly, the removal of metals, which are concentrated in the residual fraction of the crude oil, is essential. Hence, the operating conditions are much more severe and more hydrogen is required in the catalytic hydrogenation of residues.

3.4.5.3.1 Catalyst Deactivation

A crucial point in the catalytic hydroconversion of residual oils is the deactivation of the catalyst by deposition of metals. Essentially, all metals in the periodic table are present in crude oil, the major ones being nickel and vanadium. At the applied reaction conditions hydrogen sulfide is present and, as a consequence, metal sulfides rather than metals are formed. The reaction scheme is complex [38] but the overall reactions can be represented by the following simplified equations:

$$Ni - porphyrin + H_2 \rightarrow NiS + hydrocarbons \tag{3.8}$$

$$V - porphyrin + H_2 \rightarrow V_2S_3 + hydrocarbons \tag{3.9}$$

The catalyst is poisoned by these reactions. This may occur even at relatively low levels of deposition. The reason is that for a large part the deposition occurs in the outer shell of the catalyst particles. Figure 3.36 illustrates the phenomena occurring during metal deposition. Initially, the catalyst pellets are poisoned rather homogeneously, although always with some preference for the outer shell. With time, diffusion is progressively hindered and the metals are deposited more and more in the outer shell. This leads to so-called pore plugging, rendering the catalyst particle inactive although in a chemical sense the inner part of the pellets is still active.

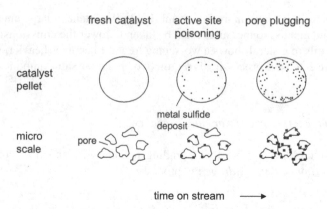

Figure 3.36   *Metal sulfide deposition on catalysts.*

3.4.5.3.2   Reactors

A wide variety of processes for residue hydroprocessing is employed or under development [5, 30, 39–44]. They can be distinguished based on the type of reactor used, that is, fixed bed (and moving bed) reactors, fluidized bed reactors (also called ebullated bed reactors), and slurry reactors. Figure 3.37 shows schematics of the three reactor types. In each case three phases are present in the reactor: gas (H_2), liquid (residual oil), and solid (catalyst).

Fixed bed reactors are generally operated as so-called trickle bed reactors, in which the gas and liquid flow cocurrently downward, although upward flow of gas and liquid is also sometimes used.

In fluidized bed reactors, gas and liquid flow upward and keep the catalyst particles in suspension. In principle, the catalyst remains in the reactor. Fresh catalyst is added during operation periodically (e.g., twice every week) and part of the spent catalyst removed. The most important problem of residue hydrotreating, namely catalyst deactivation, is solved in this way.

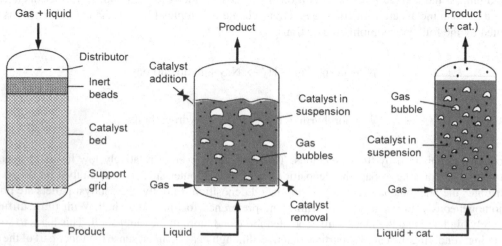

Figure 3.37   *Fixed bed (left), fluidized bed (middle), and slurry reactor (right).*

In slurry reactors, the catalyst is very finely divided and is carried through the reactor with the liquid fraction. Slurry reactors are usually mechanically stirred, but in slurry reactors for the conversion of heavy petroleum fractions suspension of the catalyst particles is no problem: the liquid/solid slurry behaves as a homogeneous phase [5].

It is interesting to compare the reactor systems. The most important advantages of the fluidized bed reactor are the excellent heat transfer properties, the option to use highly dispersed catalyst systems and the ease of addition and removal of catalyst particles. Compared to fixed bed reactors, the particle size can and must be much smaller and, as a consequence, the apparent activity and the capacity for metal removal are higher. In slurry reactors, the particle size is even smaller, with the accompanying advantages. However, separation of the fine particles from the product is often a problem. In residue processing, recovery of the catalyst is usually not practical, so that it is discarded with the unconverted residue [5].

QUESTIONS:

In which reactor type would internal diffusion rates be highest?
Give advantages and disadvantages of the three reactor types.
Discuss safety aspects of these reactors: is a "runaway" (dangerous temperature rise) possible?
In hydrocracking of certain feeds, fluidized bed reactors are also used. For which feedstocks would this be the case?

3.4.5.3.3 Processes

3.4.5.3.3.1 Processes with fixed bed reactors. Replacement of deactivated catalyst in a conventional fixed bed reactor is not possible during operation. Therefore, when processing heavy residual fractions in a fixed bed reactor, either a catalyst with a long life (highly metal resistant with accompanying low activity) is required, or a solution has to be found for easy catalyst replacement. Depending on the metal content of the feedstock various combinations can be used (Figure 3.38).

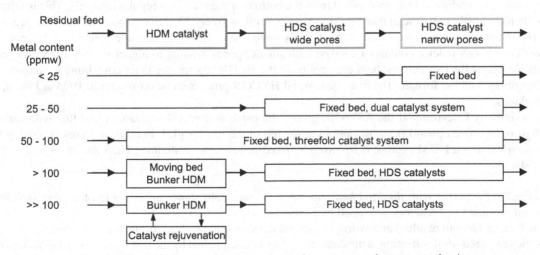

Figure 3.38 *Catalyst configuration dependence on metal content in feed.*

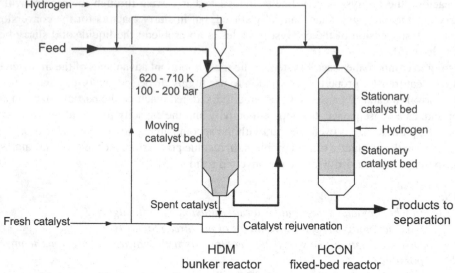

Figure 3.39   *Reactor section of the HYCON process.*

The HYCON process developed by Shell [41,42] (Figure 3.39) solves the dilemma that a fixed bed reactor is most convenient but that an active catalyst deactivates fast by applying two reactor systems in series. A reactor that enables easy catalyst replacement is used for the removal of metals (hydrodemetallization: HDM). Further conversion, that is, sulfur removal and conversion into lighter products by hydrocracking (hydroconversion: HCON), takes place in a conventional fixed bed reactor. Both reactors are operated as trickle bed reactors with the gas and liquid flowing cocurrently downward. This process is suitable for feedstocks having a metal content $\gg 100$ ppmw.

The HDM reactor is of the moving bed type, or more accurately a "bunker" reactor. Fresh catalyst, hydrocarbons, and hydrogen move concurrently downward and a special arrangement of internals inside the reactor allows withdrawal of spent catalyst from the bottom of the reactor at regular intervals. This is referred to as "bunker flow". It is crucial that the catalyst particles flow freely. Therefore, spherical particles are used. The HDM reactor contains a catalyst with relatively large pores. This catalyst is metal resistant, but not very active. The HCON reactor contains a catalyst with smaller pores, leading to higher activity but lower metal resistance. Actually, two catalyst beds are applied in the HCON reactor, the latter containing a catalyst with smaller pores than the former. The first commercial HYCON process came on stream in 1993 at Pernis, the Netherlands.

In designs by Chevron and the ASVAHL group, the catalyst moves downward, while the hydrocarbons and hydrogen move upward from the bottom to the top of the reactor [45]. In all these cases, a moving bed reactor containing a HDM catalyst is followed by fixed bed reactors containing HDS and other hydrofining catalysts.

3.4.5.3.3.2   Processes with fluidized bed reactors.   Figure 3.40 shows an example of a process scheme based on fluidized bed reactors developed by Lummus [40].

The feed (a vacuum residue) and hydrogen are preheated in separate heaters and fed to the first reactor. The velocities required for fluidization of the catalyst bed are accomplished by the usual gas recirculation, together with internal recirculation of the liquid phase. In the reactors, removal of hetero-atoms and conversion of

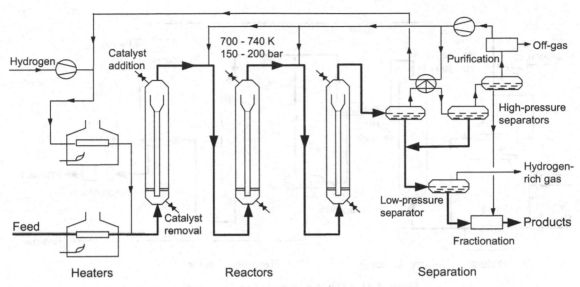

Figure 3.40   *Catalytic hydrogenation of residues in fluidized bed reactors (Chevron Lummus LC-Fining process).*

the feed take place. The reaction conditions are more severe than those in hydrotreating of distillates. The products are separated in a series of high- and low-pressure gas–liquid separators at high and low temperature.

QUESTIONS:
> *Why three reactors in series? What is the purpose of the internals in the reactors?*

3.4.5.3.3.3 Processes with slurry reactors.   Slurry processes for residue processing are normally designed with the objective of maximizing the residue conversion by hydrocracking. Downstream reactors are then used to treat the liquid products for removal of sulfur and nitrogen. Examples of slurry processes are the Veba Combi-Cracking process [43, 46] and the CANMET process [5]. Figure 3.41 shows a flow scheme of the Combi-Cracking process. Currently, many slurry processes are in the industrial demonstration stage [43].

Conversion of the residual feed takes place in the liquid phase in a slurry reactor. After separation of the residue from the products, these products are further hydrotreated in a fixed bed reactor containing a HDS catalyst. The process uses a cheap once-through catalyst, which ends up in the residue.

3.5   Current and Future Trends in Oil Refining

Current and future developments in oil refining are and will be mainly concerned with upgrading of heavy oil fractions and with environmentally more benign processes and products.

The Flexicoking process described in Section 3.3.3 and hydroprocessing of heavy residues (Section 3.4.5.3) for upgrading of heavy oil fractions are typical of the 1970s and early 1980s developments, which were aimed at maximizing the conversion of heavy oils to gasoline and middle distillate products. Although this objective is still important, the focus has shifted somewhat since the late 1980s to the development of cleaner products [6, 13, 47].

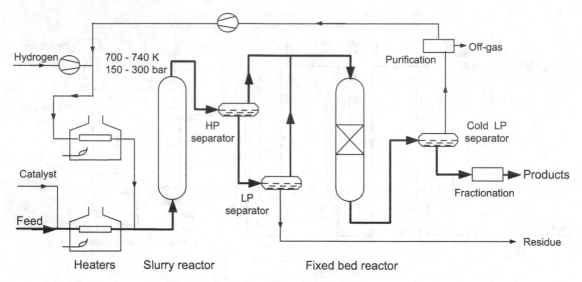

Figure 3.41   *The Veba Combi-Cracking process.*

Ever stricter environmental constraints on both products and refineries have triggered these developments.

In addition, the preferred feedstock is debated. It is expected that biomass will play a role in the production of fuels and chemicals. Some biomass-based fuels are so-called drop-in fuels and can be processed directly in existing refineries after some modest adaptations (Chapter 7).

3.5.1   Stricter Environmental Regulations

The trend of stricter environmental regulations started in the 1970s and has accelerated in recent years. Major areas of improvement include:

- Reduced particulate matter emissions. For example, better cyclones and high-efficiency electrostatic precipitators are being added to FCC units.
- Reduced SO_x and NO_x emissions. SO_x emissions from FCC units are being reduced by hydrotreating of FCC feeds (Section 3.4.5.1) and the use of SO_x-transfer additives (Section 3.4.2.5). For removal of SO_x and NO_x from other sources, flue gas scrubbing can be used. At some plants, limitations on NO_x may require the addition of post-combustion NO_x removal on boilers and heaters, for example, by selective catalytic reduction (SCR) (Chapter 8).
- Improved energy efficiency and reduced CO_2 emissions. Decreased energy consumption usually also decreases CO_2 emissions.
- Reformulated gasoline. The need for higher octane in reformulated gasoline combined with upper limits on gasoline components has a great impact on refinery operations (Section 3.5.1.1).
- Low-sulfur fuels. Recent regulations limiting the sulfur content of gasoline and diesel fuel are stimulating the installation of a large number of new hydrotreaters. This in turn is requiring an increase in sulfur recovery and hydrogen production capacity. From a process development standpoint, the need to produce low-sulfur fuels has driven the development of new processes for post-treating gasoline from FCC units.

3.5.1.1   *Reformulated Gasoline*

Over the past decades gasoline production has become much more complicated and it will become even more so. Prior to about 1973 gasoline production involved blending catalytically reformed naphtha with straight-run light naphtha (tops) and other additives (notably lead compounds) to adjust its properties, particularly the octane number, the sulfur content, and vapor pressure.

Environmental regulations, which aim at reducing the toxic and/or polluting components from car exhausts, have already had an enormous impact on gasoline production. With the advent of unleaded gasoline from 1973 onwards, oil refiners had to find other means to enhance the octane number of gasoline. For example, the isomerization of the light straight-run naphtha fraction (mainly C_5 and C_6) [3] has been gaining popularity. Today, most countries around the world use unleaded gasoline only [48]. Regulations will continue to be tightened, especially in urban areas. The need for reformulated gasoline, together with the decreasing availability of light crude oil has enormous consequences on refinery operations, not in the least from an economic viewpoint.

QUESTION:
> *Why was unleaded gasoline introduced in the 1970s?*

Current and future environmental regulations also place upper limits on sulfur, alkenes, and benzene and other lower aromatics content, while the octane number has to be preserved. Moreover, a certain minimum amount of oxygenates is required. In particular, MTBE has been added (Section 6.3.4). However, although its octane number is perfect, the detection of MTBE in groundwater has made its future questionable.

The environmentally most favorable gasoline consists of highly branched alkanes with mainly 5 to 11 carbon atoms. Such a gasoline is produced by *alkylation*, so for good quality gasoline production its capacity should greatly increase. However, alkylation is one of the most environmentally unfriendly processes in a refinery as a result of the use of the liquid sulfuric and hydrofluoric acids. As described in Section 3.4.4.3, in academia and industry large efforts have been put into developing stable solid catalysts. As a result, several solid-catalyzed alkylation processes are now available, although the industry is somewhat reluctant to incorporate these processes.

The octane number of gasoline produced by *catalytic reforming* is high owing to the large fraction of aromatics present (Table 3.7). The process will continue to play an important role as gasoline base stock, and as hydrogen producer (!), but in the long run its importance will diminish by the introduction of stricter regulations on benzene and other aromatics. In some catalytic reformers, severity is being reduced to decrease the fraction of aromatics. Others are being converted for the production of benzene, toluene, and xylenes, which are important base chemicals. Other possible solutions are the reduction of the dealkylation activity of the reforming catalyst and the removal of the C_5/C_6 fraction from the reforming feedstock (dehydrogenation of hexanes is the most important source of benzene in reformate) for isomerization.

The *fluid catalytic cracking* process contributes much to the gasoline pool. However, FCC gasoline, just like reformate, contains a large amount of aromatics. In addition, although the FCC process contributes only one-third of the gasoline, it is responsible for over 90% of the sulfur in the refinery gasoline pool. Moreover, the FCC process is a large contributor to refinery emissions. Undoubtedly, FCC gasoline will remain important, but sulfur will have to be removed to a greater extent. Hydrodesulfurization of the gasoline fraction also results in hydrogenation of the alkenes present, resulting in a decrease of the octane number. In recent years, several processes have been developed for the removal of sulfur at minimum octane loss [16].

Gasoline produced by *hydrocracking* has a more favorable composition than FCC gasoline with respect to environmental issues; it does not contain sulfur and is low in aromatics, but an additional isomerization step is required.

Table 3.13   Properties of diesel base stocks [6, 13].

Source	Sulfur (wt%)[a]	Aromatics (wt%)	Cetane number
Straight-run gas oil	1–1.5	20–40	40–50
Light cycle oil from FCC	2–2.8	>70	<25
Gas oil from thermal processes	2–3	40–70	30–50
Gas oil from hydrocracking	<0.01	<10	50–55
Fischer–Tropsch gas oil	0	≈0	>70

[a] Before hydrodesulfurization.

3.5.1.2   Diesel

The main sources of diesel fuel in a refinery are straight-run gas oil and gas oil from the hydrocracking and FCC processes. Regulations for diesel fuel mainly cover maximum sulfur and aromatics content and minimum cetane number of around 50. Table 3.13 shows these properties for a number of diesel base stocks. In the European Union, the minimum cetane number is increased, which implies a drastic reduction of the aromatics content [16].

Clearly, of the refinery products, gas oil from the hydrocracking process shows the most favorable properties. It is very low in sulfur and has a high cetane number, which is caused by the low amount of aromatics and a high content of slightly branched alkanes. Thus, hydrocracking will increasingly prevail as a diesel fuel producer. Diesel from other refinery sources can only meet current sulfur standards (Figure 3.29) when hydrotreating is applied. A lot of research has been carried out successfully to develop more active and stable desulfurization catalysts, enabling so-called deep desulfurization, that is, desulfurization down to a sulfur content of 10–15 ppmw or lower [31, 33, 34]. This research has been successful.

Similarly, to comply with the lower aromatics specifications, new hydrogenation catalysts are being developed [31, 49]. In particular upgrading of light cycle oil (LCO), which accounts for 10–20% of the FCC product, is highly desired. LCO has been used as a low-value product for heating purposes but in recent years this application has been abandoned worldwide [49]. The target reactions in upgrading of LCO are hydrogenation of the aromatics and selective ring opening of the formed naphthenes. Up to now no satisfactory catalysts seem to be available.

QUESTIONS:

How will current and future regulations concerning diesel fuel affect the hydrogen balance in refineries? Suggest means for producing hydrogen.

Fischer–Tropsch synthesis, which converts synthesis gas into long-chain alkanes (see also Chapter 6) produces a nearly ideal diesel fuel and might well become the process of the future.

3.5.2   Refinery Configurations

As mentioned before, crude oil is getting heavier worldwide. This, together with the requirement for cleaner products has increased the complexity of refineries. For instance, existing refineries that are designed to handle light crudes have been modified to handle heavy crudes.

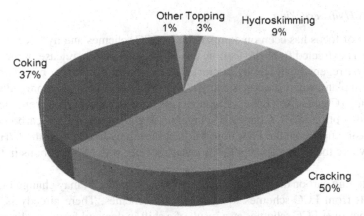

Figure 3.42   *Distribution of refinery types in 2009. Cracking includes both FCC and hydrocracking [50].*

3.5.2.1 Refinery Types

Figure 3.42 shows the capacities of the different refinery types. A topping refinery is the simplest kind of refinery. It only splits the crude oil into its main components. The amount of fuel oil produced is large. A hydroskimming refinery is a little more advanced. It has a catalytic reforming unit to produce gasoline and a hydrotreating unit for the removal of sulfur from gas oil, but still produces a large amount of fuel oil. Currently, the most common type of refinery is a cracking refinery. In such a refinery a vacuum distillation unit is added. The gas oil from this unit goes to an FCC or hydrocracking unit for the production of more gasoline and gas oil by cracking. In addition, the light alkenes from the FCC unit undergo alkylation with isobutane to produce alkylate. There is still some fuel oil production. In the most advanced refinery, a coking refinery, most of the fuel oil is upgraded to light products. It has some kind of coker to convert the vacuum residue. Delayed coking (Section 3.3.2) is employed most often.

From Figure 3.43 it is clear that the complexity of refineries has increased in the first decade of the twenty-first century. China accounted for 35% of the added coking capacity, while 40% of the added capacity was realized in the rest of Asia.

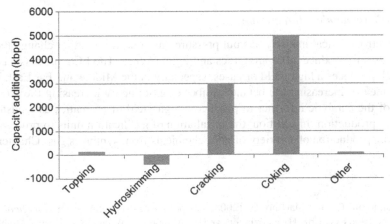

Figure 3.43   *Refinery capacity changes from 2000 to 2009. Cracking includes both FCC and hydrocracking [50].*

3.5.2.2   FCC versus Hydrocracking

In the past years a lot of focus has been on comparison of FCC schemes and hydrocracking schemes. In the FCC scheme, carbon is extracted from the feed by coke formation and its release in the form of CO_2 at low pressure in the catalyst regenerator. The hydrocracker does not produce CO_2, but the carbon extraction is shifted to the hydrogen plant, where carbon is extracted by formation of carbon monoxide and carbon dioxide during steam reforming (Chapter 5). These components are normally discharged from the hydrogen plant via the pressure swing adsorption (PSA, Chapter 6) purge gas into the flue gas, but carbon dioxide can also be captured at high pressure upstream the PSA unit. Further, during hydrocracking the C/H ratio is reduced by hydrogen addition, where the hydrogen supplier is water reacting with hydrocarbons in the steam reforming process.

The potential tightening carbon dioxide legislation (also see Chapter 5) may change the refinery landscape completely by a shift from FCC schemes to hydrocracker schemes. There already is a trend of favoring hydrocracking schemes over FCC schemes as a result of a shift in demand from gasoline to diesel. This trend has resulted in specification of grass root refineries with "mega size" hydrogen plants.

QUESTIONS: *For a fixed production rate, does the FCC process or the hydrocracking process produce the largest amount of carbon dioxide or is the overall amount of carbon dioxide produced equal for both processes?*
Which process would be preferred should carbon capture be required in the future?

3.5.2.3   Gasification/Partial Oxidation for the Production of Synthesis Gas

A trend could be to use heavy waste of the refinery (vacuum residue or coke) for the production of synthesis gas. Disposal of these refinery wastes is becoming more and more problematic due to a shift to heavier crudes as well as stricter regulations on atmospheric emissions. Gasification (Section 3.3.3, Chapter 5) using steam/oxygen is considered together with cleaning of synthesis gas for "clean disposal". This also makes it possible to convert essentially all carbon oxides to carbon dioxide for energy-efficient high-pressure carbon dioxide removal. The resulting hydrogen gas can be used in the refinery for hydrocracking/hydrotreating and the balance can be used as fuel in a gas turbine.

3.5.2.4   Refinery–Petrochemical Integration

The growth of the petrochemical industry has put pressure on refineries to either change their configuration or operating conditions to produce more aromatics and light alkenes. The FCC process has been developed to petro-FCC, which produces a high yield of gases, especially in the Middle and Far East. Furthermore, the phasing out of the idea of increasing the octane number of gasoline by increasing the aromatics content is changing the role of the catalytic reformer to produce a high yield of aromatics as feedstock for benzene, toluene, and xylenes production. In addition, the installation of gasification units to process vacuum residue opens the way for the production of a variety of petrochemicals from synthesis gas (Chapter 5).

References

[1] Sarrazin, P., Baudouin, C., and Martino, G. (2008) Perspectives in oil refining, in *Handbook of Heterogeneous Catalysis*, 2nd edn (eds G. Ertl, H. Knözinger, and J. Weitkamp). Wiley-VCH Verlag GmbH, Weinheim, pp. 2677–2695.

[2] Gary, J.H. and Handwerk, G.E. (1994) *Petroleum Refining, Technology and Economics*, 3rd edn. Marcel Dekker, New York.

[3] Meyers, R.A. (1997) *Handbook of Petroleum Refining Processes*, 2nd edn. McGraw-Hill, New York.

[4] Ditman, J.G. and Godino, R.L. (1965) Propane extraction: a way to handle residue. *Hydrocarbon Processing*, **44**, 175–178.

[5] Gray, M.R. (1994) *Upgrading Petroleum Residues and Heavy Oils*. Marcel Dekker, New York.

[6] Bousquet, J. and Valais, M. (1996) Trends and constraints of the European refining industry. *Applied Catalysis A: General*, **134**, N8–N18.

[7] Sadeghbeigi, R. (1995) *Fluid Catalytic Cracking Handbook; Design, Operation, and Trouble-Shooting of FCC Facilities*. Gulf Publishing, Houston, TX.

[8] Wilson, J.W. (1997) *Fluid Catalytic Cracking: Technology and Operation*. PennWell Publishing, Tulsa, OK.

[9] Magee, J.S. and Mitchell, M.M. Jr. (2009) *Fluid Catalytic Cracking: Science and Technology*. Elsevier, Amsterdam.

[10] Sie, S.T. (1992) *Petroleum Conversion*. Lecture Course. Delft Technical University (in Dutch).

[11] Rollmann, L.D. and Valyocisk, E.W. (1981) *Inorganic Syntheses* John Wiley & Sons, Inc., New York, p. 61.

[12] Ambler, P.A., Milne, B.J., Berruti, F., and Scott, D.S. (1990) Residence time distribution of solids in a circulating fluidized bed: Experimental and modelling studies. *Chemical Engineering Science*, **45**, 2179–2186.

[13] Martino, G., Courty, P., and Marcilly, C. (1997) Perspectives in oil refining, in *Handbook of Heterogeneous Catalysis* (eds G. Ertl, H. Knözinger, and J. Weitkamp). VCH, Weinheim, pp. 1801–1818.

[14] Rheaume, L. and Ritter, R. E. (1988) Use of catalysts to reduce SO_x emissions from fluid catalytic cracking units, in *Fluid Catalytic Cracking: Role in Modern Refining* (ed. M.L. Ocelli). ACS Symposium Series 375, American Chemical Society, Washington, DC, pp. 146–161.

[15] Soud, H. (1995) *Suppliers of FGD and NOx Control Systems*. IEA Coal Research, London.

[16] Robinson, P.R. (2006) Petroleum processing overview, in *Practical Advances in Petroleum Processing* (eds C. Hsu and P. Robinson). Springer, New York, pp. 1–78.

[17] Stratiev, D.S., Shishkova, I.K., and Dobrev, D.S. (2012) Fluid catalytic cracking feed hydrotreatment and its severity impact on product yields and quality. *Fuel Processing Technology*, **94**, 16–25.

[18] Sie, S.T. (1994) Past, present and future role of microporous catalysts in the petroleum industry, in *Advanced Zeolite Science and Applications*, (eds J.C. Jansen, M. Stöcker, H.G. Karge, and J. Weitkamp). Elsevier, Amsterdam, pp. 587–631.

[19] Von Ballmoos, R., Harris, D.H., and Magee, J.S. (1997) Catalytic cracking, in *Handbook of Heterogeneous Catalysis* (eds G. Ertl, H. Knözinger, and J. Weitkamp). VCH, Weinheim, pp. 1955–1968.

[20] Little, D.M. (1985) *Catalytic Reforming*. PennWell Publishing, Tulsa, OK.

[21] Corma, A. and Martinez, A. (1993) Chemistry, catalysts, and processes for isoparaffin-olefin alkylation: actual situation and future trends. *Catalysis Reviews: Science and Engineering*, **35**, 483–570.

[22] Lerner, H. and Citarella, V.A. (1991) Improve alkylation efficiency. *Hydrocarbon Processing*, **70**, 89–94.

[23] Lavrenov, A., Bogdanets, E., and Duplyakin, V. (2009) Solid acid alkylation of isobutane by butenes: the path from the ascertainment of the reasons for fast deactivation to the technological execution of the process. *Catalysis in Industry*, **1**, 50–60.

[24] Roeseler, C. (2004) UOP Alkylene™ process for motor fuel alkylation, in *Handbook of Petroleum Refining Processes*, 3rd edn (ed. R.A. Meyers). McGraw-Hill, New York, pp. 1.25–1.31.

[25] D'Amico, V., Gieseman, J., van Broekhoven, E., van Rooijen, E., and Nousiainen, H. (2006) The AlkyClean alkylation process – New technology eliminates liquid acids. NPRA Spring 2006 Meeting, pp. 1–9.

[26] Mukherjee, M., Nehlsen, J., Sundaresan, S., Sucio, G.D., and Dixon, J. (2006) Scale-up strategy applied to solid-acid alkylation process. *Oil & Gas Journal*, **104**(July 10), 48–54.

[27] Mukherjee, M., Sundaresan, S., Porcelli, R., and Nehlsen, J. (2007) ExSact: novel solid-acid catalyzed iso-paraffin alkylation process, in *Ultraclean Transportation Fuels* (eds O.I. Ogunsola and I.K. Gamwo). ACS Symposium Series 959, American Chemical Society, Washington, DC, pp. 181–193.

[28] Feller, A. and Lercher, J. A. (2004) Chemistry and technology of isobutane/alkene alkylation catalyzed by liquid and solid acids, in *Advances in Catalysis*, vol. **48** (ed. C.G. Bruce). Academic Press, pp. 229–295.

[29] Feller, A., Guzman, A., Zuazo, I., and Lercher, J. A. (2003) A novel process for solid acid catalyzed isobutane/butene alkylation, in *Studies in Surface Science and Catalysis. Science and Technology in Catalysis 2002 Proceedings of*

the Fourth Tokyo Conference on Advance Catalytic Science and Technology, vol. **145** (eds M. Anpo, M. Naka, and H. Amashita). Elsevier, pp. 67–72.

[30] Speight, J.G. (1981) *The Desulfurization of Heavy Oils and Residues*. Marcel Dekker, New York.

[31] Reinhoudt, H.R., Troost, R., van Langeveld, A.D., Sie, S.T., van Veen, J.A.R., and Moulijn, J.A. (1999) Catalysts for second-stage deep hydrodesulfurisation of gas oils. *Fuel Processing Technology*, **61**, 133–147.

[32] Oyama, S.T., Gott, T., Zhao, H., and Lee, Y.K. (2009) Transition metal phosphide hydroprocessing catalysts: A review. *Catalysis Today*, **143**, 94–107.

[33] Navarro, R.M., Castaño, P., Álvarez-Galván, M.C. and Pawelec, B. (2009) Hydrodesulfurization of dibenzothiophene and a SRGO on sulfide Ni(Co)Mo/Al$_2$O$_3$ catalysts. Effect of Ru and Pd promotion. *Catalysis Today*, **143**, 108–114.

[34] Stanislaus, A., Marafi, A., and Rana, M.S. (2010) Recent advances in the science and technology of ultra low sulfur diesel (ULSD) production. *Catalysis Today*, **153**, 1–68.

[35] Maxwell, I.E., Minderhoud, J.K., Stork, W.H.J., and Van Veen, J.A.R. (1997) Hydrocracking and catalytic dewaxing, in *Handbook of Heterogeneous Catalysis* (eds G. Ertl, H. Knözinger, and J. Weitkamp). VCH, Weinheim, pp. 2017–2038.

[36] Mohanty, S., Kunzru, D., and Saraf, D.N. (1990) Hydrocracking: a review. *Fuel*, **69**, 1467–1473.

[37] Bridge, A.G., Jaffe, J., Powell, B.E., and Sullivan, R.F. (1993) Isocracking heavy feeds for maximum middle distillate production. API Meeting, Los Angeles, CA, May 1993.

[38] Bonne, R.L.C., van Steenderen, P., and Moulijn, J.A. (1995) Hydrodemetalization kinetics of nickel tetraphenyl-porphyrin over Mo/Al$_2$O$_3$ catalysts. *Industrial & Engineering Chemistry Research*, **34**, 3801–3807.

[39] Quann, R.J., Ware, R.A., Hung, C., and Wei, J. (1988) Catalytic hydrodemetallation of petroleum, *Advances in Chemical Engineering*, **14**, 95–259.

[40] van Driesen, R.P. and Fornoff, L.L. (1985) Ugrade resids with LC-fining. *Hydrocarbon Processing*, **64**, 91–95.

[41] Kwant, P.B. and Van Zijll Langhout, W.C. (1985) The development of Shell's residue hydroconversion process. Proceedings of the Conference on the Complete Upgrading of Crude Oil, Siofok, 25–27 September, 1985.

[42] Scheffer, B., van Koten, M.A., Robschläger, K.W., and de Boks, F.C. (1998) The Shell residue hydroconversion process: development and achievements. *Catalysis Today*, **43**, 217–224.

[43] Liu, Y., Gao, L., Wen, L., and Zong, B. (2009) Recent advances in heavy oil hydroprocessing technologies. *Recent Patents on Chemical Engineering*, **2**, 22–36.

[44] Rana, M.S., Sámano, V., Ancheyta, J., and Diaz, J.A.I. (2007) A review of recent advances on process technologies for upgrading of heavy oils and residua. *Fuel*, **86**, 1216–1231.

[45] Ancheyta, J. (2007) Reactors for Hydroprocessing, in *Hydroprocessing of Heavy Oils and Residua* (eds J.G. Speight and J. Ancheyta). CRC Press, pp. 71–120.

[46] Wenzel, F.W. (1992) VEBA-COMBI-Cracking, a commercial route for bottom of the barrel upgrading. Proceedings of the International Symposium on Heavy Oil and Residue Upgrading and Utilization, pp. 185–201.

[47] Absi-Halabi, M., Stanislaus, A., and Qabazard, H. (1997) Trends in catalysis research to meet future refining needs. *Hydrocarbon Processing*, **76**, 45–55.

[48] UNEP (2009) Global lead phase out – Progress as of 2002 to date, http://www.unep.org/transport/pcfv/PDF/leadprogress.pdf (last accessed 17 December 2012).

[49] Upare, D., Nageswara Rao, R., Yoon, S., and Lee, C. (2011) Upgrading of light cycle oil by partial hydrogenation and selective ring opening over an iridium bifunctional catalyst. *Research on Chemical Intermediates*, **37**, 1293–1303.

[50] Hauge, K. (2009) Refining ABC, http://www.statoil.com/en/InvestorCentre/Presentations/Downloads/Refining.pdf (last accessed 17 December 2012).

4

Production of Light Alkenes

4.1 Introduction

The discovery that light alkenes can be produced in high yields from oil fractions and from alkanes present in natural gas has laid the foundation for what is now known as the petrochemical industry. The principal process used to convert the relatively unreactive alkanes into much more reactive alkenes is thermal cracking, often referred to as "steam cracking". In steam cracking, a hydrocarbon stream is thermally cracked in the presence of steam, yielding a complex product mixture. The name steam cracking is slightly illogical: cracking of steam does not occur, but steam primarily functions as a diluent and heat carrier, allowing higher conversion. A more accurate description of the process might be "pyrolysis", which stems from Greek and means bond breaking by heat.

Steam cracking occurs at much higher temperature than other thermal cracking processes (e.g., visbreaking, delayed coking, and Flexicoking discussed in Chapter 3) and in the presence of large amounts of steam. The steam cracking process mainly produces ethene, which is a very important base chemical. Valuable coproducts such as propene, butadienes, and pyrolysis gasoline (*pygas*), with benzene as the main constituent, are also produced.

Since the late 1930s when the petrochemical industry started to take shape, ethene has almost completely replaced coal-derived ethyne and now is the world's largest volume building block for the petrochemical industry, with a production exceeding 110 million t/a. Figure 4.1 shows its main uses.

Feedstocks for steam cracking range from light saturated hydrocarbons, such as ethane and propane, to naphtha and light and heavy gas oils. In North America and the Middle East ethane (from natural gas) is the primary feedstock for the production of ethene. In contrast, in Europe and Japan naphtha (from oil) is the major feedstock, which explains why steam cracking is frequently referred to as naphtha cracking.

Most steam cracking feedstocks are also feedstocks for fuel production, for example, naphtha is also converted into gasoline in the catalytic reforming process (Section 3.4.3). Furthermore, coproducts from steam cracking, such as pygas, usually find their destination in liquid fuels. On the other hand, fluid catalytic cracking (FCC, Section 3.4.2) is an important propene producer. Therefore, steam cracking is intimately connected with oil refinery operations.

Chemical Process Technology, Second Edition. Jacob A. Moulijn, Michiel Makkee, and Annelies E. van Diepen.
© 2013 John Wiley & Sons, Ltd. Published 2013 by John Wiley & Sons, Ltd.

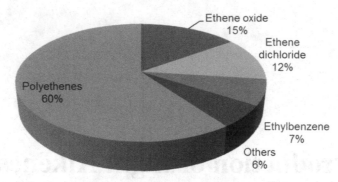

Figure 4.1   *Ethene uses (2008). Total 110 Mt/a. Source: NIC Prospectus – Jacobs Consultancy.*

There also is increasing interest in processes for the dedicated production of specific alkenes (Section 4.5), in particular propene, as a gap is developing between its supply and demand, primarily due to the high growth rate of the polypropene demand, which is one of the main end uses of propene.

4.2   Cracking Reactions

Steam cracking yields a large variety of products, ranging from hydrogen to fuel oil. The product distribution depends on the feedstock and on the processing conditions. These conditions are determined by both thermodynamic and kinetic factors.

4.2.1   Thermodynamics

In general, light alkenes, especially ethene, propene, and butadiene, are the desired products of steam cracking. Treatment of light alkanes such as ethane, propane, and butanes by steam cracking results in dehydrogenation of the alkanes to form the corresponding alkenes and hydrogen. Figure 4.2 shows the equilibrium conversions for the dehydrogenation reactions of the light alkanes ethane, propane, and isobutane represented by reactions 4.1, 4.2, and 4.3:

$$C_2H_6 \rightleftarrows C_2H_4 + H_2 \quad \Delta_r H_{298} = 137.1 \text{ kJ/mol} \tag{4.1}$$
$$C_3H_8 \rightleftarrows C_3H_6 + H_2 \quad \Delta_r H_{298} = 124.4 \text{ kJ/mol} \tag{4.2}$$
$$C_4H_{10} \rightleftarrows C_4H_8 + H_2 \quad \Delta_r H_{298} = 117.7 \text{ kJ/mol} \tag{4.3}$$

Figure 4.2 indicates that from a thermodynamic viewpoint the reaction temperature should be high for sufficient conversion. The forward reaction is also favored if the alkanes have a low partial pressure, because for every molecule converted two molecules are formed. A process under vacuum would be desirable in this respect. In practice, it is more convenient to use dilution with steam, which has essentially the same effect. Figure 4.2 also shows that the smaller the alkane the higher the temperature has to be for a given conversion.

QUESTION:
What can you conclude about optimal reaction conditions?
Why is the process not carried out as a catalytic process?
When nitrogen is available would you consider replacing steam by nitrogen?

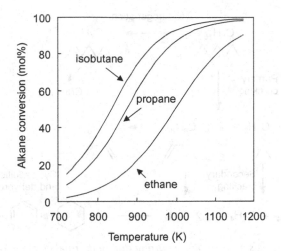

Figure 4.2   *Equilibrium conversion in the dehydrogenation of light alkanes at 1 bar as a function of temperature.*

4.2.2 Mechanism

Cracking occurs by free radical reactions (Scheme 4.1 shows the cracking of ethane). The reaction is initiated by cleavage of the C–C bond in an ethane molecule, resulting in the formation of two methyl radicals. Propagation proceeds by reaction of a methyl radical with an ethane molecule, resulting in the production of methane and an ethyl radical. The ethyl radical subsequently decomposes into ethene and a hydrogen radical, which then attacks another ethane molecule, and so on. These latter two reactions dominate in the cracking of ethane, which explains why ethene can be obtained in high yields. Termination occurs as a result of the reaction between two radicals to form either a saturated molecule or both a saturated and an unsaturated molecule. Scheme 4.1 shows that cracking of ethane besides ethene also produces methane and hydrogen. Small quantities of heavier hydrocarbons can also be formed as a result of the reaction of two radicals.

Similar, although more complex, networks apply to thermal cracking of higher alkanes (Scheme 4.2).

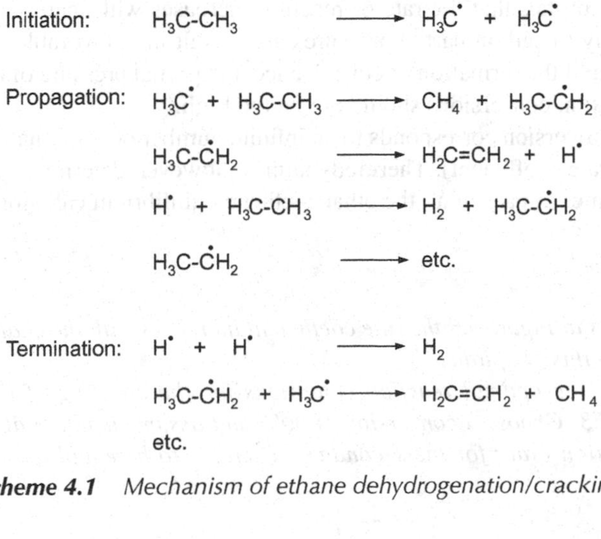

Initiation: $H_3C\text{-}CH_3 \longrightarrow H_3C^\bullet + H_3C^\bullet$

Propagation: $H_3C^\bullet + H_3C\text{-}CH_3 \longrightarrow CH_4 + H_3C\text{-}\overset{\bullet}{C}H_2$

$H_3C\text{-}\overset{\bullet}{C}H_2 \longrightarrow H_2C{=}CH_2 + H^\bullet$

$H^\bullet + H_3C\text{-}CH_3 \longrightarrow H_2 + H_3C\text{-}\overset{\bullet}{C}H_2$

$H_3C\text{-}\overset{\bullet}{C}H_2 \longrightarrow$ etc.

Termination: $H^\bullet + H^\bullet \longrightarrow H_2$

$H_3C\text{-}\overset{\bullet}{C}H_2 + H_3C^\bullet \longrightarrow H_2C{=}CH_2 + CH_4$

etc.

Scheme 4.1   *Mechanism of ethane dehydrogenation/cracking.*

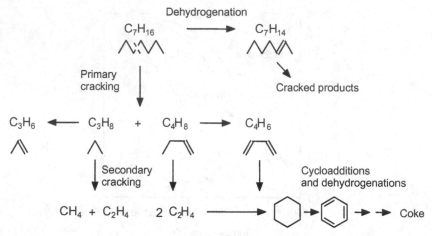

Scheme 4.2   *Examples of reactions occurring during thermal cracking of heavier alkanes.*

The primary cracking products may undergo secondary reactions such as further cracking, dehydrogenation, and condensation (combination of two or more smaller fragments to produce aromatics). These secondary reactions may also lead to the formation of coke. Coke is always formed, even when light alkanes are used as feedstock. Not surprisingly, the heavier the feedstock, the more coke is formed.

QUESTION:

> *According to Schemes 4.1 and 4.2, initiation takes place via C–C bond breaking. Is this logical? Explain your answer.*

4.2.3 Kinetics

The rate of reaction of alkanes obeys first order kinetics. Figure 4.3 shows rate coefficients for the cracking of a number of alkanes as a function of temperature. The reactivity increases with chain length. Ethane clearly shows the lowest reactivity.

The first order kinetics implies that the rate of reaction increases with increasing partial pressure of the reactants. However, high hydrocarbon partial pressures also result in unfavorable secondary reactions, such as condensation reactions and the formation of coke. Hence, the partial pressure of the hydrocarbons must be kept low. For the same reason, conversions should not be too high.

It appears that a given conversion corresponds to an infinite number of combinations of residence time and temperature (through the rate coefficient). Thermodynamics, however, determines the required temperature, and hence the residence time. In particular, the ethane–ethene equilibrium calls for a temperature as high as possible.

QUESTIONS:

> *As shown in Figure 4.3 the rate coefficient increases with the chain length. Would you have expected this? Explain.*
>
> *Estimate the optimal reaction conditions for the cracking of the alkanes referred to in Figure 4.3. Choose a conversion of 60% and assume isothermal plug flow. Do you expect the reaction order for the secondary reactions to be equal to one as well? Explain your answer.*

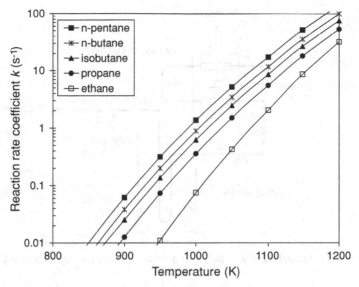

Figure 4.3   *Reaction rate coefficients of various alkanes [1].*

4.3   The Industrial Process

From the foregoing discussion a number of requirements concerning the steam cracking process can be derived:

- considerable heat input at a high temperature level;
- limitation of hydrocarbon partial pressure;
- very short residence times (<1 s);
- rapid quench of the reaction product to preserve the composition.

QUESTIONS:
> *Explain these requirements. Select the optimal temperatures.*

In industrial practice these requirements are met in the following way: a mixture of hydrocarbons and steam is passed through tubes placed inside furnaces heated by the combustion of natural gas, LPG, or fuel oil. The furnaces consist of a convection section, in which the hydrocarbon feed and the steam are preheated, and a radiation section, in which the reactions take place. The hydrocarbons undergo cracking and, subsequently, the products are rapidly quenched to prevent further reaction and thus preserve the composition. The residence time in the tubes is very short (<1 s). The temperature is chosen to be as high as possible in the sense that the properties of the construction material of the tubes determine the ceiling temperature. Figure 4.4 shows a simplified flow scheme of a steam cracker for the cracking of naphtha.

4.3.1   Influence of Feedstock on Steam Cracker Operation and Products

Steam cracking yields a complex product mixture. In practice it is crucial to choose the reaction conditions in such a way that the product distribution is optimal. Usually this means that the amount of ethene produced

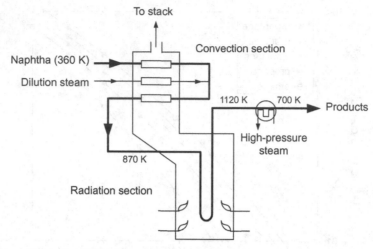

Figure 4.4   *Simplified flow scheme of a steam cracker for naphtha cracking.*

should be as high as possible. Depending on market developments and the local situation, however, other products might be more desirable. For instance, on the world market at present there is a shortage of propene.

The product distribution in ethane cracking is determined solely by the process parameters, such as temperature, pressure, residence time, and steam-to-ethane ratio. For naphtha no general product distribution can be given because naphtha is not a single compound and its composition varies with its source and refinery operating conditions and may include hydrocarbons ranging from C_3 to $\sim C_{15}$ [2]. The naphtha composition, expressed as the percentages of alkanes, naphthenes, and aromatics present (Table 4.1), has a pronounced effect on the cracking yield. This can be explained from their different reactivities and hydrogen-to-carbon ratios. Alkanes can be converted relatively easily and when decomposed produce high yields of light products such as ethene and propene as a result of their high hydrogen content. Aromatics, on the other hand, are very stable and they have a low hydrogen-to-carbon ratio, so their yield of light cracking products is negligible. Naphthenes are intermediate between alkanes and aromatics with regard to cracking behavior.

Tables 4.2 and 4.3 show typical product distributions (obtained in a pilot plant) for ethane cracking and cracking of naphtha with the composition shown in Table 4.1.

The data in these examples clearly indicate that reaction conditions are very critical and that it pays to operate the plant as close as possible to optimal conditions. These conditions depend on the feedstock and on the desired products. It is not surprising that in practice extensive simulation programs are used. Commercially available simulation packages contain several hundreds to thousands of chemical reactions.

Table 4.1   *Composition of naphtha feed for yield patterns in Table 4.3 [2].*

Hydrocarbon type	wt% of total
n-Alkanes	36.1
iso-Alkanes	36.6
Alkenes	0.2
Naphthenes	21.1
Aromatics	6.0
Total	100.0

Table 4.2 *Examples of commercial ethane cracking yield patterns for various residence times at a steam dilution ratio of 0.3 kg steam/kg feed [2].*

Residence time (s)	0.4607	0.3451	0.1860	0.1133
Conversion (kg/kg)	0.6501	0.6497	0.6501	0.6501
Yield (wt%)				
Hydrogen	4.05	4.06	4.10	0.13
Methane	3.80	3.57	3.23	2.88
Ethyne	0.44	0.47	0.54	0.75
Ethene	51.88	52.31	52.85	53.43
Ethane	34.99	35.03	34.99	34.99
C_3H_4[a]	0.02	0.02	0.02	0.02
Propene	1.22	1.13	1.06	0.97
Propane	0.12	0.12	0.12	0.13
Butadiene	0.05	0.05	0.05	0.06
Butenes	0.19	0.19	0.18	0.16
Butanes	0.21	0.21	0.21	0.22
Pyrgas[b]	1.02	0.88	0.75	0.55
Fuel oil	0.21	0.16	0.11	0.06
Total	100.00	100.00	100.00	100.00

[a]Methyl-ethyne and propadiene.
[b]Fraction boiling from 301 K (isopentane) to 478 K.

Table 4.3 *Examples of commercial naphtha cracking yield patterns for different residence times and steam dilution ratios [2].*

	Low severity cracking		Medium severity cracking	
Residence time (s)	0.4836	0.1784	0.4840	0.1828
Steam dilution (kg/kg)	0.4	0.4	0.45	0.45
P/E (kg/kg)[a]	0.65	0.65	0.55	0.55
Yield (wt%)				
Hydrogen	0.82	0.85	0.90	0.94
Methane	13.59	12.46	15.27	14.34
Ethyne	0.32	0.36	0.44	0.53
Ethene	25.23	26.08	27.95	29.24
Ethane	4.76	3.62	4.59	3.57
C_3H_4[b]	0.59	0.70	0.71	0.87
Propene	16.42	16.97	15.38	16.09
Propane	0.68	0.61	0.53	0.48
Butadiene	0.09	0.10	0.12	0.16
Butenes	4.63	5.27	4.54	5.28
Butanes	0.72	0.80	4.41	4.75
Pygas[c]	23.74	23.76	21.42	20.60
Fuel oil	2.65	2.08	3.32	2.68
Total	100.00	100.00	100.00	100.00

[a]Propene-to-ethene ratio, a measure of severity.
[b]Methyl-ethyne and propadiene.
[c]Fraction boiling from 301 K (isopentane) to 478 K.

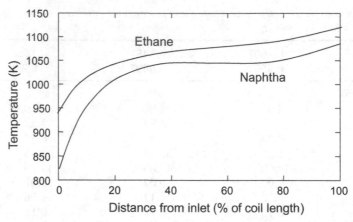

Figure 4.5 *Typical reactor temperature profiles in modern steam crackers. (Adapted from [1]. Used with permission of Wiley-VCH.)*

An obvious difference between ethane and naphtha cracking is the much wider product distribution of the latter. Ethene can be produced from ethane with a selectivity of over 70 wt%, with hydrogen and methane the second most important products (note: selectivity is defined as kg product/kg feed converted, yield as kg product/kg feed fed). In naphtha cracking the selectivity towards ethene is much lower, with large proportions of methane and propene being formed.

As a result of its stability, ethane requires higher temperatures and longer residence times than naphtha to achieve acceptable conversion. The cracking tubes are far from isothermal. The feed inlet temperature ranges from about 800 to 950 K depending on the feedstock, while the outlet temperature is between 1050 and 1200 K. Figure 4.5 shows typical temperature profiles.

QUESTIONS:

Interpret the data in the tables: compare the reaction conditions and compare ethane and naphtha.

In Table 4.3 P/E ratios are given for naphtha instead of conversions. Why? The P/E ratio is a measure of cracking severity or conversion. Does the conversion increase or decrease with increasing P/E ratio. Explain the logic of the temperature profiles in Figure 4.5. Why were the inlet temperatures for the two feedstocks chosen differently?

Compare steam cracking with FCC.

Compare the reaction conditions used in the various thermal cracking processes discussed in this book. Explain the differences.

4.3.2   Cracking Furnace

Figure 4.6 shows the cross-section of a typical two-cell cracking furnace, in which two fireboxes are connected to one convection section. The convection section contains various zones used for preheating the steam and the feed. The actual reactions take place in the radiant section, which can be as high as 15 m.

The cracking tubes, which are usually hung in a single plane down the center of the furnace, vary widely in diameter (30–200 mm) and length (10–100 m), depending on the production rate for each tube

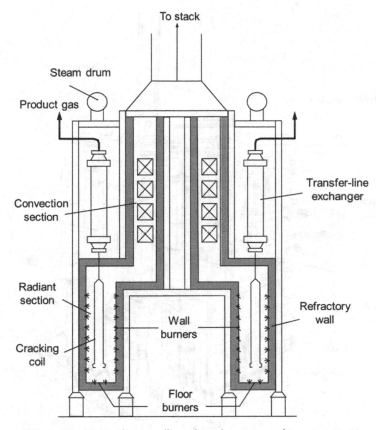

Figure 4.6   *Typical two-cell cracking furnace configuration [1–3].*

and the rate of coke deposition. For a given total feed rate a furnace either contains many short small diameter tubes or a smaller number of long large diameter tubes. The longer tubes consist of straight tubes connected by U-bends. Depending on the tube dimensions and the desired furnace capacity, the number of tubes in a single furnace may range from 2 to 180. Figure 4.7 shows some typical tube designs.

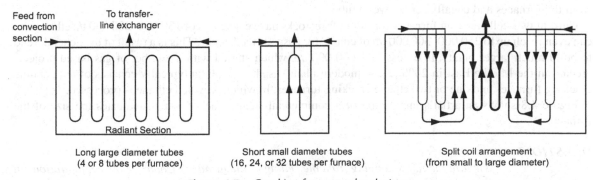

Figure 4.7   *Cracking furnace tube designs.*

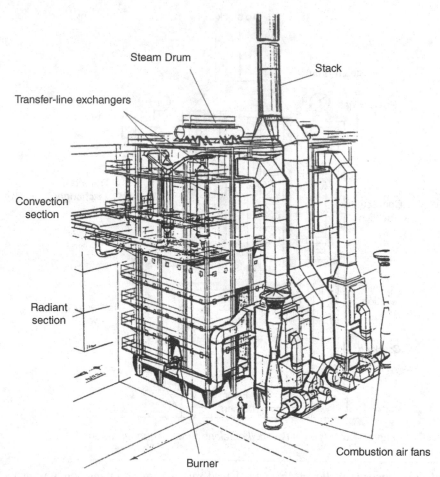

Figure 4.8   *Artist's impression of a commercial steam cracking furnace. (Courtesy of Van Schijndel [personal communication].)*

The cracked gas is quenched immediately after leaving the radiant zone in order to retain its composition as much as possible. The heat exchangers, called transfer-line exchangers, are mounted directly on the outlet from the furnaces and usually serve several tubes.

Current two-cell cracking furnaces for liquid feedstocks have capacities of 130 000–200 000 t/a ethene and can even reach up to 250 000–300 000 t/a of ethene for gaseous feedstocks. This is a result of the general trend to build much larger plants, in excess of 1 000 000 t/a, which started with a number of gas-based projects erected in the Middle East in 2000 [2]. A modern plant based on naphtha typically consists of ten naphtha cracking furnaces and one or two ethane cracking furnaces, in which recycled ethane is processed.

Figure 4.8 shows an artist's impression of a commercial steam cracker, which indicates the size of the furnace.

QUESTIONS:

Why are the tubes hanging from the ceiling of the furnace? Compare the configurations in Figure 4.6. Give the pros and cons.

> *Estimate the number of burners per furnace (capacity of a burner normally is between 0.1 and 1 MW, modern burners might have a 10 MW capacity).*
> *During start-up, the cracking tubes will expand. Estimate the increase in the tube length. Are special precautions necessary?*

During the past decades, significant improvements have been made in the design and operation of cracking furnaces. The use of high quality alloys for the cracking tubes enables operation at higher temperatures (up to about 1400 K [2]), which permits operation at shorter residence times and higher capacity. For instance, the shortest possible residence time has decreased from 0.5–0.8 seconds in the 1960s to 0.1–0.15 seconds in the late 1980s [4]. In addition, energy requirements have been reduced by heat recovery from furnace flue gas and rapid development has occurred in process control, data management, and optimization systems.

4.3.3 Heat Exchanger

The cracking products usually leave the reactor at a temperature exceeding 1070 K. They should be instantaneously cooled to prevent consecutive reactions. Quenching can be direct or indirect or a combination of both. Direct quenching involves the injection of a liquid spray, usually water or oil, and cooling can be extremely fast. Indirect cooling by *transfer-line exchangers* (TLEs) (Figure 4.9) has the advantage that valuable high-pressure steam can be generated. This steam can be used in the refrigeration compressors and the turbines driving the cracked gas (Section 4.4).

In designing transfer-line exchangers the following points are of concern:

- minimum residence time in the section between furnace outlet and TLE;
- low pressure drop;

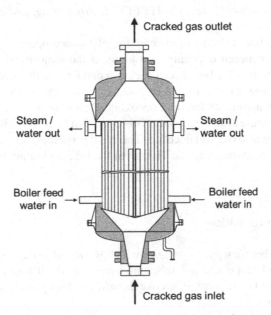

Figure 4.9   *Schematic of a transfer-line exchanger.*

- efficient heat recovery – the outlet temperature for liquid cracking should not be too low (condensation of heavy components);
- acceptable run times (fouling occurs by coke deposition and condensation of heavy components).

Usually a transfer-line exchanger is used, followed by direct quenching. In ethane cracking, where only small amounts of fuel oil are produced, cooling is mainly accomplished in the TLE. During steam cracking of heavy naphtha or gas oil, in which large amounts of fuel oil are produced, the heat exchange area of the TLE is reduced and most of the cooling is performed by a direct quench. In the extreme case of vacuum gas oils the TLEs are totally eliminated and supplanted by a direct quench.

QUESTIONS:

> *Explain the term "transfer-line exchanger."*
> *Minimization of pressure drop is important for energy saving and for product quality. Why?*
> *Discuss advantages and disadvantages of direct cooling by water injection.*
> *Why does a high pressure drop lead to lesser product quality?*
> *Why are no TLEs used when the feedstock is a vacuum gas oil?*

4.3.4   Coke Formation

The inevitable formation of coke during the cracking process causes many undesired phenomena, such as reduced heat transfer rates, increased pressure drop, lowered yields, and reduced selectivity towards alkenes.

QUESTIONS:

> *Explain these phenomena. Does the temperature inside the cracking tubes increase or decrease as a result of coke deposition?*
> *When comparing coke deposited in different processes what do you expect about its reactivity? (Compare coke deposited in FCC, hydrotreating, and ethane cracking).*

As a consequence of these issues, frequent interruption of furnace operation is necessary to remove the deposited coke. The interval between decoking operations in the majority of furnaces is 20–60 days for gaseous feedstocks and 20–40 days for liquid feedstocks, depending on the feedstock, the cracking severity (conversion), and special measures taken to reduce coke formation [5, 6] (Box 4.1). In industrial practice, always at least one furnace is usually undergoing decoking. Decoking of the cracking tubes takes place by gasification in an air/steam mixture or in steam only. This operation lasts for 5–30 hours [7]. During decoking of the reactor tubes, the TLE is also (partly) decoked. The usual procedure for complete decoking of the TLE is shutting down the furnace, disconnecting the TLE from the tube and removing the coke mechanically or hydraulically.

Box 4.1   Reducing Coke Formation

A distinction can be made between *pyrolysis coke*, which is formed in the gas phase and *catalytic coke* [8]. The cracking tubes and heat exchanger tubes are usually made of heat-resistant alloys containing iron and nickel, which unfortunately catalyze coke deposition. This process is often the largest source of coke in cracking furnaces.

Reducing the formation of catalytic coke would mean large cost savings, in particular due to reduced downtime of the cracking furnaces. Therefore, in the past decades a lot of effort has been put into finding appropriate methods or additives to this effect [2, 9].

One particularly effective method is the technology developed by Nova Chemicals, in which the cracking tubes are pretreated so that an inert barrier is created on the internal surface, preventing the catalytic influence of the tubes [10]. This technology has recently been commercialized and led to reduced coking rates of up to 90% and thus a 10-fold runtime increase.

The tubes can also be passivated by chemical treatment prior to start-up or during operation. Such passivation technologies are offered by various companies, including Atofina, ChevronPhillips, and Nova. The advantages of this process are that it is less costly and that dosing can be adjusted during operation. A disadvantage is that the chemicals used for passivation could end up in the cracking products [2].

An alternative approach is to apply a catalytic coating to the cracking tubes, which enables the removal of the deposited coke while the cracking furnace is in operation. BASF has recently started to commercialize such surface coatings (through a newly formed corporation, BASF Qtech) [11, 12].

QUESTION:

Which reaction would the catalytic surface coating developed by BASF Qtech catalyze? What are advantages and disadvantages compared to the other options?

4.4   Product Processing

Processing of cracked gas, that is, the separation into the desired components, can be performed in many different sequences that depend on the feedstock type and the degree of recovery desired for the different products. For example, with pure ethane as feedstock, the amounts of C_3 and heavier hydrocarbons are small and their recovery is often not economically feasible.

Figures 4.10 and 4.11 show examples of conventional flow schemes for downstream processing of the cracking products. Not surprisingly, the processing of products resulting from the cracking of gaseous feedstocks (ethane, propane, butane) (Figure 4.10) is less complicated than the processing of products resulting from the cracking of liquid feedstocks (naphtha, gas oil) (Figure 4.11).

The product mixture resulting from cracking of gaseous feedstocks is firstly cooled in the transfer-line exchanger with subsequent cooling by direct quench with water. Secondly, the product stream is compressed by multistage compression, typically in four to six stages with intermediate cooling. Before the last compressor stages, acid gas components (mainly hydrogen sulfide and carbon dioxide) are removed. After the last compressor stage, water removal takes place by chilling and drying over zeolites. Subsequent fractionation of the cracking products is based on cryogenic distillation (temperature <273 K) and conventional distillation under pressure (15–35 bar). Cryogenic distillation is very energy intensive due to the need for a refrigeration system.

The separation by distillation of ethene/ethane and propene/propane mixtures is difficult: the separation of the former mixture requires 80–150 trays, while in the latter case over 160 trays are needed [4]. The trays are often placed in two towers in series.

In the schemes of Figures 4.10 and 4.11, ethyne and C_3H_4 compounds are converted to ethene and propene, respectively, by selective hydrogenation. This is conventionally done in a gas phase process in catalytic fixed

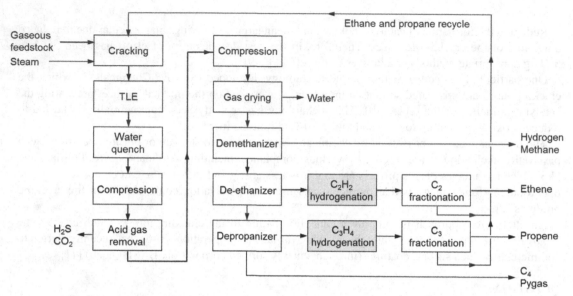

Figure 4.10 *Simplified process flow diagram for producing ethene via gas cracking.*

bed reactors. As the hydrogenation reaction is highly exothermic, temperature control may be problematic; with high hydrogen and ethyne levels, there is even a risk of thermal runaways. As a result of changed operating conditions in steam crackers during the past decades, the product streams are richer in ethyne. Therefore, currently, trickle bed reactors with an inert liquid phase are increasingly used to convert the largest part of the ethyne. To achieve the very low ethyne levels required for polymerization of ethene (Chapter 11),

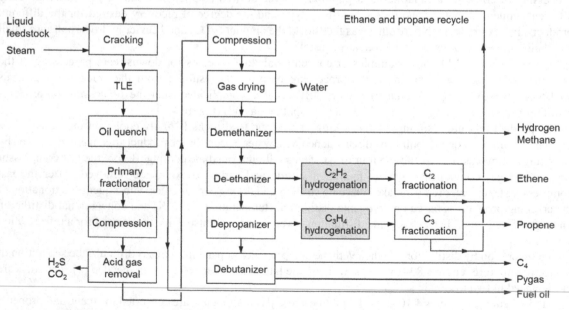

Figure 4.11 *Simplified process flow diagram for producing ethene via liquid cracking.*

Table 4.4 *Characteristics of columns in a 450 000 t/a ethene plant based on naphtha cracking [4].*

Column	Diameter (m)	Height (m)	Temperature (K)	Pressure (bar)
Primary fractionator	7.1	35.6	610	4.5
Demethanizer	2.0–3.0	64.7	131	9.0
De-ethanizer	2.7	39.7	230/360	24.5
C_2 fractionator	3.7	84.3	230	20.5
Depropanizer	2.3	22–32	370	20.0
C_3 fractionator	4.1	47.9	330	22.0
Debutanizer	2.4	34.3	410	7.5

a second step, using the conventional gas phase process, is required [13]. Occasionally, ethyne is produced and sold for welding applications.

Downstream processing of the product mixture resulting from cracking of liquid feedstocks is somewhat more complex because heavier components are present. In this case a primary fractionator is installed upstream of the compressor stages to remove fuel oil, which is produced in considerable amounts when cracking liquid feedstocks. This fuel oil is partly used in the direct quench operation, while the remainder may be sold or used as fuel for the cracking furnaces. The primary fractionator also produces a so-called pygas side stream. C_4 hydrocarbons are produced in sufficient quantities to justify their separate recovery and, therefore, usually a debutanizer is installed.

Recycle ethane and propane are normally cracked in separate furnaces, rather than including them as feedstocks to cracking furnaces built for liquid feedstocks.

QUESTIONS:

> *Why is the gas mixture cooled between the compression stages?*
> *Think of alternatives for cryogenic distillation. What are advantages and disadvantages of these alternatives?*
> *Why would separate furnaces be used for the cracking of naphtha and recycle ethane?*
> *Why are acid gas components removed prior to the final compressor stage?*
> *Why is there a trend towards product streams richer in ethyne?*
> *Ethyne and propyne are converted to the corresponding alkenes in a selective catalytic hydrogenation process. What change(s) in the operation of steam crackers would have been the most important cause for going from gas phase to liquid phase operation? Why is an additional gas phase process required for achieving very low ethyne levels?*
> *Pygas is also processed in a selective hydrogenation reactor in order to improve the product quality. Why would this operation improve the product quality?*

Table 4.4 shows typical dimensions and operating conditions of some of the columns used in the processing of cracking products from a naphtha cracker.

4.5 Novel Developments

Steam cracking plants have developed in an evolutionary way. Plant capacities have increased tremendously over the years. The first commercial ethene plants, in the 1950s, had a standard capacity of about 20 000 t/a. Since then, plant sizes have continuously been increasing to over 1 000 000 t/a at present [2]. Debottlenecking has played an important role. For example, distillation tower capacities were increased by the application of suitable internals, namely random or even structured packings. Currently, the maximum plant capacity is limited by the size of the cracked-gas compressor. Machinery manufacturers report that this capacity limit is most likely close to 2 000 000 t/a for a gas cracker, considering current technology [2].

In addition, fundamentally different processes have received attention. The most obvious disadvantage of the steam cracking process is its relatively low selectivity towards the desired products. A wide range of products is produced with limited flexibility. This is inherent to the non-catalytic nature of the process. In this section, a selection of alternative processes and feedstocks for the production of the light alkenes is presented.

In recent years the demand for propene, butenes, and butadiene has increased much more rapidly than the demand for ethene [14, 15], while many of the new steam crackers that are being built will utilize ethane as the feedstock, leading to a further mismatch with the market [15]. In addition, the growth of refinery sources (e.g., from FCC, Section 3.4.2) for these products has not kept pace with this growing demand. Therefore, dedicated production methods are receiving increasing attention. In 2011, about 57% of propene was produced as a coproduct in steam crackers, about 35% as a by-product of FCC and only about 8% by dedicated or "on purpose" technologies [16].

4.5.1 Selective Dehydrogenation of Light Alkanes

A dedicated process for the production of light alkenes by (catalytic) selective dehydrogenation of the corresponding alkanes suggests itself. Such a process is feasible. Alkane dehydrogenation is highly endothermic and high temperatures are required (Figure 4.2). At these high temperatures, however, secondary reactions such as cracking and coke formation become appreciable. Extensive deposition of coke on the catalyst takes place, so regeneration schemes are necessary. Furthermore, the thermodynamic equilibrium limits the conversion per pass, so a substantial recycle stream is required.

Several catalytic dehydrogenation processes are commercially available (Figure 4.12, Table 4.5). These processes differ in the type of catalyst used, the reactor design, the method of heat supply, and the method used for catalyst regeneration.

The UOP *Oleflex* process [17] uses radial flow, moving bed reactors in series with interstage heaters. The process is operated in the same way as the UOP continuous-regenerative reforming process (Section 3.4.3).

A key feature of the *Catofin* process [18] developed by ABB Lummus is the principle of storing the needed reaction heat in the catalyst bed. Reactors (usually three to eight in parallel) are alternately on stream for reaction and off stream for regeneration. During regeneration by burning the coke off the catalyst, the bed is heated, while this heat is consumed in the endothermic dehydrogenation reaction during the on-stream period. This technology is not new. It has been used extensively in coal gasification, where coal beds were fed alternately with air (exothermic) and steam (endothermic).

The *STeam Active Reforming* (STAR) process, originally developed by Phillips Petroleum and acquired by Uhde in 1999, uses multiple tubular reactors in a firebox, similar to the steam reformer used for the production of synthesis gas (Section 5.2). In this process a similar reaction–regeneration procedure is followed as in the Catofin process. From 2000 onwards, Uhde has added an oxydehydrogenation step (not shown in Figure 4.12) downstream of the reformer; this has significantly increased the performance [16].

The Snamprogetti *fluidized bed reactor design* (FBD) [19, 20] is similar to older FCC units (Section 3.4.2) with continuous catalyst circulation between a fluidized bed reactor and regenerator.

The selectivities in the commercial catalytic dehydrogenation processes are much higher than those in steam cracking, but the conversions are necessarily lower to limit side reactions. Still, single pass yields of 20–40 wt% can be achieved, considerably higher than the yields obtained in steam cracking (circa 16 wt% propene is produced in naphtha cracking (Table 4.3); propane and butane cracking produce similar yields [1]).

Processing of products from dehydrogenation reactors is similar to product processing in a steam cracking plant, although it is less complex as a result of the more concentrated alkene streams.

Table 4.5 *Performance of propane and isobutane dehydrogenation processes [17, 19, 21, 22].*

Process	Oleflex	Catofin	STAR[a]	FBD
Reactor	adiabatic moving beds	parallel adiabatic fixed beds with swing reactor	isothermal tubular reactors in furnace	fluidized bed
Catalyst	Pt/Al_2O_3	Cr/Al_2O_3	$Pt-Sn/Zn-Al_2O_3$	Cr/Al_2O_3
Catalyst life (y)	1–3	1–2	1–2	not available
Temperature (K)	820–890	860–920	750–890	820–870
Pressure (bar)	1–3	0.1–0.7	3–8	1.1–1.5
Conversion (%)[b]	P: 25	P: 48–65	P: 30–40	P: 40
	IB: 35	IB: 60–65	IB: 40–55	IB: 50
Selectivity (%)[b]	P: 89–91	P: 82–87	P: 80–90	P: 89
	IB: 91–93	IB: 93	IB: 92–98	IB: 91

[a]Without oxydehydrogenation reactor.
[b]P: propane dehydrogenation; IB: isobutane dehydrogenation.

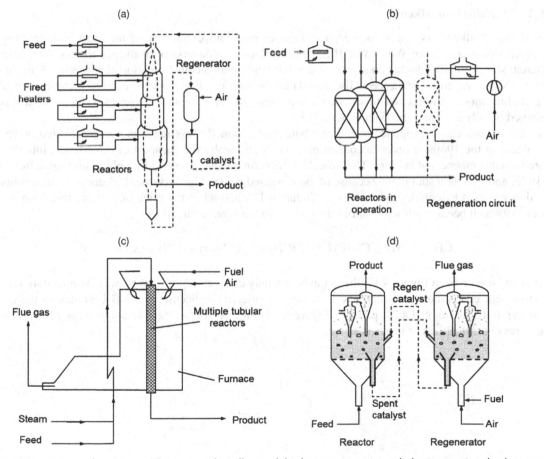

Figure 4.12 *Types of commercial reactors for alkane dehydrogenation: (a) adiabatic moving bed reactors (Oleflex); (b) parallel adiabatic fixed bed reactors (Catofin); (c) tubular fixed bed reactors in a furnace (STAR); (d) fluidized bed reactor and regenerator (FBD).*

In principle, catalytic dehydrogenation of ethane is also possible, but up till now a commercially attractive process has not been realized.

QUESTIONS:

> *Compare the four processes. Give pros and cons.*
> *What properties would the ideal catalyst have?*
> *Why is ethane a more difficult feed than heavier alkanes?*
> *The dehydrogenation reactions are endothermic equilibrium reactions (Figure 4.2). How does oxydehydrogenation in the STAR process help to increase the alkane conversion? (Hint: In which reaction does oxygen participate? What are the two effects of this reaction on the equilibrium of the dehydrogenation reaction?)*
> *Would you use air, oxygen-enriched air or pure oxygen as the source of oxygen in the STAR oxydehydrogenation reactor? Why?*
> *In view of the low conversions of the alkane dehydrogenation processes, which part(s) of the process would account for most of the capital costs?*

4.5.2   Metathesis of Alkenes

Metathesis of alkenes is one of the very few fundamentally novel and one of the most important organic reactions discovered since World War II [23]. It involves the conversion of alkenes to produce alkenes of different size. Formally, double bonds are broken with the simultaneous formation of new ones (Scheme 4.3). In detailed investigations, it has been discovered that the reaction takes place via carbene complexes resulting in a slightly more complex reaction network than presented in Scheme 4.3. However, the product mixture observed is fully in agreement with Scheme 4.3.

In the metathesis of propene, ethene and butene are formed. This reaction was discovered at Phillips Petroleum in the 1960s in order to convert propene, traditionally the cheapest of the alkenes, into the then more valuable ethene and butenes. This so-called Triolefin process operated for only a short time, from 1966 to 1972, and was then shut down because of the increased demand for propene. Metathesis is an equilibrium reaction which is essentially thermoneutral (Figure 4.13), and when the price of propene rose high enough (in the 1980s) it became attractive to run the process in the reverse direction:

$$CH_2 = CH_2 + CH_3CH = CHCH_3 \rightleftarrows 2CH_3CH = CH_2 \quad \Delta_r H_{298} \approx 0 \qquad (4.4)$$

As can be seen from Figure 4.13, the composition only changes slightly over a wide temperature range.

At present ethene/butene metathesis processes are offered for license by ABB Lummus as the Olefins Conversion Technology (OCT) process (Figure 4.14) and by IFP as the Meta-4 process [24]. Table 4.6 compares characteristics of these processes.

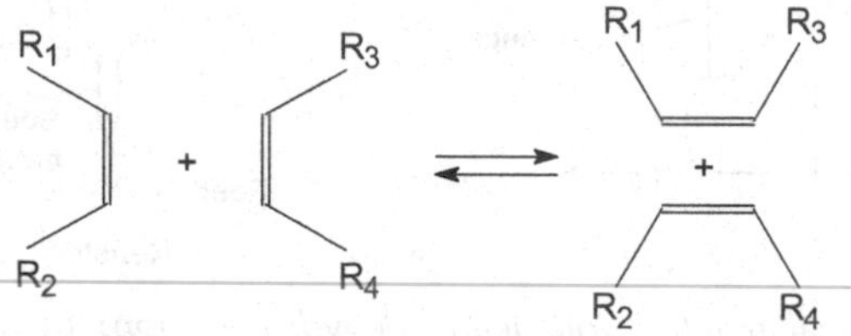

Scheme 4.3   *Metathesis of alkenes; R_1, R_2, R_3, R_4 = H, CH_3, CH_2CH_3, and so on.*

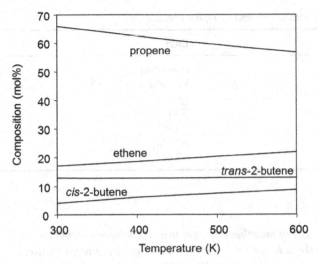

Figure 4.13 *Equilibrium composition of an ethene/propene/2-butene mixture.*

Feed and recycle ethene are mixed with feed and recycle butenes and heated prior to entering the fixed bed metathesis reactor. Most butene feeds not only contain 2-butenes, but also 1-butene (and some other C_4 components). Therefore, the metathesis reactor contains a catalyst that also catalyzes the isomerization of 1-butene to 2-butene. The products from the reactor are cooled and ethene is removed for recycle. In the propene column, high purity, polymer-grade propene is obtained by separation from the butenes, which are recycled. A small part is purged to remove butanes, isobutenes, and heavier products.

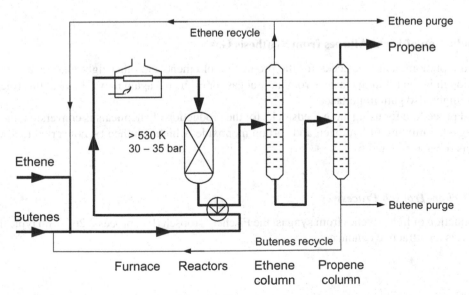

Figure 4.14 *Typical flow scheme of the OCT metathesis process. Only one of multiple reactors is shown.*

Table 4.6   *Comparison of metathesis technologies [24, 25].*

	OCT	Meta-4
Reactor	fixed bed	moving bed
Catalyst	$WO_3/SiO_2 + MgO^a$	Re_2O_7/Al_2O_3
Phase	gas	liquid
Temperature (K)	530	310
Pressure (bar)	30–35	60
Butene conversion (%)	60–75	63
Propene selectivity (%)	95	not available

[a] For double-bond isomerization.

QUESTIONS:

> *Why is part of the ethene stream from the ethene column purged?*
>
> *Some coke is formed during operation, requiring regeneration of the catalyst. What would be the procedure for this? (Hint: Often companies feel most comfortable with technologies they are experienced in.)*
>
> *Derive a reaction network for the metathesis of propene based on carbenes (present as carbene complexes) as the key intermediates.*

A number of companies have integrated the OCT process with naphtha crackers in order to increase the production of propene [24]. The propene-to-ethene yield can be increased from 0.5–0.65:1 kg/kg to as much as 1.1:1 kg/kg [25].

Metathesis is certainly not restricted to the three lightest alkenes. Among others, it has opened up new industrial routes to important petrochemicals, polymers, and specialty chemicals. An example of considerable industrial importance is the Shell Higher Olefins Process (SHOP, Section 9.4). Other examples can be found elsewhere [23, 24].

4.5.3 Production of Light Alkenes from Synthesis Gas

Currently, the predominant feedstocks for the production of ethene and other light alkenes are naphtha and ethane, although in the future methane (from natural gas) or even coal again may take over this role. Biomass conversion might also gain importance.

The usual procedure for using these feedstocks for the production of chemicals is conversion into synthesis gas (or syngas), a mixture of hydrogen and carbon monoxide, which can then be converted to a wide range of chemicals (Chapters 5 and 6).

4.5.3.1 Fischer–Tropsch Process

For the production of light alkenes from syngas, the Fischer–Tropsch (FT) process [26], with typical reactions given below, is an attractive candidate.

General:

$$nCO + 2n\,H_2 \rightarrow C_nH_{2n} + nH_2O \quad \Delta_r H_{298} \ll 0 \tag{4.5}$$

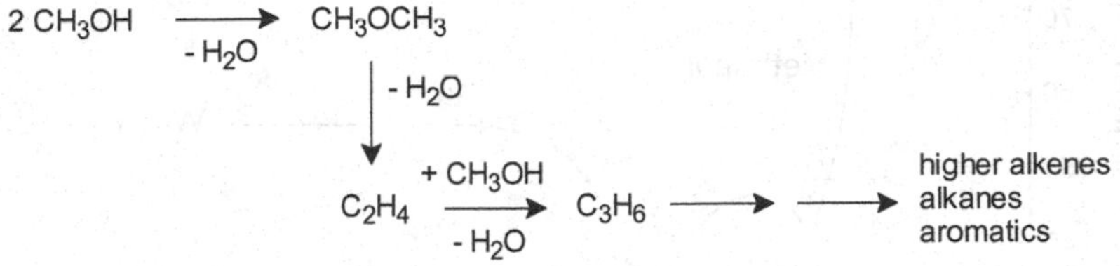

$$2\,CH_3OH \xrightarrow[-H_2O]{} CH_3OCH_3$$

$$\downarrow -H_2O$$

$$C_2H_4 \xrightarrow[-H_2O]{+CH_3OH} C_3H_6 \longrightarrow \longrightarrow \begin{array}{l}\text{higher alkenes}\\\text{alkanes}\\\text{aromatics}\end{array}$$

Scheme 4.4   *Reactions in the conversion of methanol to light alkenes and gasoline components.*

Ethene:

$$2CO + 4H_2 \rightarrow C_2H_4 + 2H_2O \quad \Delta_r H_{298} = -211 \text{ kJ/mol} \tag{4.6}$$

The Fischer–Tropsch process (Section 6.3) is capable of producing a wide range of hydrocarbons and oxygenates, depending on the catalyst, reactor, and reaction conditions. It has mainly been used for the production of liquid fuels (gasoline, diesel), but in principle the process can also be tuned for the production of light alkenes by using different catalysts. A disadvantage of such a process is the requirement of large recycle streams. Current catalysts show low selectivity towards ethene and other light alkenes, while methane formation is high. In addition, catalyst activity and life are still issues.

QUESTION:

> *Although the FT reaction has been known for a long time, there is still discussion about its mechanism. Propose one or more possible mechanisms.*

4.5.3.2 *Methanol-to-Olefins Process*

Another option for the production of light alkenes is the methanol-to-olefins (MTO) process [27–34], in which methanol is first dehydrated to form dimethyl ether (DME), which then further reacts to form alkenes (olefins). This process has evolved from Exxon Mobil's methanol-to-gasoline (MTG) process [32], which converts methanol into gasoline with alkenes as intermediates (Section 6.3). The observed reactions can be described in simplified form by Scheme 4.4.

The real reaction mechanism is rather complex and although much research has been done it is not yet fully understood. An important issue for the production of light alkenes is to suppress the formation of aromatics. This can be accomplished by using suitable zeolites [27–31]. The keyword here is shape selectivity, in this case product selectivity, that is, only small molecules can leave the pores of the catalyst, and thus aromatics cannot (Section 10.4).

Figure 4.15 shows a typical product distribution as a function of reaction time, which clearly indicates the sequential character of the reactions.

QUESTIONS:

> *Suggest a reaction pathway for the formation of ethene. Do the same for the formation of propene. What would be the mechanism for the formation of aromatics?*

Coke formation plays an important role in the MTO process. Deactivation of the catalyst by coke makes frequent regeneration necessary. Furthermore, the reaction is highly exothermic, which requires good temperature control. Therefore, it has been proposed to carry out the process in a system consisting of two fluidized

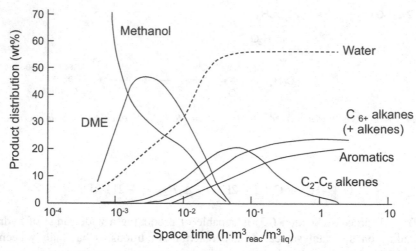

Figure 4.15   *Formation of various products from methanol as a function of space time (T = 643 K; p = 1 bar).*

bed reactors, one for the exothermic reaction and one for the exothermic catalyst regeneration [32, 33]. Figure 4.16 shows a flow scheme, including a possible configuration of the separation section.

A demonstration plant of this process has been operated successfully in Norway since 1995 and a larger demonstration plant was started up in Belgium in 2008. In this demonstration plant, the larger alkenes are cracked in a separate process (the olefin cracking process (OCP) developed by UOP/Total) to obtain larger ethene and propene yields. Commercial scale MTO plants are being constructed and planned in China and

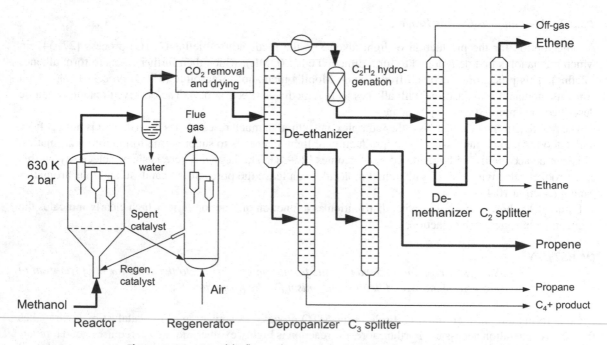

Figure 4.16   *Possible flow scheme of the UOP/Hydro MTO process.*

Table 4.7 *Comparison of product gas compositions from naphtha cracker, UOP/ Hydro MTO process and Lurgi MTP process [2, 36].*

Product (wt%)	Naphtha cracker 1070 K, 2 bar	MTO 630 K, 2 bar	MTP 730 K, 1.5 bar
Light gases[a]	27	2	1
Ethene	29	39	—
Propene	16	42	72
Butenes	5	12	—
C_5+	23	5	27
Total	100	100	100

[a]Including light alkanes.

Nigeria. The process uses a proprietary molecular sieve catalyst and operates at nearly 100 % methanol conversion.

The ratio of ethene to propene can be adjusted by altering the reaction conditions, which makes the process flexible. Whether the MTO process is attractive from an economic viewpoint depends on the local conditions, such as the availability of natural gas at low cost and transportation distance, but the catalyst activity and obtained selectivities are favorable for this process [28, 33].

An alternative to the MTO process is the methanol-to-propene process (MTP) developed by Lurgi [35]. In this process fixed bed reactors are employed. As the name suggests, the MTP process mainly produces propene. Table 4.7 compares the product gas composition of the MTO and MTP processes with that of a typical naphtha cracker.

QUESTIONS:

In the flow scheme of Figure 4.16, the first separation step after the reactor is the removal of water. Explain.

As an ethene producer using the MTO process, would you purchase your methanol on the market or build your own methanol plant with remote natural gas as feedstock? If you choose the latter, will you locate your methanol and MTO plants near each other or not? (Hint: compare total ethene and methanol production and plant sizes.)

4.5.4 Dehydration of Bioethanol

The production of ethene by the catalytic dehydration of bioethanol (e.g., produced from sugarcane, see also Chapter 7) has attracted renewed interest, particularly in Brazil and the United States [2, 37]. In the presence of a suitable catalyst and at sufficiently high temperature (600–770 K), the following endothermic reaction takes place:

$$CH_3CH_2OH \rightleftarrows CH_2 = CH_2 + H_2O \quad \Delta_r H_{298} = 45.3 \text{ kJ/mol} \quad (4.7)$$

Like in the MTO process, ethanol can be converted into its corresponding ether, which is then converted into ethene:

$$2CH_3CH_2OH \rightleftarrows CH_3CH_2OCH_2CH_3 + H_2O \quad \Delta_r H_{298} = -24.4 \text{ kJ/mol} \quad (4.8)$$

$$CH_3CH_2OCH_2CH_3 \rightarrow 2CH_2 = CH_2 + H_2O \quad \Delta_r H_{298} = 115 \text{ kJ/mol} \quad (4.9)$$

Formation of the diethyl ether is favored at temperatures of 420–570 K.

Table 4.8   *Comparison of fixed bed reactors for the production of ethene from bioethanol [37].*

	Isothermal	Adiabatic
Temperature (K)	600–650	720–770 (at inlet)
Time between catalyst regenerations (months)	1–6	6–12
Conversion (%)	98–99	99
Ethene selectivity (%)	95–99	97–99

The most important by-products are acetaldehyde and hydrogen, which are formed by the following reaction:

$$CH_3CH_2OH \rightleftarrows CH_3CHO + H_2 \quad \Delta_r H_{298} = 68.5 \text{ kJ/mol} \tag{4.10}$$

Other by-products, which are present in the ethanol feed or are formed in small quantities, include acetic acid, methanol, higher hydrocarbons, carbon monoxide, and carbon dioxide.

Temperature control is important because at too low a temperature the ethanol conversion is not sufficient and the formation of diethyl ether increases, while at too high a temperature acetaldehyde and excessive amounts of decomposition by-products, including coke, are formed. The reaction is carried out in the vapor phase over solid acid catalysts in fixed or fluidized bed reactors. Table 4.8 shows typical data for fixed bed reactors, which can be operated either isothermally or adiabatically. For isothermal operation, multitubular reactors are used with the catalyst placed in the tubes and a hot fluid circulating in the shell. With adiabatic operation, the heat of reaction is supplied by adding preheated steam to the feed. Usually three fixed beds are used with addition of ethanol between the reactors and interstage furnaces to reheat the reaction mixture.

QUESTIONS:

> *Why is the inlet temperature of the adiabatic fixed bed reactor higher than the temperature of the isothermal fixed bed reactor?*
> *Draw a possible flow scheme for the bioethanol dehydration process including separation steps.*

4.5.5   Direct Conversion of Methane

The direct conversion of methane (the largest constituent of natural gas) into valuable chemicals is a holy grail. The direct conversion of methane into ethane and ethene can take place by applying so-called 'oxidative coupling', that is, the selective catalytic oxidation of methane:

$$2CH_4 + \tfrac{1}{2}O_2 \rightarrow C_2H_6 + H_2O \quad \Delta_r H_{298} = -177 \text{ kJ/mol} \tag{4.11}$$

$$C_2H_6 + \tfrac{1}{2}O_2 \rightarrow C_2H_4 + H_2O \quad \Delta_r H_{298} = -105 \text{ kJ/mol} \tag{4.12}$$

The reaction produces ethane, which is subsequently dehydrogenated to ethene, while through sequential reactions heavier hydrocarbons are also formed in small amounts. However, large amounts of carbon monoxide and carbon dioxide are formed as by-products:

$$CH_4, CH_3CH_3, \textit{etc.} + O_2 \rightarrow CO, CO_2 \quad \Delta_r H_{298} \ll 0$$

This not only reduces the selectivity towards ethene but the high exothermicity of these oxidation reactions also presents enormous heat removal problems (runaways).

In the past three decades oxidative methane coupling has received much attention [38–40]. Since the first publication in 1982 [41] considerable progress has been made in the development of catalysts for the oxidative coupling reaction. However, as with all selective oxidations, it is a challenge to achieve both a high conversion and a high selectivity, and the best combinations of conversion and selectivity thus far achieved in laboratory fixed bed reactors, leading to maximum single-pass yields of 30–35% [42], are still well below those required for economic feasibility. Accordingly, the emphasis in oxidative coupling research has shifted somewhat to innovative reactor designs, such as staged oxygen addition and membrane reactors for combining the oxidation reaction and ethene removal [39] (see also Chapter 5). An interesting idea is to combine the exothermic oxidative coupling reaction with the endothermic steam reforming reaction (Chapter 5). This results in a higher methane conversion with the additional advantage that the heat generated by the oxidative coupling reaction is consumed in the steam reforming reaction, so that an opportunity for autothermal operation is created [40].

QUESTIONS:

Explain why it is not surprising that a commercially attractive process has not been developed yet for the oxidative coupling of methane, despite the enormous amount of research that has been carried out?

What is meant by the term "single pass yield"?

What would be the reaction temperature and the partial pressures of methane and oxygen in the reactor? Explain your answer.

This chapter presents several very different technologies for producing light alkenes. List the pros and cons. Take into account the prices of feeds and products and evaluate the results.

References

[1] Grantom, R.L. and Royer, D.J. (1987) Ethylene, in *Ullmann's Encyclopedia of Industrial Chemistry*, 5th edn, vol. **A10** (ed. W. Gerhartz). VCH, Weinheim, pp. 45–93.

[2] Zimmermann, H. and Waltzl, R. (2009) Ethylene, in *Ullmann's Encyclopedia of Industrial Chemistry*. Wiley-VCH Verlag GmbH, Weinheim, pp. 1–66.

[3] Kniel, L., Winter, O. and Stork, K. (1980) *Ethylene: Keystone to the Petrochemical Industry*. Marcel Dekker, New York.

[4] Sundaram, K.M., Shreehan, M.M., and Olszewski, E.F. (1994) Ethylene, in *Kirk-Othmer Encyclopedia of Chemical Technology*, 4th edn, vol. **9** (eds J.I. Kroschwitz and M. Howe-Grant). John Wiley & Sons, Inc., New York, pp. 877–915.

[5] Sundaram, K.M., Van Damme, P.S., and Froment, G.F. (1981) Coke deposition in the thermal cracking of ethane. *AIChE Journal*, **27**, 946–951.

[6] Wysiekierski, A.G., Fisher, G., and Schillmoller, C.M. (1999) Control coking for olefins plants. *Hydrocarbon Process*, **78**, 97–100.

[7] Chauvel, A. and Lefebvre, G. (1989) Petrochemical Processes 1. Synthesis Gas Derivatives and Major Hydrocarbons. Technip, Paris.

[8] Cai, H., Krzywicki, A., and Oballa, M.C. (2002) Coke formation in steam crackers for ethylene production. *Chemical Engineering and Processing*, **41**, 199–214.

[9] Wang, J., Reyniers, M.F., Van Geem, K.M., and Marin, G.B. (2008) Influence of silicon and silicon/sulfur-containing additives on coke formation during steam cracking of hydrocarbons. *Industrial & Engineering Chemistry Research*, **47**, 1468–1482.

[10] Gyorffy, M., Benum, L., and Sakamoto, N. (2006) Increased run length and furnace performance with Kobota and Nova Chemicals ANK 40 anticoking technology. AIChE 18th Ethylene Producers' Conference, AIChE Spring Meeting, Orlando, FL, April 23–27, 2006.

[11] Parkinson, G. (2010) Furnace tube coatings reduce carbon formation and increase efficiency in olefins plants. *Chementator*, **March 1**, 14.

[12] Ondrey, G. (2011) BASF Qtech formed to commercialize catalytic surface coatings for steam-cracker furnace tubes, *Chemical & Engineering News*, September 13.

[13] Edvinsson, R.K., Holmgren, A.M., and Irandoust, S. (1995) Liquid-phase hydrogenation of acetylene in a monolithic catalyst reactor. *Industrial & Engineering Chemistry Research*, **34**, 94–100.

[14] Zehnder, S. (1998) What are Western Europe's petrochemical feedstock options? *Hydrocarbon Processing*, **77**, 59–65.

[15] Tallman, M.J., Eng, C., Choi, S., and Park, D.S. (2010) Naphtha cracking for light olefins production. *PTQ*, **Q3**, 87–91.

[16] Uhde (2011) The Uhde STAR Process, Uhde, www.uhde.eu (last accessed 17 December 2012).

[17] Pujado, P.R. and Vora, B.V. (1990) Make C_3-C_4 olefins selectively. *Hydrocarbon Processing*, **69**, 65–70.

[18] Arora, V.K. (2004) Propylene via CATOFIN propane dehydrogenation technology, in *Handbook of Petrochemicals Production Processes* (ed. R.A. Meyers). McGraw-Hill, New York.

[19] Buonomo, F., Sanfilippo, D., and Trifirò, F. (1997) Dehydrogenation of alkanes, in *Handbook of Heterogeneous Catalysis* (eds G. Ertl, H. Knözinger, and J. Weitkamp). VCH, Weinheim, pp. 2140–2151.

[20] Sanfilippo, D., Buonomo, F., Fusco, G., Miracca, I., SpA, S., Maritano, V., Kotelnikov, G., Yarsintez and Oktyabrya, P. (1998) Paraffins activation through fluidized bed dehydrogenation: the answer to light olefins demand increase, in *Studies in Surface Science and Catalysis*, vol. **119** (ed. A. Parmaliana). Elsevier, pp. 919–924.

[21] Calamur, N. and Carrera, M. (1996) Propylene, in *Kirk-Othmer Encyclopedia of Chemical Technology*, 4th edn, vol. **20** (eds J.I. Kroschwitz and M. Howe-Grant). John Wiley & Sons, Inc., New York, pp. 249–271.

[22] Buyanov, R.A. and Pakhomov, N.A. (2001) Catalysts and processes for paraffin and olefin dehydrogenation. *Kinetics And Catalysis*, **42**, 64–75.

[23] Delaude, L. and Noels, A.F. (2000) Metathesis, in *Kirk-Othmer Encyclopedia of Chemical Technology*, John Wiley & Sons, Inc., New York, pp. 920–958. doi: 10.1002/0471238961.metanoel.a01

[24] Mol, J.C. (2004) Industrial applications of olefin metathesis. *Journal of Molecular Catalysis A*, **213**, 39–45.

[25] Parkinson, G. (2004) Mastering propylene production. *Chemical Engineering Progress*, **100**, 8–11.

[26] Fischer, F. and Tropsch, H. (1923) The preparation of synthetic oil mixtures (Synthoil) from carbon monoxide and hydrogen. *Brennstoff-Chemie*, **4**, 2776–285.

[27] Chang, C.D. (1984) Methanol conversion to light olefins. *Catalysis Reviews, Science And Engineering*, **26**, 323–345.

[28] Inui, T. and Takegami, Y. (1982) Olefins from methanol by modified zeolites. *Hydrocarbon Processing*, **61**, 117–120.

[29] Inui, T., Phatanasri, S., and Matsuda, H. (1990) Highly selective synthesis of ethene from methanol on a novel nickel-silicoaluminophosphate catalyst. *Journal of the Chemical Society, Chemical Communications*, 205–206.

[30] Kaiser, S.W. (1985) Methanol conversion to light olefins over silico-aluminophosphate molecular sieves. *Arabian Journal for Science and Engineering*, **10**, 361–366.

[31] Kumita, Y., Gascon, J., Stavitski, E., Moulijn, J.A. and Kapteijn, F. (2011) Shape selective methanol to olefins over highly thermostable DDR catalysts. *Applied Catalysis A*, **391**, 234–243.

[32] Bos, A.N.R., Tromp, P.J.J., and Akse, H.N. (1995) Conversion of methanol to lower olefins. Kinetic modeling, reactor simulation, and selection. *Industrial & Engineering Chemistry Research*, **34**, 3808–3816.

[33] Nilsen, H.R. (1997) The UOP/Hydro Methanol to Olefin process: its potential as opposed to the present application of natural gas as feedstock. Proceedings of the 20th World Gas Conference, Copenhagen, Denmark, 10–13 June.

[34] Chang, C.D. and Silvestri, A.J. (1977) The conversion of methanol and other O-compounds to hydrocarbons over zeolite catalysts. *Journal of Catalysis* , **47**, 249–259.

[35] Kvisle, S., Fuglerud, T., Kolboe, S., Olsbye, U., Lillerud, K.P. and Vora, B.V. (2008) Methanol-to-hydrocarbons, in *Handbook of Heterogeneous Catalysis*, 2nd edn, vol. **6** (eds G. Ertl, H. Knözinger, H. Schüth, and J. Weitkamp). Wiley-VCH Verlag GmbH, pp. 2950–2965.

[36] Diercks, R., Arndt, J.D., Freyer, S., Geier, R., Machhammer, O., Schwartze, J. and Volland, M. (2008) Raw material changes in the chemical industry. *Chemical Engineering & Technology*, **31**, 631–637.

[37] Morschbacker, A. (2009) Bio-ethanol based ethylene. *Polymer Reviews*, **49**, 79–84.

[38] Lunsford, J.H. (1997) Oxidative coupling of methane and related reactions, in *Handbook of Heterogeneous Catalysis* (eds G. Ertl, H. Knözinger, and J. Weitkamp). VCH, Weinheim, pp. 1843–1856.

[39] Androulakis, I.P. and Reyes, S.C. (1999) Role of distributed oxygen addition and product removal in the oxidative coupling of methane. *AIChE Journal*, **45**, 860–868.

[40] Tiemersma, T.P. (2010) Integrated Autothermal Reactor Concepts for Combined Oxidative Coupling and Reforming of Methane. PhD Thesis, Twente University, Enschede.

[41] Keller, G.E. and Bhasin, M.M. (1982) Synthesis of ethylene via oxidative coupling of methane: I. Determination of active catalysts. *Journal of Catalysis*, **73**, 9–19.

[42] Makri, M. and Vayenas, C.G. (2003) Successful scale up of gas recycle reactor-separators for the production of C_2H_4 from CH_4. *Applied Catalysis A*, **244**, 301–310.

5

Production of Synthesis Gas

5.1 Introduction

Synthesis gas (or syngas) is a general term used to designate mixtures of hydrogen and carbon monoxide in various ratios. These mixtures are used as such and are also sources of pure hydrogen and pure carbon monoxide. Table 5.1 shows some syngas applications. From the examples it can be noted that the term "syngas" is used more generally than stated above: the N_2/H_2 mixture for the production of ammonia is also referred to as syngas.

Syngas may be produced from a variety of raw materials ranging from natural gas to coal. The choice for a particular raw material depends on the cost and availability of the feedstock, and on the downstream use of the syngas. Syngas is generally produced by one of three processes, which are distinguished based on the feedstock used:

- steam reforming of natural gas or light hydrocarbons, optionally in the presence of oxygen or carbon dioxide;
- partial oxidation of (heavy) hydrocarbons with steam and oxygen;
- partial oxidation of coal (gasification) with steam and oxygen.

The name of the processes may be somewhat confusing. The term *steam reforming* is used to describe the reaction of hydrocarbons with steam in the presence of a catalyst. This process should not be confused with the *catalytic reforming* process for improving the gasoline octane number (Section 3.4.3). In the gas industry, *reforming* is commonly used for the conversion of a hydrocarbon by reacting it with oxygen-containing molecules, usually H_2O, carbon dioxide, and/or oxygen. A combination of steam reforming and partial oxidation, in which endothermic and exothermic reactions are coupled, is often referred to as *autothermal reforming*.

Partial oxidation (also called steam/oxygen reforming) is the non-catalytic reaction of hydrocarbons with oxygen and usually also steam. This process may be carried out in an autothermal or allothermal way. Catalysis is possible and the process is then referred to as catalytic partial oxidation. Gasification is the more common term to describe partial oxidation of coal or petroleum coke (petcoke).

Chemical Process Technology, Second Edition. Jacob A. Moulijn, Michiel Makkee, and Annelies E. van Diepen.
© 2013 John Wiley & Sons, Ltd. Published 2013 by John Wiley & Sons, Ltd.

Table 5.1   *Syngas applications in refining, the chemical process industry and fuel production.*

Mixtures	Main uses	Chapter of this book
H_2	Refinery hydrotreating and hydrocracking	3
$3\ H_2 : 1\ N_2{}^{a}$	Ammonia	6
$2\ H_2 : 1\ CO$	Substitute natural gas (SNG)	5
$2\ H_2 : 1\ CO$	Alkenes (Fischer–Tropsch reaction)	6
$2\ H_2 : 1\ CO$	Methanol, higher alcohols	6, 9
$1\ H_2 : 1\ CO$	Aldehydes (hydroformylation)	9
CO	Acids (formic and acetic)	6, 9

[a]*With N_2 from air.

QUESTIONS:

What is "petcoke"? Why is it used in gasification?

Most syngas today is produced by steam reforming of natural gas or light hydrocarbons up to naphtha. For light feedstocks partial oxidation is usually not an economic option, because of the high investment costs as a result of the required cryogenic air separation. Partial oxidation processes, however, are employed where feeds suitable for steam reforming are not available or in special situations where local conditions exist to provide favorable economics. Such conditions could be the availability of relatively low priced heavy feedstocks or the need for a syngas with high carbon monoxide content. Coal gasification is an important source of chemicals via syngas production in countries with abundant coal resources, such as South Africa, China, India, and the United States. It is also increasingly used for power generation.

Figure 5.1 shows general flow schemes of the main processes for syngas production. The steam reforming feed usually has to be desulfurized. Sulfur is a poison for metal catalysts because it can block active sites by the formation of rather stable surface sulfides. In steam reforming and in downstream uses in many reactors transition-metal-based catalysts are used. When sulfur is present as hydrogen sulfide (H_2S), adsorption (for instance on activated carbon), reaction with an oxide (for instance zinc oxide), or scrubbing with a solvent may be employed. If the feed contains more stable sulfur compounds, hydrotreating (Section 3.4.5) may be required.

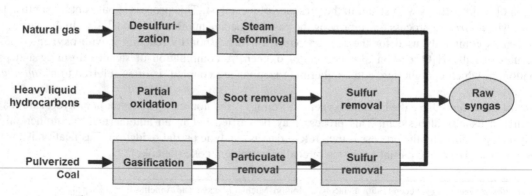

Figure 5.1   *General process flow diagrams for the production of syngas.*

Processes for the production of syngas based on coal and heavy oil fractions require removal of sulfur compounds (mainly H_2S) from the syngas. Feed purification is not possible with these raw materials, although attempts have been made using biotechnological approaches.

Depending on its production process and its downstream use (Table 5.1), the raw syngas may be treated in several ways. Syngas conditioning includes such processing steps as the water–gas shift reaction, carbon dioxide removal, methanation, and so on (Section 5.4) to achieve the right composition.

Partial oxidation of heavy hydrocarbons is very similar to coal gasification and will not be discussed further. This chapter deals with the conversion of natural gas and coal into syngas.

QUESTIONS:

In syngas production processes sulfur is generally removed. In reforming of natural gas this is done before the reforming step but in the case of heavy oil fractions and coal sulfur removal takes place after conversion (Figure 5.1). Explain.

Where in the process of producing syngas would you consider scrubbers? What solvents would you use?

The scale of the plant is an important characteristic in selecting a process for hydrogen production. What would you do when designing a plant for the production of fine chemicals? (Hint: Consider two cases, a very small plant and a larger fine chemicals plant. Take into consideration that, in principle, hydrogen can be produced from any molecule that contains hydrogen atoms.)

5.2 Synthesis Gas from Natural Gas

5.2.1 Reactions and Thermodynamics

Although natural gas does not solely consist of methane (Section 2.3.4) for simplicity it is assumed that this is the case. Table 5.2 shows the main reactions during methane conversion. When converting methane in the presence of steam the most important reactions are the steam reforming reaction 5.1 and the water–gas shift reaction 5.2.

Some processes, such as the reduction of iron ore [1] and the hydroformylation reaction (Table 5.1, Section 9.3), require syngas with a high carbon monoxide content, which might be produced from methane and carbon dioxide in a reaction known as *CO_2 reforming* (5.3). The latter reaction is also referred to as "dry reforming", obviously because of the absence of steam.

Table 5.2 *Reactions during methane conversion with steam and/or oxygen.*

Reaction	$\Delta_r H_{298}$ (kJ/mol)	(Reaction number)
$CH_4 + H_2O \rightleftarrows CO + 3\,H_2$	206	(5.1)
$CO + H_2O \rightleftarrows CO_2 + H_2$	−41	(5.2)
$CH_4 + CO_2 \rightleftarrows 2\,CO + 2\,H_2$	247	(5.3)
$CH_4 \rightleftarrows C + 2\,H_2$	75	(5.4)
$2\,CO \rightleftarrows C + CO_2$	−173	(5.5)
$CH_4 + {}^1/_2\,O_2 \rightarrow CO + 2\,H_2$	−36	(5.6)
$CH_4 + 2\,O_2 \rightarrow CO_2 + 2\,H_2O$	−803	(5.7)
$CO + {}^1/_2\,O_2 \rightarrow CO_2$	−284	(5.8)
$H_2 + {}^1/_2\,O_2 \rightarrow H_2O$	−242	(5.9)

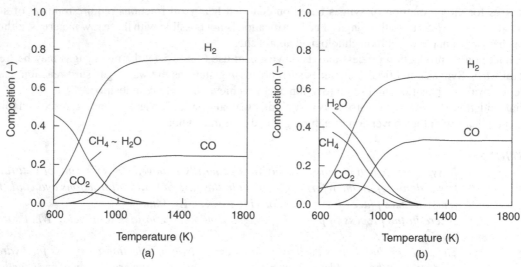

Figure 5.2   *Equilibrium gas composition at 1 bar as a function of temperature: (a) steam reforming of methane, $H_2O/CH_4 = 1$ mol/mol, H_2O curve coincides with CH_4 curve; (b) partial oxidation of methane, $O_2/CH_4 = 0.5$ mol/mol.*

The main reactions may be accompanied by coke formation, which leads to deactivation of the catalyst. Coke may be formed by decomposition of methane (5.4) or by disproportionation of carbon monoxide, the Boudouard reaction (5.5).

In the presence of oxygen, methane undergoes partial oxidation to produce carbon monoxide and hydrogen (5.6). Side reactions, such as the complete oxidation of methane to carbon dioxide and H_2O (5.7), and oxidation of the formed carbon monoxide (5.8) and hydrogen (5.9) may also occur.

The reaction of methane with steam is highly endothermic while the reactions with oxygen are moderately to extremely exothermic. Operation can hence be allothermal (steam and no or little oxygen added, required heat generated outside the reactor) or autothermal (steam and oxygen added, heat generated by reaction with oxygen within the reactor), depending on the steam/oxygen ratio.

Figure 5.2 compares the equilibrium gas compositions of the reaction of methane with steam and with oxygen at one bar as a function of temperature. Stoichiometric amounts of steam and oxygen were added for reactions 5.1 and 5.6, respectively. Both Figures 5.2a and 5.2b show essentially the same composition pattern as a function of temperature, although the product ratios are different.

This observation is not surprising, since apart from the irreversible oxidation reactions, the same reactions occur in both steam reforming and partial oxidation. The difference in product distribution is easily explained: the ratio of the elements carbon, oxygen, and hydrogen in the feed is different for the two processes. This is most obvious at high temperatures where only hydrogen and carbon monoxide are present. The molar H_2/CO ratios resulting from steam reforming and partial oxidation at these high temperatures are three and two, respectively.

The hydrogen and carbon monoxide contents of the equilibrium gas increase with temperature, which is explained by the fact that the reforming reactions 5.1 and 5.3 are endothermic. The carbon dioxide content goes through a maximum. This can be explained as follows: carbon dioxide is formed in exothermic reactions only, while it is a reactant in the endothermic reactions. Hence, at low temperature carbon dioxide is formed, while with increasing temperature the endothermic reactions in which carbon dioxide is converted (mainly reaction 5.3) become more important.

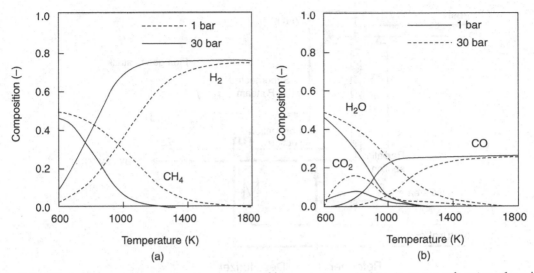

Figure 5.3 *Effect of temperature and pressure on equilibrium gas composition in steam reforming of methane, $H_2O/CH_4 = 1$ mol/mol; (a) CH_4 and H_2; (b) CO, CO_2 and H_2O.*

Both steam reforming and partial oxidation of methane are hindered at elevated pressure because the number of molecules increases due to these reactions. Figure 5.3 illustrates the effect of pressure on steam reforming of methane at a steam/methane ratio of 1 mol/mol.

QUESTIONS:

> *Draw a qualitative profile of the heat consumed or required (as a function of temperature) for both situations in Figure 5.2.*
> *In autothermal reforming both steam and oxygen are added. Sketch a composition profile in the reactor for autothermal operation.*
> *Steam reforming of ethane will result in different thermodynamic product compositions from those for steam reforming of methane shown in Figure 5.2. Guess the concentration versus temperature relationships.*

At a pressure of 30 bar, the equilibrium conversion to hydrogen and carbon monoxide is only complete at a temperature of over 1400 K. However, in industrial practice, temperatures in excess of 1200 K cannot be applied because of metallurgical constraints. It is shown later that this point is of great practical significance.

5.2.2   Steam Reforming Process

Even though steam reforming is carried out at high temperature (>1000 K) a catalyst is still required to accelerate the reaction. The reason for this is the very high stability of methane. The catalyst is contained in tubes, which are placed inside a furnace that is heated by combustion of fuel. The steam reformer (Figure 5.4) consists of two sections. In the convection section, heat recovered from the hot flue gases is used for preheating of the gas feed and process steam, and for the generation of superheated steam. In the radiant section the reforming reactions take place.

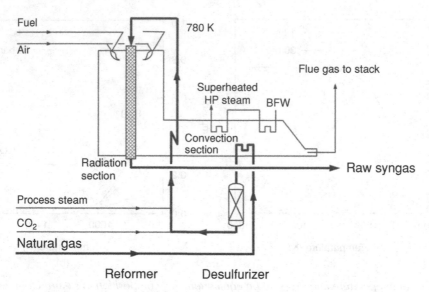

Figure 5.4   Simplified flow scheme of the steam reforming process.

After sulfur removal, the natural gas feed is mixed with steam (and optionally CO_2) and preheated to approximately 780 K before entering the reformer tubes. The heat for the endothermic reforming reaction is supplied by combustion of fuel in the reformer furnace (allothermal operation). The syngas composition leaving the reformer may be further modified in additional process steps such as secondary reforming (Section 5.2.3) and/or shift reactors that reduce the carbon monoxide content (Section 5.4).

QUESTIONS:

For which type of syngas applications would you expect CO_2 to be added? What would be the source of this CO_2?

Figure 5.5 shows the general arrangement of a typical steam reformer. The preheated process stream enters the catalyst tubes through manifolds and pigtails, and flows downwards. The product gas is collected in an outlet manifold from which it is passed upwards to the effluent chamber from which it will be further processed [2]. The tubes are carefully charged with the catalyst particles: an even distribution over the tubes is essential and the catalyst bed should be dust free. A furnace may contain 500–600 tubes with a length of 7–12 m and an inside diameter of between 70 and 130 mm.

QUESTIONS:
Explain the logic of the process conditions in Table 5.3.

5.2.2.1   Carbon Formation

Carbon formation in steam reformers must be prevented for two main reasons. Firstly, coke deposition on the active sites of the catalyst leads to deactivation. Secondly, carbon deposits grow so large that they can cause total blockage of the reformer tubes, resulting in the development of "hot spots". Hence, the reforming conditions must be chosen such that carbon formation is strictly limited.

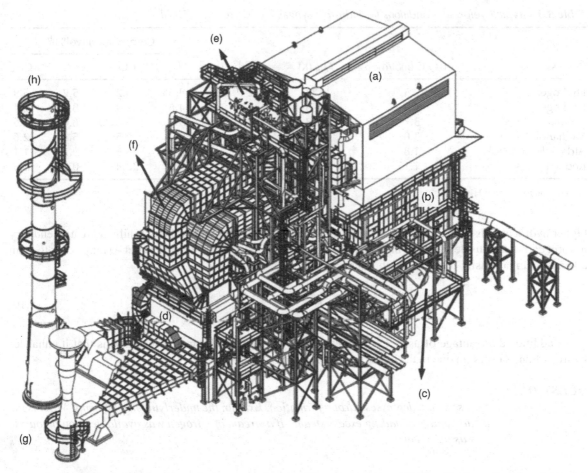

Figure 5.5 *General arrangement of a steam reformer. (Courtesy of van Uffelen, Technip Benelux B.V.) (a) reformer penthouse structure containing distribution systems for feed, fuel, and combustion air, including support systems and operator access to burners and catalyst tubes; (b) reformer radiant box with the catalyst tubes arranged in multiple rows; (c) process gas boiler for cooling of the reformer effluent in a single step to approximately 620 K, with associated steam drum in "piggyback" arrangement; (d) combustion air preheater to reduce fuel consumption; (e) steam drum to provide boiler water to boilers in the convection section; (f) convection section, for heat recovery from flue gas leaving the radiant section, with preheat coils for the process and coils for steam generation/superheating and air preheat; (g) combustion air fan with air intake; (h) stack for discharge of flue gases to safe location.*

QUESTION:

> Carbon formation changes catalytic activity and, as a consequence, the temperature profile in the reactor changes. Explain that extended carbon deposition can lead to hot spots.

Carbon forming reactions can be suppressed by adding excess steam. Therefore, it is common practice to operate the reformer at steam-to-carbon ratios of 2.5–4.5 mol H_2O per mol C, the higher limit applying to

Table 5.3   *Typical reformer conditions for industrial syngas-based processes [1, 3].*

Process	H$_2$O/C (mol/mol)	T_{exit} (K)	p_{exit} (bar)	Composition (vol%)[b]			
				H$_2$	CO	CO$_2$	CH$_4$
Hydrogen	2.5	1123	27	48.6	9.2	5.2	5.9
Hydrogen[a]	4.5	1073	27	34.6	5.3	8.0	2.4
Ammonia	3.7	1073	33	39.1	5.0	6.0	5.5
Methanol	3.0	1123	17	50.3	9.5	5.4	2.6
Aldehydes/alcohols	1.8	1138	17	28.0	25.9	19.7	1.1
Reducing gas	1.15	1223	5	70.9	22.4	0.9	1.5

[a] From naphtha; [b] Rest is H$_2$O

higher hydrocarbons such as naphtha. Compared to methane, higher hydrocarbons exhibit a greater tendency to form carbonaceous deposits. At high temperature (>920 K) steam cracking (Chapter 4) may occur to form alkenes that may easily form carbon through reaction 5.10:

$$C_nH_{2n} \rightarrow carbon + H_2 \qquad \Delta_r H_{298} < 0 \text{ kJ/mol} \qquad (5.10)$$

An additional advantage of using excess steam (a larger than stoichiometric quantity) is that it enhances hydrocarbon conversion due to the lower partial pressure of hydrocarbons.

QUESTIONS:

Excess steam depresses carbon formation. Explain the underlying mechanisms. What is the disadvantage of adding excess steam? If a stream of nitrogen was available, would you use this as a diluent?

5.2.2.2   Methane Slip

Most applications require syngas at elevated pressure (e.g., methanol synthesis occurs at 50–100 bar and ammonia synthesis at even higher pressure, Chapter 6). Therefore, most modern steam reformers operate at pressures far above atmospheric, despite the fact that this is thermodynamically unfavorable. The advantages of operating at elevated pressure are the lower syngas compression costs and a smaller reformer size. The down side is a lower methane conversion. To counterbalance the negative effect on the equilibrium, higher temperatures are applied and more excess steam is used. Figure 5.6 shows the effect of pressure (left) and excess steam (right) on the methane slip (percentage of unreacted methane).

It is generally economically advantageous to operate at the highest possible pressure and, as a consequence, at the highest possible temperature. The tube material, however, places constraints on the temperatures and pressures that can be applied: a maximum limit exists for the operating temperature at a given pressure because of the creep limit of the reformer tubes.

At typical reformer temperatures and pressures an appreciable amount of methane is still present in the syngas produced. A low methane slip is often crucial for the economics of the process. There is a trend towards higher operating temperatures (exceeding 1200 K), which is made possible by the increasing strength of the tube materials [4].

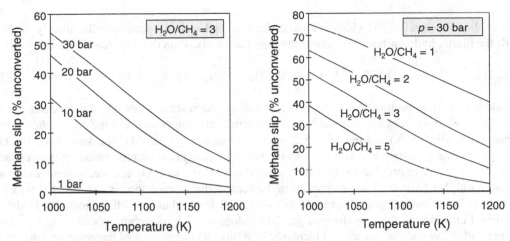

Figure 5.6 *Effect of pressure (left) and steam-to-methane feed ratio (right) on methane slip as a function of temperature.*

QUESTIONS:

> *Estimate the percentage of CH_4 in the "dry syngas" (i.e., after removal of H_2O) at the conditions given in Table 5.3.*
>
> *Compare the costs of compression for the following two cases: (i) syngas is produced at atmosperic pressure (for practical reasons slightly above 1 bar) and the product mixture is subsequently compressed to, say, 30 bar; (ii) the feed is compressed and synthesis is carried out at evelated pressure.*

5.2.2.3 Advances in Steam Reforming

Although steam reforming is a well established technology, there is continuing progress, resulting in less costly plants. Improvements include better materials for reformer tubes, better control of carbon limits, better catalysts regarding sulfur tolerance and carbon deposition, and better process concepts with high feedstock flexibility [5].

Various modifications of the basic steam reforming process (Figure 5.4) can be implemented. For instance, if the natural gas feed is rich in higher hydrocarbons or if naphtha is used as the feedstock, a so-called pre-reformer may be installed upstream the reformer furnace (Box 5.1) to convert these compounds [4]. In this way the worst carbon precursors are removed in advance, which allows operation of the actual reformer at lower steam-to-carbon ratio.

Several different process configurations for reforming of higher hydrocarbons, including circulating fluidized bed membrane reformers (CFBMRs), have been reported [6].

Box 5.1   Pre-Reforming

When higher hydrocarbons (e.g., naphtha) are used as feedstock or if the natural gas contains appreciable amounts of higher hydrocarbons, it is advantageous to install a pre-reformer upstream of the steam reformer. In the pre-reformer, which is a catalytic fixed bed reactor operating adiabatically at about 770 K, the higher hydrocarbons are converted to methane and carbon dioxide (reaction 5.11).

$$C_nH_{2n+2} + \tfrac{1}{2}\,(n-1)\,H_2O \rightarrow \tfrac{1}{4}\,(3n+1)\,CH_4 + \tfrac{1}{4}\,(n-1)\,CO_2 \; \Delta_r H_{298} < 0 \text{ kJ/mol} \qquad (5.11)$$

In addition, carbon monoxide and steam are formed by the reverse water–gas shift reaction (–5.2). In the actual reformer, which operates at higher inlet temperature than would be the case without a pre-reformer (typically 900 K), reactions 5.1, 5.2, and 5.3 subsequently proceed to the desired equilibrium.

The pre-reformer utilizes excess heat from the reformer flue gases, thereby reducing the heat load on the reformer and improving the heat recovery, both resulting in a reduced steam generation rate. Because no higher hydrocarbons are present in the feed to the steam reformer, the selected reformer catalyst can be a "natural gas reforming catalyst" for all feeds, thus providing increased feedstock flexibility. Furthermore, the pre-reformed gas will reduce the risk of carbon formation in the steam reformer and thereby allow operating at increased heat flux and increased feed temperature, since there are no C_2^+ hydrocarbons which can cause carbon formation reactions.

Syngas plants normally produce high-pressure steam that is only partly used in the plant. It is possible to benefit from the high temperature of the reformer effluent by heat exchange with additional feed, allowing reforming of part of the feed in a multitubular reactor (see also Box 5.3 in Section 5.2.3) without using a furnace. For newly built syngas plants this concept is usually not applied due to the severe corrosion ("metal dusting") inside the reactor, requiring the selection of expensive materials. However, for capacity revamps this concept is attractive, enabling typically 25% capacity increase at low investment costs [R. van Uffelen, personal communication].

QUESTIONS:

> *What is meant by "metal dusting"?*
> *Technip markets this technology under the name EHTR (Enhanced Heat Transfer Reformer). Explain the name.*
> *Starting from Figure 5.4, draw a process scheme in which an ETHR is included.*

Steam reforming of natural gas results in syngas with a relatively high H_2/CO ratio (>3 mol/mol). However, most applications call for a more carbon monoxide-rich syngas. To increase the carbon monoxide content of the produced gas, carbon dioxide can be added to the reformer inlet. Because carbon dioxide is a (mild) oxidizing agent, an additional advantage is that less steam is needed to prevent carbon deposition. Furthermore, this carbon dioxide reforming – or dry reforming – can contribute to reducing global carbon dioxide emissions (Box 5.2). Several processes have been developed in which carbon dioxide reforming is used with little or no steam addition [7–9]. In this way H_2/CO ratios in the product gas below 0.5 can be achieved.

Box 5.2   Chemical Cooling

If carbon dioxide for dry reforming is available at high temperature, it could be used in so-called "chemical cooling": the reaction of carbon dioxide from the hot process stream with methane would lead to cooling of the gas mixture in the reformer. Many process streams exist for which this principle might be used successfully [10].

Coupling of exothermic and endothermic processes has often been considered as a means to upgrade thermal energy. Examples are the use of solar energy, waste heat from nuclear energy plants, and high-temperature exhaust gases from mineral processing plants for providing heat to chemical processes. In such a way, the heat for which no good outlet would be available otherwise, is transformed into chemical energy. It is needless to say that one of the options mentioned constitutes an easy target for environmental pressure groups.

5.2.3   Autothermal Reforming Process

In autothermal reforming, heat is generated in the reactor by combustion of part of the feed with oxygen; the product distribution changes to lower H_2/CO ratios. Autothermal reforming is the combination of steam reforming and partial oxidation in a single reactor in the presence of a catalyst. Figure 5.7 shows a schematic of an autothermal reformer.

The reactor is a refractory-lined pressure vessel. Therefore, higher pressures and temperatures can be applied than in steam reforming. Part of the feed is oxidized in the combustion zone (reactions 5.6 and 5.7). In the lower part the remaining feed is catalytically reformed with the produced carbon dioxide and H_2O (reactions 5.1 and 5.3). The endothermic reforming duty is provided by the exothermic oxidation reaction.

Autothermal reforming of natural gas and light hydrocarbons is usually not applied on its own because of the high investment and operating costs. These are, for a large part, attributed to the cryogenic air separation plant needed for the production of oxygen. An autothermal reformer is frequently used, however, in the production of syngas for ammonia manufacture (Section 6.1). In that case, this so-called secondary reformer is installed downstream of the steam reformer and air is used instead of oxygen. The advantage of this

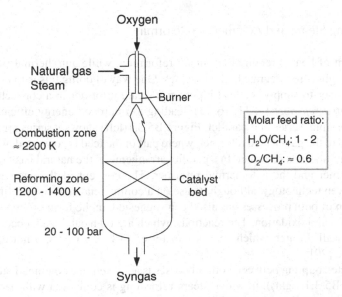

Figure 5.7   *Schematic of an autothermal reformer for the production of syngas.*

arrangement is that unreacted methane from the steam reformer is converted in the autothermal reformer, so that the steam reformer can be operated at lower temperature and higher pressure (circa 1070 K and 30 bar). These conditions are thermodynamically unfavorable. However, a high outlet pressure is advantageous because it decreases the costs for compression of the syngas to the pressure of the ammonia synthesis loop (>100 bar). Since air supplies the required nitrogen, no expensive oxygen plant is needed in this special case.

Syngas for methanol plants is also increasingly produced by combined steam and autothermal reforming, with the advantage of operation at high pressure (methanol synthesis takes place at 50–100 bar) but the disadvantage of requiring pure oxygen. Therefore, the major part of methanol production is based on steam reforming of natural gas.

Presently, a lot of interest exists in so-called catalytic partial oxidation (CPO), which differs from autothermal reforming due to the fact that no burner is used; all reactions take place in the catalyst zone [11]. The production of syngas through partial oxidation of hydrocarbons over rhodium catalysts has been studied in detail in short-contact-time reactors [12–13].

A number of reactor systems combining steam reforming and autothermal reforming have been developed in the past two decades, some concepts of which are presented in the Box 5.3. These are examples of "Process Intensification", the principles of which are discussed in more detail in Chapter 14.

QUESTIONS:

Syngas produced by autothermal reforming has a lower H_2/CO ratio than syngas produced by steam reforming. What is the reason? Why should air not be used in the production of syngas for methanol production? Why would there be so much interest in catalyzing the oxidation reactions occurring in autothermal reforming?

Why is autothermal reforming as shown in Figure 5.7 limited to light hydrocarbons? What is the reason for choosing a refractory-lined vessel?

Box 5.3   Combining Steam and Autothermal Reforming

Since a large amount of heat is required for steam reforming, while autothermal reforming produces heat, an obvious thought is to integrate both processes. One way of doing this is to use the hot gas from an autothermal reformer to supply the heat input for a steam reformer in a convective *heat exchange reformer* or *gas-heated reformer* [4, 11, 16, 17], leading to improved energy efficiency. Two types of arrangements are possible, series and parallel. Figure B5.3.1(left) shows a series arrangement. Natural gas and steam are fed to the gas-heated reformer, where part of the feed is converted, while the remainder is converted in the autothermal reformer. In a parallel arrangement, the natural gas feed is split between the gas-heated reformer and the autothermal reformer. The concept of the gas-heated reformer has already become proven technology, although its use on a commercial scale is still limited.

Further integration of both processes has also been suggested, in the form of direct heat exchange by coupling of reforming and oxidation. The reactions, which are both catalyzed, occur on opposite sides of an impermeable wall through which heat is transferred [18, 19]. It is still not clear how operable these systems will be [20].

Even more intimate coupling between both processes is achieved in a combined autothermal reforming process (Figure B5.3.1(right)), in which steam reforming is combined with partial oxidation in a single fluidized bed reactor [21]. This technology has been demonstrated at pilot scale [20].

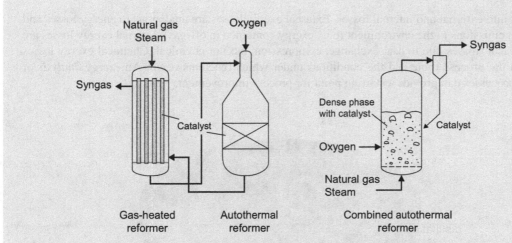

Figure B5.3.1 *Combination of a gas-heated reformer with an autothermal reformer (left) and combined autothermal reforming (right).*

5.2.4 Novel Developments

Producing syngas by conventional steam reforming is an expensive and energy intensive step in the production of chemicals. For instance, approximately 60–70% of the investment in a methanol plant is associated with the syngas plant (including compression). The developments outlined above could reduce the investment and operating costs to some extent. Catalytic partial oxidation especially has a large potential because it is attractive from the point of view of *exergy efficiency* (Box 5.4, [22–24]) and materials usage. Nevertheless, a major breakthrough would be required to make any substantial reduction in the cost of producing syngas.

One such radically different development is based on membrane technology. Two different approaches have been pursued. The first approach, intended for pure hydrogen production, makes use of a palladium or ceramic membrane to separate hydrogen from the steam reforming process *in situ*, thus pushing the equilibrium towards full conversion even at low temperatures [4, 25–27]. This has, however, only been achieved to any extent at laboratory scale [20]. The second approach combines oxygen separation from air and autothermal reforming in a single step [11, 28, 29], so-called oxygen membrane reforming (Box 5.5).

Other approaches aim at the direct conversion of methane into valuable chemicals instead of first producing syngas as an intermediate. Examples are the production of methanol by direct partial oxidation of methane (Section 6.2) and the production of ethene by oxidative coupling of methane with oxygen [4, 30–32] (Section 4.5).

Box 5.4 Exergy

Exergy is the theoretically maximum amount of work that can be obtained from a system. It is a measure for the *quality* of the energy. Exergy losses ("lost work") in a process occur because the quality of the energy streams entering the process is higher than the quality of the energy streams leaving the process (while the *quantity* of the energy remains the same: the first law of thermodynamics). Exergy losses can be divided into external and internal losses. External exergy losses are analogous to energy losses and result from emissions to the environment (e.g., exergy contained in off-gas). Internal exergy losses are either physical (losses due to heat exchange, compression, etc.) or chemical. Chemical exergy losses depend on the process route and the conditions under which reactions occur. An *exergy analysis* of chemical processes can provide a starting point for process improvement.

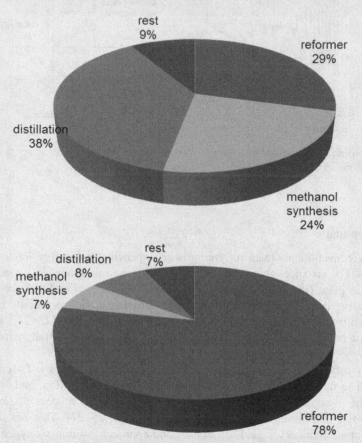

Figure B5.4.1   *Contribution of process sections in the ICI low-pressure methanol process to total energy (top) and exergy losses (bottom) [33].*

Figure B5.4.1 clearly shows the different results obtained from an energy and exergy analysis for the production of methanol by steam reforming of natural gas followed by conversion of the syngas using the low-pressure ICI process (Section 6.2). Based on an energy analysis the focus would be on improving the distillation section, while an exergy analysis points in a different direction, namely the reformer. It has been demonstrated that replacing the steam reforming process by a process combining a gas-heated reformer with an autothermal reformer (Figure B.5.3.1), results in a decrease in exergy losses of nearly 30% [33]. The main reason for this improvement in *exergy efficiency* is the fact that the irreversible combustion reaction is combined with the endothermic reforming reaction, eliminating heat requirements for a reformer furnace. A further improvement would result from eliminating the need for heat exchange, for instance by producing the syngas in a single catalytic partial oxidation reactor.

QUESTION:

> *Why would the combustion reaction have such an impact on the exergy loss of the process? What is another "exergy advantage" of using the combined reforming process? How would direct production of methanol from methane compare to the syngas route in terms of exergy efficiency?*

Box 5.5 Oxygen Membrane Reforming

Figure B5.5.1 shows the principle of oxygen membrane reforming. A ceramic membrane selectively extracts oxygen from air, which flows on one side of the membrane. The oxygen is transported through the membrane and reacts with the hydrocarbon feedstock flowing on the other side. The catalyst may be in the form of pellets or directly attached to the membrane itself [11]. Consortia led by major industrial gas suppliers have been working for about 15 years on oxygen transport membranes for oxygen separation from air. So-called ion transport membranes [34] have been developed by Air Products. These are ceramic membranes that are non-porous, multicomponent metallic oxides. They operate at high temperatures and have exceptionally high oxygen flux and selectivity. The membrane module design has been tested successfully on a commercial scale in a process development unit with a capacity of about 30 Nm^3/h syngas flow [35]. A reduction in the capital cost of syngas generation of greater than 30% is anticipated compared to a process based on autothermal reforming.

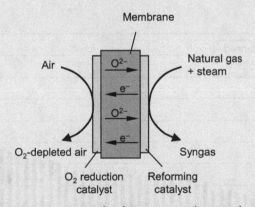

Figure B5.5.1 *Principle of oxygen membrane reforming.*

QUESTION:

> *The "ion transport membranes" were developed for air separation, based on selective permeation of O$_2$ (minimal permeation of N$_2$). In view of the reaction pathway presented here, what could be the underlying mechanism of the selectivity?*

5.3   Coal Gasification

Gasification of coal to produce syngas (or coal gas or town gas) dates back to the end of the eighteenth century. During the mid-1800s, coal gas was widely used for heating and lighting in urban areas. The development of large-scale processes began in the late 1930s and the process was gradually improved. Following World War II, however, interest in coal gasification dwindled because of the increasing availability of inexpensive oil and natural gas. In 1973, when oil and gas prices increased sharply, interest in coal gasification was renewed and, especially in the last 10–15 years, much effort has been put in improving this process (Figure 5.8). The main reason for this interest is the dramatic increase in oil and gas prices, which seems to be a continuing trend, together with the wide availability of coal compared to oil and natural gas (Section 2.3).

Although coal gasification, like steam reforming of natural gas, produces syngas, the incentive of the two processes is different: The goal of steam reforming is the *production* of carbon monoxide and hydrogen for chemical use, whereas coal gasification was primarily developed for the *conversion* of coal into a gas, which happens to predominantly contain carbon monoxide and hydrogen.

In 2007, 49% of syngas produced by coal gasification was used to produce Fischer–Tropsch liquids (Section 6.3), 32% as a raw material for the chemical industry, 11% for power generation, and 8% for gaseous fuels [27]. The application of coal gasification for power generation in a so-called integrated gasification combined cycle (IGCC) is a relatively recent development. The name stems from the fact that gasification is combined with power generation by a gas turbine driven by the gases coming from the gasifier with the waste heat being used in a steam turbine (Section 5.3.6).

5.3.1   Gasification Reactions

In a broad sense, coal gasification is the conversion of the solid material coal into gas. The basic reactions, as given in Table 5.4, are very similar to those that take place during steam reforming (Table 5.2).

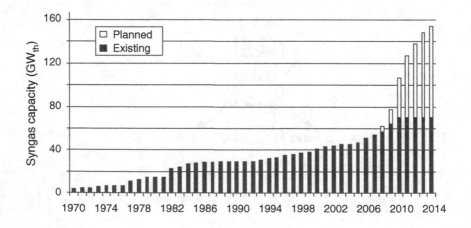

Figure 5.8   *Worldwide gasification capacity (2007 data); MW$_{th}$ refers to thermal power produced [36]. Note that this capacity also includes gasifier feeds other than coal (heavy oil fractions, gas, petcoke, and biomass). In 2007, coal contributed approximately 55%.*

Table 5.4 *Reactions occurring during coal gasification.*

Reaction	$\Delta_r H_{800}$ (kJ/mol)	(Reaction number)
Heterogeneous reactions		
$C + H_2O \rightleftarrows CO + H_2$	136	(5.12)
$C + CO_2 \rightleftarrows 2\,CO$	173	(−5.5)
$2\,C + O_2 \rightleftarrows 2\,CO$	−222	(5.13)
$C + O_2 \rightleftarrows CO_2$	−394	(5.14)
$C + 2\,H_2 \rightleftarrows CH_4$	−87	(−5.4)
Homogeneous reactions		
$2\,CO + O_2 \rightleftarrows 2\,CO_2$	−572	(5.8)
$CO + H_2O \rightleftarrows CO_2 + H_2$	−37	(5.2)

The first two reactions with carbon are endothermic, whereas the latter three are exothermic. It is common practice to perform coal gasification in an "autothermal" way. The reaction is carried out with a mixture of O_2 (or air) and H_2O, so that combustion of part of the coal produces the heat needed for heating up to reaction temperature and for the endothermic steam gasification reaction. Besides these heterogeneous reactions, homogeneous reactions occur.

When coal is burned in air, strictly speaking also gasification occurs, but this process is usually not referred to as gasification. Coal is a complex mixture of organic and mineral compounds. When it is heated to reaction temperature (depending on the process, between 1000 and 2000 K), various thermal cracking reactions occur. Usually, the organic matter melts, while gases evolve (H_2, CH_4, aromatics, etc.). Upon further heating the organic mass is transformed into porous graphite-type material, called "char". This process is called pyrolysis. Figure 5.9 illustrates the complex pyrolysis network.

When reactive gases are present in the gas phase, several succeeding reactions occur. This is the case in coal combustion in particular. In a combustor, coal particles entering the flame are heated very fast (>10 000 K/s). Immediately, pyrolysis reactions occur, but the product mixture is not complex, because the products are essentially completely burned. When less than stoichiometric amounts of oxygen are present, as in coal gasification, the atmosphere is overall reducing and the product contains a complex mixture of organic compounds.

Coal is not unique in this sense: many hydrocarbons and related compounds, for instance plastics, sugar, biomass, and heavy oil, exhibit similar behavior. For instance, heating of sugar leads to melting followed by complex sequences of reactions leading to volatiles and a solid material that might be referred to as caramel (mild conditions) or sugar char (severe conditions).

Depending on the temperature of the gasifier, the mineral (inorganic) matter will be released as a liquid or a solid. When the mineral matter is released as a liquid, the gasifier is called a "slagging" gasifier.

5.3.2 Thermodynamics

It will be clear that the structure of coal and its products is very complex. For convenience, in thermodynamic calculations we assume that coal solely consists of carbon. Furthermore, under practical gasification conditions, it is usually assumed that only the following species are present in the gas phase: H_2O, CO, CO_2, H_2, and CH_4. Of course, other hydrocarbons are also present, but their concentrations are relatively small and in the thermodynamic calculations these are all represented by CH_4. Figure 5.10 shows the equilibrium composition resulting from gasification of coal in an equimolar amount of steam as a function of temperature and pressure.

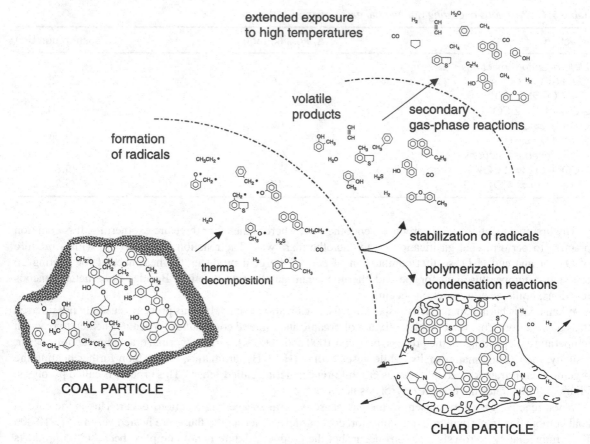

Figure 5.9   *Pyrolysis network [38].*

With increasing temperature, the CO and H_2 mol fractions increase due to the increasing importance of the endothermic H_2O gasification reaction. Accordingly, H_2O decreases with temperature. Both CO_2 and CH_4 go through a maximum as a result of the exothermicity of their formation and the endothermicity of their conversion. As can be seen from Figure 5.10, a low pressure is favorable for CO and H_2 forming reactions due to the increase in number of molecules. The CO_2 and CH_4 maxima shift to higher temperature with increasing pressure.

The heat required for coal gasification may either be supplied outside the reactor ("allothermal" gasification) or within the reactor by adding oxygen ("autothermal" gasification). Addition of oxygen not only supplies the necessary energy, but also changes the composition of the product gas. This can be seen from Figure 5.11, which shows the equilibrium composition resulting from autothermal gasification as a function of temperature and pressure.

QUESTION:

The partial pressure of H_2 goes through a maximum (Figure 5.11a). What is the explanation?

The equilibrium composition in autothermal gasification shows the same trend concerning temperature and pressure as in allothermal steam gasification. However, the fractions of CO and CO_2 are much higher, while

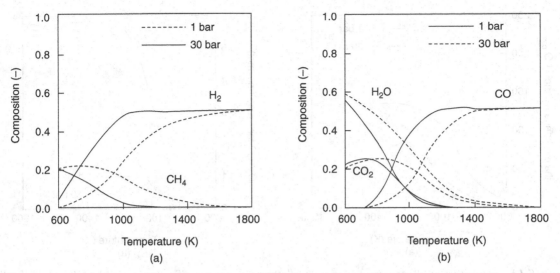

Figure 5.10   *Effect of temperature and pressure on equilibrium gas composition in steam gasification of coal (represented as C); $H_2O/C = 1$ mol/mol: (a) CH_4 and H_2; (b) CO, CO_2, H_2O [39].*

those of H_2 and CH_4 are lower. This obviously is the result of the higher oxygen content (lower H/O ratio) of the feed, particularly at high temperature.

Figure 5.12a shows the enthalpy change (energy required for reactions) as a result of allothermal gasification and Figure 5.12b shows the O_2/H_2O ratio required for autothermal operation.

The enthalpy change reflects the total change in reaction enthalpies of all reactions involved. Thus, at 1600 K, 135 kJ is required per mole of H_2O converted. This is the reaction enthalpy of the steam gasification reaction, the predominant reaction at high temperature. At low temperature exothermic reactions play a

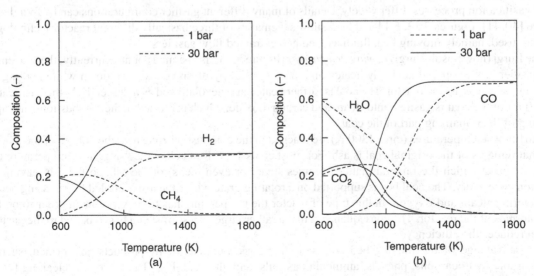

Figure 5.11   *Effect of temperature and pressure on equilibrium gas composition in autothermal gasification of coal (C): (a) CH_4 and H_2; (b) CO, CO_2, H_2O; $H_2O/O_2/C$ ratio depends on temperature [39].*

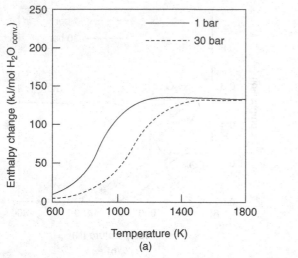
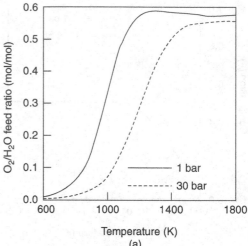

Figure 5.12   *(a) Enthalpy change associated with allothermal steam gasification and (b) O_2/H_2O feed ratio (mol/mol) in autothermal gasification [39].*

role, but at higher temperature the endothermic reactions become increasingly important, resulting in larger enthalpy changes. These enthalpy changes determine the required O_2/H_2O ratio for the system to become autothermal. At higher temperature more oxygen is required to compensate for the energy consumed in the endothermic steam gasification reactions.

5.3.3   Gasification Technologies

Coal gasification processes differ widely. Details of many different gasifier configurations can be found elsewhere [40,41]. Figures 5.13, 5.14 and 5.15 show schematics of three basically different reactor technologies that are used, namely moving bed, fluidized bed and entrained flow gasifiers.

The Lurgi (now Sasol–Lurgi) *moving bed gasifier* (Figure 5.13), is operated countercurrently; the coal enters the gasifier at the top and is slowly heated and dried (partial pyrolysis) on its way down while cooling the product gas as it exits the reactor. The coal is further heated and devolatilized as it descends. In the gasification zone, part of the coal is gasified into steam and carbon dioxide, which is formed in the combustion zone upon burning of the remaining part of the coal.

The highest temperatures (circa 1300 K) are reached in the combustion zone near the bottom of the reactor. All that remains of the original coal is ash. For most coals 1300 K is below the "slagging" temperature (the temperature at which the mineral matter becomes sticky or even melts) of the ash, so the ash leaving the reaction zone is dry. The coal bed is supported on a rotating grate where the ash is cooled by releasing heat to the entering steam and oxygen. In this type of reactor the temperature has to be kept low in order to protect the internals of the reactor. The consequence is that a rather large excess of steam has to be fed to the gasifier, which reduces the efficiency.

A disadvantage of the moving bed reactor is that large amounts of by-products are formed, such as condensable hydrocarbons, phenols, ammonia tars, oils, naphtha, and dust. Therefore, gas cleaning for the Lurgi gasification process is much more elaborate than for other gasification processes. On the other hand, the countercurrent operation makes the process highly energy efficient.

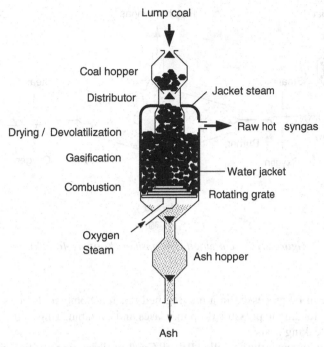

Figure 5.13 *Moving bed gasifier (Lurgi).*

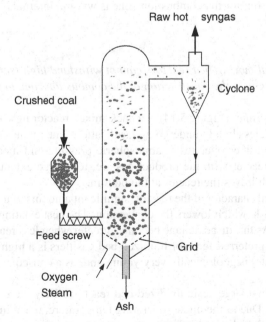

Figure 5.14 *Fluidized bed gasifier (Winkler, high temperature).*

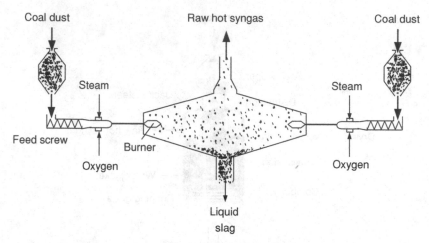

Figure 5.15   *Entrained flow gasifier (Koppers–Totzek).*

Not all types of coal can be processed in a moving bed reactor. Some coals get sticky upon heating and form large agglomerates, leading to pressure drop increases and even plugging of the reactor. Coals showing this behavior are called "caking coals".

A relatively novel development is the so-called British Gas/Lurgi slagging gasifier, in which the temperature in the combustion zone is above 2000 K, resulting in slagging of the ash. To enable the use of such a high temperature, the construction of the reactor had to be changed. The conventional Lurgi reactor contains a grate but in the slagging Lurgi reactor the combustion zone is without internals. Steam consumption is much lower in the slagging gasifier.

QUESTIONS:

> *Conventional moving bed reactors cannot withstand high temperatures in the combustion zone. Why? Guess the construction of the combustion zone in a slagging Lurgi reactor.*

The Winkler *fluidized bed gasifier* (Figure 5.14) is a back-mixed reactor in which coal particles in the feed are well mixed with coal particles already undergoing gasification. The intensive mixing results in excellent mass and heat transfer. This gasifier operates at atmospheric pressure and moderate uniform temperature. Char particles that leave the reactor with the product gas are, to a large extent, recovered in cyclones and recycled to the reactor. Dry ash leaves the reactor at the bottom.

As a result of the back-mixed character of the reactor, significant amounts of unreacted carbon are removed with the product gas and the ash, which lowers the conversion. The best existing fluidized bed reactors offer a carbon conversion of 97%. As this disadvantage becomes greater when less reactive coal or coal with a low ash melting point is used, the preferred feed for fluidized bed gasifiers is a highly reactive coal [42]. Brown coal (often referred to as lignite) is geologically very young and, as a consequence, is highly functionalized and, therefore, very reactive.

Winkler reactors were the first large-scale fluidized bed reactors. They were used in Germany for brown coal gasification in the 1930s. Due to the higher operating temperature, the Winkler gasifier produces much lower quantities of impurities than the Lurgi gasifier.

Other fluidized bed gasification technologies have been developed by various companies [40]. Many of these developments were aimed at increasing the operating pressure, and thus temperature.

Because of the limited carbon conversion in fluidized bed gasifiers, the ash particles contain a relatively large amount of unconverted carbon. They are fed to a boiler system, where the carbon is burned, providing heat for steam generation. This additional step is required in all fluidized bed reactor systems.

QUESTION:

> *Why would one want to perform gasification at elevated pressure?*

The Koppers–Totzek *entrained flow gasifier* (Figure 5.15) is a plug-flow system in which the coal particles react cocurrently with steam and oxygen at atmospheric pressure. The residence time in the reactor is a few seconds. The temperature is high in order to maximize coal conversion. At this high temperature, mainly carbon monoxide and hydrogen are formed and hardly any by-products. Ash is removed as molten slag. The entrained flow gasifier is the only gasifier that can handle all types of coal.

QUESTION:

> *Sketch the axial temperature profile for the entrained flow gasifier. Is the statement that the reactor can be described as a plug-flow reactor correct for the whole reactor?*

In fluidized bed and entrained flow reactors the coal particles are heated very quickly, typically at rates over 1000 K/s. During heating extensive pyrolysis occurs and when mixing is poor considerable amounts of volatiles, coke, and so on will be formed. The process design should be such that this will not lead to lowered efficiency and problems downstream of the reactor.

Table 5.5 compares the general characteristics of the three reactor types. The lowest exit gas temperature is obtained in the moving bed reactor because of the countercurrent mode of operation.

The cold gas efficiency, a key measure of the efficiency of coal gasification, represents the chemical energy in the syngas relative to the chemical energy in the incoming coal. It is based on the heating values at standard temperature (at which H_2O is condensed, explaining the term "cold" gas efficiency). The cold gas efficiencies of moving bed and fluidized bed reactors are high compared to that of the entrained flow reactor. This is explained by the fact that the product gas of the former two processes contains methane and higher hydrocarbons, which have a higher heating value than carbon monoxide and hydrogen.

An important aspect of coal gasification is the state of the ash produced. When it melts (as in the slagging Lurgi and Koppers–Totzek process), it forms a vitrified material that only exhibits very limited leaching upon

Table 5.5 *Characteristics of moving bed, fluidized bed and entrained flow coal gasifiers.*

	Moving bed	Fluidized bed	Entrained flow
Preferred coal type	Non-caking	Reactive	Any
Coal size, d_p (mm)	5–50	<5 mm	<0.1 mm
Gasifier temperature (K)	1250–1350	1250–1400	1600–2200
Outlet temperature (K)	700–900	1150–1300	1200–1850
Throughput (t/d)	500–1500	3000	5000
Carbon conversion (%)	99	95–97	99
O_2 consumption (kg/kg coal)	0.43 (dry ash) 0.53 (slagging)	0.71	0.8–0.9
Cold gas efficiency (%)	84 (dry ash) 88 (slagging)	81–85	74–81

disposal. In a low-temperature process, the ash does not go through a molten phase and, as a consequence, it is less inert causing extensive leaching upon disposal.

QUESTIONS:

> *Explain why the conventional Lurgi reactor is not fit for all types of coal. Why does the moving bed reactor (usually) have the highest cold gas efficiency? Why is the cold gas efficiency not a good parameter for comparing combined power generation processes?*

The three gasifier types result in quite different gas compositions. This is mainly due to their different operating temperature, which in turn depends on the reactor design, and is controlled by the amount of oxygen and steam fed to the gasifier. For instance, the maintenance of a relatively low combustion temperature is essential for moving bed gasifiers, since if the ash melts it cannot be handled on a supporting grate. Therefore, excess steam and very little oxygen are added. In an entrained flow reactor with its small residence time, on the other hand, high temperatures are required to ensure high reaction rates. Hence, the oxygen-to-steam ratio is high.

Figure 5.16 indicates the composition of the gas phase, as derived from thermodynamic equilibrium data of carbon gasification, as a function of temperature. The operating ranges of industrial gasifiers are also indicated.

The gas composition resulting from the low-temperature moving bed process is the most complex. The hydrocarbon conversion is far from complete and carbon dioxide is present. In fact, the composition is more complex than shown because many other hydrocarbons (in the figure only CH_4 is considered) are present. Furthermore, the product from moving bed reactors also contains large amounts of tars, oils, and phenolic liquors due to the countercurrent mode of operation, thus requires extensive and expensive syngas cleaning.

An entrained flow reactor is operated at higher oxygen/coal ratio and lower amount of steam, thus lower H/O ratio. This type of gasifier produces a gas with higher carbon monoxide and lower hydrogen content, while virtually no hydrocarbons are present in the product.

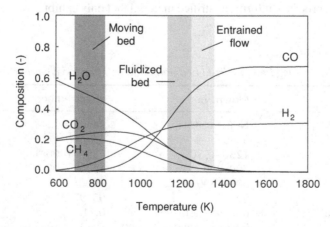

Figure 5.16   *Equilibrium exit gas composition of autothermal carbon gasification at 30 bar as a function of temperature; the range of the outlet temperatures of the gas from industrial gasifiers is indicated [39].*

Table 5.6 *Syngas composition from different gasification processes with oxygen as the oxidant (Coal type: Illinois#6) [43].*

Composition (vol%)	Moving bed	Fluidized bed	Entrained flow
H_2O	5.1	4.4	2.0
H_2	52.2	27.7	26.7
CO	29.5	54.6	63.1
CO_2	5.6	4.7	1.5
CH_4	4.4	5.8	0.03
COS	0.04	0.1	0.1
H_2S	0.9	1.3	1.3
NH_3 + HCN	0.5	0.08	0.02
N_2 + Ar	1.5	1.7	5.2

Table 5.6 shows measured compositions, including trace impurities, of syngas obtained by gasification of a standard grade of coal often used as a reference employing different gasification processes.

5.3.4 Recent Developments in Gasification Technology

Coal gasification is currently under consideration for many applications. The majority of the most successful coal gasification processes developed after 1950 are entrained flow processes operating at pressures of 20–70 bar. The various designs differ in their coal feed systems, configurations for introducing the reactants, type of reactor wall, and the ways in which heat is recovered from the syngas, as shown in Figure 5.17 for some of the most important processes, which all use a different combination of these features. Table 5.7 summarizes major characteristics of these processes.

The Shell, General Electric (GE), and ConocoPhillips (COP) gasifier technologies can be considered proven technologies that are applied commercially in several coal gasification plants. The Mitsubishi Heavy

Table 5.7 *Characteristics of some important entrained flow processes.*

Process	Shell	GE[a]	COP[b]	MHI
Feed	Dry	slurry, 60–70 wt% coal	slurry, 50–70 wt% coal	dry
Configuration	1-stage upflow	1-stage downflow	2-stage upflow	2-stage upflow
Reactor wall	Membrane	Refractory	Refractory	Membrane
Syngas cooling	Gas quench and syngas cooler	Water quench or syngas cooler	Convective	Convective
Oxidant	Oxygen	Oxygen	Oxygen	(enriched) air or oxygen
Outlet temperature (K)	1600–1900	1500–1800	~1300	not available
Pressure (bar)	30–40	30–80	20–40	>25
Cold gas efficiency (%)	80–83	69–77	71–80	70–75

[a] Acquired from ChevronTexaco in 2004, originally Texaco.
[b] Originally developed by Dow, acquired by COP in 2003.

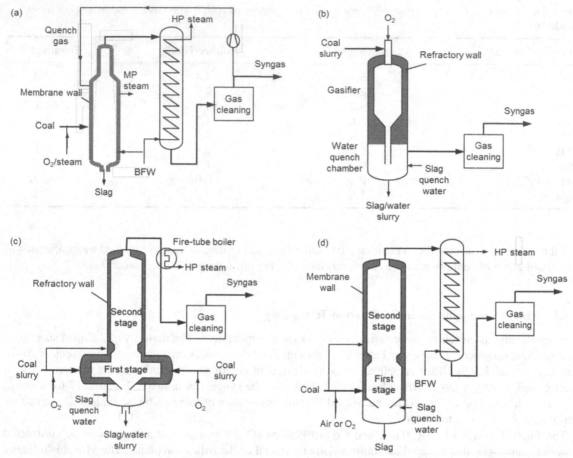

Figure 5.17 *Schematics of entrained flow gasifiers showing reactor, feed, and syngas-cooling characteristics: (a) Shell; (b) GE; (c) COP; (d) MHI.*

Industries (MHI) technology is currently still in the demonstration phase. The reason that it is mentioned here is that it combines the dry feed system and two-stage operation in one design.

The coal feed system is a very important aspect in gasifier design. In the GE and COP processes, a slurry of coal in water is added to the reactor. The reliability of this feeding system is very high. However, evaporation of the water requires energy, resulting in lower cold gas efficiency. Therefore, in many processes dry feeding systems are applied.

In two-stage gasifiers, part of the coal is fed with oxygen to the first stage, where oxidation takes place. In the second stage, more coal is fed to the hot gas stream flowing upward, but no additional oxygen is supplied. Here the gasification reactions occur. The main advantage of two-stage gasification, which is used in the COP and MHI gasifiers, is the lower oxygen demand due to the fact that only part of the coal is gasified at the high temperatures required for slagging.

Refractory-wall gasifiers require less investment costs than membrane-wall gasifiers, but they are high in maintenance due to the need to replace refractories at 6–9 month intervals. Therefore, to ensure sufficient availability, a spare gasifier should be used, which negates the low investment costs to a large extent [40].

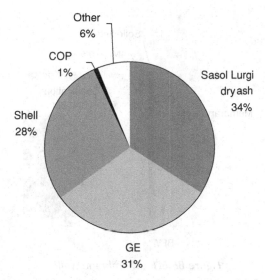

Figure 5.18   *Gasification capacity in 2007 according to gasification technology; total 56 238 MW$_{th}$ [37].*

Membrane walls (Box 5.6) require much less maintenance but have a larger heat loss through the wall. An additional drawback is the higher investment cost.

QUESTIONS:

The cooling of the product gas is carried out in quite different ways. Evaluate them.
Discuss dry versus slurry feeding. Which do you prefer?
At the time of writing this chapter, the outlet temperature of the MHI gasifier was unknown. In what range do you expect it to be?

In 2007 at least 15 different gasification technologies (using coal or other feedstocks) were in operation in plants around the world [37]. However, three commercial technologies were dominant, holding >90% of the 2007 world market (Figure 5.18). According to the same source [37], Shell was expected to achieve 100% of the projected capacity growth (17 000 MW$_{th}$) in 2008–2010, mainly realized in a huge Fischer–Tropsch plant in Qatar based on gasification (rather called partial oxidation) of natural gas, but also in a number of coal gasification projects for the production of chemicals in China.

Box 5.6   Membrane Wall

A membrane wall (Figure B5.6.1) consists of high-pressure tubes in which steam is generated. Part of the molten slag forms a layer coating the inner surface of the gasifier. Thus, the liquid slag does not come into contact with the wall, avoiding corrosion and erosion problems.

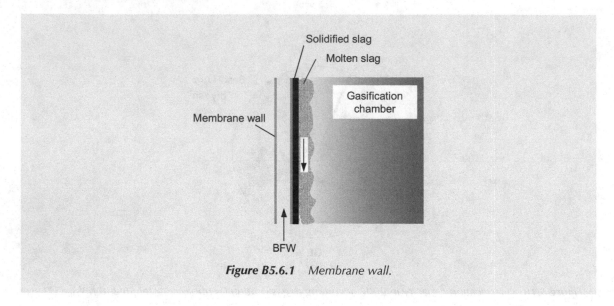

Figure B5.6.1   *Membrane wall.*

5.3.5   Applications of Coal Gasification

The use of coal as the feedstock for a "coal refinery", analogous to an oil refinery, has been considered. This idea is feasible in principle but at present coal cannot compete with oil in this respect. Figure 5.19 shows a number of applications of coal gasification.

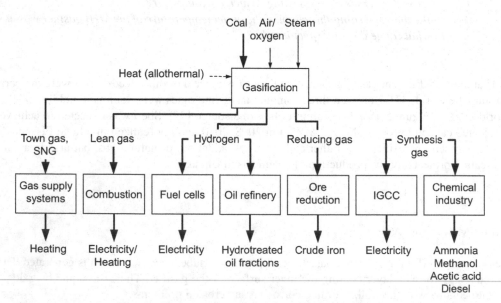

Figure 5.19   *Coal gasification applications. IGCC = integrated gasification combined cycle.*

When coal is gasified in air/steam mixtures low calorific gas is produced ("lean gas"), which is usually used for heating or power generation. Coal-derived syngas is also applied in gas supply systems as "Substitute Natural Gas" or "Synthetic Natural Gas" (SNG) (Section 5.4):

$$coal \rightarrow CO/H_2 \rightarrow CH_4 \tag{5.15}$$

This application has received much interest but developments are currently on hold due the discovery and exploitation of very cheap shale gas (Section 2.3). In principle, there is also a direct route from coal to methane (Box 5.7). Another important application of coal gasification is as a raw material for the chemical industry and for the production of liquid transportation fuels, for instance in Fischer–Tropsch type reactions (Section 6.3):

$$coal \rightarrow CO/H_2 \rightarrow hydrocarbons + alcohols \tag{5.16}$$

Coal gasification used to be the primary technology for the production of hydrogen:

$$coal \rightarrow CO/H_2 \rightarrow water\text{-}gas\ shift \rightarrow H_2 \tag{5.17}$$

Hydrogen can be used in a wide variety of processes applied in oil refining and in the chemical industry. However, as a result of the much higher investment costs for a coal gasification plant compared to a methane steam reforming plant, coal gasification is no longer used for hydrogen generation, except occasionally in situations where natural gas is not available. At present, about 95% of hydrogen production is based on steam reforming.

Most applications of coal gasification are not unique to coal. Furthermore, the technology for coal gasification in principle is also applicable to other hydrocarbon resources. Gasification is used for producing syngas from gas oils and heavy residues. With these feedstocks the term partial oxidation is usually used. Probably in the future gasification will also be used on a large scale to convert biomass, plastic waste, and so on.

Box 5.7 A Thermoneutral Process for Methane Production

In principle, coal can be converted directly into methane by reaction with steam, so-called hydromethanation:

$$2C + 2H_2O \rightleftharpoons CH_4 + CO_2 \qquad \Delta_r H_{800} = 11.4\ kJ/mol \tag{5.18}$$

An attractive aspect of this reaction is that the process is nearly autothermal without adding oxygen. From the thermodynamic data (Figure 5.10) it is clear that suitable process conditions are a relatively low temperature (<600–800 K) and a pressure depending on the temperature; higher temperature requires higher pressure for maximum CH_4 selectivity, as apparent from Figure 5.10.

However, coal is not sufficiently reactive at these temperatures. Processes have been suggested and developed based on the use of a catalyst, high pressure and recycle of produced hydrogen and carbon monoxide. Figure B5.7.1 shows a block diagram of a process developed by Exxon (now ExxonMobil) [44,45].

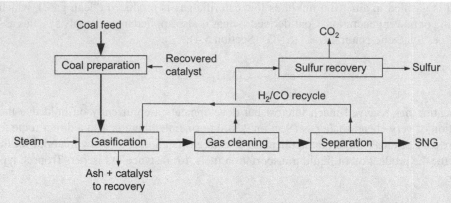

Figure B5.7.1   *Block diagram of the Exxon catalytic gasification process.*

The process appeared to be feasible technically, but not economically. At the time that the process was being developed (late 1970s/early 1980s), it was concluded that there would be a good market for it. Later, large amounts of natural gas were discovered and the process lost its economic appeal before the intended large-scale plants had been built.

The original patents of the Exxon process have now expired and currently GreatPoint Energy, founded in 2005, is commercializing this technology, called Bluegas™ [46].

QUESTIONS:
> *When would conversion of coal into methane be useful?*

5.3.6   Integrated Gasification Combined Cycle

In a conventional pulverized coal (PC) power plant, steam is generated by heat exchange of boiler feed water with the flue gas from a pulverized coal combustor. This steam then drives a steam turbine to generate electricity.

A relatively recent development in power generation is the integrated gasification combined cycle (IGCC) system, which consists of a coal gasifier and a combination of a gas turbine and a steam turbine for power generation. The world's first IGCC power plant of semi commercial size (253 MW), presently owned by Nuon, has been successfully operated at Buggenum, The Netherlands, since 1994. In 2002 trials with biomass cogasification started and currently the gasifier feed can contain up to 30 wt% biomass. The plant was a technological success and has been running until 2012.

In 2010, six IGCC plants operating on coal as the primary feedstock were in operation (Table 5.8) and more are being constructed or planned.

Figure 5.20 shows a simplified flow scheme of the Buggenum IGCC process. The coal gasification process is based on an oxygen-blown, dry feed, entrained flow gasifier developed by Shell. Oxygen for the gasifier is obtained in a cryogenic air separation unit. The gas from the gasifier flows into a gas cooler, after which particulates are removed in a cyclone and candle filter. Then the gas is sent to a gas cleaning unit, where, among other things, sulfur compounds are converted into sulfur and removed. The seemingly small gas cleaning unit in fact features a COS hydrolysis and H_2S absorption unit (Section 5.4), a Claus plant (Section 8.3), a gas saturator, and a water cleaning system.

Table 5.8   *Coal-based IGCC plants in operation (2010).*

Item	Location	Gasifier	Feedstock[a]	Net Output (MW$_e$)	Efficiency (%) (LHV)	Start up
Nuon	Buggenum, Netherlands	Shell	coal/biomass	253	43.1	1994
Wabash River	Illinois, USA	COP	coal/petcoke	262	39.2	1995
Polk Power Station	Florida, USA	GE	coal/petcoke	250	41.2	1996
SUAS	Vresova, Czech Republic	Lurgi	lignite	351	44	1996
				430	41	2005
Elcogas	Puertollano, Spain	Uhde[b]	coal/petcoke	298	45	1997
Nakoso Power Station	Iwaki City, Japan	MHI	coal	250	42.5	2007

[a]Petcoke, or petroleum coke, is coke derived from oil refinery coker units or other cracking units.
[b]Similar to Shell gasifier. Gasifier is called Prenflo (pressurized entrained flow) reactor.

The purified gas is sent to the gas turbine, where it is burned with compressed air to provide a stream of hot, high-pressure gas. The gas expands and conducts work on the turbine blades to turn the shaft that drives both the compressor and an electricity generator. The exhaust gases from the gas turbine are sent to a waste heat boiler, where high-pressure steam is generated by heat exchange with boiler feed water. This steam is used in the steam turbine for the generation of additional electricity.

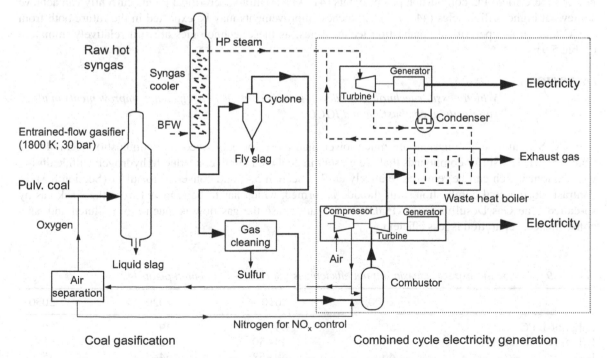

Figure 5.20   *Flow scheme of integrated gasification combined cycle power generation.*

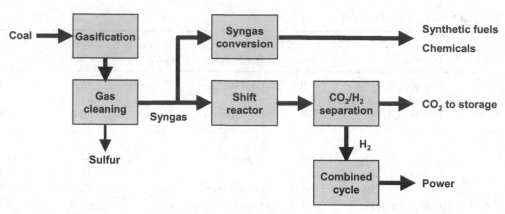

Figure 5.21 *Example of polygeneration in an IGCC power plant with carbon capture.*

5.3.7 Why Gasify, Not Burn for Electricity Generation?

Important advantages of IGCC technology compared to power generation based on pulverized coal combustion are the high energy efficiency, inherently low environmental impact, and ability to produce electricity as well as synthetic fuels, chemicals and marketable by-products such as sulfur (polygeneration) (Figure 5.21).

Pulverized coal combustion power plants can be categorized by their steam pressure as subcritical, super-critical or ultra-supercritical plants [47]. Current IGCC power plant efficiencies (40–45%) are much better than those of conventional subcritical PC combustion power plants (37–39%) and comparable to those of the newer supercritical PC combustion power plants (41–43%). Ultra-supercritical plants currently can achieve somewhat higher efficiencies (44–47%). Efficiency improvements may be expected in the future both from IGCC and ultra-supercritical combustion technologies, as these technologies are still relatively immature (Table 5.9).

QUESTIONS:

> *Which possible technological developments could lead to efficiency improvements in ultra-supercritical combustion and IGCC?*

IGCC SO_2 and NO_x emissions are much lower than those for pulverized coal combustion technologies (Table 5.10). The sulfur compounds that are present in coal are mainly converted to hydrogen sulfide during gasification, which can be removed relatively easily (Section 5.4) and converted to sulfur (Section 8.3). In contrast, during coal combustion sufur dioxide is formed, which has to be removed from the stack gas by so-called Flue Gas Desulfurization (FGD). In the latter case the gas flow is much more diluted and, as a consequence, separation is less efficient.

Table 5.9 *Actual and expected maximum net efficiencies (% LHV) for coal-fired plants [48].*

	2000	2010	2020	2050
Subcritical PC	39	39	39	39
(Ultra-)supercritical PC	44	48–50	50–53	51–55
IGCC	45	50–52	54–56	55–60

Table 5.10   *Comparison of emissions from coal-fired power plants [48].*

Emission (kg/MWh)	Ultra-supercritical PC combustion	IGCC
SO_2	0.6	0.06
NO_x	1.2	0.4
CO_2	760	740

During gasification, nitrogen and ammonia are formed. nitrogen is harmless and ammonia is more easily removed from syngas than NO_x from the stack gases of a coal combustor (usually done by selective or non-selective catalytic reduction, SCR or NSCR, Section 8.4). NO_x is formed in the combustor of the gas turbine in an IGCC plant, but its concentration can be limited to currently acceptable levels by diluting the feed to the combustor with nitrogen from the air separation plant and with steam.

The lower amount of harmful emissions from gasification versus combustion is one of the major incentives for considering this expensive technology, especially if carbon capture and storage, also known as carbon capture and sequestration (CCS), needs to be employed (Section 5.3.8).

QUESTIONS:

Why is ammonia in gasifier gas more easily removed than NO_x in stack gases? How does nitrogen addition to the combustor in Figure 5.20 help to control NO_x emissions from the process?

Why is sulfur more efficiently removed in gasification as compared to combustion? Compare oil and coal with respect to the preferred technologies for reducing sulfur emissions.

Extensive research and development studies are being carried out with the aim of cleaning the gases from the gasifier at high temperature rather than cooling them to enable scrubbing at low temperature. What would be the reason?

5.3.8   Carbon Capture and Storage (CCS)

Significant research, development, and demonstration efforts are ongoing with respect to CCS as a greenhouse gas mitigation option. The idea is to capture large-scale carbon dioxide emissions and store the gas underground so that it is not released into the atmosphere. Instead of carbon dioxide storage, some of the captured carbon dioxide could be used as a chemical feedstock, for example in the production of methanol or urea.

In coal-based technologies, there are basically three routes to carbon capture:

- post-combustion (recovering CO_2 from flue gases);
- oxy-fuel combustion (combustion in oxygen instead of air); and
- pre-combustion (extracting CO_2 before the fuel is burned).

Post-combustion is the route that would be applied to conventional pulverized-coal combustion power plants. Carbon dioxide is captured from the flue gases that mostly consist of nitrogen and water vapor, with only 15% or less being carbon dioxide. Carbon dioxide capture can be done by absorption in methanol or in amine-based solutions (Section 5.4). In principle, post-combustion capture is applicable to all types of power plants. However, the low pressure, large volume, and low concentration of carbon dioxide lead to huge capture systems with large energy consumption.

In oxy-fuel combustion, the fuel, in this case coal, is burned in (nearly) pure oxygen instead of air, eliminating the large amount of nitrogen from the flue gas, which mostly consists of carbon dioxide and water vapor. The latter is condensed through cooling, resulting in an almost pure stream of carbon dioxide. The main problems with this technology are the high investment costs and the large energy consumption in the production of oxygen.

Pre-combustion carbon capture has been integrated with gasification for several years, but more as a hydrogen purification step in the production of syngas for the manufacture of chemicals and synfuels than as a carbon capture step. The most common method is converting (part of the) carbon monoxide in the syngas into carbon dioxide and hydrogen via the water–gas shift reaction, followed by absorption of carbon dioxide in methanol or amine-based solutions (Section 5.4). Possible alternatives are the use of membrane technology or pressure swing adsorption (Section 6.1.6). The resulting carbon dioxide emissions can be pulled off in a relatively pure stream, while the formed hydrogen can be combusted or used in the production of chemicals.

QUESTION:

> *What are advantages and disadvantages of the three routes? Which technological problems can you think of for these routes?*

Employing CCS leads to increased capital costs and a lower energy efficiency, and thus a higher cost of electricity (Figure 5.22). CCS results in efficiency losses of up to 15% for pulverized-coal combustion plants and up to 8% for IGCC plants.

As can be seen from Figure 5.22, the impact of CCS on pulverized-coal combustion plants is much larger than that on IGCC plants, making the latter even more economically attractive than the former.

QUESTIONS:

> *Why does carbon capture add much more to the capital cost and the cost of electricity of a pulverized-coal combustion plant than to an IGCC plant?*

Worldwide, there are several industrial-scale CCS demonstration projects [50].

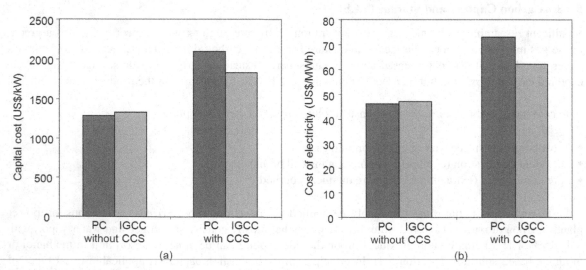

Figure 5.22 *Impact of adding carbon capture on coal-fired technologies: (a) capital cost; (b) cost of electricity. Capture costs include CO$_2$ compression but not transport and storage [49].*

5.4 Cleaning and Conditioning of Synthesis Gas

It is self-evident that the gas produced in a reformer or gasifier in general does not have the desired composition; impurities are present and the H_2/CO ratio usually needs to be adjusted. The required cleaning and conditioning of syngas depends on its source and use. In conventional gas cleaning processes, cooling of the raw syngas to low temperatures is usually required. Particulate matter present in syngas from a coal gasification process can then be removed in cyclones, fabric filters, wet scrubber systems (these also remove trace quantities of chlorides, NH_3, and HCN), gravity settlers, and electrostatic precipitators [43]. Some further cleaning and conditioning processes are discussed in the following sections.

5.4.1 Acid Gas Removal

The removal of acid gases, such as carbon dioxide, hydrogen sulfide, and other sulfur compounds (e.g., COS, CS_2, and mercaptans) from gas streams is often necessary for various reasons.

One of the most common processes to eliminate acid gases from gas streams is absorption in the liquid phase, either in chemical or physical solvents. Some of these processes allow the simultaneous removal of hydrogen sulfide and carbon dioxide, which is convenient if a sour water–gas shift (Section 5.4.2) is used.

Chemical solvents are used in the form of aqueous solutions to react with hydrogen sulfide and carbon dioxide reversibly and form products that can be regenerated by a change of temperature or pressure or both [51]. Commonly used chemical absorbents are alkanolamines (Figure 5.23). Advantages of these amines are their high solubility in water (solutions of 15–30 wt% are commonly used) and their low volatility. Amine processes offer good reactivity at relatively low cost and good flexibility in design and operation.

Physical solvents are attractive at high concentrations of hydrogen sulfide and carbon dioxide. Examples of physical solvents are methanol and propene carbonate. Regeneration is often possible by a simple pressure swing.

5.4.1.1 Removal of CO_2

In hydrogen and ammonia production, carbon dioxide is present in large quantities after the water–gas shift reactors (Section 5.4.2, Chapter 6). Table 5.11 shows a comparison of the energy requirements of various carbon dioxide removal processes. The advantage of MEA (monoethanolamine) and DEA (diethanolamine) solutions is that they exhibit high mass transfer rates. However, their regeneration is energy intensive. Therefore, at present solvents requiring lower heats of regeneration are being used.

QUESTION:
> *Explain the trends in Table 5.11.*

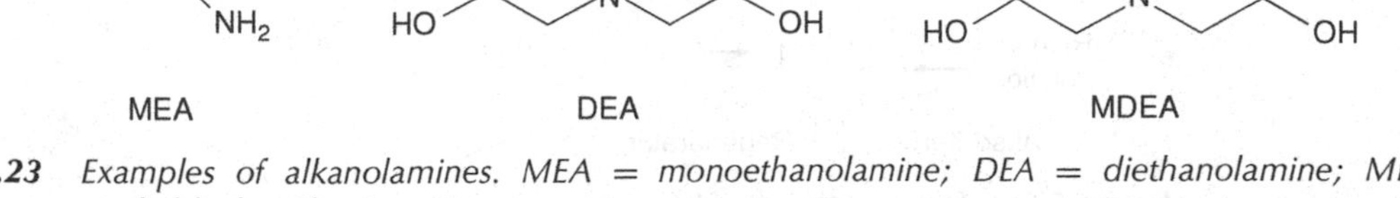

Figure 5.23 *Examples of alkanolamines. MEA = monoethanolamine; DEA = diethanolamine; MDEA = methyldiethanolamine.*

Table 5.11   *Solvent concentrations and energy requirements for CO$_2$ removal systems [2, 51].*

Removal system	Amine concentration (wt%)	Energy requirement (GJ/mol CO$_2$)
MEA	15–20	210
MEA with inhibitors	25–35	93–140
K$_2$CO$_3$ with additives	20–35	62–107
MDEA with additives	40–55	40–60

MEA forms corrosive compounds on decomposition and, therefore, its concentration has to be relatively low. Adding a compound that prevents MEA from decomposing, called an inhibitor, permits higher amine concentrations, thus allowing lower circulation rates and, consequently, lower regeneration duty.

The carbon dioxide removal processes based on hot potassium carbonate (K$_2$CO$_3$) require even less regeneration energy because the bonding between carbon dioxide and the carbonate is relatively weak. Various additives are used to improve mass transfer rates and to inhibit corrosion.

MDEA (methyldiethanolamine) has gained wide acceptance for carbon dioxide removal. MDEA requires very low regeneration energy because the CO$_2$–amine bonding is weaker than that for the other solvents. The additives mainly serve to improve mass transfer, as MDEA does not form corrosive compounds.

5.4.1.2   *Removal of H$_2$S and COS*

Figure 5.24 shows a typical process scheme for the removal of hydrogen sulfide from acid gas streams. Raw gas is fed to the bottom of the absorber containing typically 20–24 trays or an equivalent packing. Clean gas leaves the top. The regenerated amine (lean amine) enters at the top of this tower and the two streams are contacted countercurrently. Hydrogen sulfide (and CO$_2$) are absorbed into the amine phase by chemical reaction.

The rich amine solution loaded with hydrogen sulfide (and CO$_2$) leaving the bottom of the absorber is preheated and fed to the regenerator, where the acid gases are stripped with steam generated in a reboiler. The acid gases are taken from the top and fed to a condenser, where most of the steam condenses and is returned to the regeneration column.

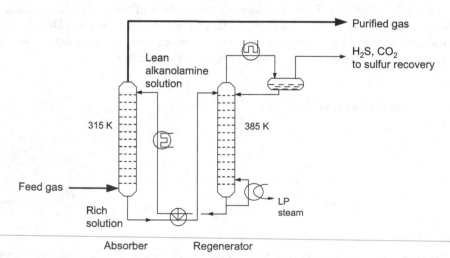

Figure 5.24   *Flow scheme for H$_2$S removal by amine absorption.*

The concentrated hydrogen sulfide is frequently converted to elemental sulfur by the "Claus" process [52, 53] (Section 8.3).

Carbonyl sulfide (COS) is present in minor amounts in the raw syngas from coal gasifiers. It needs to be removed because of its adverse effects on subsequent processes. In order to remove COS, prior catalytic hydrolysis to hydrogen sulfide may be necessary:

$$COS + H_2O \rightleftarrows H_2S + CO_2 \qquad \Delta_r H_{298} = -34 \text{ kJ/mol} \qquad (5.19)$$

QUESTION:

Why are the acid gas components not removed by scrubbing with amines directly, without prior COS hydrolysis?

5.4.2 Water–Gas Shift Reaction

The water–gas shift reaction serves to increase the H_2/CO ratio of the syngas for use in, for example, the production of various chemicals, Fischer–Tropsch liquids and SNG (Table 5.1), or to enable the complete removal of carbon monoxide and carbon dioxide, for example in syngas for ammonia synthesis or hydrogen production and in IGCC power plants with carbon capture.

In the water–gas shift reaction, carbon monoxide is converted to carbon dioxide:

$$CO + H_2O \rightleftarrows CO_2 + H_2 \qquad \Delta_r H_{298} = -41 \text{ kJ/mol} \qquad (5.2)$$

Although the reaction is only moderately exothermic, the temperature has a great impact on the equilibrium constant, as illustrated in Figure 5.25.

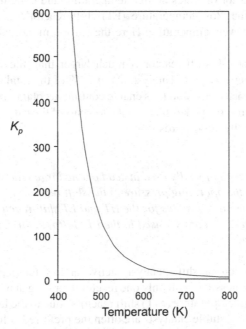

Figure 5.25 *Water–gas shift equilibrium constant; $K_p = p_{H_2} p_{CO_2} / p_{H_2O} p_{CO}$.*

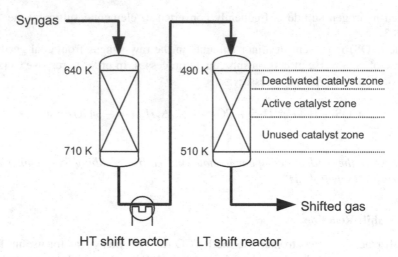

Figure 5.26   *Water–gas shift reactors with typical temperatures and catalyst profile in the LT-shift reactor.*

Clearly, the temperature has to be low! In the past, catalysts were only active at temperatures above about 600 K. Later, more active catalysts were developed which are already active at temperatures of around 500 K. These catalysts, however, require extremely pure gases because they are very sensitive to poisoning. Moreover, they are not stable at higher temperatures. For many practical purposes, the water–gas shift reaction is carried out using two adiabatic fixed bed reactors with cooling between the two reactors (Figure 5.26). In the first reactor, the "high-temperature (HT) shift reactor", most of the carbon monoxide is converted. This reactor operates at high temperature and contains a classical iron-oxide-based catalyst. The second reactor, the "low-temperature (LT) shift reactor", contains a more active copper-based catalyst and operates at lower temperature. Here the carbon monoxide content of the gas is further reduced.

The amount of catalyst in the LT shift reactor is much larger than the amount needed to achieve the required carbon monoxide conversion; extra catalyst, about 70% of the total catalyst volume, is included to compensate for the inevitable deactivation and thus enable continuous plant operation for at least two or three years. Initially, almost all of the reaction takes place in the top part of the catalyst bed, but deactivation causes the reaction zone to move gradually downwards [54].

QUESTION:

> *Why are two reactors usually used instead of one large reactor with a superior LT catalyst?*
> *What would be the operating pressure in the shift reactors?*
> *Sketch the temperature profiles for the HT and LT shift reactors.*
> *Why is so much extra catalyst used in the LT shift reactor, instead of replacing the catalyst load more often?*

Breakthroughs in the production of shift catalysts active at low temperature, but also stable at high temperature, have made schemes possible with only one reactor operating at a low inlet temperature; these are generally referred to as "medium-temperature (MT) shift reactors" due to the larger temperature increase over the reactor. These new temperature-stable catalysts are often the preferred solution for hydrogen production.

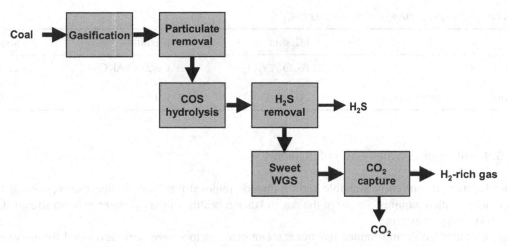

Figure 5.27   *Sweet water–gas shift (WGS) in IGCC with carbon capture.*

They enable operation of the steam reformer at lower steam-to-carbon ratio, thereby reducing the total specific energy consumption for hydrogen production.

Syngas from a coal gasifier may be shifted before or after sulfur removal, referred to as sour or sweet water–gas shift. Schematics for both options are given in Figures 5.27 and 5.28. Table 5.12 compares properties of the different types of shift reactors.

The sweet water–gas shift is carried out in two stages as discussed above. In the HT shift reactor, the carbon monoxide concentration is reduced to approximately 7–8 mol%. In the LT shift reactor, the carbon monoxide concentration is further reduced to around 0.3 mol%.

In the sour water–gas shift, a sulfur-tolerant sulfided cobalt–molybdenum catalyst is used. In an IGCC plant, the use of two or three intercooled adiabatic reactors results in a final carbon monoxide concentration of 1.6–0.8 mol%. A side reaction is the hydrolysis of COS, so a separate hydrolysis reactor is not necessary [45, 55]. Box B.5.8 describes alternative methods for carrying out the water–gas shift reaction.

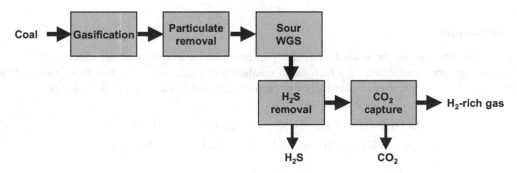

Figure 5.28   *Sour water–gas shift (WGS) in IGCC with carbon capture.*

Table 5.12 *Some properties of shift reactors [54].*

	HT shift	LT shift	Sour shift
Catalyst	Fe_3O_4/Cr_2O_3	$Cu/ZnO/Al_2O_3$	CoMoS
Temperature (K)	620–770	460–550	520–770
Maximum sulfur content (ppm)	20	0.1	>1000

Box 5.8 Alternative Water–Gas Shift Reactors

A disadvantage of currently available carbon dioxide removal processes is that they operate at low temperature, which requires cooling of the gas and then reheating of the hydrogen product stream. This is not very energy efficient.

As for other equilibrium-limited reactions, membrane reactors have been developed for the water–gas shift reaction. The membranes used are either permeable to hydrogen or carbon dioxide, thus shifting the reaction equilibrium to the product side. An alternative method to achieve this is the so-called sorption-enhanced water–gas shift (SEWGS) reaction, developed by Air Products and ECN [56] (Figure B5.8.1). The CO_2 formed during the reaction is adsorbed on the sorbent and the reverse reaction cannot occur. The SEWGS process is a batch process: CO_2 accumulates on the sorbent and at some point the sorbent becomes saturated with CO_2. The sorbent then has to be regenerated, either by reducing the pressure (pressure-swing adsorption) or by increasing the temperature (temperature-swing adsorption), see Section 6.1.6.

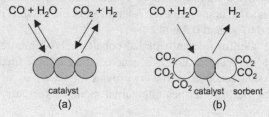

Figure B5.8.1 *Principle of sorption-enhanced WGS: (a) ordinary WGS; (b) SEWGS.*

5.4.3 Methanation

The methanation process has been used extensively in commercial ammonia plants (Section 6.1), where it is the final syngas purification step in which small residual concentrations of carbon monoxide and carbon dioxide (< 1 mol%) are removed catalytically by reaction with hydrogen over a nickel catalyst:

$$CO + 3\,H_2 \rightleftarrows CH_4 + H_2O \qquad \Delta_r H_{298} = -206 \text{ kJ/mol} \qquad (5.20)$$

$$CO_2 + 4\,H_2 \rightleftarrows CH_4 + 2\,H_2O \qquad \Delta_r H_{298} = -165 \text{ kJ/mol} \qquad (5.21)$$

This step is necessary because even very small amounts of these carbon oxides are poisonous to the iron-based ammonia synthesis catalyst.

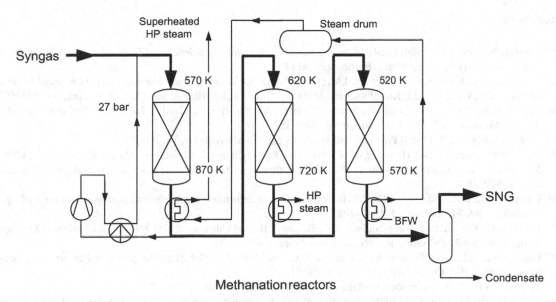

Figure 5.29   *Simplified flow scheme of the TREMP™ methanation process [57, 58].*

The primary methanation reaction, reaction (5.20), is the reverse of the steam reforming reaction. Both reactions (5.20) and (5.21) are highly exothermic, but this is not a problem because of the small concentrations of carbon monoxide and carbon dioxide present.

Methanation is also used in the production of SNG from coal-derived syngas, but in this case the concentrations of carbon monoxide and carbon dioxide are much higher. As a result, the temperature rise in the reactor would be much higher. Two basically different approaches can be used to circumvent this problem, namely the use of two or more adiabatic fixed bed reactors with interstage cooling or a reactor design with improved heat transfer properties, such as a fluidized bed reactor [57].

Figure 5.29 shows a simplified flow scheme of a process in which the former approach is used: the TREMP™ (Topsøe Recycle Energy-efficient Methanation Process), developed by Haldor Topsøe [58]. Part of the methane-rich exit gas from the first reactor is recycled in order to control the temperature rise in the first methanation reaction, which is still rather high. The heat that is generated during the exothermic process is recovered as steam and hot water, resulting in high energy efficiency.

QUESTIONS:

> *Why would a methanation step be preferred over additional water–gas shift conversion in preparing syngas for ammonia production?*
> *Note the temperatures given in the flow scheme of Figure 5.29. Comment upon the logic of the values.*
> *What steps would you use in the processing of raw syngas for methanol synthesis from steam reforming and coal gasification, respectively?*
> *Will water–gas shift reactors be present in a coal gasification process for the production of SNG?*
> *In Hybrid Electric Vehicle applications, CO also often has to be removed in order to protect the catalyst. This is done by selective oxidation. Why would this technology not be attractive in ammonia production?*

References

[1] Rostrup-Nielsen, J.R. (1984) Catalytic steam reforming, in *Catalysis, Science and Technology* (eds J.R. Anderson and M. Boudart). Springer-Verlag, Berlin, pp. 1–117.

[2] Czuppon, T.A., Knez, S.A., and Rovner, J.M. (1992) Ammonia, in *Kirk-Othmer Encyclopedia of Chemical Technology*, 4th edn., vol. **2** (eds J.I. Kroschwitz and M. Howe-Grant). John Wiley & Sons, Inc., New York, pp. 638–691.

[3] Pearce, B.B., Twigg, M.V., and Woodward, C. (1996) Methanation, in *Catalyst Handbook*, 2nd edn. (ed. M.V. Twigg). Manson Publishing Ltd, London, pp. 340–383.

[4] Rostrup-Nielsen, J.R. (1993) Production of synthesis gas. *Catalysis Today*, **18**, 305–324.

[5] Aasberg-Petersen, K., Bak Hansen, J.-H., Christensen, T.S., Dybkjaer, I., Christensen, P.S., Stub Nielsen, C., Winter Madsen, S.E.L., and Rostrup-Nielsen, J.R. (2001) Technologies for large-scale gas conversion. *Applied Catalysis A*, **221**, 379–387.

[6] Chen, Z. and Elnashaie, S.S.E.H. (2005) Optimization of reforming parameter and configuration for hydrogen production. *AIChE Journal*, **51**, 1467–1481.

[7] Dibbern, C., Olesen, P., Rostrup-Nielsen, J.R., Tottrup, P.B., and Udengaard, N.R. (1986) Make low H_2/CO syngas using sulfur passivated reforming. *Hydrocarbon Process*, **65**, 71–74.

[8] Udengaard, N.R., Bak Hansen, J.H., Hanson, D.C., and Stal, J.A. (1992) Sulfur passivated reforming process lowers syngas H_2/CO ratio. *Oil Gas Journal*, **90**, 62–67.

[9] Teuner, S. (1987) A new process to make oxo-feed. *Hydrocarbon Process*, **66**, 52.

[10] Xu, X.D. and Moulijn, J.A. (1996) Mitigation of CO_2 by chemical conversion: plausible chemical reactions and promising products. *Energy Fuels*, **10**, 305–325.

[11] Dybkjaer, I., Rostrup-Nielsen, T., and Aasberg-Petersen, K. (2009) Synthesis gas and hydrogen, in *Encyclopaedia of Hydrocarbons*, **II** (ed. G. Treccani). Istituto Della Enciclopedia Italiana Fondata da Giovanni Treccani, Rome, Italy, pp. 469–500.

[12] Bharadwaj, S.S. and Schmidt, L.D. (1995) Catalytic partial oxidation of natural gas to syngas. *Fuel Processing Technology*, **42**, 109–127.

[13] Hohn, K.L. and Schmidt, L.D. (2001) Partial oxidation of methane to syngas at high space velocities over Rh-coated spheres. *Applied Catalysis A*, **211**, 53–68.

[14] Schmidt, D., Klein, J., Leclerc, A., Krummenacher, J., and West, N. (2002) Syngas in millisecond reactors: higher alkanes and fast lightoff. *Chemical Engineering Science*, **58**, 1037–1041.

[15] Beretta, A. and Forzatti, P. (2004) Partial oxidation of light paraffins to synthesis gas in short contact-time reactors. *Chemical Engineering Journal*, **99**, 219–226.

[16] Pena, M.A., Gómez, J.P., and Fierro, J.L.G. (1996) New catalytic routes for syngas and hydrogen production. *Applied Catalysis A*, **144**, 7–57.

[17] Aasberg-Petersen, K., Christensen, T.S., Dybkjaer, I., Sehested, J., Østberg, M., Coertzen, R.M., Keyser, M.J., and Steynberg, A.P. (2004) Synthesis gas production for FT synthesis, in *Studies in Surface Science and Catalysis. Fischer-Tropsch Technology*, Vol. **152** (eds A.P. Steynberg and M.E. Dry). Elsevier, pp. 258–405.

[18] Frauhammer, J., Eigenberger, G., Hippel, L., and Arntz, D. (1999) A new reactor concept for endothermic high-temperature reactions. *Chemical Engineering Science*, **54**, 3661–3670.

[19] Piga, A. and Verykios, X.E. (2000) An advanced reactor configuration for the partial oxidation of methane to synthesis gas. *Catalysis Today*, **60**, 63–71.

[20] Stitt, E.H. (2005) Reactor technology for syngas and hydrogen, in *Sustainable Strategies for the Upgrading of Natural Gas: Fundamentals, Challenges, and Opportunities*, **191** (eds E.G. Derouane, V. Parmon, F. Lemos, and F.R. Ribeiro). Springer, The Netherlands, pp. 185–216.

[21] Eisenberg, B., Fiato, R.A., Mauldin, C.H., Say, G.R., and Soled, S.L. (1998) Exxon's advanced gas-to-liquids technology, in *Studies in Surface Science and Catalysis. Natural Gas Conversion V, Proceedings of the 5th International Natural Gas Conversion Symposium*, Vol. **119**, (ed. A. Parmaliana). Elsevier, pp. 943–948.

[22] Smith, J.M. and Van Ness, H.C. (1987). *Introduction to Chemical Engineering Thermodynamics*, 4th edn. McGraw-Hill, New York.

[23] Kotas, T.J. (1985) *The Exergy Method of Thermal Plant Analysis*. Buttersworth, London.

[24] De Swaan Arons, J. and Van der Kooi, H.J. (1993) Exergy analysis. Adding insight and precision to experience and intuition, in *Precision Process Technology* (eds M.P.C. Weijnen and A.A.H. Drinkenburg). Kluwer, Dordrecht, The Netherlands, pp. 89–113.

[25] Sogge, J. and Ström, T. (1997) Membrane reactors – a new technology for production of synthesis gas by steam reforming, in *Studies in Surface Science and Catalysis. Natural Gas Conversion IV*, Vol. **107**, (ed. M. de Pontes). Elsevier, pp. 561–566.

[26] Lu, G.Q., Diniz da Costa, J.C., Duke, M., Giessler, S., Socolow, R., Williams, R.H., and Kreutz, T. (2007) Inorganic membranes for hydrogen production and purification: a critical review and perspective. *Journal of Colloid and Interface Science*, **314**, 589–603.

[27] Rostrup-Nielsen, J.R. and Rostrup-Nielsen, T. (2002) Large-scale hydrogen production. *CATTECH*, **6**, 150–159.

[28] Balachandran, U., Dusek, J.T., Maiya, P.S., Ma, B., Mieville, R.L., Kleefisch, M.S., and Udovich, C.A. (1997) Ceramic membrane reactor for converting methane to syngas. *Catalysis Today*, **36**, 265–272.

[29] Carolan, M.F., Chen, C.M., and Rynders, S.W. (2003) Development of the ceramic membrane ITM Syngas/ITM hydrogen process in *Fuel Chemistry Division Preprints*, **48**. American Chemical Society, Division of Fuel Chemistry, Washington, DC, pp. 343–344.

[30] Lunsford, J.H. (1997) Oxidative coupling of methane and related reactions, in *Handbook of Heterogeneous Catalysis* (eds G. Ertl, H. Knözinger, and J. Weitkamp). VCH, Weinheim, pp. 1843–1856.

[31] Androulakis, I.P. and Reyes, S.C. (1999) Role of distributed oxygen addition and product removal in the oxidative coupling of methane. *AIChE Journal*, **45**, 860–868.

[32] Keller, G.E. and Bhasin, M.M. (1982) Synthesis of ethylene via oxidative coupling of methane: I. Determination of active catalysts. *Journal of Catalysis*, **73**, 9–19.

[33] Stougie, L., Dijkema, G.P.J., Van der Kooi, H.J., and Weijnen, M.P.C. (1994) Vergelijkende exergie-analyse van methanolprocessen, *NPT Procestechnologie*, **Aug**, 19–23 (in Dutch).

[34] Bose, A.C., Stiegel, G.J., Armstrong, P.A., Halper, B.J., and Foster, E.P. (2009) Progress in ion transport membranes for gas separation applications, in *Inorganic Membranes for Energy and Environmental Applications* (ed. A.C. Bose). Springer, New York, pp. 3–26.

[35] Allam, R.J. (2007) Improved Oxygen Production Technologies, 2007/14, IEA Environmental Projects, 1–84.

[36] Simbeck, D. (2007) Gasification Opportunities in Alberta. Paper presented at Seminar on Gasification Technologies sponsored by Alberta Government and GTC, Edmonton.

[37] NETL (2007) Gasification World Database 2007, NETL, US Department of Energy, http://www.netl.doe.gov/technologies/coalpower/gasification/database/Gasification2007_web.pdf, pp. 32 (last accessed 17 December 2012).

[38] Tromp, P.J.J. and Moulijn, J. A. (1988) Slow and rapid pyrolysis in coal, in *New Trends in Coal Science* (ed. Y. Yürüm). Kluwer, Dordrecht, the Netherlands, pp. 305–338.

[39] Van Diepen, A.E. and Moulijn, J.A. (1998) Effect of process conditions on thermodynamics of gasification, in *Desulfurization of Hot Coal Gas*, **G42** (eds A.T. Atimtay and D.P. Harrison). Springer-Verlag, Berlin, pp. 57–74.

[40] Higman, C. and van der Burgt, M. (2007) *Gasification*, 2nd edn. Elsevier.

[41] University of Stavanger (2010) IGCC State-of-the-Art Report. University of Stavanger, Norway, pp. 91. http://www.h2-igcc.eu/Pdf/State-of-the-art%20IGCC%20_2010-04-29.pdf (last accessed 6 December 2012).

[42] Cornils, B. (1987) Syngas via coal gasification, in *Chemicals From Coal: New Processes* (ed. K.R. Payne). John Wiley & Sons, Inc., New York, pp. 1–31.

[43] Garcia Cortés, C., Tzimas, E., and Peteves, S.D. (2009) Technologies for Coal Based Hydrogen and Electricity Co-Production Power Plants With CO2 Capture. JRC Scientific and Technical Reports, pp. 74. http://www.energy.eu/publications/a05.pdf (last accessed 6 December 2012).

[44] Gallagher, J.E. and Euker, C.A. (1980) Catalytic coal gasification for SNG manufacture. *International Journal of Energy Research*, **4**, 137–147.

[45] Teper, M., Hemming, D.F., and Ulrich, W.C. (1983) The Economics of Gas From Coal. EAS report E2/80, IEA Coal Research, Economic Assessment Service, London.

[46] NETL (2010) GreatPoint Energy. http://www.netl.doe.gov/technologies/coalpower/gasification/gasifipedia/4-gasifiers/4-1-3-2_gpe.html (last accessed 6 December 2012).

[47] MIT (2007) The Future of Coal. Massachusetts Institute of Technology, pp. 192. http://web.mit.edu/coal/The_Future_of_Coal.pdf (last accessed 6 December 2012).

[48] Lako, P. (2004) Coal-Fired Power Technologies ECN Project, ECN-C-04-076, pp. 38.

[49] Rubin, E.S., Chen, C., and Rao, A.B. (2007) Cost and performance of fossil fuel power plants with CO_2 capture and storage. *Energy Policy*, **35**, 4444–4454.

[50] MIT (2010) Power Plant Carbon Dioxide Capture and Storage Projects, Massachusetts Institute of Technology, http://sequestration.mit.edu/tools/projects/index.html (last accessed 6 December 2012).

[51] Abdel-Aal, H.K., Fahim, M.A., and Aggour, M. (2003) Sour gas treating, in *Petroleum and Gas Field Processing*. CRC Press, pp. 251–284.

[52] Nehr, W. and Vydra, K. (1994) Sulfur, in *Ullman's Encyclopedia of Industrial Chemistry*, 5th edn, **A25** (ed. W. Gerhartz). VCH, Weinheim, pp. 507–567.

[53] Piéplu, A., Saur, O., Lavalley, J.C., Legendre, O., and Nédez, C. (1998) Claus catalysis and H_2S selective oxidation. *Catalysis Reviews, Science and Engineering*, **40**, 409–450.

[54] Lloyd, L., Ridler, D.E., and Twigg, M.V. (1996) The water-gas shift reaction, in *Catalyst Handbook*, 2nd edn (ed. M.V. Twigg). Wolfe Publishing Ltd, London, pp. 283–339.

[55] EPRI (2006) Feasibility Study for an Integrated Gasification Combined Cycle Facility at a Texas Site, pp. 238. http://www.scdhec.gov/environment/baq/docs/SanteeCooper/Comments/psd_Santee_exhibit_F.feasibility.pdf (last accessed 6 December 2012).

[56] van Selow, E.R., Cobden, P.D., van den Brink, R.W., Hufton, J.R., and Wright, A. (2009) Performance of sorption-enhanced water-gas shift as a pre-combustion CO_2 capture technology. *Energy Procedia*, **1**, 689–696.

[57] Kopyscinski, J., Schildhauer, T.J., and Biollaz, S.M.A. (2010) Production of synthetic natural gas (SNG) from coal and dry biomass – A technology review from 1950 to 2009. *Fuel*, **89**, 1763–1783.

[58] Haldor Topsoe (2009) From Solid Fuels to Substitute Natural Gas (SNG) Using TREMPTM. Haldor Topsoe, Denmark, pp. 8. http://www.topsoe.com/business_areas/gasification_based/~/media/PDF%20files/SNG/Topsoe_TREMP.ashx (last accessed 6 December 2012).

6

Bulk Chemicals and Synthetic Fuels Derived from Synthesis Gas

The chemical process industry to a large extent aims at producing value added chemicals from various raw materials. A convenient route is via synthesis gas or syngas, the production of which has been discussed in Chapter 5. Historically the most important processes have been ammonia and methanol production. As these processes have been developed along a path of instructive innovations they are examined here in detail. Some of their main derivatives, namely urea from ammonia and formaldehyde and methyl *tert*-butyl ether (MTBE) from methanol, are also briefly discussed.

Syngas is a convenient intermediate in the conversion of natural gas and coal into transportation fuels (gasoline, diesel, and jet fuel) via, for instance, the Fischer–Tropsch (FT) synthesis or the methanol-to-gasoline (MTG) process. Although these processes are currently only used in special cases, in the future they will probably gain importance, as natural gas and, especially, coal reserves well exceed those of oil.

6.1 Ammonia

6.1.1 Background Information

Ammonia is a major product of the chemical industry. Already at the beginning of the nineteenth century its synthesis was the subject of much research. The Industrial Revolution and the related growth of the population generated a large demand for nitrogen fertilizers. Natural resources of nitrogen-containing fertilizers were saltpeter (KNO_3), Chile saltpeter ($NaNO_3$), and guano (seabird droppings). At the beginning of the twentieth century, ammonia was produced as by-product in coke ovens and gas works. In these industries, ammonia is formed during the distillation of coal for the production of town gas (syngas with a high methane and nitrogen content). This source of ammonia is no longer important today, but it explains why industries that originally were coal-based often are still ammonia producers.

It was recognized as early as the turn of the twentieth century that fertilizer supplies were not sufficient for agricultural needs. Moreover, the explosives industry was developing into a large volume industry due to the beginning of World War I.

Chemical Process Technology, Second Edition. Jacob A. Moulijn, Michiel Makkee, and Annelies E. van Diepen.
© 2013 John Wiley & Sons, Ltd. Published 2013 by John Wiley & Sons, Ltd.

Table 6.1   *Energy data for nitrogen.*

	Value for N_2 (kJ/mol)	Compare with (kJ/mol)
Bond dissociation energy	945	C–H in CH_4: 439
Ionization enthalpy	1503	O_2: 1165
Electron affinity	34900	O_2: 43

QUESTIONS:

> *What is meant by "distillation of coal"? Compare distillation of coal, the production of coke for blast furnaces, and "coking" processes in the oil refinery (Section 3.3). In what respects are they similar?*

Understandably, the direct synthesis of ammonia from nitrogen in the air was attempted. However, the difficulty in the synthesis of ammonia is that nitrogen is very stable and inert. This can be understood from the data on the bond dissociation energy, the ionization enthalpy, and the electron affinity of nitrogen (Table 6.1).

All values are very high. The bond dissociation energy is extremely high and, as a consequence, dissociation is expected to be very difficult. Activation by ionizing the molecule is also expected to be nearly impossible: the value for the electron affinity of nitrogen is exceptionally high, as illustrated by comparison with the value for oxygen, which is about 1000 times lower.

The history of ammonia synthesis is very interesting because it was the first large-scale synthesis in the chemical industry at high pressure (>100 bar) and high temperature (circa 700 K). It is remarkable that scientists and engineers succeeded in realizing a safe and reliable process in a short time, even to modern standards. The reaction to be carried out is:

$$N_2 + 3\,H_2 \rightleftarrows 2\,NH_3 \qquad \Delta_r H_{298} = -91.44 \text{ kJ/mol} \qquad (6.1)$$

A crucial question concerned thermodynamics. Precise thermodynamic data were not known and, as a consequence, a strong controversy existed as to where the equilibrium is located at the conditions at which attempts to synthesize ammonia were made.

Around 1910 it was well documented that at atmospheric pressure in a mixture of nitrogen, hydrogen and ammonia hardly any ammonia is present. For instance, Haber found that at 1290 K the fraction of NH_3 in an equilibrium mixture of N_2, H_2, and NH_3 ($N_2/H_2 = 1/3$ mol/mol) was only 0.01%. These data convinced many experts, in first instance including Haber, that industrial synthesis based on nitrogen would never be economically feasible.

Meanwhile, more reliable thermodynamic data became available. Haber extrapolated these data to lower temperatures and concluded that an industrial process was feasible, if suitable catalysts could be developed. Interestingly, Haber also pointed out that an industrial process is possible even if conversion is not complete. He proposed a recycle loop under pressure and also suggested the use of a feed-effluent heat exchanger. These concepts are still the basis of modern ammonia synthesis plants.

In the unbelievably short period of five years (1908–1913), Haber developed a commercial process in operation (30 t/d) in cooperation with Bosch and Mittasch at BASF. The group at BASF tested over 6500 (!) catalysts and discovered that a (multipromoted) iron-based catalyst exhibits superior activity. The catalysts used today do not differ essentially from this catalyst, although recently a new (noble-metal-based) catalyst has been commercialized by Kellogg [1].

Figure 6.1 shows that ammonia production has grown exponentially, although especially in Europe saturation, and even some decline, has now occurred. Capacities of modern ammonia plants are high (Figure 6.1)

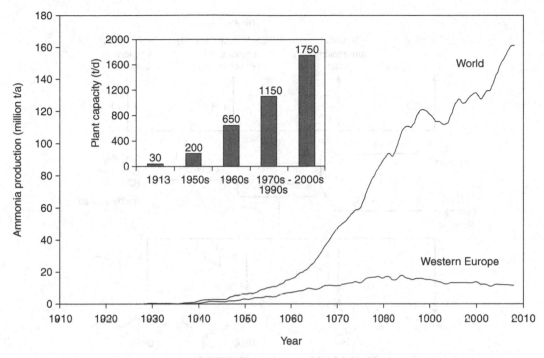

Figure 6.1   *Growth of ammonia production and average single-train plant capacity (insert) [3, 4].*

and are still increasing. In the early 2000s, the maximum attainable single-train capacity was 2200 t/d, and designs for plants with sizes of 3000–>4000 t/d are already available [2].

Figure 6.2 shows a number of products manufactured from ammonia. At present, about 85% of ammonia production is used for nitrogen fertilizers. Direct application of ammonia represents the largest single consumption (about 30%). Of the solid fertilizers, urea is the most important, accounting for about 40% of ammonia usage. Other solid nitrogen fertilizers are ammonium nitrate, ammonium sulfate, and ammonium phosphates. Industrial applications of ammonia include the production of amines, nitriles, and organic nitrogen compounds for use as intermediates in the fine chemicals industry.

Ammonia is becoming increasingly important in environmental applications, such as removal of NO_x from flue gases of fossil fuel power plants.

The production of ammonia requires a mixture of hydrogen and nitrogen in a ratio of 3:1. The source of nitrogen is invariably air, but hydrogen can be produced from a variety of fossil fuels and biomass. Steam reforming of natural gas followed by secondary reforming with air is most often employed in ammonia plants and accounts for over 80% of ammonia production. Section 6.1.5 describes a complete ammonia plant based on natural gas, but first the ammonia synthesis reaction is considered.

6.1.2   Thermodynamics

Because of the high practical relevance, very reliable thermodynamic data are available on the ammonia synthesis reaction. Figure 6.3 shows the ammonia content in equilibrium syngas as a function of temperature and pressure.

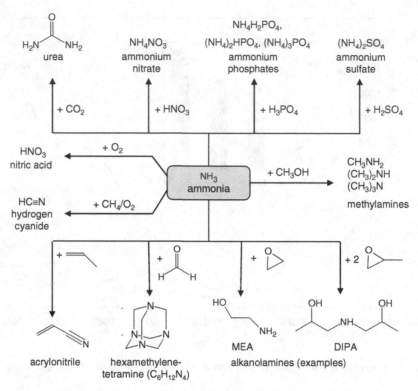

Figure 6.2 *Fertilizers and chemicals from ammonia.*

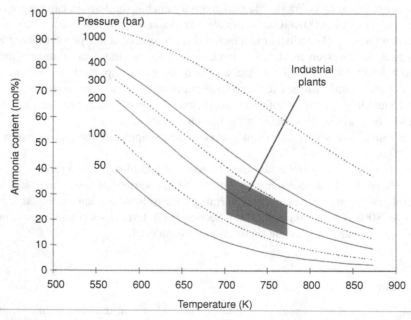

Figure 6.3 *Ammonia content in equilibrium syngas ($H_2/N_2 = 3$ mol/mol) [5].*

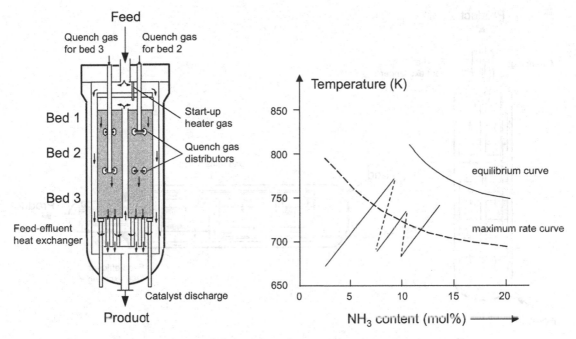

Figure 6.4 ICI quench reactor and temperature–concentration profile.

Clearly, favorable conditions are low temperature and high pressure. However, kinetic limitations exist: at temperatures below about 670 K the rate of reaction is very low. Therefore, the temperature to be used is the result of an optimization procedure. Similarly, regarding pressure, the optimal situation is a compromise between thermodynamically favorable conditions and minimal investment. Typical reaction conditions are:

Temperature : 675 K (inlet); 720–770 K (exit)
Pressure : 100–250 bar

QUESTION:

Estimate the maximum attainable (single-pass) conversion under industrial conditions. (Hint: keep in mind that the reaction is associated with a large heat of reaction.)

6.1.3 Commercial Ammonia Synthesis Reactors

Ammonia synthesis reactors are classified by flow type (axial, radial or cross flow) and cooling method. Temperature control is crucial in ammonia synthesis: the reaction is exothermic and the heat produced needs to be removed. Two methods are applied: (1) direct cooling and (2) indirect cooling:

1. In the case of direct cooling in so-called quench reactors, cold feed gas is added at different heights in the reactor.
2. Indirect cooling is achieved with heat exchangers that are placed between the catalyst beds.

An example of a reactor with direct cooling is the ICI quench reactor [5,6], which is shown in Figure 6.4 together with the temperature–concentration profile.

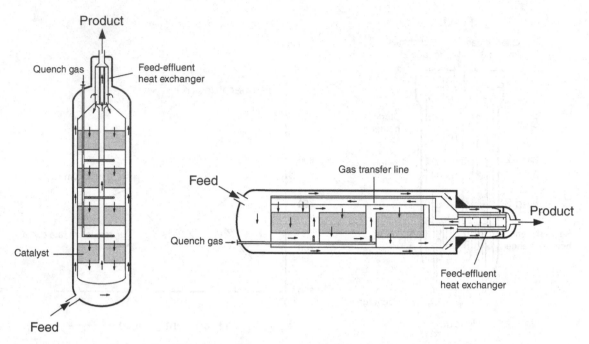

Figure 6.5   *Kellogg quench reactors; vertical (left) and horizontal (right).*

QUESTIONS:

Explain the temperature profile in Figure 6.4. Why are the dashed lines not parallel to the temperature axis? Are the solid lines parallel? Why or why not?

Part of the cold feed is introduced at the top and passes downward through the annular space between the catalyst container and the converter shell. This is a feature used in most ammonia converter designs. It serves to keep the reactor shell, which is made of carbon steel or a low-alloy steel, at a relatively low temperature in order to prevent hydrogenation of the carbon in the steel, which would result in embrittlement of the shell.

The gas is then heated in the feed-effluent heat exchanger before entering the catalyst bed. The remainder of the feed is injected at the inlet of the second and third catalyst zones resulting in a reduction of the temperature. The single catalyst bed consists of three zones separated only by gas distributors, facilitating catalyst unloading. The effluent gas undergoes heat exchange with the feed and leaves the reactor at approximately 500 K. An additional feature of this reactor is a separate inlet to which preheated feed gas can be fed to start up the reactor.

Another quench reactor is the Kellogg vertical quench converter (Figure 6.5) [5, 6]. The Kellogg vertical reactor consists of four beds held on separate grids. Quench gas distributors are placed in the spaces between the beds. The feed enters the reactor at the bottom and flows upward between the catalyst bed and shell to be heated in a feed-effluent exchanger located at the top of the vessel. The product gas outlet is also located at the top.

QUESTIONS:

Compare the ICI and Kellogg quench reactors. Which reactor is most convenient concerning catalyst loading and unloading?

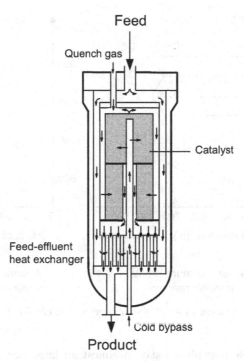

Feed

Quench gas

Catalyst

Feed-effluent
heat exchanger

Cold bypass

Product

Figure 6.6   *Haldor Topsøe radial flow reactor.*

It is attractive to employ catalyst particles that are as small as possible to increase the efficiency of the catalyst. However, smaller particles increase the pressure drop in the catalyst beds. Kellogg has designed a horizontal converter [5, 6] for plant capacities in excess of 1700 t/d (Figure 6.5), in which the gas flows through two or more catalyst beds in a cross-flow pattern. A larger cross-sectional area is possible than in axial flow reactors, providing a larger catalytic area per unit volume of reactor and permitting the use of smaller catalyst particles without large pressure drop.

Another design achieving this effect is a radial flow reactor. An example is the reactor designed by Haldor Topsøe [5, 7] (Figure 6.6), in which two annular catalyst beds are used through which the gas flows in radial direction. After passing downward through the annulus between shell and internals, then up again through a heat exchanger located at the bottom, the feed gas passes up through a central pipe, from which it flows radially through the top catalyst bed. The gas then passes downward around the bottom bed to flow inward radially together with the quench gas.

Both the Kellogg horizontal converter and the Haldor Topsøe radial flow reactor are also available as indirectly cooled reactors.

Indirect cooling [1,5] has advantages (and disadvantages) compared to direct cooling. Figure 6.7 shows an example of an indirectly cooled reactor. The cooling medium can be the feed gas or high-pressure steam can be generated. In the latter case, the catalyst beds and heat exchangers are often not installed together inside a single pressure vessel, but the individual catalyst beds are accommodated in separate vessels and have separate heat exchangers. Figure 6.7 shows a typical temperature profile and a temperature–concentration diagram. The advantage of this type of reactor is that heat is recovered at the highest possible temperature. However, the investment cost is higher than that of quench reactors due to the cost of the heat exchangers.

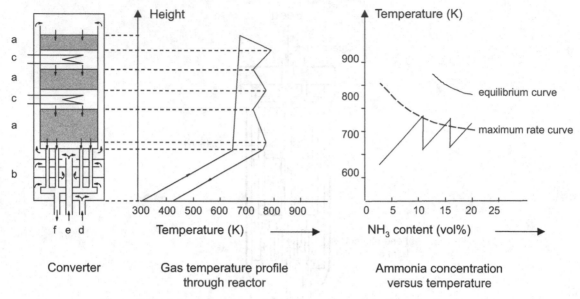

Figure 6.7   *Schematic of a multibed converter with indirect cooling [1, 5].*

The indirect cooling principle is applied today in almost all large new ammonia plants. Also, in the modernization (revamping) of older plants an increasing number of quench converters are modified to the indirect cooling mode.

QUESTIONS:

Compare the reactor designs given. What are the similarities and what are the differences? What are advantages and disadvantages of the different designs?
Compare the ammonia concentration profiles of Figures 6.4 and 6.7.

6.1.4   Ammonia Synthesis Loop

Because of the far from complete single-pass conversion (20–30% per pass), unconverted syngas leaves the reactor together with ammonia. Fortunately, a simple separation is possible. Ammonia is condensed and the unconverted syngas is recycled to the reactor. Various synthesis loop arrangements are used (Box 6.1).

Box 6.1   Synthesis Loop Designs

Figure B.6.1.1 shows some typical arrangements of synthesis loops. In all four cases the mixture leaving the reactor is separated in a condenser that removes the ammonia in the liquid state. The arrangements differ with respect to the points in the loop at which the feed gas is added and the location of ammonia condensation. If the feed gas is free of catalyst poisons, such as water, carbon dioxide, carbon monoxide, and perhaps some hydrogen sulfide, it can flow directly to the reactor (Figure B.6.1.1a). This represents the most favorable arrangement from a kinetic and an energy point of view: it results in the lowest ammonia content at the entrance of the reactor and the highest ammonia concentration for condensation.

If the feed gas is not sufficiently pure, the condensation stage is partly or entirely located between the feed gas supply and the reactor in order to remove the impurities before the gas enters the reactor. Figure B.6.1.1b shows the simplest form of such an arrangement. It has the disadvantage that the ammonia produced in the reactor has to be compressed together with the recycle gas. This disadvantage is avoided by using the scheme in Figure B.6.1.1c, in which the recycle gas is compressed after ammonia separation. In Figure B.6.1.1c, however, ammonia is condensed at lower pressure, resulting in a higher ammonia concentration in the gas stream, and thus at the reactor inlet. Another way to avoid compression of ammonia is to split the ammonia condensation as shown in Figure B.6.1.1d. The disadvantage of this scheme is the need for an additional condenser.

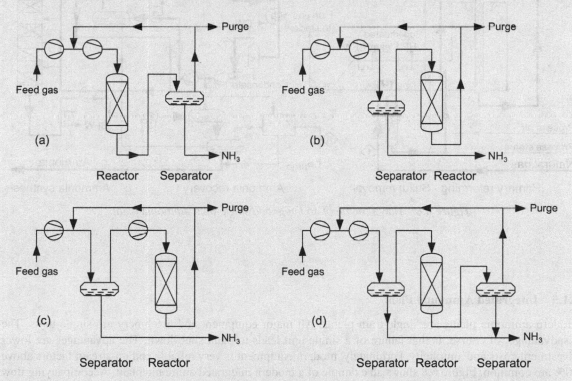

Figure B.6.1.1 *Typical ammonia synthesis loop arrangements: (a) pure and dry feed gas; (b) ammonia recovery after recycle compression; (c) ammonia recovery before recycle compression; (d) two stages of product condensation.*

QUESTIONS:

> *Why is scheme (a) in Figure B.6.1.1 the most favorable from a kinetic point of view (at equal condensing temperature)? In this scheme draw the heat exchangers that you feel should be placed.*
>
> *Analyze the four arrangements with respect to energy consumption.*

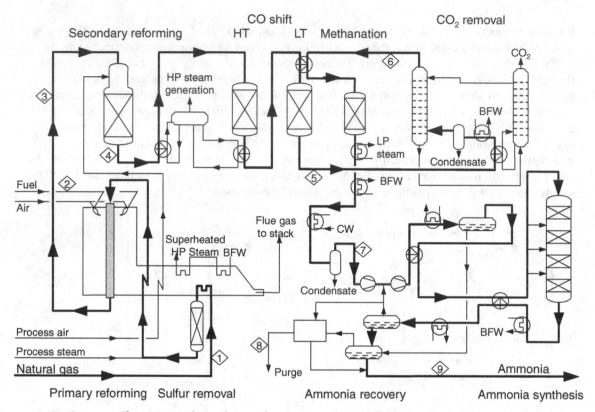

Figure 6.8 *Flow scheme of an integrated single-train ammonia plant.*

6.1.5   Integrated Ammonia Plant

Modern ammonia plants are single-train plants. All major equipment and machinery are single units. The disadvantage, of course, is that failure of a single unit leads to plant shut down. The advantages are lower investment costs and simplicity. Fortunately, modern equipment is very reliable and on-stream factors above 90% are common. Figure 6.8 shows an example of a modern integrated ammonia plant. Accompanying flow rates, stream compositions and process conditions are given in Table 6.2.

After desulfurization the natural gas is fed to the steam reformer (or primary reformer), where part of the feed is converted. Air containing the required amount of nitrogen for ammonia synthesis is introduced in an autothermal reformer (the secondary reformer). The catalyst in the secondary reformer is similar to that employed in the primary reformer.

Carbon oxides are highly poisonous to the iron-based ammonia synthesis catalyst. Therefore, the reformed gas is shifted and scrubbed for carbon dioxide removal (Section 5.4). The remaining carbon dioxide and carbon monoxide are removed by reaction with hydrogen to produce methane and water in a methanation step (Section 5.4). After compression, the syngas is converted to ammonia in the synthesis reactor.

Table 6.3 shows the approximate catalyst volumes of the reactors shown in Figure 6.8.

Table 6.2   *Compositions and flow rates of streams in a 1500 t/d integrated ammonia plant (Figure 6.8) [8].*

Stream no. Composition (mol%)[a]	1	2	3	4	5	6	7	8	9
CH_4	97.04	97.04	14.13	0.60	0.53	0.65	1.16	24.34	0.12
CO_2	1.94	1.94	10.11	7.38	18.14	0.01			
CO			9.91	13.53	0.33	0.40			
Ar				0.28	0.25	0.30	0.37	7.21	0.01
H_2			65.52	54.57	59.85	73.08	73.54	21.72	0.03
N_2	1.02	1.02	0.33	23.64	20.90	25.56	24.93	46.55	0.02
NH_3								0.18	99.82
Flow rate (kmol/h)									
Dry gas	1713.7	434.43	5296.4	8414.2	9520.7	7764.0	8041.1	319.9	3676.6
H_2O			3520.6	4086.1	2979.6	22.8	13.3	0.2	0.9
p (bar)	50	2.5	39.5	39.0	36.1	34.3	32.3	2.5	25.0
T (K)	298	298	1081	1249	502	323	308	311	293

[a]On a dry basis.
[b]Including 5.80 mol% higher hydrocarbons.

QUESTIONS:

Which reactions take place in the reactors in Figure 6.8 (give reaction equations)?
Why would carbon oxides be poisonous to the iron catalyst? Much more CO is converted in the HT shift reactor than in the LT shift reactor. Nevertheless, the catalyst volumes are approximately equal. What is the explanation?
At many places heat is exchanged. Make an analysis with respect to the production of HP and LP steam, and so on.
Referring to Box 6.1, which synthesis loop arrangement is applied in the flow scheme of Figure 6.8?
Do you expect that for the ammonia synthesis process further improvements are possible? Estimate the efficiency of the process.
List possible energy saving measures.

Table 6.3   *Approximate catalyst volumes used in the major catalytic stages in ammonia production for a 1500 t/d ammonia plant [9].*

Stage	Catalyst volume (m³)
Desulfurization	76
Primary reforming	25
Secondary reforming	50
HT shift	86
LT shift	92
Methanation	38
Ammonia synthesis	92

Table 6.4   *Specific energy requirements of ammonia processes [1, 5, 8].*

Process	GJ /t ammonia (LHV)[a]
Classical Haber–Bosch (coke)	80–90
Reformer pressure 5–10 bar (1953–1955)	47–53
Reformer pressure 30–35 bar (1965–1975)	33–42
Low-energy concepts (1975–1984)	27–33
State of the art (since 1991)	24–26
Theoretical minimum	20.9

[a]LHV = lower heating value.

Table 6.4 shows specific energy requirements for various generations of ammonia processes. The classical Haber–Bosch ammonia process was based on coal as feedstock. This process is rather inefficient for the production of hydrogen because of the low H/C ratio in coal. Also, gas purification is not very efficient in this process. The introduction of steam reformers operating on a natural gas feed much improved the energy efficiency because of the high H/C ratio in the feed and the higher purity of the raw syngas. During the 1960s high-pressure reformers came on stream, reducing the syngas compression power required (and investment costs). At that time also low-temperature shift and methanation catalysts were introduced, resulting in a significant improvement in gas quality, and thus higher ammonia yields per amount of natural gas processed.

The most important improvement of that time, however, was the use of centrifugal instead of reciprocating compressors. Centrifugal compressors have many advantages over reciprocating compressors, such as low investment and maintenance costs, high reliability, and low space requirement. Centrifugal compressors are very efficient at high flow rates, making them well suited for use in large capacity single-train ammonia plants.

Low-energy concepts resulted in better heat integration; for example, recovery of most of the reaction heat by preheating boiler feed water and raising high-pressure steam. With state-of-the-art technology, using different ammonia converter designs, even lower energy consumptions are possible.

6.1.6   Hydrogen Recovery

Off-gases from units such as fluid catalytic crackers and hydrotreaters and purge gas in, for example, ammonia plants contain considerable amounts of hydrogen. It is often economical to recover this hydrogen instead of using it as fuel. Hydrogen is a valuable commodity and is becoming more so because of the increasing need for hydroprocessing of feeds and products of refinery processes.

Several methods are available for gas purification and recovery. The most important are cryogenic distillation, absorption, adsorption, and membrane separation. Cryogenic distillation is, for example, used for the separation of the product stream from naphtha and ethane crackers (Chapter 4). This process requires a large amount of energy, because very low temperatures are required (circa 120 K). Absorption is discussed in Chapter 5 for the removal of hydrogen sulfide and carbon dioxide from gas streams, but it is not suitable for obtaining hydrogen at the required purity (80–95 vol%). At present, adsorption and membrane separation are the preferred technologies for the recovery of hydrogen from purge gas from the ammonia synthesis loop.

Table 6.5 *Advantages and disadvantages of membranes.*

Advantages	Disadvantages
Low energy consumption (no phase transfer).	Fouling.
Mild conditions.	Short lifetime.
Low pressure drop (compared to "classical" filtration).	Often low selectivity.
No additional phase required.	No economies of scale (scale-up factor ≈ 1).
Continuous separation.	
Easy operation.	
No moving parts (except recycle compressor in gas separation applications).	

6.1.6.1 Adsorption

Gas phase adsorption on solids is widely used in industry for air drying, nitrogen production, hydrogen purification, and so on. It is based on the selective adsorption of one or more components from a gas stream. Hydrogen is practically not adsorbed, and is thus easy to purify by this method. Adsorption processes for recovery of hydrogen are usually carried out in a transient mode and consist of two steps:

1. impurities are adsorbed yielding a purified gas stream containing the hydrogen;
2. the adsorbent bed is regenerated either by raising the temperature ("Temperature Swing Adsorption", TSA) or by reducing the pressure ("Pressure Swing Adsorption", PSA) (Box 6.2).

6.1.6.2 Membrane Separation

A membrane is a selective barrier that allows separation of compounds on the basis of molecular properties such as molecular size and strength of adsorption on or solubility in the membrane material. Although osmosis through membranes has been known since the 1830s, and has been applied for liquid separation since the 1970s, the application of membranes for the separation of gas mixtures is more recent. Examples of gas separations using membranes are hydrogen recovery from purge streams in hydrotreating and ammonia synthesis, and removal of carbon dioxide from natural gas.

Membranes usually are prepared from polymers, although inorganic membranes are receiving increasing attention. The most important practical properties required for membranes are high selectivity, high permeability, high mechanical stability, thermal stability, and chemical resistance. For inorganic membranes, the selectivity (degree of separation) depends on the adsorption properties and the pore size of the surface layer that is in contact with the fluid (gas or liquid). For polymer membranes, solubility plays a major role. The permeability (flow rate through the membrane) is also determined by the membrane thickness. Table 6.5 summarizes advantages and disadvantages of membranes in separation processes.

The performance of most pressure-driven membranes is hindered by progressive fouling of the membrane. Solutions are found either by changing the surface characteristics of the membrane or by hydrodynamic changes aimed at reducing the concentration of materials that can cause fouling at the surface of the membrane.

QUESTIONS:

Why would membranes not exhibit favorable economies of scale?
Membranes are especially used in the purification of purge streams. Why?

Box 6.2 Pressure Swing Adsorption (PSA)

PSA processes were developed as replacement for adsorption processes involving thermal regeneration. Figure B.6.2.1 shows the principle of PSA for a hydrogen purification process. Adsorption is carried out at elevated pressure (typically 10−40 bar), while regeneration is carried out at low, often atmospheric, pressure. In industry several units (at least two, but usually four or more) are placed in parallel.

High-pressure gas is fed to adsorber 1. Here the adsorbable components are adsorbed. The purified hydrogen gas exits the adsorber. When an adsorber is taken out of the adsorption stage (in this case adsorber 4), it expands the gas (mainly hydrogen, which is contained in the voids of the adsorber packing) into two vessels at lower pressure. The first is the vessel in which adsorption is to take place in the next step, but which is still at low pressure (adsorber 2). The remaining available pressure is expanded in the remaining vessel (3), which is in its regeneration (purging) mode; the feed impurities are desorbed from the bed.

Since the gas from adsorber 4 alone cannot bring adsorber 2 to the required pressure, some of the purified hydrogen is added in order to do so. Now, adsorption can take place in adsorber 2, while adsorber 1 undergoes depressurization, and so on. Figure B.6.2.2 shows the cycle sequence of the PSA system. In an elegant way, the pressures of the four adsorbers are varied such that a quasi-continuous process is realized.

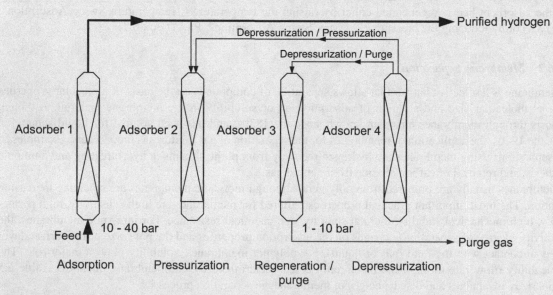

Figure B.6.2.1 *Hydrogen purification using PSA; one step shown, not all piping shown.*

QUESTIONS:

> *Why is the PSA process not carried out in a configuration of two parallel adsorption beds (similar to absorption processes, for instance H_2S separation by amine absorption, Figure 5.24)?*

What would be the pressure of columns 4 and 2 in Figure B.6.2.1 after the depressurization/pressurization? Draw a schematic of the next step after the situation in the figure.

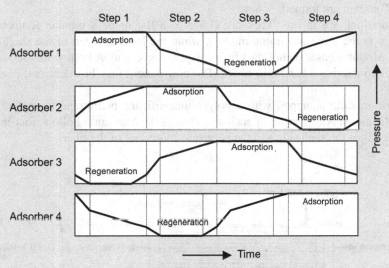

Figure B.6.2.2 *Cycle sequence in PSA with four adsorbers [10].*

The central part of a membrane plant is the module: the technical arrangement of the membranes. An example is the Monsanto module (Box 6.3), which is used in gas separation applications.

Figure 6.9 shows a flow scheme of a membrane-based process for the recovery of hydrogen from the purge stream in an ammonia plant. The hydrogen permeates preferentially through the membrane. Usually, a two-stage process such as shown in Figure 6.9 is employed. The two membrane units each consist of several modules. After compression the permeate streams combine with the main recycle stream and enter the ammonia synthesis reactor. The retentate stream is sent to the reformer.

QUESTIONS:

Why are two sets of membrane modules used?
What is the purpose of the scrubber?
Every process consumes energy. Analyze this scheme in terms of energy consumption.

6.1.7 Production of Urea

Urea was first synthesized in 1828 from ammonia and cyanic acid. This was the first synthesis of an organic compound from an inorganic compound [12]. Commercially, urea is produced by reacting ammonia and carbon dioxide. The latter is a by-product of the ammonia production process.

Box 6.3   Membrane Modules

Membrane modules are produced on an industrial scale. Most modules are so-called three-end modules (Figure B.6.3.1). The feed stream enters the module. Because of the action of the membrane, the composition inside the module will change. A "permeate" stream, which passes through the membrane, and a "retentate" stream are formed.

Many configurations are in use. They are either of a flat or of a tubular geometry. An example of the former is the plate-and-frame module, while the latter is often of the shell-and-tube heat exchanger type. Modules based on hollow fibers are quite successful. A hollowfiber module consists of a pressure vessel containing a bundle of fibers. The flow pattern may be cocurrent, countercurrent or cross-flow.

A hollow fiber module equipped with a polysulfone-silicone membrane is applied in many gas permeation applications. An example of such a module is the Monsanto hollow fiber module (Figure B.6.3.1). The feed is at the shell side and the permeate comes out of the fibers.

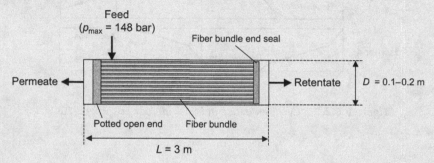

Figure B.6.3.1   *Configuration of the Monsanto hollow-fiber module and some characteristics. Adapted from Ref. [11].*

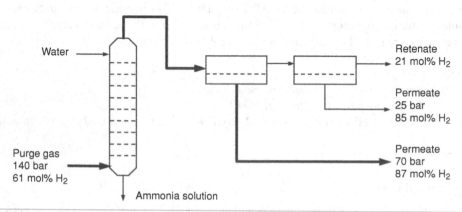

Figure 6.9   *Flow scheme of hydrogen recovery from purge gas in ammonia synthesis [11].*

QUESTIONS:

> *Because of environmental concerns, many research groups are attempting to develop processes based on the conversion of CO_2. What could be attractive processes? For every practical process, the availability of feedstocks is crucial. List processes that could provide the CO_2 feedstock for such a process.*

6.1.7.1 Reactions

The formation of urea occurs through two uncatalyzed equilibrium reactions. The first reaction, the formation of ammonium carbamate (liquid at reaction conditions, solid at standard temperature and pressure) is exothermic while the second reaction, the dehydration of the carbamate, is endothermic. Since more heat is produced in the first reaction than is consumed in the second reaction, the overall reaction is exothermic.

$$2\,NH_3 \;+\; CO_2 \;\overset{fast}{\rightleftharpoons}\; H_2NCONH_4 \qquad\qquad \Delta_r H_{298} = -\,159\;kJ/mol$$

ammonium carbamate

$$(6.2)$$

$$H_2NCONH_4 \;\overset{slow}{\rightleftharpoons}\; H_2NCNH_2 \;+\; H_2O \qquad\qquad \Delta_r H_{298} = 31.4\;kJ/mol$$

urea

$$(6.3)$$

Thermodynamically, both the carbon dioxide conversion and the urea yield as a function of temperature go through a maximum, which usually lies between 450 and 480 K at practical conditions. Increasing the NH_3/CO_2 ratio leads to a higher carbon dioxide conversion. The yield on urea has a maximum at an NH_3/CO_2 ratio of somewhat above the stoichiometric ratio. A large excess of ammonia results in a reduced yield. In current practice, an NH_3/CO_2 ratio of 3–5 mol/mol is used.

QUESTIONS:

> *What is the influence of temperature and pressure on the two reactions?*
> *Why do the CO_2 conversion and the urea concentration go through a maximum as a function of temperature?*
> *What is the effect of water on the CO_2 conversion and urea yield?*

From a kinetic viewpoint, both carbamate formation and urea formation proceed faster with increasing temperature. The carbamate to urea conversion is much slower than carbamate formation; temperatures of over 420 K are required for a sufficiently high reaction rate. At these temperatures, pressures of over 130 bar are required to prevent dissociation of carbamate into ammonia and carbon dioxide, that is, the reverse of reaction 6.2. At economically and technically feasible conditions, the conversion of carbon dioxide to urea is only between 50 and 75%.

6.1.7.2 Challenges in the Design of Urea Processes

Because of the equilibrium, the reactor effluent will always contain a considerable amount of carbamate (Figure 6.10). This carbamate needs to be separated from the urea solution.

Figure 6.10   *Schematic of urea production.*

To be able to separate the carbamate, it is decomposed back into carbon dioxide and ammonia (the reverse of reaction 6.2) to separate it from the urea solution. The key challenges in the design of a urea plant are efficient handling of the unconverted carbon dioxide and ammonia and minimizing the water concentration.

The very first commercial urea processes were *once-through* processes (Figure 6.11), in which carbamate decomposition was effected by reducing the pressure of the reactor effluent by flashing through an expansion valve. The unconverted ammonia was then neutralized with, for example, nitric acid, to produce ammonium nitrate. However, a major disadvantage of this type of plant is that the economics of the plant are not only dependent on the market of the main product, that is, urea, but also on that of the coproducts [13, 14].

Once-through processes were soon replaced by *total recycle* processes, in which essentially all of the unconverted ammonia and carbon dioxide are recycled to the urea reactor. Recycling of the NH_3/CO_2 mixture is not as straightforward as it may seem. The reason is that recompression to reaction pressure is required, which would lead to the recombination of ammonia and carbon dioxide to form liquid carbamate droplets (or solid carbamate crystals at lower temperature) that can damage compressor components.

There are many different approaches to circumvent this problem, which is illustrated by the remarkably large number of urea processes that have been developed [13, 15]. In the following, the so-called stripping processes will be discussed.

6.1.7.3   Stripping Processes

In conventional urea processes, decomposition of carbamate is accomplished by a combination of pressure reduction and heating as described in the previous section. However, lowering of the carbon dioxide or ammonia concentration in the solution leaving the reactor also leads to decomposition of the carbamate. This can be accomplished by stripping the mixture from the urea reactor with either carbon dioxide (Stamicarbon and Toyo Engineering) (Figure 6.12, Box 6.4) or ammonia (Snamprogetti) [12, 14, 16].

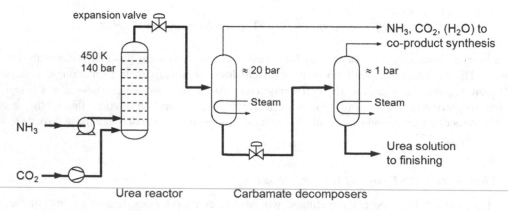

Figure 6.11   *Once-through process for urea production.*

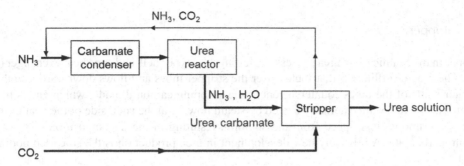

Figure 6.12   *Simplified block diagram of a CO_2 stripping process for urea production.*

Carbon dioxide contacts the effluent from the reactor in a stripper which operates at the same pressure as the reactor, while heat is supplied at the same time. Unconverted ammonia is stripped from the solution, resulting in the decomposition of most of the ammonium carbamate. The gases leaving the stripper are sent to the carbamate condenser, in which part of the gas is condensed to form ammonium carbamate, and this mixture is fed to the reactor bottom.

The major advantage of stripping processes is that the unconverted reactants are mainly recycled via the gas phase; this eliminates a large water recycle to the reactor section, which is typical of the older carbamate-solution-recycle processes. Therefore, stripping processes operate at lower NH_3/CO_2 ratios than conventional total recycle processes, typically between 3 and 3.7 mol/mol.

QUESTIONS:

> *Explain why stripping with ammonia or carbon dioxide works.*
> *Why is a large water recycle to the reaction section, typical of carbamate-solution-recycle processes, a disadvantage?*
> *The carbamate condenser, the urea reactor, and the stripper all operate at a pressure of 140 bar. What is the reason?*
> *What is the function of the demister at the top of the stripper (Box 6.4)?*

6.1.7.4   *Corrosion*

Solutions containing ammonium carbamate are very corrosive. Some of the measures to limit corrosion of the construction materials are:

- proper selection of construction materials, at least stainless steel [17];
- addition of a small amount of air to the carbon dioxide feed – by reaction of oxygen with the stainless steel, the protective metal oxide layer is kept intact;
- minimal exposure of materials to the corrosive carbamate solution – stripping processes are preferred over carbamate-solution-recycle processes from this point of view.

Box 6.4   Stripper

The stripper in the Stamicarbon urea process is a falling-film type shell-and-tube heat exchanger (Figure B.6.4.1). The reactor effluent is distributed over the stripper tubes and flows downward as a thin film on the inner walls of the tubes countercurrent to the stripping carbon dioxide, which enters the tubes at the bottom. The tubes are externally heated by steam flowing at the shell side of the heat exchanger. Unconverted ammonia is stripped from the solution, resulting in the decomposition of most of the ammonium carbamate. A relatively new development in urea production is the so-called pool reactor (Box 6.5).

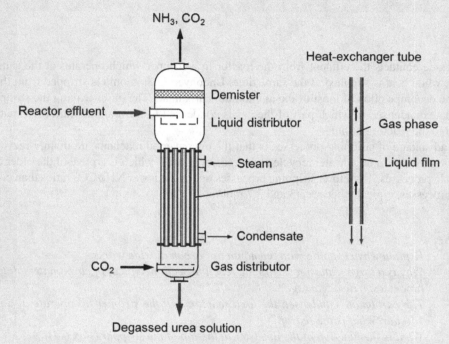

Figure B.6.4.1   *Schematic of the stripper used in the Stamicarbon CO₂ stripping process.*

6.1.7.5   *Biuret Formation*

The most important side reaction in the urea process is the formation of biuret:

$$2\ H_2NCNH_2 \underset{\text{slow}}{\rightleftharpoons} H_2NCNCNH_2 + NH_3 \quad \text{(slightly endothermic)}$$

(6.4)

Box 6.5 Pool Reactor

The development by Stamicarbon of the so-called "pool reactor" in the 1990s [12, 14], has greatly simplified the reactor section and, thereby, reduced the investment costs. The pool reactor combines the condensation of carbamate from ammonia and carbon dioxide with the dehydration of carbamate into urea in one vessel (Figure B.6.5.1) and can thus be considered an example of process intensification.

The gases leaving the stripper and the ammonia feed are introduced into the pool reactor. The continuous liquid phase is agitated by the gases from the stripper entering through a gas divider, which enhances heat and mass transfer compared to a conventional falling-film condenser. An advantage of pool condensation specific for urea production is the long residence time of the liquid phase, so that besides condensation conversion of the carbamate to urea also takes place. The heat of condensation is partly used to supply the heat for this endothermic urea forming reaction and partly for raising steam in the tube bundle.

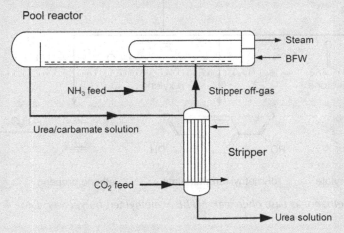

Figure B.6.5.1 *Pool reactor synthesis section for the Stamicarbon CO_2 stripping process.*

Even at very low concentrations, biuret is detrimental to crops, so its formation should be minimized [17]. In the urea reactor itself this is not a problem because of the high ammonia concentration, which shifts the reaction to the left. However, in downstream processing, ammonia is removed from the urea solution, thereby creating a driving force for biuret formation. To minimize biuret formation, it is essential to keep the residence times of urea-containing solutions at high temperature as short as possible.

6.2 Methanol

6.2.1 Background Information

The major application of methanol is in the chemical industry, where it is used as a solvent or as an intermediate. It is also increasingly used in the energy sector. Figure 6.13 shows a survey of reactions involving methanol.

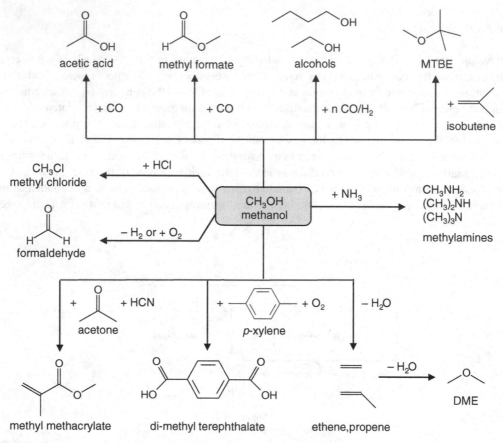

Figure 6.13   *Methanol as base chemical; MTBE = methyl* tert-*butyl ether; DME = dimethyl ether.*

Methanol synthesis was the second large-scale process involving catalysis at high pressure and temperature (BASF, 1923). The same team that developed the ammonia synthesis process also developed a commercial process for the production of methanol based on syngas.

6.2.2   Reactions, Thermodynamics, and Catalysts

The main reactions for the formation of methanol from syngas are:

$$CO + 2\,H_2 \rightleftarrows CH_3OH \qquad \Delta_r H_{298} = -90.8 \text{ kJ/mol} \qquad (6.5)$$

$$CO_2 + 3\,H_2 \rightleftarrows CH_3OH + H_2O \qquad \Delta_r H_{298} = -49.6 \text{ kJ/mol} \qquad (6.6)$$

The two methanol forming reactions are coupled by the water–gas shift reaction, which has been discussed in Chapter 5:

$$CO + H_2O \rightleftarrows CO_2 + H_2 \qquad \Delta_r H_{298} = -41 \text{ kJ/mol} \qquad (6.7)$$

Table 6.6 CO and CO_2 equilibrium conversion data [a] [18].

Temp. (K)	CO conversion Pressure (bar)			CO_2 conversion Pressure (bar)		
	50	100	300	50	100	300
525	0.524	0.769	0.951	0.035	0.052	0.189
575	0.174	0.440	0.825	0.064	0.081	0.187
625	0.027	0.145	0.600	0.100	0.127	0.223
675	0.015	0.017	0.310	0.168	0.186	0.260

[a]Feed composition: 15% CO, 8% CO_2, 74% H_2, and 3% CH_4

Table 6.6 shows typical equilibrium data for the methanol forming reactions. Clearly, the temperature has to be low. Note that the carbon dioxide conversion increases with temperature as a result of the reverse water–gas shift reaction.

The original catalysts were only active at high temperature. Therefore, the pressure had to be very high (250–350 bar) to reach acceptable conversions. Until the end of the 1960s basically these original catalysts were used in what are now call classical methanol processes. More active catalysts (based on copper) were known, but these were not resistant to impurities in the feed such as sulfur. In the late 1960s, the ability to produce sulfur-free syngas allowed the use of these very active catalysts and this has led to a new generation of plants, the "low-pressure plants".

QUESTION:
> *Which development(s) made it possible to produce sulfur-free syngas?*

Figure 6.14 shows the equilibrium conversion of carbon monoxide to methanol as a function of temperature and pressure. The typical temperature ranges for the classical and modern methanol processes are indicated

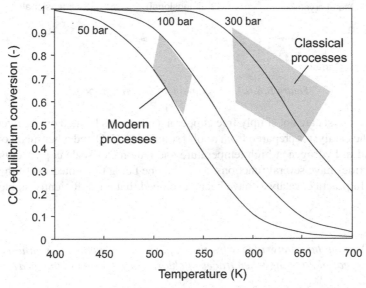

Figure 6.14 *Equilibrium CO conversion to methanol (feed $H_2/CO = 2$ mol/mol).*

in the figure. The figure illustrates that the development of catalysts that are active at lower temperature made it possible to operate at lower pressure (50–100 bar), while maintaining the same conversion as in the classical process. Temperature is critical. A low temperature is favorable from a thermodynamic point of view. Moreover, the high-activity catalysts are sensitive to "sintering" (Box 6.6), which increases progressively with temperature.

Catalyst deactivation will always occur. The usual practice for restoring activity, increasing temperature, is normally employed to maintain the rate of reaction, but pressure increase is also employed. The temperature should not exceed 570 K because then unacceptable sintering (Box 6.6) of the catalyst will occur [19].

Compared to ammonia synthesis, catalyst development for methanol synthesis was more difficult because, besides activity, *selectivity* is crucial. It is not surprising that in the hydrogenation of carbon monoxide other products, such as higher alcohols and hydrocarbons, can be formed. Thermodynamics show that this is certainly possible. Figure 6.15 shows thermodynamic data for the formation of methanol and some possible by-products resulting from the reaction of carbon monoxide with hydrogen.

Clearly, methanol is thermodynamically less stable and, in that sense, less likely to be formed from carbon monoxide with hydrogen than other possible products, such as methane, which can be formed by

Box 6.6   Sintering

Metal catalysts often consist of metal crystallites deposited on a porous carrier. Sintering of these catalysts occurs by the growth of the metal crystallites, resulting in the loss of active surface area (Figure B.6.6.1). Sintering is usually irreversible. Sintering phenomena in most cases are a result of too high temperatures, that is, they lead to thermal deactivation. However, the chemical environment can also play a role. For instance, the presence of steam or traces of chlorine often accelerates sintering.

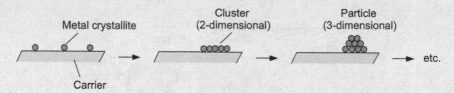

Figure B.6.6.1   *Schematic of sintering process.*

Industrial copper catalysts are not simply free copper nanoparticles located on the internal surface of a porous support. The catalyst is prepared from a solid solution of CuO and other metal oxides. This solid solution is reduced in hydrogen at high temperature and nanometer-scale copper crystals are formed. The other metal oxides have several functions, a crucial one being the protection of the copper crystals against sintering. In practice, catalysts have been developed that are sufficiently stable to be active for several years.

QUESTION:

> *Do you have any idea why the copper-based methanol synthesis catalyst shows increased sintering when traces of chlorine are present in the feed?*

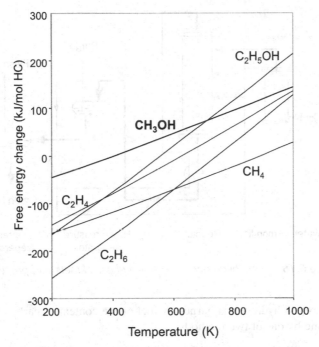

Figure 6.15 *Products from CO and H$_2$; standard free energy of formation.*

the methanation reaction (Section 5.4). Therefore, the catalyst needs to be very selective. The selectivity of modern copper-based catalysts is over 99%, which is remarkable considering the large number of possible by-products [19].

QUESTION:

> *Why can in methanol synthesis, in theory a poor catalyst (having a low selectivity), lead to a "run-away"?*

6.2.3 Synthesis Gas for Methanol Production

Nowadays, most syngas for methanol production is produced by steam reforming of natural gas. The production of syngas has been dealt with in Chapter 5. The ideal syngas for methanol production has a H$_2$/CO ratio of about 2 mol/mol. A small amount of carbon dioxide (about 5%) increases the catalyst activity. A H$_2$/CO ratio lower than 2 mol/mol leads to increased by-product formation (higher alcohols, etc.), a higher ratio results in a less efficient plant due to the excess hydrogen present in the syngas, which has to be purged.

The composition of syngas depends on the feedstock used. When naphtha is the raw material, the stoichiometry is approximately right. When methane is used, however, hydrogen is in excess.

QUESTIONS:

> *Why, despite the better stoichiometry of naphtha reforming, would natural gas (consisting mainly of methane) still be the preferred feedstock for methanol production?*
>
> *In the purification of syngas for methanol, would a methanation reactor be included? And a shift reactor?*

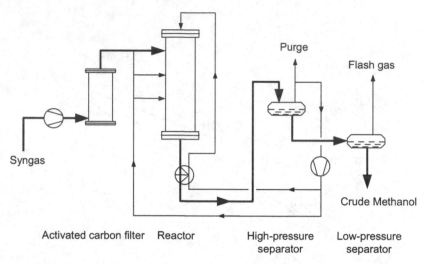

Activated carbon filter   Reactor   High-pressure   Low-pressure
                                       separator   separator

Figure 6.16   *Simplified flow scheme of a classical methanol process.*

In practice, either the excess hydrogen is burned as fuel or the content of carbon oxides in the syngas is increased. This can be done by one of two methods:

1. When available, carbon dioxide addition to the process is a simple and effective way to balance the hydrogen and carbon oxides content of the syngas. Carbon dioxide addition can be implemented by injecting it either in the reformer feed stream or in the raw syngas. In both cases the stoichiometric ratio for methanol synthesis is achieved, although the compositions will be somewhat different.

2. Installing an oxygen-fired secondary reformer downstream of the steam reformer, as discussed in Chapter 5, or using only autothermal reforming. In the latter case, the syngas is too rich in carbon oxides ($H_2/CO \approx 1.7-1.8$ mol/mol), so adjustment of the gas is required. This can be done by removing carbon dioxide or by recovering hydrogen from the synthesis loop purge gas and recycling it to the reactor feed.

QUESTION:

> *Why will the syngas compositions be different for the injection of carbon dioxide in the feed and in the syngas?*

6.2.4   Methanol Synthesis

The first industrial plants were based on a catalyst that was fairly resistant to impurities but not very active nor selective. To achieve a reasonable conversion to methanol, these plants were operated at high pressure. Figure 6.16 shows a simplified flow scheme of such a high-pressure plant.

The heart of the process is the reactor with a recirculation loop according to the same principle as in ammonia synthesis. All methanol processes still use this principle. The crude methanol is distilled to separate the methanol from water and impurities. The conversion per pass is low, which facilitates good temperature control.

Nevertheless, special reactor designs are required. Quench reactors and cooled multitubular reactors are most commonly applied.

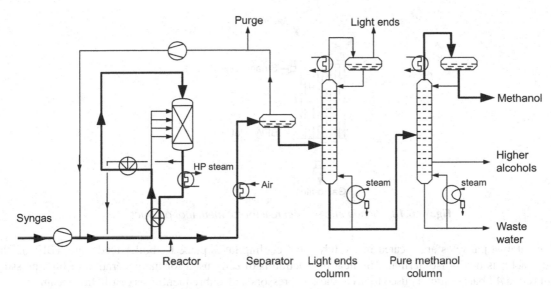

Figure 6.17 *Flow scheme of the ICI low-pressure methanol process.*

The high operating pressure (circa 300 bar) of this classical methanol process results in high investment costs and high syngas compression costs. In addition, large amounts of by-products are formed due to the low selectivity of the catalyst. These by-products include ethers, hydrocarbons, and higher alcohols.

A major breakthrough came in the late 1960s with the development of low-pressure processes using the more active and selective copper-based catalysts. All modern processes are low-pressure processes with plant capacities ranging from 150 to 6000 t/d, although plants using remote natural gas may have a capacity as large as 10 000 t/d. The plants differ mainly in reactor design, and, interrelated with this, in the way the heat of reaction is removed.

In the ICI process [20], an adiabatic reactor is used with a single catalyst bed. The reaction is quenched by adding cold reactant gas at different heights in the catalyst bed. The temperature in the bed has a saw-tooth profile (compare Figure 6.4). Figure 6.17 shows a flow scheme of the ICI process.

Fresh syngas is compressed and mixed with recycle gas. The mixture is heated with reactor effluent. Subsequently, about 40% of the stream is sent to the reactor after undergoing supplementary preheating also by the reactor effluent. The rest is used as quench gas. The reactor effluent is cooled by heat exchange with the feed and with water required for high-pressure steam generation, and then by passage through an air-cooled exchanger in which the methanol and water are condensed. Gas/liquid separation is carried out in a vessel under pressure. The gas is recycled after purging a small part to keep the level of inerts (mainly argon) in the loop within limits. The crude methanol is purified in two columns. The first column removes gases and other light impurities, the second separates methanol from heavy alcohols (side stream) and water.

QUESTIONS:

> *What is the source of argon in the syngas?*
> *The crude methanol is purified in two distillation columns. Are other configurations possible apart from the one given in Figure 6.17? Which one do you prefer and why?*

The Lurgi process is very similar to the ICI process. The most important difference is the reactor. In the Lurgi process, a cooled tubular reactor is used (Figure 6.18) [21, 22].

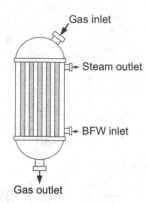

Figure 6.18   *Schematic of Lurgi reactor for methanol production.*

The catalyst particles are located in the tubes and cooling takes place by boiler feed water (BFW). The Lurgi reactor is nearly isothermal. The heat of reaction is directly used for the generation of high-pressure steam (circa 40 bar), which is used to drive the compressors and, subsequently, as distillation steam.

In the Haldor Topsøe process (Figure 6.19), several adiabatic reactors are used, arranged in series [23]. Intermediate coolers remove the heat of reaction. The syngas flows radially through the catalyst beds, which results in reduced pressure drop compared to axial flow. Crude methanol purification is the same as in the other processes.

The choice of methanol process design depends on various factors, such as type of feedstock, required capacity, energy situation, local situation, and so on. Box 6.7 shows a process for methanol production based on a slurry reactor, while Box 6.8 shows and alternative conversion route.

QUESTIONS:

> *Compare the reactor designs shown in Figures 6.17, 6.18, and 6.19. Give advantages and disadvantages. Which reactor configuration has the smallest total catalyst volume for the same methanol production and which the largest? Explain.*

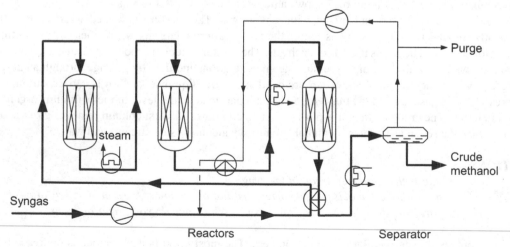

Figure 6.19   *Flow scheme of the reaction section of the Haldor Topsøe methanol process.*

Box 6.7 Methanol Process Based on Slurry Reactor

A relatively new development in methanol synthesis is a *three-phase* process invented by Chem Systems and developed by Air Products [24, 25] (Figure B.6.7.1). This process uses a slurry reactor, in which the solid catalyst is suspended in an inert hydrocarbon liquid, while the syngas passes through the bed in the form of bubbles. The catalyst remains in the reactor and the hydrocarbon liquid, after separation from the gas phase, is recycled via a heat exchanger.

The main advantage of this process is that the presence of the inert hydrocarbon limits the temperature rise, as it absorbs the heat liberated by the reaction, while it also keeps the temperature profile in the reactor uniform. Therefore, a higher single-pass conversion can be achieved than in conventional processes, reducing the syngas compression costs.

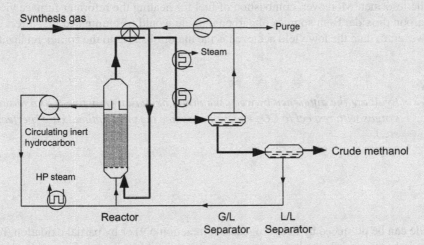

Figure B.6.7.1 *Three-phase methanol process.*

QUESTION:

It has been found that the slurry process tolerates feedstocks with high CO content, such as those produced in coal gasification without carrying out a water–gas shift. Explain this observation.

6.2.5 Production of Formaldehyde

One of the largest applications of methanol, accounting for about 35% of methanol consumption, is the synthesis of formaldehyde, which is based on one of the following reactions:

$$CH_3OH \rightleftarrows CH_2O + H_2 \qquad \Delta_r H_{298} = 85 \text{ kJ/mol} \qquad (6.9)$$

$$CH_3OH + \tfrac{1}{2} O_2 \rightarrow CH_2O + H_2O \qquad \Delta_r H_{298} = -158 \text{ kJ/mol} \qquad (6.10)$$

Box 6.8   The Direct Conversion of Methane to Methanol

The direct conversion of methane to methanol has been the subject of many academic research efforts during the past decades:

$$CH_4 + 0.5\ O_2 \rightleftarrows CH_3OH \qquad \Delta_r H_{298} = -126\ kJ/mol \qquad (6.8)$$

Such a process would be a spectacular event. In principle, the efficiency would be increased enormously, compared to the conventional process of steam reforming followed by methanol synthesis. In addition, a real contribution to the reduction of the greenhouse effect would be realized. This is linked to the formation of carbon dioxide, a major greenhouse gas, from methane in the reformer feed and fuel [26]: during steam reforming of hydrocarbons (Chapter 5) carbon dioxide is formed by the water–gas shift reaction in the reformer. Moreover, combustion of fuel for heating the reformer furnace yields a large amount of carbon dioxide. Both sources of carbon dioxide would be eliminated in a direct conversion process. However, to date the low yield achieved is the major obstacle to the commercialization of this route [27].

QUESTION:

    Evaluate the difference between the direct oxidation (reaction 6.8) and the route via syngas with respect to CO$_2$ emission. Assume the most optimal case (perfect catalysts, etc.).

So, formaldehyde can be produced by dehydrogenation (reaction 6.9) or by partial oxidation (reaction 6.10). The former is endothermic, while the latter is exothermic. As should be expected, several side reactions take place.

In industrial practice dehydrogenation and partial oxidation are often carried out over a silver-based catalyst in a single reactor in which the heat produced by the exothermic partial oxidation reaction supplies the heat for the endothermic dehydrogenation reaction [28–30]. Figure 6.20 shows a flow scheme of such a process.

Fresh and recycle methanol is evaporated in a vaporizer into which air is sparged to generate a feed mixture that is outside the explosion limits. The resulting vapor is combined with steam and heated to the reaction temperature. Reaction takes place in a shallow bed of catalyst, after which the product gases are immediately cooled in a waste heat boiler, thereby generating steam. After further cooling the gases are fed to an absorber. Subsequently, distillation yields a 40–55 wt% formaldehyde solution in water. Methanol is recycled to the vaporizer.

The overall reaction is highly exothermic, despite some heat consumption by the endothermic dehydrogenation reaction. Because it occurs at essentially adiabatic conditions, good temperature control is critical. This is mainly achieved by the use of excess methanol, which is recycled, and by the addition of steam to the feed.

A process of about equal industrial importance is based on only the partial oxidation reaction over a metal oxide catalyst [28–30]. In contrast to the silver-catalyzed process, this process operates with excess air. Therefore, it has the advantage that the distillation column for methanol recovery can be omitted, because in this case methanol conversion is over 99%.

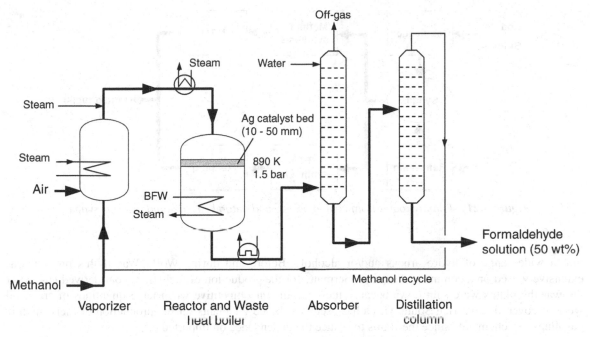

Figure 6.20   *Flow scheme of typical formaldehyde process.*

QUESTIONS:

Why would the catalyst bed for the combined dehydrogenation/oxidation process be so shallow? What would be the composition of the off-gas? (What are possible side reactions?) What percentage of methanol would have to be converted by the partial oxidation reaction for autothermic operation? Would a fully balanced process (in terms of heat consumption) be possible in practice? (Hint: consider differences in the expected rates of the dehydrogenation and the oxidation reactions.)

According to the flow scheme of Figure 6.20, in the methanol/formaldehyde distillation methanol goes over the top. Compare the boiling points of methanol and formaldehyde: they suggest the opposite! Explain this counterintuitive behavior of the distillation column. What type of reactor would you choose for the partial oxidation process and why? What would be advantages and disadvantages of this process compared to the process described above?

6.3   Synthetic Fuels and Fuel Additives

Processes for the conversion of syngas derived from coal or natural gas into liquid fuels such as gasoline and diesel (Figure 6.21) have been considered on and off for many years, usually as an alternative for oil-based fuels. These processes are known as coal-to-liquids (CTL) and gas-to-liquids (GTL) processes. More recently, biomass-to-liquids (BTL) processes have been receiving more and more attention.

The best known of such processes is the Fischer–Tropsch (FT) process, named after F. Fischer and H. Tropsch, the German coal researchers who discovered, in 1923, that syngas can be converted catalytically

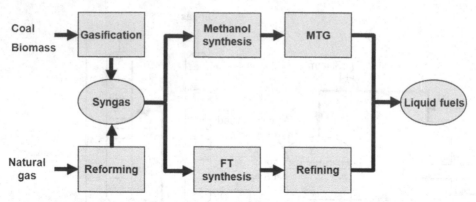

Figure 6.21   *Primary routes from coal, biomass, and natural gas to liquid fuels via syngas.*

into a wide range of hydrocarbons and/or alcohols. Before and during World War II this process was extensively used on a commercial scale in Germany for the production of fuels from coal-derived gas. After the war the plants were shut down because they became uncompetitive when large quantities of crude oil were discovered. Coal-rich South Africa has used coal-based FT plants for the production of fuels (mainly gasoline) and chemicals since the 1950s to reduce the dependence on imported oil.

A less known route is the conversion of syngas to gasoline via methanol. This methanol-to-gasoline (MTG) process [31–35] was developed by Mobil (now ExxonMobil) during the 1970s and 1980s in response to the critical energy situation in the Western world, which triggered the search for non-oil-based processes for fuel production. A natural-gas-based MTG plant was in operation in New Zealand from 1985 to 1997. Currently, the plant only produces methanol.

Until recently, these synthetic fuels have never been able to compete economically with oil-based fuels, unless in special cases like those mentioned above. However, in recent years these processes have come into the picture again, for instance as a means to convert natural gas from remote gas fields into liquid fuels. The transportation of the gas to possible consumer markets, either by pipeline or as liquefied natural gas (LNG) in special tankers, is costly and logistically difficult. An interesting option then may be to convert this gas into more readily transportable liquids, such as the bulk chemicals ammonia and methanol, or liquid fuels. The latter have a much larger market and are, therefore, more attractive from the viewpoint of economy of scale.

6.3.1   Fischer–Tropsch Process

6.3.1.1   *Reactions and Catalysts*

Although the chemistry of the Fischer–Tropsch synthesis is rather complex, the fundamental aspects can be represented by the generalized stoichiometric relationships in Table 6.7. Characteristic of FT reactions is their high exothermicity; the formation of one mole of —CH_2— is accompanied by a heat release of 145 kJ, which is an order of magnitude higher than that of typical catalytic reactions in oil refining [36].

The hydrocarbons are formed by a chain growth process (Box 6.9). The selective production of hydrocarbons other than methane has not yet been achieved; a mixture of hydrocarbons with various chain lengths and thus molecular weights is always formed ([37–39], Box 6.9). Also, from Figure 6.15 in Section 6.2 it is clear that thermodynamically both hydrocarbons and alcohols can be formed. For these reasons, the choice of

Table 6.7 *Major overall reactions in Fischer–Tropsch synthesis.*

Main reactions		
Alkanes	$n\,CO + (2n + 1)\,H_2 \rightarrow C_n\,H_{2n+2} + n\,H_2O$	(6.11)
Alkenes	$n\,CO + 2n\,H_2 \rightarrow C_nH_{2n} + n\,H_2O$	(6.12)
Side reactions		
Water–gas shift	$CO + H_2O \rightleftarrows CO_2 + H_2$	(6.7)
Alcohols	$n\,CO + 2n\,H_2 \rightarrow H(CH_2)_nOH + (n - 1)\,H_2O$	(6.13)
Boudouard reaction	$2\,CO \rightarrow C + CO_2$	(6.14)

catalyst and process conditions is very important [37]; an appropriate choice of catalyst, temperature, pressure, and H_2/CO ratio enables the value of the chain growth probability (α) to be shifted, enabling different product ranges to be produced (Figure 6.22).

Current catalysts used for FT synthesis are based on cobalt or iron. Cobalt catalysts are more active and exhibit a relatively large chain growth probability α. Thus, they are more suitable for the production of diesel and waxes. For the production of gasoline or linear alkenes, the use of iron catalysts at high temperature is the best option [45]. Iron catalysts are also preferred for use with coal-derived syngas, which has a relatively low H_2/CO ratio (circa 0.7–1). The reason for this is that iron promotes the water–gas shift reaction, leading to a higher H_2/CO ratio in the reactor. Cobalt catalysts do not have this high activity for the water–gas shift reaction and, therefore, must operate at higher H_2/CO ratios, thus making them more suitable for syngas obtained from natural gas.

QUESTIONS:

Referring to Table 6.7, what would be the ideal syngas composition for FT synthesis?

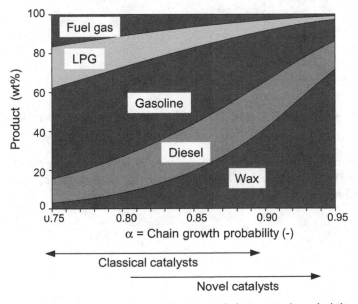

Figure 6.22 *Product distribution as function of chain growth probability [38].*

Box 6.9   Chain Growth Probability in Fischer–Tropsch Synthesis

Various mechanisms have been proposed for the FT synthesis reactions [39–43]. However, for an understanding of the product distribution only the basics of the reaction mechanism need to be considered. No matter what the exact mechanism is, growth of a hydrocarbon chain occurs by stepwise addition of a one-carbon segment derived from CO at the end of an existing chain (Scheme B.6.9.1). For a certain minimum chain length, say $n \geq 4$, it is plausible that the relative probability of chain growth and chain termination (α and $1-\alpha$, respectively) is independent of the chain length and hence constant [39]. Often it is assumed that this applies to all values of n. Then, the carbon number distribution of FT products can be represented by a simple statistical model, the so-called Anderson–Flory–Schulz (AFS) distribution [37–39, 44]. α is dependent on the catalyst used. In practice this is a fortunate catalyst characteristic allowing simple tuning of the process to obtain the maximum yield of the desired product, usually either diesel or gasoline. The H_2/CO ratio and temperature are also parameters that can be used to tune the selectivity. α becomes smaller with increasing temperature.

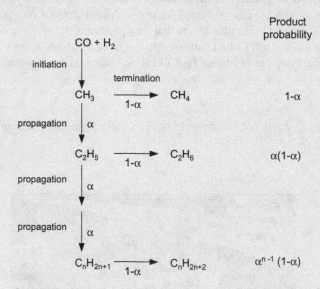

Scheme B.6.9.1   *Chain growth mechanism for FT synthesis; Anderson–Flory–Schulz kinetics [38].*

QUESTIONS:

> *Based on the model shown in Scheme B.6.9.1, what would be the effect of an increase in the H_2/CO ratio in the feed on the average chain length of the produced hydrocarbons? Complete Figure 6.22 for lower values of α.*

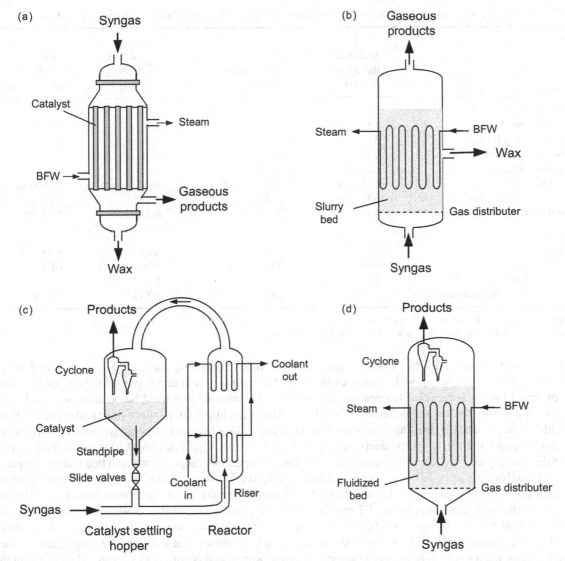

Figure 6.23 *Reactors used for Fischer–Tropsch synthesis: (a) multitubular fixed bed reactor (Arge); (b) slurry-phase reactor; (c) circulating fluidized bed reactor (Synthol), (d) fixed fluidized bed reactor (Sasol Advanced Synthol, SAS) (BFW = boiler feed water).*

6.3.1.2 Reactors Used in Fischer–Tropsch Synthesis

Due to the high exothermicity of the reactions, efficient removal of the heat of reaction is a major consideration in the design of FT reactors. Many reactor types have been proposed and developed for proper heat management [39]. Currently used reactors are shown in Figure 6.23, while main characteristics are presented in Table 6.8.

Multitubular fixed bed reactors and slurry-phase reactors are classified as low-temperature FT reactors, while fluidized bed reactors are high-temperature FT reactors.

Table 6.8 Typical dimensions and operating conditions of Fischer–Tropsch reactors (Fe catalyst) [36, 48, 49].

	Multitubular fixed bed reactor	Slurry-phase reactor	Circulating fluidized bed reactor	Fixed fluidized bed reactor
Dimensions				
Reactor height (m)	12	22	46	22
Reactor diameter (m)	3	5	2.3	8
Tube diameter (m)	0.05	—	—	—
Number of tubes	>2000	—	—	—
Catalyst size	1–3 mm	10–150 μm	40–150 μm	5–100 μm
Reactor capacity (b/d)	600	2500	1500	11 000
Potential reactor capacity (b/d)	3000	30 000	8000[a]	20 000[b]
Conditions				
Inlet T (K)	496	533	593	610
Outlet T (K)	509	538	598	615
Pressure (bar)	27	15	22	25
H_2/CO feed ratio (mol/mol)	1.7–1.8	≥0.7	2.5–3	2.5–3

[a]Capacities of CFB reactors at the Sasol II and III plants were 7500 b/d.
[b]Potential capacity has now been attained with a reactor diameter of 10.7 m.

In the multitubular fixed bed reactor, small diameter tubes containing the catalyst are surrounded by circulating boiler feed water, which removes the heat of reaction. A high linear gas velocity is applied and unconverted syngas is recycled to enhance heat removal. The multitubular fixed bed reactor is suitable for operation at relatively low temperature (490–530 K). Above the upper temperature limit carbon deposition would become excessive, leading to blockage of the reactor. This reactor can be considered a trickle flow reactor, since a large part of the products formed are liquids (waxes: C_{19+}) that trickle down the catalyst bed.

In the slurry-phase reactor, which operates in the same temperature range as the fixed bed reactor, a finely divided catalyst is suspended in a liquid medium, usually the FT wax product, through which the syngas bubbles upwards. This also provides agitation of the reactor contents. As the gas flows through the slurry, it is converted into more wax by the FT reaction. The fine particle size of the catalyst reduces diffusional mass and heat transfer limitations. The heat generated by the reaction is removed by internal cooling coils. The liquid medium surrounding the catalyst particles greatly improves heat transfer. The temperature in the slurry reactor must not be too low, because then the liquid wax would become too viscous, while at too high temperature hydrocracking takes place (Section 3.4.5).

Separation of the product wax from the suspended catalyst has been a challenge. A heavy product is produced, so distillation and flashing are not options. The successful development of an effective and relatively cheap liquid–solid separation step was crucial to the development of the slurry-phase reactor.

The key difference between the reactors described above and the fluidized bed reactors is that there is no liquid phase present in the latter. The formation of wax must be prevented, because this would condense on the catalyst particles, causing agglomeration of the particles and thus defluidization [36]. Therefore, these reactors are preferably used at temperatures above 570 K. The upper temperature limit of fluidized bed reactors is not much higher in order to avoid excessive carbon formation.

In the circulating fluidized bed (CFB) reactor, designed by Kellogg, the syngas circulates continuously together with powdered catalyst. In the improved design by Sasol, cooling coils remove a significant part of the heat of reaction, while the remainder is removed by the recycle and product gases, which are then

Table 6.9   *Typical selectivity data for Fischer–Tropsch processes (Fe catalyst) [43, 50, 51].*

Selectivity (carbon atom basis), wt%	Low temperature Multitubular fixed bed reactor	Low temperature Slurry-phase reactor	High temperature Circulating and fixed fluidized bed reactor
Methane	2.0	3.3	7.0
C_2–C_4 alkenes	5.6	8.2	24.1
C_2–C_4 alkanes	5.2	3.9	5.8
C_5–C_{11} (gasoline)	18.0	13.7	36.5
C_{12}–C_{18} (diesel)	14.0	17.0	12.5
C_{19+} (waxes)	52.0	49.5	9.0
Oxygenates	3.2	4.4	5.1

separated from the catalyst in cyclones. The main disadvantage of this type of reactor is the erosion of the walls of the reactor due to the flow of solid catalyst particles.

In the advanced fixed fluidized bed (FFB) reactor, circulation of the catalyst is eliminated, leading to simpler construction, lower cost, and more efficient operation. One might wonder why this design was not used at Sasol in the first place. This has to do with technical problems encountered when FFB Fischer–Tropsch technology was tried in the 1940s and 1950s. These problems had to do with the aforementioned defluidization [46,47].

QUESTIONS:

> *Give advantages and disadvantages of the four reactor types. (Consider operability, economics, ease of catalyst replacement, product flexibility, possibility of runaway, ease of scale-up, etc.)*
>
> *H_2S is a strong poison for the FT catalysts. Therefore, syngas purification is necessary. In the case of an upset in the syngas purification plant, causing a sudden higher H_2S content in the syngas feed than usual, which reactor would suffer least from activity loss and why? (Hint: think of flow characteristics, mixed or plug flow.)*

The different reactor types yield different product distributions (Table 6.9). As a result of the relatively low temperature in the fixed bed and in the slurry-phase reactor, in these reactors the selectivity towards heavy products (waxes) is high. In the fluidized bed reactors, which operate at much higher temperature, gasoline is the major product. In addition, these reactors produce a larger quantity of light products, such as methane and lower alkenes.

QUESTIONS:

> *Would the products be mainly linear or branched? What is desired for gasoline and diesel fuel, respectively, and why?*
>
> *Why is much more gasoline produced in fluidized bed reactors than in a fixed bed or slurry-phase reactor?*
>
> *Estimate the chain growth probability α for each reactor.*
>
> *Why does a low H_2/CO ratio lead to a high selectivity to liquid products?*
>
> *Catalyst particles are used in various shapes: spheres, cylinders, trilobes, quadrilopes, etc. Which would you prefer for each of the reactors in Figure 6.23? The pores in catalyst particles can be in the macro-, the meso- and the microsize range. What is the optimal size range?*

Table 6.10   *Currently operating and planned Fischer–Tropsch plants.*

Company	Location	Feedstock	Reactor (catalyst)	Operation	Current liquids output (b/d)
Sasol (I)	Sasolburg, South Africa	initially coal, currently natural gas	fixed bed (Fe) CFB (Fe) slurry (Fe)	1955 – present 1957–1993 1993 – present	5000
Sasol (Sasol Synfuels[a])	Secunda, South Africa	coal, currently supplemented by natural gas	CFB (Fe) FFB (Fe)	1980–2000 1995 – present	160 000
PetroSA[b]	Mossel Bay, South Africa	natural gas	CFB (Fe)	1991 – present	30 000
Shell	Bintulu, Malaysia	natural gas	fixed bed (Co)	1993 – present	14 700
Sasol-Chevron/Qatar Petroleum (Oryx)	Ras Laffan, Qatar	natural gas	slurry (Co)	2007 – present	34 000
Shell/Qatar Petroleum (Pearl)	Qatar	natural gas	fixed bed (Co)	2011/2012	140 000
Sasol-Chevron	Escravos, Nigeria	natural gas	slurry (Co)	2013	34 000

[a] Formerly Sasol II and Sasol III.
[b] Formerly Mossgas.

The Fischer–Tropsch process produces one of the most desirable diesel fuels: FT diesel has a high cetane number and negligible sulfur content (see also Section 3.5).

6.3.1.3   Removal of Carbon Deposits

As in most catalytic processes involving hydrocarbon reactions at high temperature, deposition of carbon on the catalyst is inevitable. Fluidized bed and slurry-phase reactors allow catalyst to be replaced during operation, but this is impossible in multitubular fixed bed reactors. Usually, removal of deposited carbon is carried out periodically (e.g., after a year of operation) once signs of catalyst deactivation, such as loss in conversion, become evident. The alternatives are to replace the deactivated catalyst with fresh catalyst or to regenerate it with hydrogen:

$$C + 2\,H_2 \rightarrow CH_4 \qquad \Delta_r H_{298} = -75\ kJ/mol \qquad\qquad (6.15)$$

6.3.1.4   Overview of Fischer–Tropsch Processes

Various companies are active in FT technology [44, 49], but only a few processes are used commercially (Table 6.10).

6.3.1.5   Sasol Process

Box 6.10 illustrates that the original Sasol I plant, which came on stream in 1955, can be seen as a "coal refinery" analogous to an oil refinery. The plant was highly complex, combining both low temperature and high temperature processes. Later, the CFB reactors were replaced by a slurry-phase reactor with a production

Box 6.10 Sasol I Plant

Figure B.6.10.1 shows a simplified block diagram of the Sasol I plant.

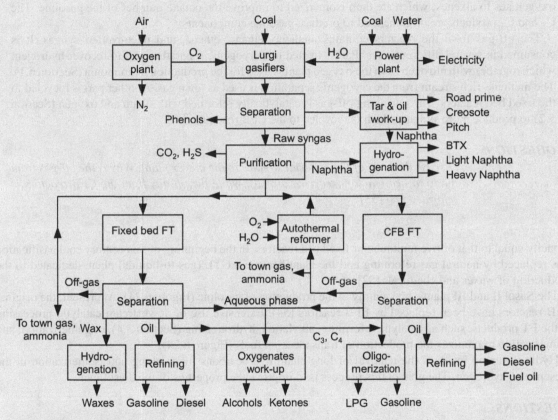

Figure B.6.10.1 *Simplified block diagram of the original set-up of the Sasol I Fischer–Tropsch process [37, 40].*

The syngas feed for the low-temperature FT reactors is produced by gasification of coal with steam and oxygen in 13 Lurgi gasifiers (Section 5.3). The required oxygen is produced by cryogenic air separation. The raw syngas is cooled to remove water and tars. The aqueous phase is treated to recover phenols. The tar is distilled to produce a range of products.

The syngas is then purified to remove hydrogen sulfide, carbon dioxide, and hydrocarbons in the naphtha boiling range. The latter, together with the naphtha from the tar work-up, are hydrotreated and the products are either blended into the gasoline pool or sold as aromatic solvents (benzene, toluene, and xylenes: BTX).

The purified syngas is sent to the reactor sections. The effluent from the reactors is cooled and water and oil are condensed. The water is sent to the oxygenate recovery and refining plant, where various alcohols and ketones are produced. The hydrocarbon product from the low-temperature FT reactors is distilled to produce gasoline, diesel, and waxes. The latter are distilled in a vacuum column to produce medium (590–770 K) and hard wax (>770 K), which are then hydrogenated to remove the remaining oxygenated compounds and alkenes.

The oil fraction from the high-temperature FT reactors is treated over an acid catalyst to convert the oxygenates to alkenes, which are then isomerized to improve the octane number of the gasoline. The C_3 and C_4 products are oligomerized to produce gasoline components.

The off-gas from the reactors contains methane, ethane, ethene, and unconverted syngas. It is consumed in several different ways. Part is treated in a cryogenic separation unit to recover hydrogen, which together with nitrogen from the oxygen plant is used for the production of ammonia (Section 6.1). The methane-rich stream from the cryogenic separation is used as town gas. Another part is recycled to the fixed bed reactors. The remaining off-gas is catalytically reformed with steam and oxygen (Section 5.2) to produce more syngas, which is recycled to the CFB reactors.

QUESTIONS:

Discuss the logic of the material streams in the power plant. Why is the off-gas from the fixed bed reactors recycled directly, while the off-gas from the CFB reactors is reformed first [53]?

capacity equal to that of five multitubular fixed bed reactors. In the beginning of this century coal gasification was replaced by natural gas reforming and the plant became a GTL (gas-to-liquids) plant, dedicated to the production of waxes and chemicals [52].

The Sasol II and III plants, aim mainly at the production of gasoline (Figure 6.24). At present, the original CFB reactors have been replaced by FFB reactors [54]. Extensive use of downstream catalytic processing of the FT products, such as catalytic reforming, alkylation, hydrotreating (Section 3.4), oligomerization, and isomerization maximizes the production of transportation fuels (Figure 6.24).

Hydrodewaxing involves the removal of long-chain hydrocarbons by cracking and isomerization in the presence of hydrogen. The goal of this process is to improve the properties of the diesel fuel.

QUESTIONS:

What is the purpose of catalytic reforming? Why are pentanes and hexanes subjected to a different process (isomerization)? What is an alternative process for the production of gasoline from the C_3/C_4 fraction? How can CO_2 be removed?

6.3.1.6   Shell Middle Distillate Synthesis (SMDS) Process

Shell has been the first to build an FT plant for the production of middle distillates based on remote natural gas. The Shell Middle Distillate Synthesis (SMDS) [38, 39] plant using natural gas from offshore fields has been in operation in Malaysia since 1993.

The syngas is primarily produced by partial oxidation of natural gas with oxygen. The H_2/CO ratio is about 1.7, which is below the 2.15 required for the cobalt-based catalyst used in the FT section. The ratio is raised by adding hydrogen-rich gas produced by catalytic steam reforming of the methane produced in the FT

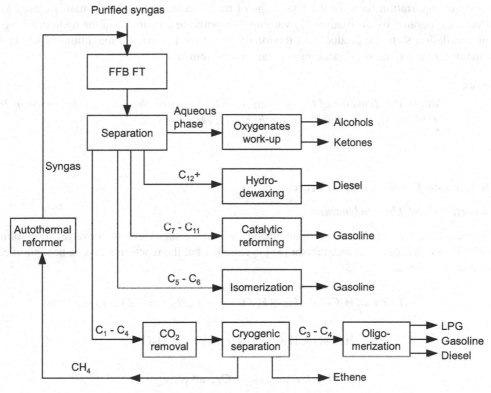

Figure 6.24   *General layout of the Sasol Synfuels plant (Sasol II and III).*

reactors [53]. Hydrocarbon synthesis takes place in a modernized version of the classical Fischer–Tropsch synthesis (Figure 6.25). The reactors used are multitubular fixed bed reactors containing over 10 000 tubes filled with a cobalt catalyst. Thus, the synthesis of long-chain waxy alkanes is favored. The heavy alkanes are subsequently converted in a mild hydrocracking (Section 3.4.5) step, to produce the desired products,

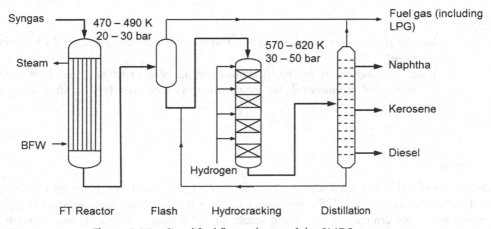

Figure 6.25   *Simplified flow scheme of the SMDS process.*

predominantly transportation fuels. In the final stage of the process, the products; mainly diesel, kerosene, and naphtha, are separated by distillation. By varying the operating conditions in the hydrocracking and the subsequent distillation step, the product distribution can be shifted towards a maximum kerosene mode or towards a maximum diesel mode, depending on market requirements.

QUESTIONS:

> Would the diameter of the tubes in the FT reactor be the same as that given in Table 6.8 (0.05 m)? If not, would it be smaller or larger and why?

6.3.2   Methanol-to-Gasoline (MTG) Process

6.3.2.1   Reactions and Thermodynamics

The MTG process converts methanol into light alkenes, which are subsequently converted to gasoline-range hydrocarbons in one process. The mechanism [31] is complex but the reaction path can be simplified by the following sequence of reactions [33]:

$$2\ CH_3OH \rightleftarrows H_3C\text{--}O\text{--}CH_3 + H_2O \qquad \Delta_r H_{298} = -23.6\ \text{kJ/mol} \tag{6.16}$$

$$H_3C\text{--}O\text{--}CH_3 \rightarrow C_2\text{--}C_5\ \text{alkenes} + H_2O \tag{6.17}$$

$$C_2\text{--}C_5\ \text{alkenes} \rightarrow C_{6+}\ \text{alkenes} \tag{6.18}$$

$$C_{6+}\ \text{alkenes} \rightarrow C_{6+}\ \text{alkanes, cycloalkanes, aromatics} \tag{6.19}$$

The initial step is the dehydration of methanol to form dimethyl ether (DME). This is an equilibrium-limited reaction but the subsequent conversion of DME into light alkenes, and then to other products, drives the dehydration reaction to the right. Figure 6.26 shows a typical product distribution as function of space time, which clearly indicates the sequential character of the reactions.

QUESTIONS:

> Are the data shown in Figure 6.26 in agreement with the sequential character of the reactions?
>
> Apparently, depending on the conditions various product mixtures can be produced. If it was suggested to produce diesel, would you agree with this? Why or why not?

6.3.2.2   Processes

Mobil (now ExxonMobil) discovered a very effective catalyst for converting methanol into high octane gasoline and still dominates this field [32]. Important features of the MTG process from a technological point of view are the high exothermicity of the overall reaction (-55.6 kJ/mol methanol) and the relatively fast deactivation of the catalyst (cycle time of a few months) by coke formation.

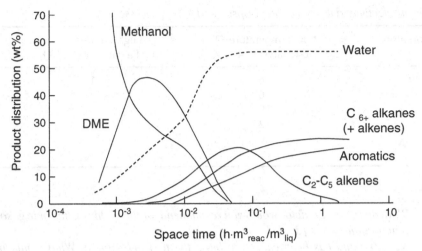

Figure 6.26 *Formation of various products from methanol as function of space time (T = 643 K; p = 1 bar).*

Figure 6.27 shows a simplified flow scheme of the ExxonMobil fixed bed MTG process. In the DME reactor crude methanol (containing about 17% water) is converted to an equilibrium methanol/DME/water mixture over an acidic catalyst. This mixture is combined with recycle gas and fed to the gasoline synthesis reactors containing the ZSM-5 catalyst. Recycling of cold product gas limits the temperature rise caused by the exothermic DME production.

As a result of deactivation of the zeolite catalyst by coke deposition, the composition of the reactor product is not constant. Therefore, the gasoline synthesis section consists of five reactors in parallel, of which always one is being regenerated. Regeneration takes place by burning off deposited coke.

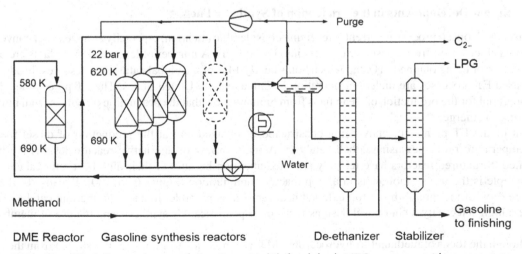

Figure 6.27 *Simplified flow scheme of the ExxonMobil fixed bed MTG process with one reactor being regenerated.*

Table 6.11   *Typical selectivity data for Fischer–Tropsch and MTG processes.*

Selectivity (carbon atom basis), wt%	Low-temperature FT Co catalyst @ 470 K	High-temperature FT Fe catalyst @ 570 K	MTG
C_{2-} (fuel gas)	6	15	1
C_3–C_4 (LPG)	6	23	10
C_5–C_{11} (gasoline)	19	36	89
C_{12}–C_{18} (diesel)	22	16	–
C_{19+} (waxes)	46	5	–
Oxygenates	1	5	–

QUESTIONS:

> *What alternative reactor would you recommend for the highly exothermic gasoline synthesis reaction and why? [32,34]*
> *New Zealand has been chosen to build the first MTG plant. What would have been the reason? (Hint: what is the source of methanol?)*

In the fixed bed MTG process, the main part of the carbon in methanol is converted to gasoline-range hydrocarbons (Table 6.11), the other products are light alkenes and alkanes and coke. Essentially no hydrocarbons larger than C_{11}, which corresponds to the end point of gasoline, are produced. This is a result of the so-called shape selectivity (Section 10.4) of the zeolite catalyst; hydrocarbons with carbon numbers outside the gasoline range cannot escape the pores [35].

An interesting alternative to the MTG process is the TIGAS (Topsøe integrated gasoline synthesis) process (Box 6.11)

DME can also be used as a fuel alternative as such. It shows a good performance in diesel engines with lower SO_x, NO_x, and carbon dioxide emissions and soot-free operation. It can also be blended with LPG up to 20 wt% without any modifications to the engine [59].

6.3.3   Recent Developments in the Production of Synthetic Fuels

The Fischer–Tropsch process is currently receiving considerable attention, especially as a means of converting natural gas located far from consumer markets into liquid fuels. A number of large-scale FT plants are under construction or being planned in Qatar, which holds nearly 14% of the global natural gas reserves. In addition, coal-based FT processes are under consideration in China and the United States. The FT process also has a high potential for the production of clean fuels from biomass, now that the production of syngas from biomass is starting to emerge.

Most of the FT plants currently being constructed or planned aim at the production of diesel fuel via low-temperature processes using cobalt catalysts. A wide variety of alternative reactor designs are being evaluated. Structured reactors have recently attracted attention because of their high heat removal capacity. An example is the use of monoliths in a loop reactor configuration (Figure 6.28) [60]. Because of the low pressure drop of the monolith reactor, external heat removal is possible. In addition, the thin catalyst layers enable a high selectivity. Other novel designs focus on supercritical FT synthesis and the use of membranes [61].

Although the focus of methanol-to-hydrocarbon (MTH) conversion technologies has shifted from the MTG process to the methanol-to-olefins (MTO) process (Section 4.5.3), at present there is still some activity in the fuel area.

Box 6.11 TIGAS Process

The TIGAS process (Figure B.6.11.1) was developed to reduce the capital and energy costs of producing gasoline by integrating methanol synthesis with the methanol-to-gasoline (MTG) step into a single loop without isolating methanol as an intermediate [55]. The process was designed for use in remote areas for the recovery of low-cost natural gas.

In this process, syngas is converted to a mixture of methanol and DME in one process step at 510–550 K and 30–80 bar over a bifunctional catalyst or a physical mixture of a classical methanol catalyst and a dehydration catalyst [56]. In this way, the thermodynamic limitations of methanol synthesis are eliminated [57]. The methanol/DME mixture is converted to gasoline in a similar way as in the MTG process.

In the mid-1980s, a demonstration plant based on steam reforming of natural gas was in operation in Houston, Texas. Recently, plans have been announced to demonstrate a biomass-based TIGAS process [58].

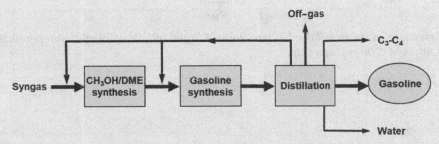

Figure B.6.11.1 *Simplified block diagram of the TIGAS process.*

QUESTION:

> *What is the optimal H_2/CO feed ratio for this process? (Hint: besides methanol formation followed by conversion to dimethyl ether, the water–gas shift reaction is also important.)*

6.3.4 Fuel Additives – Methyl Tert-Butyl Ether

Among the various gasoline additives, ethanol, methyl *tert*-butyl ether (MTBE), and ethyl *tert*-butyl ether (ETBE) are the most frequently used fuel oxygenates worldwide. MTBE has a high octane number (118 RON); it has been used as an octane booster in gasoline since the 1970s and later also for cleaner-burning gasoline. It has been the fastest growing chemical in the bulk chemical industry of the latter part of the 1980s and the 1990s (Figure 6.29). More recently, however, its environmental acceptability has been questioned and in the United States MTBE has gradually been phased out from 2000 onward. By 2006 it was no longer used in gasoline and completely replaced by bioethanol (Chapters 7 and 13). In Europe, MTBE is gradually being replaced by ETBE produced from bioethanol. In other parts of the world the market for MTBE is still growing [62, 63]. Nowadays, ethanol is the largest added oxygenate.

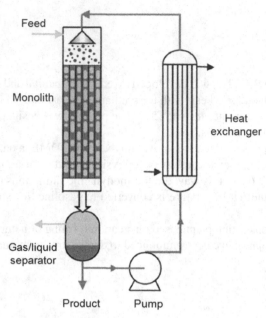

Figure 6.28   *Schematic of the monolithic loop reactor with liquid recycle. Reprinted with permission from [60] Copyright (2003) Elsevier Ltd*

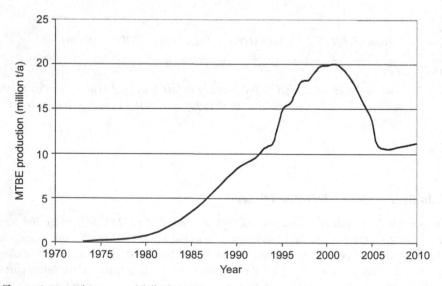

Figure 6.29   *The rise and fall of world MTBE production (data from various sources).*

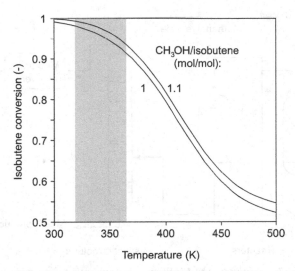

Figure 6.30   *Thermodynamic equilibrium conversion of isobutene to MTBE in the liquid phase as a function of temperature and methanol/isobutene mole ratio at a pressure of 20 bar (shaded area represents typical operating temperatures).*

MTBE is produced by reacting methanol and isobutene over an acid catalyst:

$$CH_3OH \;+\; \underset{H_3C}{\overset{H_3C}{>}}C=CH_2 \;\rightleftharpoons\; H_3C-\underset{CH_3}{\overset{CH_3}{\underset{|}{\overset{|}{C}}}}-O-CH_3 \qquad \Delta_r H_{298,liq} = -37.5 \text{ kJ/mol}$$

(6.20)

The isobutene feed usually is a mixed C_4 hydrocarbon stream containing only about 20–30% isobutene. The other constituents are mainly *n*-butane, isobutane, and *n*-butene, which are inert under typical reaction conditions. The reaction takes place in the liquid phase in two or three reactors in series.

With conventional processes employing fixed bed reactors and up to 10% excess methanol it is possible to obtain isobutene conversions of 90–96%, slightly less than the thermodynamic attainable conversion (Figure 6.30).

Separation of MTBE from the unconverted C_4-fraction, consisting of butanes, *n*-butenes and some isobutene, and methanol is achieved by distillation. Methanol is recovered for recycle (Figure 6.31). Because separation of isobutene from the other hydrocarbons is difficult, it usually leaves the process with the inert hydrocarbons. Therefore, an isobutene loss of 3–10% must be accepted. This is not the case if catalytic distillation is used (Section 14.3.1).

QUESTIONS:

What are possible sources of the C_4 feed? Why is a mixture processed instead of pure isobutene? Part of the cooled effluent from the first reactor is commonly recycled to the reactor inlet (Figure 6.31). Why? Would you use the same operating temperatures in both reactors? Why or why not?

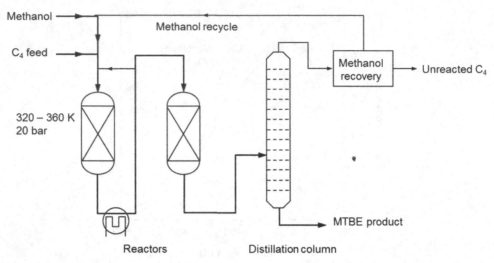

Figure 6.31 *Production of MTBE by the conventional process [63–65].*

References

[1] Appl, M. (1999) *Ammonia. Principles and Industrial Practice*, Wiley-VCH, Weinheim.

[2] Maxwell, G. (2005) Ammonia synthesis, in *Synthetic Nitrogen Products* (ed. G.R. Maxwell), Springer US, New York, pp. 163–198.

[3] DiFranco, C.A., Kramer, D.A., and Apodaca, L.E. (2009) Nitrogen (Fixed) – Ammonia Statistics. U.S. Geological Survey (USGS), http://minerals.usgs.gov/ds/2005/140/nitrogen.pdf (last accessed 8 December 2012).

[4] USGS. (2010) Bureau of Mines Mineral Yearbook (1932–1993). http://minerals.usgs.gov/minerals/pubs/usbmmyb.html (last accessed 8 December 2012).

[5] Bakemeier, H., Huberich, T., Krabetz, R., Liebe, W., Schunck, M., and Mayer, D. (1985) Ammonia, in *Ullmann's Encyclopedia of Industrial Chemistry*, 5th edn, Vol. **A2** (ed. W. Gerhartz), VCH, Weinheim, pp. 143–242.

[6] Jennings, R.J. and Ward, S.A. (1996) Ammonia synthesis, in *Catalyst Handbook*, 2nd edn. (ed. M.V. Twigg), Manson Publishing Ltd, London, pp. 384–440.

[7] Chauvel, A. and Lefebvre, G. (1989) Petrochemical Processes 1. Synthesis Gas Derivatives and Major Hydrocarbons. Technip, Paris.

[8] Appl, M. (2000) Ammonia, in *Ullmann's Encyclopedia of Industrial Chemistry*. Wiley-VCH Verlag GmbH, Weinheim.

[9] Pearce, B.B., Twigg, M. V., and Woodward, C. (1996) Methanation, in *Catalyst Handbook*, 2nd edn. (ed. M.V. Twigg), Manson Publishing Ltd, London, pp. 340–383.

[10] Watson, A.M. (1983) Use pressure swing adsorption for lowest cost hydrogen. *Hydrocarbon Processing*, 91–95.

[11] Rautenbach, R. and Albrecht, R. (1989) *Membrane Processes*, John Wiley & Sons, Inc., New York, pp. 422–454.

[12] Mavrovic, I., Shirley, A.R., and Coleman, G.R. (2000) Urea, in *Kirk-Orhmer Encyclopedia of Chemical Technology*, John Wiley & Sons, Inc., New York.

[13] Van den Berg, P.J. and de Jong, W.A. (1980) *Introduction to Chemical Process Technology*, Delft University Press, Delft.

[14] Meessen, J.H. (2000) Urea, in *Ullmann's Encyclopedia of Industrial Chemistry*, Wiley-VCH Verlag GmbH, Weinheim.

[15] UN Industrial Development Organization. (1998) *Fertilizer Manual*, 3rd edn. Springer, pp. 615.

[16] Morikawa, H. and Sakata, E. (2008) Toyo's ACES24: advanced urea production technology. In IFA Technical Symposium, 10–14 March 2008, Sao Paulo, Brazil, pp. 14.

[17] Meessen, J.H. and Petersen, H. (1996) Urea, in *Ullmann's Encyclopedia of Industrial Chemistry*, 5th edn, Vol. **A27** (ed. W. Gerhartz), VCH, Weinheim, pp. 333–365.

[18] Chang, T., Rousseau, R.W., and Kilpatrick, P.K. (1986) Methanol synthesis reactions: calculations of equilibrium conversions using equations of state. *Industrial and Engineering Chemistry Process Design and Development*, **25**, 477–481.

[19] Hansen, J.B. (1997) Methanol synthesis, in *Handbook of Heterogeneous Catalysis* (eds G. Ertl, H. Knözinger, and J. Weitkamp), VCH, Weinheim, pp. 1856–1876. See also 2nd edn.: http://dx.doi.org/10.1002/9783527610044.hetcat0148 (last accessed 8 December 2012).

[20] Royal, M.J. and Nimmo, N.M. (1969) Why LP methanol costs less. *Hydrocarbon Processing*, **48**, 147–153.

[21] Hiller, H. and Marchner, F. (1970) Lurgi makes low pressure methanol. *Hydrocarbon Processing*, **49**, 281–285.

[22] Supp, E. (1981) Improved methanol process. *Hydrocarbon Processing*, **60**, 71–75.

[23] (1983) Methanol – Haldor Topsoe A/S. *Hydrocarbon Processing*, **62**, 111–175.

[24] Sherwin, M.B. and Frank, M.E. (1976) Make methanol by 3 phase reaction. *Hydrocarbon Processing*, **55**, 122–124.

[25] Brown, D.M., Leonard, J.J., Rao, P., and Weimer, R.F. (1990) US Patent 4,910,227. Assigned to Air Products and Chemicals.

[26] Rostrup-Nielsen, J.R. (1993) Production of synthesis gas. *Catalysis Today*, **18**, 305–324.

[27] Cheng, W.H. and Kung, H.H. (1994) *Methanol, Production and Use*. Marcel Dekker, New York, Chapter 1.

[28] Davies, P., Donald, R.T., and Harbord, N.H. (1996) Catalytic oxidations, in *Catalyst Handbook*, 2nd edn (ed. M.V. Twigg), Manson Publishing Ltd, London, pp. 490–502.

[29] Gerberich, H.R. and Seaman, G.C. (1994) Formaldehyde, in *Kirk-Othmer Encyclopedia of Chemical Technology*, 4th edn, Vol. **11** (eds J.I. Kroschwitz and M. Howe-Grant), John Wiley & Sons, Inc., New York, pp. 929–951.

[30] Reuss, G., Disteldorf, W., Grundler, O., and Hilt, A. (1988) Formaldehyde, in *Ullmann's Encyclopedia of Industrial Chemistry*, 5th edn (ed. W. Gerhartz), VCH, Weinheim, pp. 619–651.

[31] Derouane, E.G. (1984) Conversion of methanol to gasoline over zeolite catalysts. I. Reaction mechanisms, in *Zeolites: Science and Technology* (eds F.R. Ribeiro, A.E. Rodrigues, L.D. Rollmann, and C. Naccache), Martinus Nijhoff, The Hague, pp. 515–528.

[32] MacDougall, L.V. (1991) Methanol to fuels routes - the achievements and remaining problems. *Catalysis Today*, **8**, 337–369.

[33] Chang, C.D. and Silvestri, A.J. (1977) The conversion of methanol and other O-compounds to hydrocarbons over zeolite catalysts. *Journal of Catalysis*, **47**, 249–259.

[34] Blauwhoff, P.M.M., Gosselink, J.W., Kieffer, E.P., Sie, S. T., and Stork, W. H. J. (1999) Zeolites as catalysts in industrial processes, in *Catalysis and Zeolites: Fundamentals and Applications* (eds J. Weitkamp and L. Puppe), Springer-Verlag, Berlin, pp. 437–537.

[35] Gabelica, Z. (1984) Conversion of methanol to gasoline over zeolite catalysts. II. Industrial processes, in *Zeolites: Science and Technology* (eds F.R. Ribeiro, A.E. Rodrigues, L.D. Rollmann, and C. Naccache), Martinus Nijhoff, The Hague, pp. 529–544.

[36] Steynberg, A.P., Dry, M.E., Davis, B.H., and Breman, B.B. (2004) Fischer-Tropsch Reactors, in *Studies in Surface Science and Catalysis. Fischer-Tropsch Technology*, Vol. **152** (eds A.P. Steynberg and M.E. Dry), Elsevier, pp. 64–195.

[37] Baldwin, R.M. (1993) Liquefaction, in *Kirk-Othmer Encyclopedia of Chemical Technology*, 4th edn, Vol. **6** (eds J.I. Kroschwitz and M. Howe-Grant), John Wiley & Sons, Inc., New York, pp. 568–594.

[38] Sie, S.T., Senden, M.M.G., and Van Wechem, H.M.H. (1991) Conversion of natural gas to transportation fuels via the Shell middle distillate synthesis process (SMDS). *Catalysis Today*, **8**, 371–394.

[39] Sie, S.T. (1998) Process development and scale up: IV. Case history of the development of a Fischer-Tropsch synthesis process. *Reviews in Chemical Engineering*, **14**, 109–157.

[40] Dry, M.E. (1981) The Fischer-Tropsch synthesis, in *Catalysis Science and Technology*, Vol. **1** (eds J.R. Anderson and M. Boudart), Springer-Verlag, Berlin, pp. 159–255.

[41] Biloen, P. and Sachtler, W.M. H. (1981) Mechanism of hydrocarbon synthesis over F-T catalysts, in *Advances in Catalysis* (eds D.D. Eley, H. Pines and P.B. Weisz), vol. **30**. Academic Press, New York, pp. 165–216.

[42] Ponec, V. (1978) Some aspects of the mechanism of methanation and Fischer-Tropsch synthesis. *Catalysis Reviews – Science and Engineering*, **18**, 151–171.

[43] Dry, M.E. (1990) The Fischer-Tropsch process – commercial aspects. *Catalysis Today*, **6**, 183–206.

[44] Van der Laan, G.P. (1999) Kinetics, selectivity and scale up of the Fischer-Tropsch synthesis. PhD Thesis, University of Groningen, http://irs.ub.rug.nl/ppn/181518880 (last accessed 8 December 2012).

[45] Dry, M.E. (2008) The Fischer-Tropsch (FT) synthesis processes, in *Handbook of Heterogeneous Catalysis*, 2nd edn, Vol. **6** (eds G. Ertl, H. Knözinger, H. Schüth, and J. Weitkamp), Wiley-VCH Verlag GmbH, Weinheim, pp. 2965–2994.

[46] Jager, B., Dry, M.E., Shingles, T., and Steynberg, A.P. (1990) Experience with a new type of reactor for Fischer-Tropsch synthesis. *Catalysis Letters*, **7**, 293–301.

[47] Collings, J. (2002) Mind Over Matter. The Sasol Story: A Half-Century of Technological Innovation. Sasol, Johannesburg, South Africa.

[48] Derbyshire, F. and Gray, D. (1986) Coal Liquefaction, in *Ullmann's Encyclopedia of Industrial Chemistry*, 5th edn, Vol. **A7** (ed. W. Gerhartz), VCH, Weinheim, pp. 197–243.

[49] Davis, B.H. (2005) Fischer–Tropsch synthesis: overview of reactor development and future potentialities. *Topics in Catalysis*, **32**, 143–168.

[50] Steynberg, A.P., Espinoza, R.L., Jager, B., and Vosloo, A.C. (1999) High temperature Fischer-Tropsch synthesis in commercial practice. *Applied Catalysis A*, **186**, 41–54.

[51] Kaneko, T., Derbyshire, F., Makino, E., Gray, D., and Tamura, M. (2000) Coal Liquefaction, in *Ullmann's Encyclopedia of Industrial Chemistry*. Wiley-VCH Verlag GmbH, Weinheim.

[52] Leckel, D. (2009) Diesel production from Fischer–Tropsch: the past, the present, and new concepts. *Energy Fuels*, **23**, 2342–2358.

[53] Dry, M.E. (2002) The Fischer-Tropsch process: 1950–2000. *Catalysis Today*, **71**, 227–241.

[54] Dancuart, L.P. and Steynberg, A.P. (2007) Fischer-Tropsch based GTL technology: a new process? in *Studies in Surface Science and Catalysis. Fischer-Tropsch Synthesis, Catalyst and Catalysis*, Vol. **163** (eds B.H. Davis and M.L. Ocelli), Elsevier, pp. 379–399.

[55] Keil, F.J. (1999) Methanol-to-hydrocarbons: process technology. *Microporous And Mesoporous Materials*, **29**, 49–66.

[56] Ramos, F.S., Farias, A.M.D., Borges, L.E.P., Monteiro, J.L., Fraga, M.A., Sousa-Aguiar, E.F., and Appel, L.G. (2005) Role of dehydration catalyst acid properties on one-step DME synthesis over physical mixtures. *Catalysis Today*, **101**, 39–44.

[57] Shen, W.J., Jun, K.W., Choi, H.S., and Lee, K.W. (2000) Thermodynamic investigation of methanol and dimethyl ether synthesis from CO_2 Hydrogenation. *Korean Journal of Chemical Engineering*, **17**, 210–216.

[58] Joensen, F., Nielsen, P., and Palis Sørensen, M. (2011) Biomass to green gasoline and power. *Biomass Conversion and Biorefinery*, **1**, 85–90.

[59] Erdener, H., Arinan, A. and Orman, S. (2011) Future fossil fuel alternative; di-methyl ether (DME) a review. *International Journal of Renewable Energy Technology*, **1**, 252–258.

[60] de Deugd, R.M., Chougule, R.B., Kreutzer, M.T., Meeuse, F.M., Grievink, J., Kapteijn, F., and Moulijn, J.A. (2003) Is a monolithic loop reactor a viable option for Fischer–Tropsch synthesis? *Chemical Engineering Science*, **58**, 583–591.

[61] Remans, T.J., Jenzer, G., and Hoek, A. (2008) Gas-to-liquids, in *Handbook of Heterogeneous Catalysis*, 2nd edn, Vol. **6** (eds G. Ertl, H. Knözinger, H. Schüth, and J. Weitkamp), Wiley-VCH Verlag GmbH, Weinheim, pp. 2994–3010.

[62] Winterberg, M., Schulte-Körne, E., Peters, U., and Nierlich, F. (2000) Methyl tert-butyl ether, in *Ullmann's Encyclopedia of Industrial Chemistry*, Wiley-VCH Verlag GmbH, Weinheim.

[63] Krause, A.O. and Keskinen, K.I. (2008) Etherification, in *Handbook of Heterogeneous Catalysis*, 2nd edn, Vol. **6** (eds G. Ertl, H. Knözinger, H. Schüth, and J. Weitkamp), Wiley-VCH Verlag GmbH, Weinheim, pp. 2864–2881.

[64] DeGarmo, J.L., Parulekar, V.N., and Pinjala, V. (1992) Consider reactive distillation. *Chemical Engineering Progress*, **88**, 43–50.

[65] Järvelin, H. (2004) Commercial production of ethers, in *Handbook of MTBE and Other Gasoline Oxygenates* (eds H. Hamid and M.A. Ali), CRC Press, pp. 194–212.

7

Processes for the Conversion of Biomass

7.1 Introduction

Biomass is the oldest source of energy and currently provides roughly 10% of total energy demand. Traditionally, biomass in the form of fuel wood used for heating and cooking is the main source of bioenergy, but liquid biofuel production has shown rapid growth during the last decade. Similar to crude oil, biomass can be and is processed in several ways. Figure 7.1 gives an overview showing different approaches and processes for biomass conversion [1].

Biomass can be gasified at high temperature in the presence of a substoichiometric amount of oxygen and the produced synthesis gas (Chapter 5) can be further processed to obtain the "normal" product spectrum, including, for example, methanol and Fischer–Tropsch liquids [2] (Chapter 6). During pyrolysis, a simple process similar to coking in the oil refinery, which takes place at intermediate temperature in the absence of oxygen, biomass is converted into a liquid phase, referred to as bio-oil [3]. This process is robust and can be used with a large variety of feedstocks but the bio-oil produced is a low-quality fuel with low stability and the yield is modest.

Not all biomass conversion processes are analogous to oil-related processes. Completely different approaches also suggest themselves. The underlying reason is the completely different structure of biomass compared to crude oil. Figures 2.22–2.24 in Chapter 2 show that biomass has a highly functionalized structure and contains a large amount of oxygen, in sharp contrast with crude oil. The highly functionalized structure has resulted in an extensive biotechnological industry, the most important process being the production of ethanol by fermentation of sugars obtained from sugar cane and corn. Another category of potential processes involves dissolution and depolymerization (by hydrolysis) of lignocellulosic biomass into the corresponding monomers. Such processes are hot topics in R&D. Obvious monomers are sugars, the most important being glucose from cellulose and xylose from hemicellulose. The sugars may subsequently be fermented or converted to useful chemicals by (bio)catalytic means.

Thus, biomass conversion processes range from high-temperature thermochemical processes, such as combustion, gasification, and pyrolysis, to more subtle (bio)chemical processes in the liquid phase, such as hydrolysis and fermentation. The former category is very robust because the detailed structure of the biomass only plays a minor role and the complete organic part of the biomass is converted into a large pool of chemical compounds of various nature. The latter category is different due to the possibility of selective conversion

Chemical Process Technology, Second Edition. Jacob A. Moulijn, Michiel Makkee, and Annelies E. van Diepen.
© 2013 John Wiley & Sons, Ltd. Published 2013 by John Wiley & Sons, Ltd.

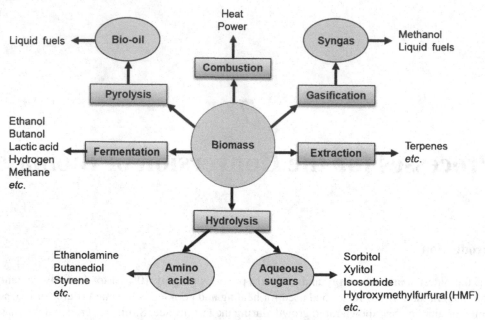

Figure 7.1   *Routes and products in biomass conversion processes.*

routes under milder conditions. For this class, the biomass structure offers the potential of efficient processes with high yields of target products.

QUESTIONS:

> *Referring to Figure 7.1, discuss the main process aspects. In which processes does catalysis play a role? Is the separation of the product(s) an issue? What are the reaction conditions (temperature, pressure)?*

It might be wondered whether it would not be advisable to process biomass in the existing industrial infrastructure, such as in oil refineries. The simplest way would be to defunctionalize the biomass, so that the stoichiometry of the products approaches that of crude oil, and subsequently utilize these products as feedstocks for the refinery. It has been proposed to do this by hydrodeoxygenation (HDO), similar to other hydrotreating processes such as hydrodesulfurization (HDS) described in Section 3.4.5. However, it can easily be shown that for the major sources of biomass the economics of such a process are questionable.

QUESTIONS:

> *Estimate the economic feasibility of the production of hydrocarbons by HDO based on lignocellulosic biomass. (Hints: Assume a simple composition for the biomass and the product. Do not take into account the processing costs, but limit yourself to the prices of reactants and products. Follow an optimistic scenario and assume 100% yields and no waste.)*
>
> *Compare the temperature profiles in the reactors for a typical HDS process and HDO of lignocellulosic biomass.*

Apart from the cost factor, there is the question what the optimal products of biomass conversion are. *A priori* there is no need to copy the hydrocarbon industry. In principle, for several reasons it is most attractive to produce oxygen-containing products. Firstly, less oxygen atoms need to be removed, saving process steps and hydrogen (typically used for removing oxygen). Secondly, the product quality might be better with oxygen atoms present. For instance, in the case of diesel fuel, oxygen-containing molecules often result in smokeless combustion and, as a result, less pollution[1]. Thirdly, when less oxygen is removed less mass is lost.

A fundamental point is the following: Intuitively, it is clear that it would be attractive for the products to have a structure similar to that of the feedstock monomers. The consequence of making such products would be that only a minimal number of bond breaking and forming steps would be required. Usually this leads to an efficient process as far as the chemistry is concerned. Therefore, in deciding upon the desired products, it makes sense to take into account the structure of the feedstock. One of the great challenges in biomass applications is to identify attractive molecular structures.

QUESTIONS:

> *In analogy with crude oil applications, a possible product spectrum from biomass consists of fuels and chemicals. Would this be the optimum product spectrum or is there a priori a preference for either fuels or chemicals? Take into consideration the market volumes and prices.*
>
> *Oxygen-containing molecules in diesel fuel often result in less pollution, as suggested by the smokeless combustion. However, under some conditions health effects still could be significant. Explain.*

7.2 Production of Biofuels

Biofuels are growing tremendously in volume; Figure 7.2 shows the global trend. The major biofuels are bioethanol and biodiesel. Bioethanol has the greater global importance, but biodiesel is the most common biofuel in the European Union (EU) (Figure 7.3).

Bioethanol was first used as an oxygen-containing additive for gasoline and later also as a primary transportation fuel, mainly in Brazil, the United States, Canada, and some European countries. Bioethanol can also be used as a feedstock for the production of ethyl *tert*-butyl ether (ETBE), which blends more easily with gasoline and has a high octane number (118). Currently, the use of biodiesel is limited to low concentration mixtures with oil-derived diesel.

QUESTIONS:

> *What fractions of gasoline and diesel are based on biomass in the world and in the EU (see also Chapter 2)?*
>
> *The contribution of bioethanol to the Greenhouse Effect is said to be lower than that of oil-based gasoline. Is this statement always correct?*

Table 7.1 shows the status of various technologies used for the production of biofuels. Current commercial technologies for producing bioethanol and biodiesel are based on food crops that can be easily converted. A

---

[1] Smokeless combustion not necessarily means cleaner combustion. Although the total weight of particulate matter in the exhaust gas is lower, the number of particles may be the same or even higher (but of smaller size).

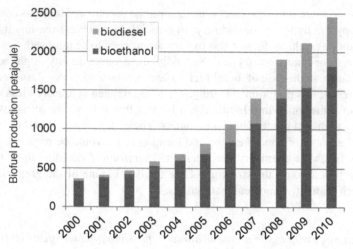

Figure 7.2   *Global biofuel production [4].*

major disadvantage is the direct competition with food supplies. Biogas, a gas mainly consisting of methane, is also produced on a commercial scale by anaerobic digestion (Section 13.4.2).

Technologies for the production of more advanced liquid biofuels, based on lignocellulosic biomass, are currently at the large-scale demonstration stage. The lignocellulosic-biomass derived biofuels produced by thermochemical methods, such as pyrolysis and gasification, are much more similar in composition and fuel value to oil-derived fuels than conventional biofuels. They may offer a really sustainable alternative as transportation fuels; they have lower requirements for land and fertilizer, and often there is no competition with usage as food.

7.2.1   Bioethanol and Biobutanol

At present, virtually all bioethanol is produced from food crops, with about 75% being produced from sugar crops, including sugarcane, sugar beet, and molasses; the remainder is mainly produced from corn. Bioethanol

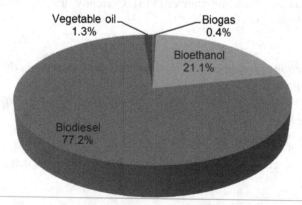

Figure 7.3   *Breakdown of EU biofuel consumption for transportation by type (total: 13.9 million metric tons oil equivalent) [5].*

Table 7.1 Status of the technology used for the production of biofuels [6].

Biofuel	Feedstock	Process	Development stage
Bioethanol	Sugar and starch crops	Hydrolysis and fermentation	Commercial
	Lignocellulosic biomass	Pretreatment, hydrolysis and fermentation	Demonstration
Biodiesel	Vegetable oils, animal fats	Transesterification	Commercial
(FAME)	Microalgae	Transesterification	Basic and applied R&D
Green diesel	Vegetable oils	Hydrotreating	Early commercial
BTL[a] diesel	Lignocellulosic biomass	Gasification + FT[b]	Demonstration
Biomethanol	Lignocellulosic biomass	Gasification + methanol synthesis	Early commercial
Bio-DME[c]	Biomethanol	Methanol dehydration	Demonstration
Biobutanol	Sugar and starch crops	Hydrolysis and fermentation	Early commercial
	Lignocellulosic biomass	Pretreatment, hydrolysis and fermentation	Demonstration
Various	Lignocellulosic biomass	Pyrolysis	Basic and applied R&D
Furanics	Sugar and starch crops Biomass	(pretreatment,) hydrolysis and chemical conversion	Basic and applied R&D
Biogas/Biomethane	Organic waste	Anaerobic digestion	Commercial
Hydrogen	Biogas	Steam reforming	Demonstration
	Lignocellulosic biomass	Gasification	Demonstration
	Various novel routes	Various	Basic and applied R&D

[a] Biomass-to-liquids.
[b] Fischer–Tropsch synthesis.
[c] Dimethyl ether.

is produced by fermentation of sugars using microorganisms such as yeasts. This process can be applied to a variety of feedstocks, provided the structure is made accessible to fermentation. For starchy crops this is done by enzymatic hydrolysis to produce mainly glucose. For the more difficult lignocellosic biomass, the technology is more complex (Figure 7.4). Processing of lignocellulosic biomass is currently the topic of a large research effort. The most important challenges being addressed are the current need for a costly pretreatment of the biomass to free the cellulose and hemicellulose and the fact that hemicellulose consists of pentose sugars that are not easily fermentable. However, in recent years much progress has been made in the development of microorganisms suitable for the fermentation of these pentose sugars.

During pretreatment, the feedstock is subjected to conditions that disrupt the fibrous matrix structure of the lignocellulose, resulting in the separation of the hemicelluloses from the polymerous cellulose chains and the interwoven lignin that binds them together (Section 2.3.4). Numerous pretreatment strategies have been developed to enhance the reactivity of cellulose and to increase the yield of fermentable sugars [7, 8]. Hydrolysis releases the individual sugar monomers (glucose) from cellulose. The cellulose and hemicellulose sugars can be fermented to ethanol (see also Section 13.4.1) by yeasts that have been modified to ferment both hexose and pentose sugars and have been adapted to deal with the inhibitory materials that are produced during pretreatment of the biomass [7,8]. Distillation and dehydration of the aqueous ethanol solution produces anhydrous ethanol of 99.9 wt% purity, acceptable for mixing with gasoline [7]. Currently, a discussion is taking place about the suitability of 96 wt% ethanol (the azeotrope of ethanol and water) for blending with gasoline. A range of coproducts, such as various chemicals, lignin, and heat or electricity may also be produced.

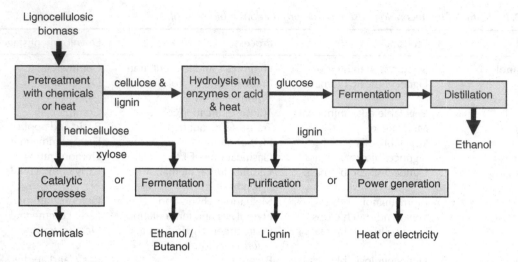

Figure 7.4   *Biochemical production of ethanol and butanol from lignocellulosic biomass.*

QUESTIONS:

> *Estimate the maximum yield of bioethanol that can be produced.*
> *What are the largest contributors to the cost of producing bioethanol from lignocellulosic biomass? How could these be reduced?*

There is an increasing interest in the use of biobutanol as a transportation fuel. Butanol better tolerates water contamination, is less corrosive than ethanol, has a higher energy content, and is more suitable for distribution through existing pipelines for gasoline. In blends with diesel or gasoline, butanol behaves more favorably with respect to the formation of stable mixtures than ethanol in case of contamination of the fuel with water. Blends of 85% butanol with gasoline can be used in unmodified gasoline engines.

Butanol production differs from ethanol production primarily in the fermentation process (different microorganism) and there are minor changes in the downstream distillation section (Section 13.4.1).

The number of announcements for commercial biobutanol plants is increasing rapidly. Currently, several bioethanol plants are being converted to biobutanol plants.

QUESTIONS:

> *Why does biobutanol tolerate water contamination better than bioethanol? What are disadvantages of biobutanol as a fuel?*

7.2.2   Diesel-Type Biofuels

7.2.2.1   *Transesterification*

So-called biodiesel is a mixture of methyl or ethyl esters of fatty acids (FAMEs and FAEEs). The ester group increases the oxygen content of diesel–biodiesel blends, improving the combustion efficiency of the conventional fossil-fuel-based diesel. Biodiesel is produced by catalytic transesterification of vegetable oils with low molecular weight alcohols like methanol or ethanol [9–13]. The most employed vegetable oils in

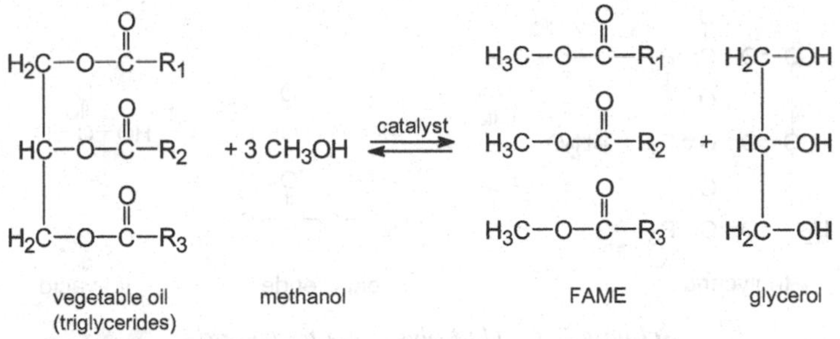

Scheme 7.1 *Transesterification of vegetable oil with methanol. FAME = fatty acid methyl esters. R_1, R_2, and R_3 are long hydrocarbon chains (with an uneven number of carbon atoms).*

Europe are rapeseed, soya bean, and sunflower oils. On the other hand, palm oil is considered an excellent feedstock for biodiesel production in tropical countries such as Malaysia and Thailand.

The preferred (homogeneous) catalysts are sodium hydroxide (NaOH) and sodium methoxide (NaOCH$_3$). The reaction can also be catalyzed by acids, but these are much less active. Scheme 7.1 shows the transesterification reaction of a vegetable oil with methanol. The reaction is reversible and excess methanol is used to shift the equilibrium towards the formation of (methyl) esters.

It is common for oils and fats to contain small amounts of water and free fatty acids (FFAs) – fatty acids that are not attached to the glycerol backbone – which give rise to side reactions. The free fatty acids (RCOOH) react with the alkaline catalyst and form soap (Scheme 7.2), as a result of which part of the catalyst is neutralized and is no longer available for transesterification. Furthermore, the presence of soap may interfere with subsequent processing steps. Therefore, the FFA content should be smaller than 0.5 wt%. If water is present, it can hydrolyze the triglycerides and form a free fatty acid (Scheme 7.3), thereby increasing the tendency towards soap formation. The water content should be less than 0.2 wt%.

Figure 7.5 shows a simplified block diagram for the production of biodiesel using an alkaline catalyst. The triglycerides are first pretreated by "degumming" for removal of phospholipids, drying, and, if necessary, removal of free fatty acids. Phospholipids are triglycerides with two fatty acid chains and one side chain formed by a phosphate ester [14]. If the triglycerides contain too much free fatty acids, these are converted to the corresponding methyl esters in the presence of an acid catalyst. Alternatively, the free fatty acids are separated from the feed for disposal or separate treatment in an acid esterification unit.

After pretreatment, the tryglicerides are subjected to transesterification with methanol in the presence of an alkaline catalyst. The methyl esters are subsequently separated from the heavier glycerol phase by phase separation. The catalyst is then neutralized by adding an acid, for example, hydrochloric acid, after which washing to remove minor amounts of by-products and drying yields a biodiesel ready for use.

Excess methanol is recycled. The resulting crude glycerol (circa 80–85 vol.%) can be used as such or be purified further by chemical treatment, evaporation, distillation, and bleaching to yield glycerol of pharmaceutical quality (>99.5 vol.%). Today, the crude glycerol that is not purified is largely burned, which must be considered as a tragic waste of a potentially very useful organic raw material [15], as shown in Section 7.3.2.

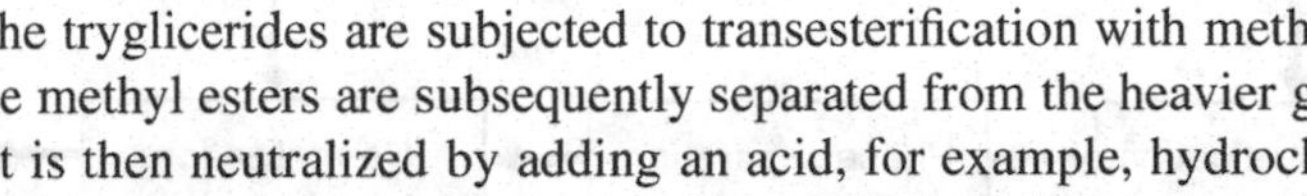

Scheme 7.2 *Soap formation in the presence of a NaOH catalyst.*

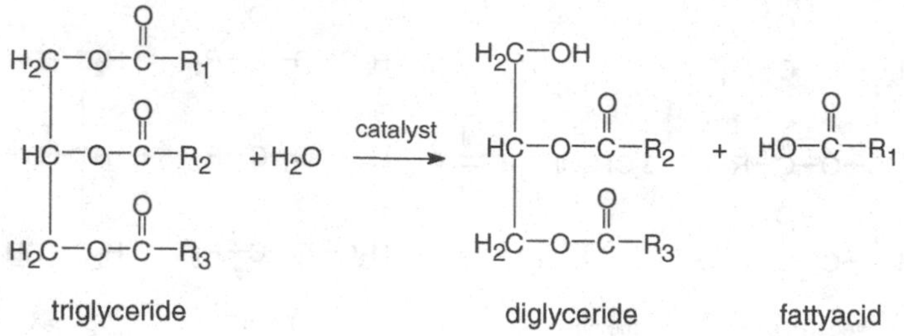

$$H_2C-O-\overset{\displaystyle O}{\overset{\|}{C}}-R_1$$

$$HC-O-\overset{\displaystyle O}{\overset{\|}{C}}-R_2 \quad + H_2O \quad \xrightarrow{\text{catalyst}} \quad HC-O-\overset{\displaystyle O}{\overset{\|}{C}}-R_2 \quad + \quad HO-\overset{\displaystyle O}{\overset{\|}{C}}-R_1$$

$$H_2C-O-\overset{\displaystyle O}{\overset{\|}{C}}-R_3 \qquad\qquad H_2C-O-\overset{\displaystyle O}{\overset{\|}{C}}-R_3$$

triglyceride   diglyceride   fattyacid

Scheme 7.3   *Hydrolysis of a triglyceride.*

QUESTIONS:

What is the reason for carrying out the transesterification process instead of using the oils and fats directly as fuels?
The concentration of water has to be kept low. Why?
Calculate the atom economy and E-factor (see Section 12.2) of the transesterification process. How is the phase separation done? What would be the impurities in the crude glycerol?
Why and how would glycerol be bleached?
Obviously, glycerol is produced as a by-product. Are the amounts produced large (compare with the world market for glycerol)?
What would be markets for crude glycerol and refined glycerol?
Suggest conversion routes for the production of marketable products from glycerol.

Although chemical transesterification using a process involving an alkaline catalyst results in high conversions of triglycerides to their corresponding methyl esters in short reaction times, the reaction has several drawbacks. The process is energy intensive, recovery of the glycerol by-product is difficult, and neutralization of the catalyst produces a large amount of waste salts.

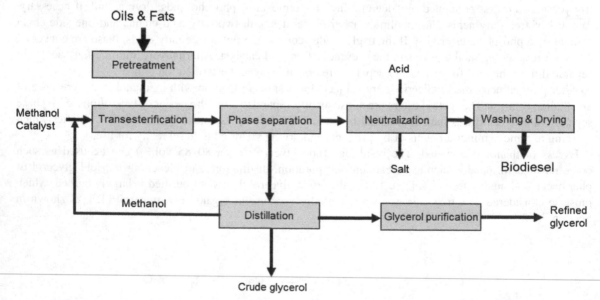

Figure 7.5   *Simplified block diagram for the production of biodiesel using an alkaline catalyst.*

Table 7.2 *Comparison of alkali- and enzyme-catalyzed transesterification processes [13].*

	Alkaline catalyst	Lipase catalyst
Reaction temperature (K)	320–350	305–315
Yield of FAME	Normal	Higher
FFA in feedstock	Saponified	Converted to FAME
Water in feedstock	Interference with reaction	No influence
Recovery of glycerol	Difficult	Easy
Purification of FAME	Repeated washing	None
Production cost of catalyst	Low	High
Catalyst separation	Difficult	Difficult

QUESTIONS:

> *Explain why the chemical transesterification process is energy intensive.*

New processes have been proposed, such as a catalyst-free process based on supercritical methanol [16] and an enzyme-based process [10, 13, 17]. Enzymes like lipases are able to effectively catalyze the transesterification of triglycerides in both aqueous and non-aqueous systems [13] (Table 7.2). Enzyme processes outperform the classical process in several aspects. In particular, glycerol can be easily removed and free fatty acids can be completely converted to the alkyl esters, which makes the process also suitable for processing cheap waste oils and fats. On the other hand, the production cost of a lipase catalyst is significantly greater than that of an alkaline one. In addition, the need for a difficult catalyst separation step remains.

QUESTIONS:

> *Draw a process scheme for transesterification based on a lipase catalyst. Why can a neutralization step be avoided?*

As mentioned above, it is not possible to perform an alkali-catalyzed transesterification process for oils with a high FFA content. For such oils the use of solid acid catalysts is recommended, because these catalysts can simultaneously catalyze the transesterification of triglycerides and the esterification of free fatty acids to methyl esters. Solid acid catalysts have a strong potential to replace homogeneous catalysts, eliminating separation, corrosion, and environmental problems [10, 18]. Figure 7.6 shows a simplified block diagram for a solid-acid-catalyzed transesterification process.

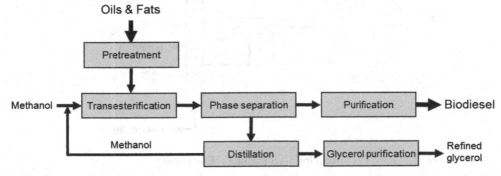

Figure 7.6 *Simplified block diagram for the production of biodiesel using a solid acid catalyst.*

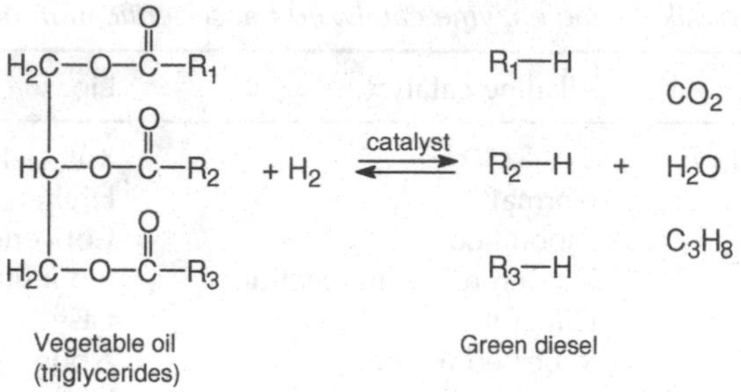

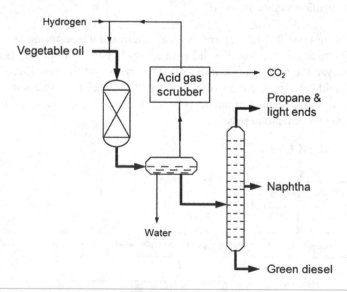

Scheme 7.4   *Hydrodeoxygenation of vegetable oil to produce green diesel.*

QUESTIONS:

Figure 7.6 includes a purification step for the biodiesel product stream. What would be the major impurity? How would you perform the purification?

7.2.2.2   Hydrodeoxygenation

An alternative to transesterification is the catalytic hydrodeoxygenation (HDO) of vegetable oils [19–24] (Scheme 7.4). The primary product is a diesel fraction consisting of branched alkanes, referred to as green diesel or renewable diesel. The molecular structure of green diesel molecules is indistinguishable from that of conventional oil-derived diesel molecules. Coproducts include propane and naphtha, while carbon oxides and water are formed as by-products.

Figure 7.7 shows a simplified flow scheme for a HDO process developed by UOP/Eni. Vegetable oil is combined with fresh and recycle hydrogen and fed to the catalytic reactor, where it is converted into a branched-alkane-rich diesel fuel. The water and carbon dioxide formed by the deoxygenation reactions

Figure 7.7   *Simplified flow scheme of the UOP/Eni Ecofining process for the production of green diesel.*

Table 7.3 *Comparison of diesel fuel properties [21, 23].*

	Oil-based ultra-low-sulfur diesel	Biodiesel	Green diesel
Oxygen (wt%)	0	11	0
Sulfur (ppm)	<10	<1	<1
Heating value (MJ/kg)	43	38	44
Distillation range (K)	470–620	610–625	535–590
Cetane number	40	50–65	70–90
Stability	Good	Poor	Good
CO_2 emissions (kg/MJ)	0.08	0.06	0.04

are separated from the fully deoxygenated hydrocarbon product. The deoxygenated liquid product is then fractionated to remove the small amount of light fuel coproducts. The excess hydrogen is recovered and recycled back to the reactor.

QUESTIONS:

What reactions occur in the UOP/Eni process? What would be the process conditions (temperature, pressure)?

Compare the HDO process with the transesterification process. List advantages and disadvantages.

Why can the economics of HDO of triglycerines be favorable while those of HDO of lignocellulosic biomass generally are not?

Table 7.3 compares the properties of green diesel with those of biodiesel and ultra-low-sulfur diesel produced from crude oil.

One of the most important advantages of the HDO route from an economic viewpoint is that the bio-based feedstocks can be processed at refineries using existing equipment, thereby minimizing capital costs [22]. The produced fuels are good examples of "drop-in" fuels.

7.3 Production of Bio-based Chemicals

Just as the petrochemical industry is based on a relatively small number (circa 20) of so-called base chemicals derived from oil components (Chapter 2), a bio-based industry could be envisioned based on a small number of building blocks, so-called platform molecules, derived from biomass components. *A priori* it is to be expected that biomass-based platform molecules are quite different form oil-based base chemicals. Several such molecules have been identified (Figure 7.8) [25, 26]. These molecules have multiple functional groups and possess the potential to be transformed into numerous commercially relevant value-added chemicals. The chemicals produced from these platform molecules may be the same as those produced from oil-derived chemicals or different but with the same function, or they may even be completely new.

QUESTIONS:

When considering the group of platform molecules proposed, do you feel that this list is complete and that all molecules proposed are good choices? Suggest additional platform

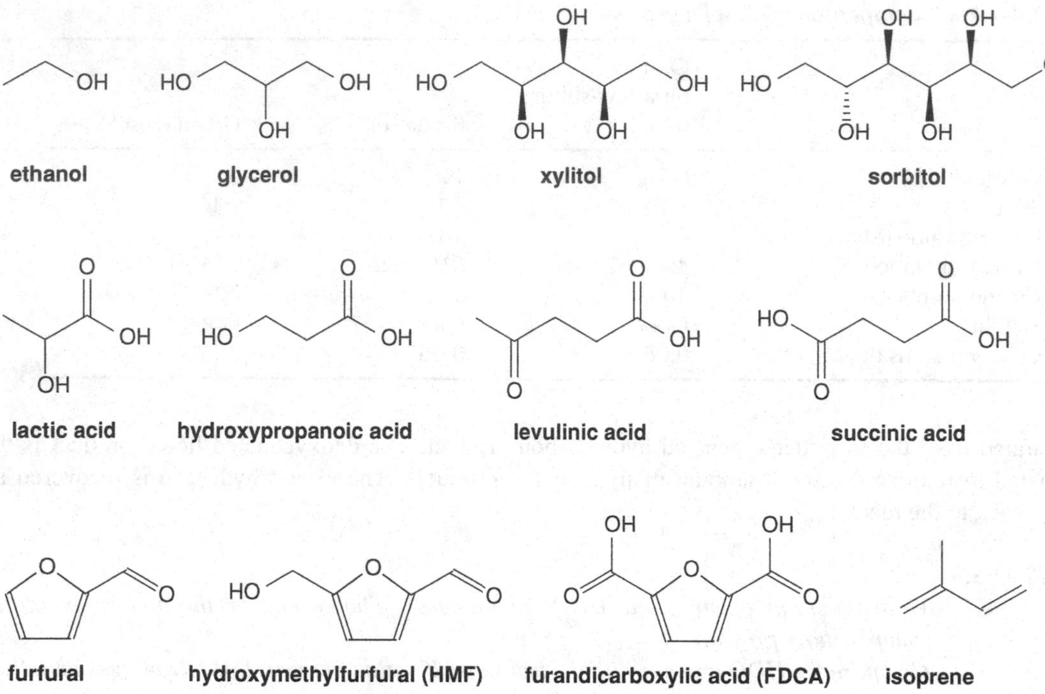

Figure 7.8   *Bio-based platform molecules [26].*

molecules. *(Hint: start from the composition of biomass and minimize the number of reactions needed.)*
What is the main general difference of the platform molecules as compared to oil-based base chemicals?

7.3.1   Ethanol

Ethanol has an enormous potential as a feedstock for the production of bulk chemicals (Figure 7.9). Steam reforming of ethanol [27] using a metal-based catalyst can serve as a renewable source of hydrogen. This route has been widely investigated in recent years. The process is similar to steam reforming of methane (Chapter 5). The reforming reaction (7.1) is highly endothermic and requires temperatures of 870 to 1200 K. These high temperatures may result in the formation of by-products such as carbon monoxide and methane. In addition, deactivation of the catalyst by the formation of coke deposits often has to be coped with.

$$C_2H_5OH + 3H_2O \rightleftarrows 4H_2 + 2CO \qquad \Delta_r H_{298} = 256\,kJ/mol \tag{7.1}$$

The production of ethene by the catalytic dehydration of bioethanol has been discussed in Section 4.5.4. High crude oil prices and the rapid development of fermentation technologies have increased the attractiveness of the bioethene route in comparison with the classical steam cracking process. Ethene is an important starting material for the production of polymers (Chapter 11).

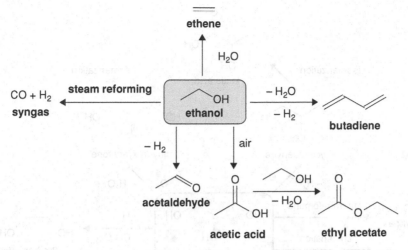

Figure 7.9 *Ethanol as a platform molecule for the production of chemicals.*

QUESTIONS:

In steam reforming of ethanol, which reactions cause the formation of carbon monoxide and methane? What are other possible by-products?

How does processing scale influence the choice of process for ethene production?

7.3.2 Glycerol

The growing production of biodiesel by transesterification of vegetable oils with methanol or ethanol has led to the surplus production of glycerol. This observation has stimulated an immense research effort into possible (new) uses. Glycerol is an intermediate in the synthesis of a large number of compounds used in industry. Figure 7.10 shows some of the most important and novel routes for the conversion of glycerol into useful chemicals.

Like ethanol, glycerol can be converted to synthesis gas by reforming. Because the process operates at relatively low temperature, reaction 7.2 is the most important reaction.

$$C_3H_5(OH)_3 + 3H_2O \rightleftharpoons 7H_2 + 3CO_2 \qquad \Delta_r H_{298} = 129 \text{ kJ/mol} \qquad (7.2)$$

This reaction takes place in the liquid phase and is termed "aqueous phase reforming". This process offers a number of advantages as compared to steam reforming of ethanol. Firstly, the higher reactivity of glycerol allows the reforming process to be carried out at milder temperatures (typically 470–550 K). Unwanted reactions are kinetically restrained under these conditions. Secondly, the process is more versatile and reaction conditions and catalysts can be selected to transform concentrated aqueous solutions of glycerol (as those produced in biodiesel production) into either syngas or hydrogen-enriched streams by coupling the reforming process with water–gas shift processes [28].

Currently, one of the most promising routes for upgrading glycerol is the production of C_3 diols by dehydration and subsequent hydrogenation of the formed intermediate. Propene glycol (1,2-propanediol), the main product obtained by this route, is used in the production of polyester resins, antifreeze, pharmaceuticals, and paints [28].

Glyceraldehyde and dihydroxyacetone can be readily formed by mild oxidation of aqueous glycerol in air. These intermediates can be transformed into lactic acid by isomerization reactions using zeolites, leading to

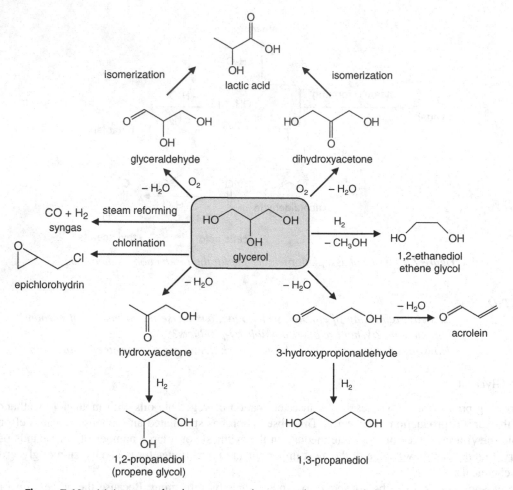

Figure 7.10   *Main routes for the aqueous phase transformation of glycerol into chemicals.*

a mixture of L- and D-lactic acid. This route offers a simple, clean, and non-enzymatic alternative to classical fermentation for the large-scale production of lactic acid, which is an important chemical that is used for production of biodegradable polymers and solvents [28,29]. It should be noted that the enzymatic route leads to either L- or D-lactic acid.

QUESTION:

> *For which applications would the selective production of either L- or D-lactic acid be interesting?*

7.3.3 Succinic Acid

At present, succinic acid is produced by the hydrogenation of petrochemically-derived maleic anhydride (Section 10.5), from which a wide range of chemicals is produced. Several bacteria can produce succinic acid as a major fermentation product; biomass-derived succinic acid could serve as an attractive replacement

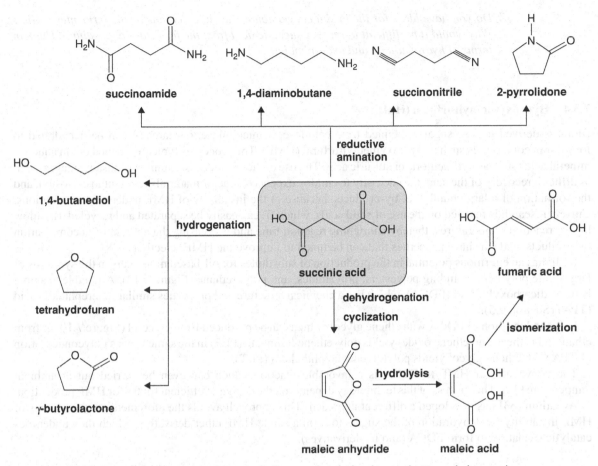

Figure 7.11 *Succinic acid as a platform molecule for the production of chemicals.*

for maleic anhydride. In addition, succinic acid could be used as a monomer for the production of polymers that are currently derived from butane. The use of succinic acid as a platform molecule is summarized in Figure 7.11. Promising derivatives from succinic acid are tetrahydrofuran, γ-butyrolacetone, and butanediol, as well as various pyrrolidones [30].

Major issues for the practical use of succinic acid are its high recovery and purification costs. The separation of by-products, such as acetic acid, formic acid, and lactic acid, has a crucial effect on process costs. Several methods for purification of succinic acid, including electrodialysis, precipitation, and extraction, have been reported [31]. A lot of attention is given to genetic manipulations and process improvements aiming at suppressing the formation of by-products. For the use of succinic acid as a starting material for further conversion, an elegant technology would be one that eliminates the need for the recovery of succinic acid by performing the conversions *in situ* in the fermentation mixture [30].

QUESTIONS:

> *What is electrodialysis? Why is it applicable, in principle, in the purification of bio-based succinic acid?*

Do you have ideas for the in situ conversion of succinic acid inside the fermenter? (Hint: Why would it be difficult to recover succinic acid from the fermentation medium? Think in terms of hydrophilicity and hydrophobicity.)

7.3.4   Hydroxymethylfurfural (HMF)

Biomass-derived hexose sugars, obtained from cellulose, hemicellulose, or starches, can be dehydrated to form furan compounds such as hydroxymethylfurfural (HMF). The process is typically carried out in aqueous mineral acids such as hydrochloric or sulfuric acid. The use of these acids has a number of disadvantages, such as difficult recovery of the acid, the necessity to employ expensive reactor materials resistant to corrosion, and the formation of a large number of by-products, because of the instability of HMF under acidic conditions. Current research is focused on the use of solid acids, which (i) can easily be separated and recycled, (ii) allow higher reaction temperatures, thereby shortening reaction time and decreasing the amount of decomposition by-products, and (iii) have properties that can be tuned to improve the HMF selectivity [28].

HMF has an enormous potential in the production of substitutes for oil-based monomers in the synthesis of large-scale polymers including polyesters, polyamides, and polyurethane (Figure 7.12). A notable example is furan dicarboxylic acid (FDCA), which has a chemical structure and properties similar to terephthalic acid (TPA) (Section 9.5).

Polycondensation of FDCA with ethene glycol (which can be produced from glycerol (Figure 7.10) or from ethanol via ethene and ethene oxide) yields poly-ethene furanoate (PEF) in the same way as polycondensation of TPA with ethene glycol yields poly(ethene terephthalate) (PET).

The conversion of HMF to FDCA is a favorable reaction which can even be carried out at ambient temperature [32]. The success of this technology depends on the design of efficient routes for HMF production.

Avantium [33] has developed a different approach. This approach avoids the aforementioned problems of HMF instability by dehydration of the sugars to form a stable HMF ether derivative, which then undergoes catalytic oxidation to form FDCA (and its derivatives).

7.4   The Biorefinery

7.4.1   Biorefinery Design Criteria and Products

Biomass is a feedstock that can be used in many different applications. In general, multiple products will be produced and in recent years the concept of biorefineries, analogous to oil refineries, has emerged. The term biorefinery refers to a facility (or group of facilities) that combines the production of transportation fuels and chemicals with energy production. By producing several products, a biorefinery takes advantage of the various components in biomass and their intermediates, thereby maximizing the value derived from the biomass feedstock.

Figure 7.13 gives a schematic listing of the main criteria for the design of a biorefinery and a possible product spectrum to be expected. For economic reasons, in general biomass processing will be performed in large-scale processes. It has not been attempted to give a full list of potential products here; this can be found in the literature, for example [34].

The demand for renewable biofuels, which can only be produced from biomass, is growing rapidly. As a consequence, the main challenge for biorefinery development seems to be the efficient and cost-effective production of transportation biofuels, while from the coproduced biochemicals additional economic and environmental benefits can be gained [19].

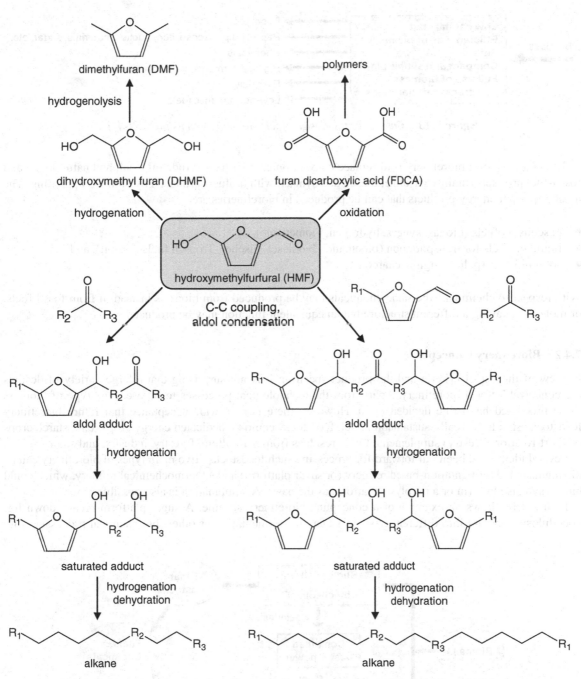

Figure 7.12 *HMF as a platform molecule for the production of chemicals and biofuels.*

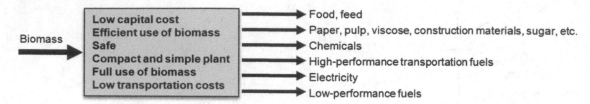

Figure 7.13   *Criteria for the design of a biorefinery with its products [1].*

The objective of a biorefinery is to replace conventional fossil fuels (crude oil, coal, and natural gas) and fossil-fuel products (mainly gasoline, diesel, and syngas) with biofuels produced by biomass upgrading. The most important energy products that can be produced in biorefineries are:

- gaseous biofuels (biogas, syngas, hydrogen, biomethane),
- liquid biofuels for transportation (bioethanol, biodiesel, Fischer–Tropsch fuels, bio-oil), and
- solid biofuels (pellets, lignin, charcoal).

With respect to chemicals, the same chemicals may be produced from biomass instead of from fossil fuels, or molecules having a different structure but an equivalent function may be produced.

7.4.2   Biorefinery Concepts

In view of the list of criteria and the product spectrum it is not surprising that a large variety of designs are conceivable for a biorefinery. Apart from the technological processes to be used, the type of biomass to be processed has to be decided upon. However, there is now wide acceptance that future biorefinery feedstocks should be really sustainable. Such feedstocks could be dedicated energy crops (e.g., starch crops or short-rotation forestry) supplemented with residues from agriculture, forestry, industry, and so on.

Several ideas have been put forward for processing such feedstocks. Two main types of biorefinery can be distinguished: a fermentation-based refinery (or sugar platform) and a thermochemical refinery, which could have a syngas platform or a pyrolysis platform as the basis. A combination is also possible.

Figure 7.14 shows an example of a conceptual biorefinery scheme. A sugar platform breaks down lignocellulosic biomass into different types of sugars for fermentation or other (bio)chemical processing into

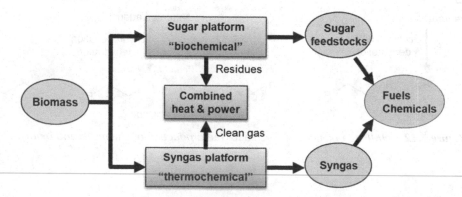

Figure 7.14   *Concept of a biorefinery [35].*

various fuels and chemicals. In a syngas platform, biomass is transformed into synthesis gas by gasification. The synthesis gas can then be converted into a range of fuels and chemicals by catalytic chemical processes.

A biorefinery should preferably produce [19, 35]:

- at least one high-value chemical or material, in addition to low-value, high-volume products (like animal feed and fertilizers);
- at least one energy product besides heat and electricity; the production of at least one biofuel (liquid, solid, or gaseous) is then required.

7.4.3 Core Technologies of a Thermochemical Biorefinery

7.4.3.1 Pyrolysis

Pyrolysis is the thermal decomposition of organic material in the absence of (molecular) oxygen. A number of pyrolysis processes have been described in previous chapters (Chapters 3–5). Pyrolysis can also be an important conversion technology within a biorefinery [3, 36]. It can be applied as a primary unit to convert biomass into bio-oil as feedstock for downstream processes. In biorefineries that first fractionate the biomass into hemicellulose, cellulose, and lignin, pyrolysis can play a role as processing unit for one or more of these fractions. Finally, pyrolysis can be used as a more peripheral unit, for example, to treat side streams that originate from other processes within the biorefinery.

Figure 7.15 shows a simple representation of a biomass pyrolysis process.

QUESTIONS:

Pyrolysis is the key technology in many large-scale processes. List the most important ones and evaluate them with respect to optimal conditions.

Why does Figure 7.15 include the term "hydrophilic organics" instead of just "organics"?

Table 7.4 summarizes the main pyrolysis technologies and their products. Slow pyrolysis is designed to maximize the yield of solid char, while fast pyrolysis (also referred to as flash pyrolysis) maximizes the yield of liquid bio-oil. In general, the goal of pyrolysis in biorefineries is to maximize the yield of bio-oil.

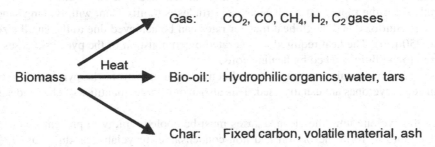

Figure 7.15 *Simple representation of a biomass pyrolysis process.*

Table 7.4   *Modes of pyrolysis and major products.*

Heating rate	Conditions	Typical product composition (wt%)
Fast	Reactor temperature 770–820 K, very high heating rates, >1000 K/s, hot vapor residence time 1–2 s solids residence time <5 s	Char: 12 Bio-oil: 75 Gas: 13
Intermediate	Reactor temperature 720–770 K, heating rate range 1–1000 K/s, hot vapor residence time 10–30 s solids residence time hours–days	Char: 25 Bio-oil: 50 Gas: 25
Carbonization (slow)	Reactor temperature <720 K, heating rate up to 1 K/s, hot vapor residence time hours solids residence time hours–days	Char: 35 Bio-oil: 30 Gas: 35
Torrefaction[a] (slow)	Reactor temperature ~560 K, solids residence 10–60 min	Solid: 80 Bio-oil: 0 Gas: 20

[a] Torrefaction is a mild pyrolyis process that converts biomass into a denser, more brittle, solid form.

QUESTIONS:

The properties (and therefore the performance) of the bio-oil produced depend on the reaction conditions. Would a low or a high heating rate result in bio-oil with the best performance? Do you expect upgrading reactions to be required? (Hint: evaluate the properties of pyrolysis gasoline.)

The technology, and in particular reactor design, will depend on the reaction conditions. Suggest suitable reactor configurations for the four cases listed in Table 7.4.

Fast pyrolysis using fluidized bed reactor technology is the preferred method, because this technology is well understood and can therefore be scaled up relatively easily. Figure 7.16 shows a typical configuration.

During fast pyrolysis, the usually finely ground and dry biomass (typically 10 wt% moisture) is rapidly heated to temperatures around 800 K. This causes the release of a wide variety of vapor phase thermal degradation products that are quenched within a few seconds to produce bio-oil, a mixture of condensed organic compounds and water. In addition, char and non-condensable gases are formed.

A simple method for the rapid heating of biomass particles is to mix them with moving sand particles in a high-temperature fluidized bed. High heat transfer rates can be achieved due to the small size of the sand particles (about 250 μm). The heat required is generated by combustion of the pyrolysis gases, and/or char, and is transferred to the fluidized bed by heating coils.

Char acts as a cracking catalyst, so rapid and effective separation from the pyrolysis product vapors is essential. To this end, cyclones are usually used. This also avoids large quantities of char and sand ending up in the bio-oil.

After leaving the cyclone unit, the cleaned gases must be cooled rapidly to prevent secondary reactions from taking place and to form the bio-oil and non-condensable, recyclable product gases. Quenching in product bio-oil or in an immiscible hydrocarbon solvent is widely practiced.

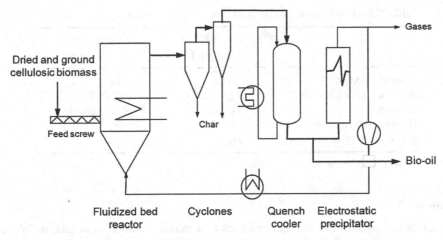

Dried and ground
cellulosic biomass

Feed screw

Char

Gases

Bio-oil

Fluidized bed
reactor   Cyclones   Quench   Electrostatic
                              cooler   precipitator

Figure 7.16   *Pyrolysis of lignocellulosic biomass in a fluidized bed reactor.*

Subsequently, electrostatic precipitators are used for coalescence and collection of what are referred to as aerosols. These are incompletely depolymerized lignin fragments which appear to exist as a liquid with a substantial molecular weight [3].

QUESTIONS:

How would you classify the pyrolysis processes discussed in previous chapters. What is the size of the biomass particles used in a fluidized bed pyrolysis reactor?
Name possible reactions that would take place if no quenching were performed.
How does an electrostatic precipitator work?
Careful handling and storage of the char is required. Why?

Bio-oil is composed of a complex mixture of oxygenated hydrocarbons containing an appreciable amount of water from the original feedstock and formed during pyrolyis. It cannot be used as such but has to be upgraded before it can be used as a transportation fuel or for the production of chemicals.

One option to upgrade the bio-oil is to gasify it (Section 7.4.3.2) and convert the produced syngas to, for example, Fischer–Tropsch liquids.

Catalytic upgrading of bio-oil for the production of transportation fuels or oil refinery feedstocks involves full or partial deoxygenation. The oxygen is either rejected as water, for instance by hydrotreating (HDO), or as carbon dioxide by catalytic cracking on zeolites (Section 3.4.2) [3, 36].

In an industrial process, the char and gas by-products (both 10–20 wt%) would be used primarily as a fuel for the generation of the required process heat (including feedstock drying). However, active carbon or carbon black could be produced from the char left. The char has also been proposed as a soil improver.

Although the bio-oil produced during fast pyrolysis can be upgraded to transportation fuels and chemicals, its high oxygen content makes it corrosive and unstable. It also contains metals and nitrogen that cause deposits on refinery catalysts, and the upgrading process requires large amounts of expensive hydrogen due to the extremely aromatic nature of the bio-oil.

A catalytic approach to pyrolysis could produce an improved bio-oil. By using multifunctional catalysts to remove the oxygen and contaminants, a bio-oil could be produced that is highly carbon efficient, requires less hydrogen to upgrade, and can be easily integrated into existing refineries [36, 37].

Table 7.5   *Comparison of coal and biomass gasification.*

	Coal gasification	Biomass gasification
Plant size (MW$_e$)[a]	>100	<80
Reactor temperature (K)	1250–2200	1070–1120
Pressure	10–40 bar	Atmospheric
Gasification agent (typical)	Oxygen	Air
Sulfur content of gas	High	Low

[a] MW$_e$ refers to electric power.

7.4.3.2   *Gasification*

Gasification technology aims to convert the feedstock to a maximum amount of gas, mainly consisting of carbon monoxide, carbon dioxide, hydrogen, and nitrogen. In Chapter 5, the gasification of coal has been discussed. Table 7.5 compares the characteristics of the gasification of coal and biomass. Coal gasification plants are generally large, while the output of biomass gasification plants is restricted. Biomass is typically gasified at lower temperatures than coal; also unlike coal, bio-gasification typically occurs at atmospheric pressure [38] and with air rather than with pure oxygen [39]. The relatively low content of sulfur and other contaminants requires less investment in gas cleaning.

QUESTIONS:

> *Explain the differences between biomass and coal gasification.*

One of the major problems in biomass gasification is how to deal with tar formed during the process. Biomass is much more prone to producing tars than coal. Tar is a complex mixture of condensable hydrocarbons, which includes single ring to five ring aromatic compounds along with other oxygen-containing hydrocarbons and complex polycyclic aromatic hydrocarbons [40]. Tar can be damaging to the gasifier, as it tends to stick to the walls and clog the entrance and exit ports [38]. Furthermore, tar is undesirable because of various problems associated with condensation, formation of tar aerosols, and polymerization to form more complex structures that cause problems in engines and turbines using the produced gas [41].

Control technologies for tar reduction can broadly be divided into two approaches [40]: (i) prevent tar build up by manipulating the gasifier conditions (primary methods) and (ii) hot gas cleaning after the gasifier (secondary methods, Figure 7.17).

Secondary methods may be chemical or physical treatment [40]. The former consists of thermal or catalytic tar cracking, the latter uses mechanical methods, such as cyclones, filters, rotating particle separators,

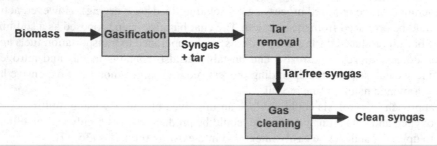

Figure 7.17   *Concept of tar reduction by secondary methods.*

electrostatic precipitators, and scrubbers. Although secondary methods have proved to be effective, treatment inside the gasifier is gaining increasing attention due to economic benefits.

In primary methods, operating parameters, such as temperature, gasifying agent, residence time, and catalytic additives, play an important role in the formation and decomposition of tar.

QUESTIONS:

Do you expect more are less tar formation in biomass gasification than in partial oxidation of crude oil or gasification of coal?

Transport and handling of biomass is a challenge. Continuous feeding of biomass calls for special technology, in particular because it contains a large amount of water and has a highly fibrous structure. After pretreatment by fast pyrolysis or torrefaction, handling of the biomass is much simpler [38].

Pretreatment by fast pyrolysis is the first stage in the overall gasification process (Section 5.3). The produced bio-oil can easily be poured into the gasifier bed of most types of gasifier.

Torrefaction (for definition see Table 7.4) destroys the fibrous nature of biomass. The product of torrefaction is a conglomeration of pellet-like chips called torrefied biomass, also sometimes called "bio-coal." In fact, torrefied biomass closely resembles coal in appearance, and greatly approaches its properties, as well. Torrefaction can be used for most types of biomass. Its most suitable usage is in pretreatment for use in entrained-flow gasifiers. Torrefied biomass is extremely brittle and can easily be crushed into particles of the small size (<0.1 mm) required for processing in these gasifiers [38].

At present, a few small industrial scale and pilot plant biomass gasification plants are in operation. In addition, some power plants cogasify coal and biomass. Advantages of cogasification over stand-alone biomass gasification are that the economies of scale inherent in coal conversion can be exploited and that biomass utilization can be greatly improved due to a synergistic effect [38,42]. Currently, also, the cofeeding of (pyrolysis) bio-oil is under investigation.

QUESTIONS:

Does the density of biomass differ from that of conventional fossil fuels? If so, are there any consequences regarding processing of the biomass?
Compare cofeeding of pyrolysis bio-oil with cofeeding of biomass as such. Which option do you prefer and why?

7.4.4 Existing and Projected Biorefineries

Table 7.6 shows examples of biorefineries at different stages of development. It illustrates the large R&D efforts required in realizing processes for the utilization of lignocellulosic feedstocks.

QUESTIONS:

Compare the various feedstocks in the examples. When you would want to build a new plant as soon as possible, what feedstock would you select? In your discussion, include the product(s) you would want to produce.

7.4.5 Possibility of Integrating a Biorefinery with Existing Plants

A logical question is whether a biorefinery can be integrated with existing plants such as oil refineries or coal gasification plants. It is instructive to evaluate the processes in Figure 7.1 with respect to the potential of merging them with oil refinery processes. For example, biomass can be used directly in an oil refinery for combustion purposes.

Table 7.6   *Examples of biorefineries [43].*

	Platform	Feedstocks	Products	Status
Crop Energies AG (Germany)	C_6 sugars	Sugar and starch crops (sugar, cereals)	Bioethanol, animal feed, electricity	Commercial
Permolex (Canada)	C_6 sugars	Starch crops (wheat, barley, corn)	Bioethanol, food (bakery, flower, gluten), animal feed	Commercial
Avantium Furanics (Netherlands)	C_6/C_5 sugars and lignin	Starch crops, lignocellulosic crops or residues	Synthetic biofuels, chemicals and polymers, electricity and heat	Pilot
Sofiproteol (France)	Transesterification and other chemical processing	Oil crops (rape, and sunflower seeds)	Biodiesel, glycerol, chemicals and polymers, animal feed	Commercial
Inbicon IBUS (Denmark)	C_6/C_5 sugars and lignin	Straw	Bioethanol, animal feed, electricity and heat	Pilot
Lignol (Canada)	C_6/C_5 sugars and lignin	Lignocellulosic crops or residues	Bioethanol, chemicals, biomaterials	Pilot
Ensyn (Canada)	Pyrolysis	Lignocellulosic residues	Synthetic biofuels, flavorings, polymers, fuels, heat	Commercial
European Bio-Hub Rotterdam (Netherlands)	Syngas C_6/C_5 sugars, lignin and protein	Lignocellulosic biomass Biofuel process residues	Synthetic biofuels, chemical Feed, chemicals, fuels	Demonstration
Green Biorefinery (Austria)	Biogas and organic solution	Grasses	Organic acids, biomaterials, fertilizers, biomethane or electricity and heat	Demonstration
M-Real Hallein AG (Austria)	C_6 sugars	Lignocellulosic crops or residues	Bioethanol, biomaterials (paper products), electricity and heat	Conceptual
Neste Oil (Finland)	Hydrodeoxygenation	Vegetable oils, animal fats	Green diesel	Commercial
Zellstoff Stendal GmbH (Germany)	Lignin	Lignocellulosic crops or residues	Biomaterials (paper products), electricity and heat	Commercial
Lenzing AG (Austria)	C_5 sugars and lignin	Lignocellulosic crops or residues	Biomaterials, chemicals, food, electricity and heat	Commercial
Highmark Renewables (Canada)	C_6 sugars and biogas	Starch crops and organic residues (manure)	Bioethanol, animal feed, fertilizer, electricity and heat	Commercial
WUR Micro-algae Biorefinery (Netherlands)	Oil and organic solution	Micro-algae	Chemicals, biodiesel, electricity and heat	Fundamental and applied research, pilot
Roquette Frères (France)	C_6 sugars	Starch crops	Starch, chemicals, animal feeds	Commercial
Virdia (USA)	C_6/C_5 sugars and lignin	Lignocellulosic biomass	Refined C_6/C_5 sugars and clean lignin	Pilot

The pyrolysis process also nicely fits within an oil refinery: a bio-oil is formed that can be separated from the hydrophilic biomass and can be used as a feedstock for further upgrading in an existing refinery [3]. If used in this manner, such a feedstock is referred to as a "drop-in fuel"; the use of drop-in fuels allows maximum integration with the existing refinery infrastructure without major changes.

Gasification of biomass is another example of a process that can be implemented relatively easily. In fact, some coal gasification plants already cogasify a large amount of biomass. In addition, if the syngas produced is upgraded to liquid fuels, these can be used in the blending section of an oil refinery.

A recent concept that has attracted much interest is the decentralized production of bio-oil for transportation to a central process plant for gasification and synthesis of hydrocarbon transportation fuels or alcohols [3].

The implementation of depolymerization (by hydrolysis) and fermentation processes in an oil refinery is hardly realistic, except for some final products such as bioethanol derivatives that can be blended with existing refinery products. In addition, during biomass conversion specific molecules can be produced and blending them with streams in the oil refinery results in downgrading of the bio-based product stream. Thus, in most cases a dedicated plant is needed for efficient biomass conversion by these routes.

7.4.6 Biorefinery versus Oil Refinery

The biorefinery differs profoundly from the oil refinery in terms of feedstock properties, process technology, product spectrum, investments, and environmental impact. Table 7.7 compares a number of characteristics. The oil refinery is associated with a high feedstock price and low-cost processing, whereas for the biorefinery the opposite is the case: a low-cost feedstock combined with high processing costs. The latter is due to several reasons, such as a low energy density, high water content, and high separation costs.

The variety in composition of biomass feedstocks is both an advantage and a disadvantage. An advantage is that biorefineries can make more classes of products than can oil refineries, and they can rely on a wider range of feedstocks. A disadvantage is that a relatively large range of processing technologies is needed, and most of these technologies are still at a pre-commercial stage [34].

For the production of biofuels, the high oxygen content of biomass is a disadvantage, because it reduces the heat content of molecules and usually gives them high polarity, which hinders blending with conventional fuels [44]. Therefore, expensive deoxygenation is required. In contrast, for the production of bio-based chemicals the presence of oxygen may be regarded as an advantage, because oxygen provides valuable physical and chemical properties, and oxidation reactions, which are often difficult and environmentally harmful, may be avoided. Instead, there could be a shift towards "greener" reduction reactions.

Table 7.7 *Comparison of a biorefinery and an oil refinery.*

	Biorefinery	Oil refinery
Feedstock	Highly varying composition	Rather constant composition
	High O content, low S content	Hydrocarbons, some S
	Low density (high volumes)	High density
	Lignocellulosics low cost, oils expensive	High cost
Process technology	Large range of processes	Relatively simple
Conversion steps	Bio and chemical	Mainly chemical
Product spectrum	Complex	Relatively simple
Investments	High	Low
Environment	Footprint can be low	Large footprint

7.5 Conclusions

Biomass has a high potential both as an energy carrier and as a feedstock for the chemical industry. Lignocellulosics are abundantly available and are the preferred biomass feedstocks for the production of chemicals. Instead of just substituting present-day fossil-fuel-derived fuel components and chemicals by bio-based analogues, it is better to judge the potential product spectrum of biomass conversion taking into account the very different structure of biomass compared to fossil fuel feedstocks.

In view of the non-homogeneous composition of biomass, a biorefinery that integrates several technological processes is often the best option. Because biomass is highly functionalized, subtle technology in the liquid phase that is adapted to the specific feedstock is most appealing. When dealing with a well-defined feedstock a dedicated plant could be justified. In practice, several technologies will be used in parallel. For instance, in all plants waste streams that must be cleaned are present. For this, robust, non-specific technology is required. Thus, conventional technologies such as combustion, gasification, and anaerobic fermentation will be applied.

Economic evaluation is a crucial phase in R&D programs for biomass conversion processes. In general, compared with fossil fuels, biomass is characterized by a combination of a relatively cheap feedstock requiring comparatively complex processing, resulting in high investment and operation costs.

QUESTIONS:

> *Make a table with prices of the various feedstocks. Compare these data with the market prices of the products. What are your conclusions?*

References

[1] Sanders, J.P.M., Clark, J.H., Harmsen, G.J., Heeres, H.J., Heijnen, J.J., Kersten, S.R.A., van Swaaij, W.P.M., and Moulijn, J.A. (2012) Process intensification in the future production of base chemicals from biomass. *Chemical Engineering and Processing*, **51**, 117–136.
[2] Puig-Arnavat, M., Bruno, J.C., and Coronas, A. (2010) Review and analysis of biomass gasification models. *Renewable Sustainable Energy Reviews*, **14**, 2841–2851.
[3] Bridgwater, A.V. (2012) Review of fast pyrolysis of biomass and product upgrading. *Biomass Bioenergy*, **38**, 68–94.
[4] Lamers, P., Hamelinck, C., Junginger, M., and Faaij, A. (2011) International bioenergy trade: a review of past developments in the liquid biofuel market. *Renewable Sustainable Energy Reviews*, **15**, 2655–2676.
[5] Eurobserv, E.R. (2011) Biofuels barometer. *Systèmes Solaires – Le Journal des énergies Renouvelables*, **204**, 68–93.
[6] IEA (2011) Technology Roadmap Biofuels for Transport. International Energy Agency, pp. 52. www.iea.org (last accessed 13 December 2012).
[7] Stephen, J.D., Mabee, W.E., and Saddler, J.N. (2012) Will second-generation ethanol be able to compete with first-generation ethanol? Opportunities for cost reduction. *Biofuels, Bioproducts and Biorefining* **6**, 159–176.
[8] Brodeur, G., Yau, E., Badal, K., Collier, J., Ramachandran, K.B., and Ramakrishan, S. (2011) Chemical and physicochemical pretreatment of lignocellulosic biomass: a review. *Enzyme Research*, **2011**, Article ID 787532, 17.
[9] Ma, F. and Hanna, M.A. (1999) Biodiesel production: a review. *Bioresource Technology*, **70**, 1–15.
[10] Helwani, Z., Othman, M.R., Aziz, N., Kim, J., and Fernando, W.J.N. (2009) Solid heterogeneous catalysts for transesterification of triglycerides with methanol: A review. *Applied Catalysis A*, **363**, 1–10.
[11] Murzin, D.Y., Mäki-Arvela, P., and Simakova, I.L. (2012) Triglycerides and oils for biofuels, in *Kirk-Othmer Encyclopedia of Chemical Technology*, John Wiley & Sons, Inc.
[12] Meher, L.C., Vidya Sagar, D., and Naik, S.N. (2006) Technical aspects of biodiesel production by transesterification: a review. *Renewable Sustainable Energy Review*, **10**, 248–268.
[13] Fukuda, H., Kondo, A., and Noda, H. (2001) Biodiesel fuel production by transesterification of oils. *Journal of Bioscience And Bioengineering*, **92**, 405–416.

[14] Lei, X. and Levente, D. (2004) Degumming, in *Nutritionally Enhanced Edible Oil Processing*, AOCS Publishing.

[15] Katryniok, B., Paul, S., Belliere-Baca, V., Rey, P. and Dumeignil, F. (2010) Glycerol dehydration to acrolein in the context of new uses of glycerol. *Green Chemistry*, **12**, 2069–2280.

[16] Ngamprasertsith, S. and Sawangkeaw, R. (2011) Transesterification in supercritical conditions, in *Biodiesel – Feedstocks and Processing Technologies* (eds M. Stoytcheva and G. Montero), Intech, pp. 247–268. Available from: http://www.intechopen.com/articles/show/title/transesterification-in-supercritical-conditions (last accessed 13 December 2012).

[17] Fjerbaek, L., Christensen, K.V., and Norddahl, B. (2009) A review of the current state of biodiesel production using enzymatic transesterification. *Biotechnology and Bioengineering*, **102**, 1298–1315.

[18] Naik, S.N., Goud, V.V., Rout, P.K., and Dalai, A.K. (2010) Production of first and second generation biofuels: a comprehensive review. *Renewable Sustainable Energy Review*, **14**, 578–597.

[19] Cherubini, F. (2010) The biorefinery concept: using biomass instead of oil for producing energy and chemicals. *Energy Conversion and Management*, **51**, 1412–1421.

[20] Kubicková, I. and Kubicka, D. (2010) Utilization of triglycerides and related feedstocks for production of clean hydrocarbon fuels and petrochemicals: a review. *Waste Biomass Valorization*, **1**, 293–308.

[21] Serrano-Ruiz, J.C., Ramos-Fernandez, E.V., and Sepulveda-Escribano, A. (2012) From biodiesel and bioethanol to liquid hydrocarbon fuels: new hydrotreating and advanced microbial technologies. *Energy & Environmental Science*, **5**, 5638–5653.

[22] Choudhary, T.V. and Phillips, C.B. (2011) Renewable fuels via catalytic hydrodeoxygenation. *Applied Catalysis A*, **397**, 1–12.

[23] Kalnes, T., Marker, T., and Shonnard, D.R. (2007) Green diesel: a second generation biofuel. *International Journal of Chemical Reactor Engineering*, **5**, A48. Available at: http://www.bepress.com/ijcre/vol5/A48 (last accessed 13 December 2012).

[24] Lavrenov, A., Bogdanets, E., Chumachenko, Y., and Likholobov, V. (2011) Catalytic processes for the production of hydrocarbon biofuels from oil and fatty raw materials: Contemporary approaches. *Catalysis in Industry*, **3**, 250–259.

[25] Werpy, T. and Petersen, G. (2004) Top Value Added Chemicals from Biomass: Volume 1 – Results of Screening Potential Candidates from Sugar and Synthesis Gas. Pacific Northwest National Laboratory and the National Renewable Energy Laboratory, US Department of Energy, pp. 1–76. http://www.nrel.gov/docs/fy04osti/35523.pdf (last accessed 13 December 2012).

[26] Bozell, J.J. and Petersen, G.R. (2010) Technology development for the production of biobased products from biorefinery carbohydrates-the US Department of Energy's "Top 10" revisited. *Green Chemistry*, **12**, 539–554.

[27] Ni, M., Leung, D.Y.C., and Leung, M.K.H. (2007) A review on reforming bio-ethanol for hydrogen production. *International Journal of Hydrogen Energy*, **32**, 3238–3247.

[28] Serrano-Ruiz, J.C., Luque, R., and Sepulveda-Escribano, A. (2011) Transformations of biomass-derived platform molecules: from high added-value chemicals to fuels via aqueous-phase processing. *Chemical Society Reviews*, **40**, 5266–5281.

[29] Taarning, E., Saravanamurugan, S., Spangsberg Holm, M., Xiong, J., West, R., and Christensen, C.H. (2009) Zeolite-catalyzed isomerization of triose sugars. *ChemSusChem*, **2**, 625–627.

[30] Cukalovic, A. and Stevens, C.V. (2008) Feasibility of production methods for succinic acid derivatives: a marriage of renewable resources and chemical technology. *Biofuels, Bioproducts and Biorefining*, **2**, 505–529.

[31] Bechthold, I., Bretz, K., Kabasci, S., Kopitzky, R., and Springer, A. (2008) Succinic acid: a new platform chemical for bio-based polymers from renewable resources. *Chemical Engineering & Technology*, **31**, 647–654.

[32] Gorbanev, Y.Y., Klitgaard, S.K., Woodley, J.M., Christensen, C.H., and Riisager, A. (2009) Gold-catalyzed aerobic oxidation of 5-hydroxymethylfurfural in water at ambient temperature. *ChemSusChem*, **2**, 672–675.

[33] Avantium (2012) Why YXY? http://www.avantium.com.

[34] Kamm, B., Gruber, P.R., and Kamm, M. (2006) *Biorefineries – Industrial Processes and Products*, Wiley-VCH Verlag GmbH, Weinheim.

[35] NREL (2009) What is a Biorefinery? National Renewable Energy Laboratory (NREL), http://www.nrel.gov/biomass/biorefinery.html (last accessed 13 December 2012).

[36] Venderbosch, R.H. and Prins, W. (2010) Fast pyrolysis technology development. *Biofuels, Bioproducts and Biorefining* **4**, 178–208.

[37] RTI (2012) Second-Generation Biofuels Technology. http://www.rti.org (last accessed 13 December 2012).

[38] Long, H.A. and Wang, T. (2012) Case study for biomass/coal co-gasification in IGCC applications. ASME Turbo Expo 2011, Vancouver, 6–10 June 2011.

[39] Bridgwater, A.V. (1995) The technical and economic feasibility of biomass gasification for power generation. *Fuel*, **74**, 631–653.

[40] Devi, L., Ptasinski, K.J. and Janssen, F.J.J.G. (2003) A review of the primary measures for tar elimination in biomass gasification processes. *Biomass Bioenergy*, **24**, 125–140.

[41] Balat, M., Balat, M., Kırtay, E., and Balat, H. (2009) Main routes for the thermo-conversion of biomass into fuels and chemicals. Part 2: Gasification systems. *Energy Conversion And Management*, **50**, 3158–3168.

[42] Liu, G., Larson, E.D., Williams, R.H., Kreutz, T.G., and Guo, X. (2010) Making Fischer–Tropsch fuels and electricity from coal and biomass: performance and cost analysis. *Energy Fuels*, **25**, 415–437.

[43] de Jong, E., Langeveld, H., and Van Ree, R. (2009) IEA Bioenergy Task 42 Biorefineries: Co-production of Fuels, Chemicals, Power and Materials from Biomass. International Energy Agency, pp. 1–28.

[44] Lange, J.P. (2007) Lignocellulose conversion: an introduction to chemistry, process and economics. *Biofuels, Bioproducts and Biorefining*, **1**, 39–48.

8

Inorganic Bulk Chemicals

8.1 The Inorganic Chemicals Industry

Inorganic chemicals are chemicals that are derived from minerals and from molecular oxygen and nitrogen in the atmosphere. In fact, they constitute the bulk of the base chemicals, compare the data shown in Appendix A (Table A3).

Nitrogen and oxygen are produced from air via liquefaction followed by fractional distillation. Helium and argon, in particular, are produced as valuable by-products (Table 8.1). The production volumes of minerals are huge, as illustrated in Table 8.2.

The sulfur, nitrogen, phosphorus, and chlor-alkali industries are the main producers of base inorganic chemicals. In this chapter, the production of a few key inorganic bulk chemicals, namely, sulfuric acid, nitric acid, and chlorine, is examined. The production processes for these chemicals present interesting technological concepts and innovative features.

In addition, one section is devoted to the production of sulfur, which is used as feedstock for the production of sulfuric acid. The production and the availability of sulfur have changed much over the last two decades; in the past mining was the main source of this element but nowadays most sulfur is produced as a by-product of oil refining.

Another important inorganic chemical is ammonia, the production of which is discussed in Chapter 6 on the production of chemicals and fuels from synthesis gas (H_2 and CO).

Sulfuric acid (H_2SO_4) is the largest-volume chemical produced (Table 8.3, Table A.3 in Appendix A). It is used in the production of all kinds of chemicals, of which fertilizers are the most important. Other major uses are in alkylation in oil refining (Chapter 3), copper leaching, and the pulp and paper industries.

Nitric acid (HNO_3) is mainly used for the production of ammonium nitrate for fertilizer use. Ammonium nitrate is also used for chemicals, explosives, and various other uses. Other relatively large uses of nitric acid are in the production of cyclohexanone, dinitrotoluene, and nitrobenzene.

Chlorine is widely used in the chemical industry (Figure 8.1). It is used to produce vinyl chloride, the starting material for the manufacture of poly(vinyl chloride) (PVC). Hydrochloric acid can be prepared by the direct reaction of chlorine and hydrogen gas. It is used as a chlorinating agent for metals and organic compounds. Chlorine is also used in water purification, in the food industry, and in household cleaning products.

Chemical Process Technology, Second Edition. Jacob A. Moulijn, Michiel Makkee, and Annelies E. van Diepen.
© 2013 John Wiley & Sons, Ltd. Published 2013 by John Wiley & Sons, Ltd.

Table 8.1   *Noble gases in the earth's atmosphere [1].*

Noble gas	Abundance (ppm)	Price (US $/m^3)
He	5.2	4.5 (industrial grade)
		33 (laboratory grade)
Ne	18.2	90
Ar	9340	5
Kr	1.1	450
Xe	0.09	4500

QUESTIONS:

Why would Ne be more expensive than He, whereas the abundance data in Table 8.1 suggest otherwise?

What is meant by the term "calcining"?

An important side product of chlorine production is sodium hydroxide or "caustic soda" (NaOH). It is widely used in the chemical industry. It is used together with chlorine to make sodium hypochlorite (NaClO) for use in disinfectants and bleaches. Considerable amounts of NaOH are used in the manufacture of paper.

8.2   Sulfuric Acid

Sulfuric acid production involves the catalytic oxidation of sulfur dioxide (SO_2) into sulfur trioxide (SO_3). The process has developed via the "lead chamber" process, which was used in the eighteenth and nineteenth centuries and used a homogeneous catalyst (nitrogen oxides), to the modern "contact process". The contact process originally used a supported platinum catalyst, but from the 1920s on this has been gradually replaced by a vanadium catalyst. The main reasons were the high price of platinum and its sensitivity to poisons. Nowadays the life time of an active catalyst is 10–20 years. Another significant change occurred in the early 1960s with the introduction of the so-called double absorption process, in which part of the formed sulfur trioxide is removed between two conversion stages.

Table 8.2   *World production of minerals in 2010 [2].*

Source	Production (10^6 t)	Main examples of uses
Lime[a] (CaO or Ca(OH)$_2$)	310	Steelmaking, flue gas desulfurization (FGD)
Phosphate rock	176	Ammonium phosphate fertilizers, detergents
Potassium compounds	33	Fertilizers
Salt (mainly NaCl)	270	Chlorine, caustic soda (NaOH) production
Sodium carbonate (Soda ash, Na$_2$CO$_3$)	46	Glass, chemicals
Sulfur	68	Sulfuric acid production

[a]CaO, also referred to as quicklime, is obtained by calcining limestone ($CaCO_3$), which is a sedimentary rock; Ca(OH)$_2$, or hydrated lime, is formed by reaction of CaO with water.

Table 8.3   *World production of inorganic chemicals in 2010.*

Chemical	Production (10^6 t)	Main examples of uses
H_2SO_4	190	Fertilizers
NH_3	160	Fertilizers
HNO_3	70	Fertilizers
Cl_2	65	PVC
NaOH	60	Paper manufacture

Sulfur dioxide can be obtained from a wide variety of sources, including elemental sulfur, spent sulfuric acid (contaminated and diluted), and hydrogen sulfide. In the past, iron pyrites were also often used, while large quantities of sulfuric acid are also produced as a by-product of metal smelting. Nowadays, elemental sulfur is by far the most widely used source. Most of this elemental sulfur is a by-product of oil refining (Chapter 3). It is produced by conversion of hydrogen-sulfide-rich gas streams (Section 8.3).

Sulfuric acid is commercially produced in various acid strengths, ranging from 33.33 to 114.6 wt%. Sulfuric acid with a strength of over 100 wt% is referred to as oleum, which consists of sulfuric acid with dissolved sulfur trioxide. The concentration of oleum is expressed as wt% dissolved sulfur trioxide ("free SO_3") in 100 wt% sulfuric acid.

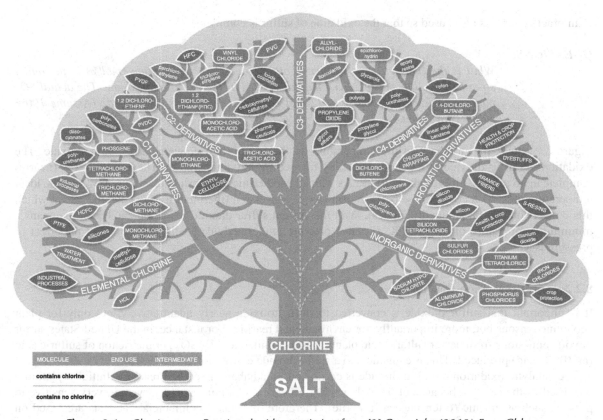

Figure 8.1   *Chorine tree. Reprinted with permission from [3] Copyright (2012) Euro Chlor.*

QUESTIONS:

> *The lead chamber process is one of the few examples of homogeneous catalysis in the gas phase. Do you know another example?*
> *Why is the process called the "lead chamber process"?*
> *What are sources of H_2S-rich gases?*
> *The vanadium catalyst used appears to be a liquid under reaction conditions. How is this engineered in practice?*

8.2.1   Reactions and Thermodynamics

The reactions involved in the production of sulfuric acid from elemental sulfur involve generation of sulfur dioxide from elemental sulfur (reaction 8.1), followed by catalytic oxidation of sulfur dioxide (reaction 8.2) and absorption of the formed sulfur trioxide in water (reaction 8.3):

$$S\,(l) + O_2\,(g) \rightarrow SO_2\,(g) \qquad \Delta_r H_{298} = -298.3 \text{ kJ/mol} \tag{8.1}$$

$$SO_2\,(g) + 1/2\,O_2\,(g) \rightleftarrows SO_3\,(g) \qquad \Delta_r H_{298} = -98.5 \text{ kJ/mol} \tag{8.2}$$

$$SO_3\,(g) + H_2O\,(l) \rightarrow H_2SO_4\,(l) \qquad \Delta_r H_{298} = -130.4 \text{ kJ/mol} \tag{8.3}$$

In practice, excess air is used so that the oxidation of sulfur is complete.

QUESTIONS:

> *What is the (theoretically) maximum attainable SO_2 concentration (mol%) in the outlet stream of a sulfur burner, given that air is used as the oxidizing medium? The actual SO_2 content varies from about 7 to 12 mol%. How much excess air is used? (Assume 100% sulfur conversion.)*

Figure 8.2 shows the influence of the temperature on the equilibrium conversion of sulfur dioxide. The oxidation of sulfur dioxide is thermodynamically favored by low temperature. The conversion is (nearly) complete up to temperatures of about 700 K. As with all exothermic equilibrium reactions, however, the ideal temperature must be a compromise between achievable conversion (thermodynamics) and the rate at which this conversion can be attained (kinetics). With the current sulfur dioxide oxidation catalysts this means a minimum temperature of 680–715 K. Elevated pressure is thermodynamically favorable, but the effect of pressure is small.

8.2.2   SO₂ Conversion Reactor

It is important that sulfur dioxide is converted to sulfur trioxide nearly quantitatively, not only for plant economic reasons but, more importantly, for environmental reasons. For instance, in the United States sulfur dioxide emissions from newer sulfuric acid plants must be limited to 2 kg SO_2 per metric ton of sulfuric acid (as 100% acid) produced. This is equivalent to a sulfur dioxide conversion of 99.7% [5].

The catalytic oxidation of sulfur dioxide is carried out in adiabatic fixed bed reactors. Initially, platinum was used as the catalyst because of its high activity. However, platinum is expensive and very sensitive to poisons such as arsenic (often present in pyrite). Therefore, nowadays commercial catalysts are based on vanadium oxide supported on a porous inorganic support.

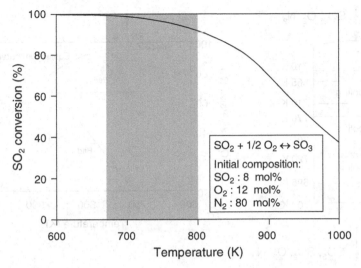

Figure 8.2 *Equilibrium conversion of SO_2 to SO_3 (p = 1 bar; shaded area indicates range of practical operating temperature) [4].*

QUESTION:

> *The vanadium-based catalyst is supported on a ceramic support. Porous silica and alumina are extensively used in heterogeneous catalysis. If you had to choose between these two supports, which would you select and why?*

As discussed in the previous section, the oxidation of sulfur dioxide is an exothermic equilibrium reaction. Consequently, with increasing conversion the temperature rises, leading to lower attainable equilibrium conversions. For example, the equilibrium conversion at 710 K is about 98% but, due to the adiabatic temperature rise, conversions of only 60–70% are obtainable in a single catalyst bed. In practice, this is overcome by using multiple catalyst beds (usually four) with intermediate cooling as shown in Figure 8.3. Cooling can be achieved by heat exchange or by quenching with air.

QUESTIONS:

> *Give examples of other processes using this type of reactor configuration.*
> *What do the diagonal lines and horizontal lines represent? Are the diagonal lines parallel to each other? Explain.*

As can be seen from Figure 8.3, the major part of the conversion is obtained in the first bed. The inlet temperature of the first bed is around 700 K and the exit temperature is 865 K or more, depending on the concentration of sulfur dioxide in the gas. The successive lowering of the temperature between the beds ensures an overall conversion of 98–99%. Still, this is not enough to meet current environmental standards. Therefore, modern sulfuric acid plants use intermediate sulfur trioxide absorption after the second or, more commonly, the third catalyst bed.

The intermediate removal of sulfur trioxide from the gas stream enables the conversion of sulfur dioxide "beyond thermodynamic equilibrium". Other examples where the principle of carrying out a reaction beyond its thermodynamic equilibrium is applied are discussed in Chapter 14.

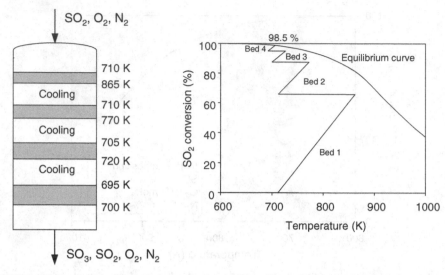

Figure 8.3 *SO₂ oxidation in reactor with four catalyst beds with intermediate cooling (initial composition same as in Figure 8.1).*

Figure 8.4 shows the equilibrium curves for sulfur dioxide conversion with and without intermediate removal of sulfur trioxide. It illustrates that an overall conversion exceeding 99% is obtained.

8.2.3   Modern Sulfuric Acid Production Process

Figure 8.5 shows a schematic of a modern sulfuric acid plant. Three main sections can be distinguished, that is, combustion of sulfur to produce sulfur dioxide (sulfur burner), catalytic oxidation of sulfur dioxide to form sulfur trioxide, and absorption of sulfur trioxide in concentrated sulfuric acid (absorption towers). Because of the presence of two absorption towers it is often referred to as a "double absorption process".

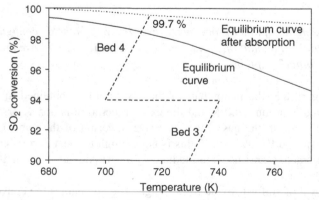

Figure 8.4 *Effect of interstage absorption of SO₃ on SO₂ oxidation (initial composition: 11.5 mol% SO₂; 9.5 mol% O₂).*

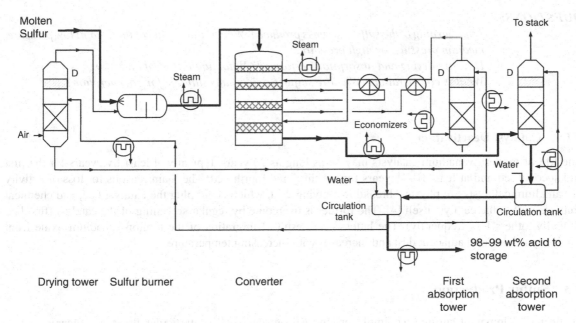

Figure 8.5 *Flow scheme of a sulfuric acid plant with two absorption towers. D = demister.*

Dried air and atomized molten sulfur are introduced at one end of the sulfur burner, which is a horizontal, brick-lined combustion chamber. A high degree of atomization and good mixing are key factors in producing efficient combustion. Atomization typically is accomplished by pressure spray nozzles or by mechanically driven spinning cups. Some sulfur burner designs contain baffles or secondary air inlets to promote mixing. Handling of molten sulfur requires keeping the temperature between 410 and 425 K, where its viscosity is lowest. Therefore, it is transported through heated lines.

Most of the heat of combustion is used to generate high-pressure steam in a waste heat boiler, which reduces the gas temperature to the desired inlet temperature of the first catalyst bed. The heat of reaction of sulfur dioxide oxidation raises the temperature in the first bed significantly. The hot exit gases are cooled before entering the second bed, where further conversion takes place. The gas leaving this bed is cooled by heat exchange with cold gas leaving the first absorption tower. It subsequently undergoes further conversion in the third bed, after which it is cooled and fed to the first absorption tower. Here, most of the sulfur trioxide produced is absorbed in a circulating stream of sulfuric acid. The gas leaving the absorption tower is reheated by heat exchange with gas from the third and second converter beds and then enters the fourth catalyst bed, where most of the remaining sulfur dioxide is converted to sulfur trioxide (overall conversion >99.7%). The final absorption tower removes this sulfur trioxide from the gas stream before release to the atmosphere.

Small amounts of sulfuric acid mists or aerosols are always formed in sulfuric acid plants when gas streams are cooled or sulfur trioxide reacts with water below the sulfuric acid dew point. Formation of sulfuric acid mists is highly undesirable because of corrosion and process stack emissions. Therefore, the absorbers in sulfuric acid plants are equipped with demisters, consisting of beds of small-diameter glass beads or Teflon fibers.

Sulfuric acid plants can be operated as cogeneration plants. Much of the heat produced in the combustion of sulfur is recovered as high-pressure steam in waste heat boilers, while some of the heat produced in the catalytic sulfur dioxide oxidation is also recovered by steam production in so-called economizers. Steam production in modern large sulfuric acid plants exceeds 1.3 t/t of sulfuric acid produced.

QUESTIONS:

> *After burning of the sulfur, steam is produced. What is the quality of this steam: low pressure, medium pressure, or high pressure?*
> *How are drying and absorption performed? What liquid is used and why?*
> *What is cogeneration? Why are sulfuric acid plants suitable for cogeneration?*

8.2.4   Catalyst Deactivation

The life of modern vanadium catalysts may be as long as 20 years, typically at least five years for the first and second bed and at least 10–15 years for the third and fourth bed. The main reasons for loss of activity of vanadium catalysts are physical breakdown giving dust, which could plug the catalyst bed, and chemical changes within the catalyst itself [4]. The former is overcome by regular screening of the catalyst (first bed annually, others less frequently). The latter is the result of migration of the molten vanadium oxide from catalyst particles into adjacent dust and increases with increasing temperature.

8.3   Sulfur Production

In the past, elemental sulfur was simply produced from deposits, in particular those of volcanic origin. Nowadays sulfur is nearly completely produced as a by-product of the processing of sulfur-containing feedstocks, which produce an off-gas rich in hydrogen sulfide.

Hydrogen sulfide obtained by amine absorption (Section 5.4) from hydrogen sulfide-containing gas such as sour natural gas, and off-gas from hydrotreating units (Section 3.4.5) or coal gasification plants (Section 5.3), is frequently converted to elemental sulfur by the "Claus" process [6, 7]. By this process 95–97% of the hydrogen sulfide is converted. To obtain greater conversions several technologies are in use. Some are already proven technology, others are under development [6].

In the Claus process, elemental sulfur is produced by partial oxidation of hydrogen sulfide in a furnace. The overall reaction may be represented simply by the highly exothermic reaction:

$$2\,H_2S + O_2 \rightleftarrows S_2 + 2\,H_2O \qquad \Delta_r H_{298} = -444 \text{ kJ/mol} \qquad (8.4)$$

but in reality the chemistry is much more complex [6, 7]. In practice, the overall reaction is accomplished in two steps. One third of the hydrogen sulfide is oxidized to sulfur and water, and then the remaining hydrogen sulfide reacts with the formed sulfur dioxide converting it into sulfur:

$$2\,H_2S + 3\,O_2 \rightarrow 2\,SO_2 + 2\,H_2O \qquad \Delta_r H_{298} = -1058 \text{ kJ/mol} \qquad (8.5)$$

$$2\,H_2S + SO_2 \rightleftarrows 3/2\,S_2 + 2\,H_2O \qquad \Delta_r H_{298} = -147 \text{ kJ/mol} \qquad (8.6)$$

At the high temperature of the furnace, oxidation rates are very high and in agreement with thermodynamics sulfur dioxide is formed. Because only a substoichiometric amount of oxygen is fed to the plant, sulfur can be formed in a consecutive reaction. Sulfur recovery in this thermal step is limited to 50–70% due to the equilibrium, which shifts to the left at higher temperature.

QUESTIONS:

> *Why does the overall reaction take place in two steps?*

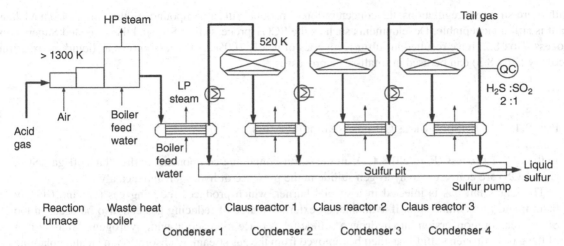

Figure 8.6   *Flow scheme of a typical Claus process. QC = quality control.*

To increase the conversion, catalytic converters (usually two or three in series) are added, which operate at relatively low temperature. In practice, about 95% of the hydrogen sulfide is converted to sulfur. Figure 8.6 shows a flow scheme of a typical Claus process.

The hydrogen sulfide-containing gas is fed to an oxidation chamber where it is partially combusted with air to form sulfur dioxide, which then reacts with the remaining hydrogen sulfide to form elemental sulfur. The heat of reaction is recovered in a waste heat boiler by raising high-pressure steam. Gas leaving the combustion chamber is cooled to about 450 K for sulfur condensation. The gas is then reheated (to 520 K) and passed into the first Claus reactor, where the Claus equilibrium is re-established and additional sulfur is formed. After each reactor sulfur is removed by condensation. In the condensers low-pressure steam is generated. The tail gas, which besides residual hydrogen sulfide and sulfur dioxide contains small amounts of carbon disulfide (CS_2) and carbonyl sulfide (COS), is either further processed (Boxes 8.1 and 8.2) or incinerated and emitted to the atmosphere, depending on emission regulations.

In order to ensure optimal conversion in the Claus plant, the H_2S:SO_2 ratio should be 2:1. The ratio is measured in the tail gas from the final condenser, and the air flow to the furnace is adjusted to achieve this ratio.

The sulfur produced is a salable product. The Claus process is widely applied in refineries and natural gas processing plants. Worldwide, about 1500 plants with an average capacity of 100 t/d sulfur are in use.

QUESTIONS:

What would be the temperature profile in the reaction furnace? In older process designs, one third of the H_2S feed entered the burner and was completely converted into SO_2. The other two thirds was added to the Claus reactor(s). What is the advantage of the configuration in Figure 8.6?

In principle, removal of H_2O from the gas mixture could be applied to enhance the conversion of H_2S. How could this be done? Why would the conversion be higher?

Why is incineration of the residual gas necessary? What compound is formed? What should an operator do when the H_2S:SO_2 ratio measured in the tail gas is lower than 2:1?

Large amounts of sulfur are produced in Claus plants. Compare the capacity with the sulfur consumed in the production of sulfuric acid.

With more stringent regulations, the concentration of residual sulfur compounds in the tail gas from a Claus plant is still unacceptable. Developments such as the SCOT process (Box 8.1) and the so-called SuperClaus process (Box 8.2) have resulted in sulfur recoveries of nearly 100%. A process known as liquid redox sulfur recovery (Box 8.3) can be used in small-scale processes.

Box 8.1   SCOT (Shell Claus Off-gas Treating) Process

In the SCOT process (Figure B.8.1.1), all the sulfur-containing components in the Claus off-gas (SO_2, CS_2, COS) are converted to hydrogen sulfide in the presence of hydrogen and a catalyst.

The Claus tail gas is injected in a special burner, which produces reducing gas (H_2 and CO) by incomplete fuel combustion. If available, an external source of reducing gas can also be used. In the SCOT reactor, conversion into hydrogen sulfide takes place by reaction with hydrogen. After cooling of the gas, hydrogen sulfide can then be removed from the gas stream by absorption in an alkanolamine solution (see also Section 5.4.1).

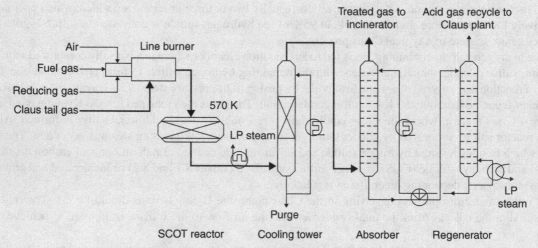

Figure B.8.1.1   *Flow scheme of the SCOT process.*

An appreciable amount of carbon dioxide is present in the gas stream to the absorber, so hydrogen sulfide absorption should be selective. This is achieved by choosing the conditions in the absorber such that the bonding of hydrogen sulfide to the amine is strong, while reaction of carbon dioxide with the amine hardly takes place.

QUESTIONS:

> *What is the origin of the CO_2 in the gas? Why would its co-absorption have to be avoided?*

Box 8.2 SuperClaus Process

The SuperClaus process has been developed by Stork–Comprimo BV, in cooperation with Gastec NV and The University of Utrecht, The Netherlands. The process is based on the use of a selective oxidation catalyst in the last reactor stage, increasing the sulfur recovery to about 99%. The catalyst catalyzes the direct oxidation of hydrogen sulfide to elemental sulfur. Due to the low operating temperature, the equilibrium is shifted to the right, so a high conversion to sulfur can be obtained. It is essential that the catalyst exhibits a high selectivity; it should produce only very small amounts of sulfur dioxide.

Figure B.8.2.1 shows a flow scheme of the SuperClaus process. The most important differences with the Claus process are the use of a selective oxidation reactor and a less sensitive, more flexible air: acid gas control.

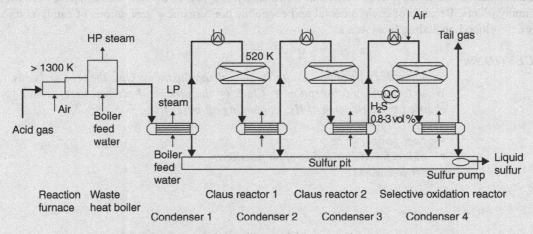

Figure B.8.2.1 *Flow scheme of the SuperClaus process.*

The acid gas feed is burned in the thermal stage in a substoichiometric amount of oxygen so that the $H_2S:SO_2$ ratio is much higher than the conventional value of 2:1. Hence, after the two Claus reactors nearly all sulfur dioxide has been converted, while hydrogen sulfide is still present. This stream is mixed with excess air before it enters the oxidation reactor. After the oxidation reactor and the final sulfur condensation step, the tail gas is incinerated.

Sulfur recovery in the SuperClaus plant is not as high as can be achieved in SCOT plants because sulfur compounds such as carbon disulfide and carbonyl sulfide are not converted. However, it is possible to convert these compounds to hydrogen sulfide by hydrogenation prior to selective oxidation. It is claimed that the sulfur recovery can then be increased to the level achieved in the SCOT process.

QUESTION:

> *Compare the SuperClaus with the combination of the Claus and SCOT processes. What are advantages and disadvantages of both options?*

Box 8.3   Liquid Redox Sulfur Recovery

The processes discussed previously are suited for large-scale operation. In small-scale processes (say <10 t/d sulfur) the gas flow rate and composition are usually not constant. This alone makes the above technologies less applicable. Moreover, Claus plants require a large investment, which is not economical for small-scale applications.

An alternative is to add a catalyst to the liquid used for absorption of hydrogen sulfide from off-gases and to convert the hydrogen sulfide directly into elemental sulfur:

$$2\,H_2S + O_2 \rightarrow S_2 + 2\,H_2O \qquad \Delta_r H_{298} = -444 \text{ kJ/mol}$$

In this respect, one might speak of a multifunctional reactor. Several processes have been developed based on various catalyst systems. Vanadium-based catalysts appeared to work well and have been used in many plants. Because of environmental and economic concerns new generations of catalysts have been developed, including biosystems.

QUESTIONS:

Evaluate this type of process for large-scale applications. Give the pros and cons. What technology (absorption + Claus or the process described here) would you choose for the cleaning of H_2S containing off-gases from:

- *coal gasifiers;*
- *plants for hydrotreating of oil fractions;*
- *sewage works?*

8.4   Nitric Acid

The history of nitric acid production goes back to the eighth century. From the Middle Ages onwards, nitric acid was primarily produced from saltpeter (potassium nitrate, KNO_3) and sulfuric acid. In the nineteenth century, Chile saltpeter (sodium nitrate, $NaNO_3$) largely replaced saltpeter. At the beginning of the twentieth century new production technologies were introduced. The rapid development of a process for the production of ammonia (first from coal, later from natural gas or naphtha, Section 6.1) made possible a commercial route for the production of nitric acid based on the catalytic oxidation of ammonia with air. This route, although modernized, is still used for all commercial nitric-acid production.

8.4.1   Reactions and Thermodynamics

Nowadays, all nitric acid production processes are based on the oxidation of ammonia, for which the overall reaction reads:

$$NH_3 + 2\,O_2 \rightarrow HNO_3 + H_2O \qquad \Delta_r H_{298} = -330 \text{ kJ/mol} \tag{8.7}$$

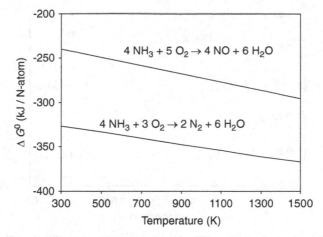

Figure 8.7 *Gibbs free energy of formation ΔG^0 in the oxidation of ammonia at 1 bar [8].*

Many reactions are involved in the oxidation, a simplified representation of which is the following sequence of reactions: burning of ammonia to nitric oxide (reaction 8.8), oxidation of the nitric oxide (reaction 8.9), and absorption of nitrogen dioxide in water (reaction 8.10):

$$4\,NH_3 + 5\,O_2 \rightarrow 4\,NO + 6\,H_2O \qquad \Delta_r H_{298} = -907 \text{ kJ/mol} \qquad (8.8)$$

$$2\,NO + O_2 \rightarrow 2\,NO_2 \qquad \Delta_r H_{298} = -113.8 \text{ kJ/mol} \qquad (8.9)$$

$$3\,NO_2 + H_2O \rightarrow 2\,HNO_3 + NO \qquad \Delta_r H_{298} = -37 \text{ kJ/mol} \qquad (8.10)$$

8.4.1.1 Ammonia Oxidation

Ammonia oxidation (reaction 8.8) is highly exothermic and very rapid. However, the reaction system has a strong tendency towards the formation of nitrogen:

$$4\,NH_3 + 3\,O_2 \rightarrow 2\,N_2 + 6\,H_2O \rightarrow \Delta_r H_{298} = -1261 \text{ kJ/mol} \qquad (8.11)$$

Figure 8.7 illustrates this by presenting the Gibbs energy of formation of reactions 8.8 and 8.11. Clearly, the formation of nitrogen is thermodynamically much more favorable. This can be overcome by using a selective catalyst and short residence time (10^{-4}–10^{-3} s) at high temperature.

QUESTION:

Why is a combination of short residence and high temperature chosen and not a larger residence time and a lower temperature?

NO formation is catalyzed by a solid catalyst (a Pt-Rh alloy). What type of reactor would you choose?

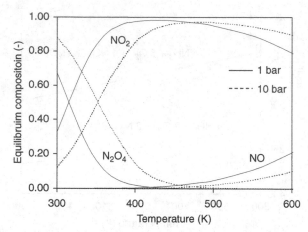

Figure 8.8   *Influence of temperature and pressure on the equilibrium composition (mol/mol total N compounds) of a mixture of NO/NO₂/N₂O₄ (starting from a stoichiometric mixture of NO and O₂).*

8.4.1.2   *Oxidation of Nitric Oxide and Absorption in Water*

The oxidation of nitric oxide (NO) to nitrogen dioxide (NO₂) (reaction 8.9) is a non-catalyzed reaction. The thermodynamic data in Figure 8.8 show that the conversion of nitric oxide is favored by low temperature and high pressure. Nitrogen dioxide (NO₂)is in equilibrium with its dimer dinitrogen tetroxide (N₂O₄).

The oxidation of nitric oxide is a famous reaction because it seems to be one of the few third order reactions known:

$$r_{NO} = -k \cdot p_{NO}^2 \cdot p_{O_2} \tag{8.12}$$

Moreover, a peculiarity of the oxidation is the fact that its reaction rate coefficient increases with decreasing temperature. Therefore, this reaction is favored by low temperature not only thermodynamically but also kinetically. The data suggest a stepwise reaction mechanism.

QUESTION:
> *Suggest a reaction mechanism that can explain the observed behavior of k.*

The absorption process of nitrogen dioxide in water (reaction 8.10) is quite complex because several reactions can occur, both in the liquid phase and in the gas phase. For practical purposes the simplified reaction (8.10) is satisfactory.

A peculiar point is that in the reaction not only is the desired product (HNO₃) formed but also part of the reactant (NO) is formed back. In addition, as reaction 8.10 takes place in the liquid phase, while NO oxidation (reaction 8.9) occurs in the gas phase, the rate of mass transfer between the gas and liquid phase plays a major role.

8.4.2   Processes

The optimal conditions for ammonia oxidation are high temperature, short residence time (in order to minimize side and following reactions) and low pressure. In practice, these conditions are realized by employing a reactor

Table 8.4 *Typical data for nitric acid plants [4].*

Parameter	Atmospheric pressure	Intermediate pressure		High pressure	
		single	dual	single	dual
Converter pressure (bar)	1–1.4	3–6	0.8–1	7–12	4–5
Absorption pressure (bar)	1–1.3	3–6	3–5	7–12	10–15
Acid strength (wt%)	49–52	53–60	55–69	52–65	60–62
NH_3 to HNO_3 conversion (%)	93	88–95	92–96	90–94	94–96
NH_3 to NO conversion (%)	97–98	96	97–98	95	96
Pt loss, g/t HNO_3	0.05	0.10	0.05	0.30	0.1
Catalyst temperature, average (K)	1100	1150	1100	1200	1150
Typical catalyst life, months	8–12	4–6	8–12	1.5–3	4–6

consisting of several layers of a woven catalyst net, composed of platinum strengthened with rhodium [4, 9]. Due to crystallization and erosion at the high gas velocity applied, some platinum is lost with the product gas. In practice, this is a significant cost factor. The lost platinum is often recovered by traps, consisting of a gauze pack of palladium. The catalyst life varies from several months to a year (Table 8.4).

The choice of the reactor pressure is not trivial. In practice, higher pressure plants are most common, because of reduced equipment size, and lower capital costs, and because the absorption of nitrogen dioxide is a process that operates best at high pressure. However, a higher pressure lowers the oxidation efficiency (percentage ammonia converted to NO). Therefore, a dilemma exists: should a low or a high pressure be applied? In practice, several options are used; both the oxidation and the absorption are carried out at atmospheric pressure, or a combination of low pressure for the oxidation reactor and high pressure for the absorption process is used. Table 8.4 shows some typical data on the various plant types. Currently, atmospheric pressure plants are obsolete.

QUESTIONS:

How do process conditions influence the acid strength?
Can you explain the differences in the catalyst temperature?
It has been discovered that the Pt losses are due to the formation of relatively volatile Pt oxides. How do you think the Pd trap works?
Which type of wire gauzes do you expect to be better, wire gauzes composed of porous or non-porous wires?

8.4.2.1 Single-Pressure Process

Figure 8.9 shows a simplified flow scheme of a single-pressure nitric acid process. Most single-pressure processes operate at high pressure (7–12 bar). The higher catalyst temperature (Table 8.4) and operating pressure enable more efficient energy recovery.

Both ammonia and air are filtered to remove impurities such as rust and other particulates present in the feedstocks; rust particles promote the decomposition of ammonia. The compressed air and gaseous ammonia are mixed such that there is excess oxygen (circa 10 vol.% ammonia in mixture) and are passed over a stack of several gauzes of platinum catalyst to produce nitric oxide (NO) and water. The resulting gases are rapidly cooled, thus generating steam that can be used to power the steam turbine driving the compressors.

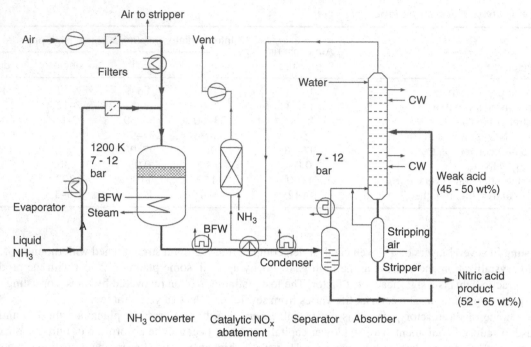

Figure 8.9   *Flow scheme of a single-pressure nitric acid plant (high pressure) using catalytic NO_x abatement (SCR). BFW = boiler feed water, CW = cooling water.*

Upon cooling, nitric oxide is further oxidized to form nitrogen dioxide (NO_2) in equilibrium with its dimer, dinitrogen tetroxide (N_2O_4). In the condenser, condensation of (weak) nitric acid takes place. The choice of materials is critical because of the highly acidic conditions. The weak acid stream is subsequently separated from the gas mixture and fed to the absorption column. The gases leaving the separator are mixed with secondary air (from stripper) to enhance oxidation of nitric oxide, and then fed to the bottom of the absorber, where the equilibrium mix of nitrogen dioxide and dinitrogen tetroxide is absorbed into water, producing nitric acid. The nitric acid leaving the bottoms of the absorber is contacted with air in a stripper to strip dissolved NO_x (NO and NO_2) from the nitric acid product.

The gas leaving the top of the absorber column contains residual nitrogen oxides (NO_x), which have to be removed for environmental reasons before venting to the atmosphere. Figure 8.9 shows NO_x removal by means of so-called selective catalytic reduction (SCR) (Section 8.4.3).

QUESTIONS:

> *What would be the optimum concentration of ammonia in the air/ammonia mixture from the point of view of reaction stoichiometry (see reaction 8.8)? Why is the concentration used (circa 10 vol.%) lower than this value?*

8.4.2.2   Dual-Pressure Process

Figure 8.10 shows a simplified flow scheme of a dual-pressure nitric acid plant (high pressure). The main change in the flow scheme is the addition of a compressor between the ammonia conversion stage and the

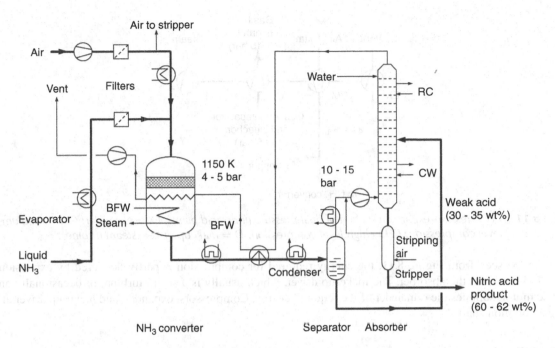

Figure 8.10   *Flow scheme of a dual-pressure nitric acid plant using extended absorption for NO$_x$ abatement. BFW = boiler feed water, CW = cooling water, RC = refrigerated cooling.*

absorption stage. Another difference in this scheme is the way in which NO$_x$ abatement is achieved, that is, by extended absorption. Note, however, that the NO$_x$ abatement methods are interchangeable between the single-pressure and dual-pressure processes.

QUESTIONS:

> *Explain the term "extended" absorption. How is this performed in Figure 8.10?*
> *The gas flow leaving the separator is cooled. Why?*

Extended absorption reduces NO$_x$ emissions by increasing the absorption efficiency (i.e., acid yield). This option can be implemented by installing a single large absorption tower, extending the height of an existing tower, or by adding a second tower in series with the existing tower. The increase in the volume and the number of trays in the absorber results in more NO$_x$ recovered as nitric acid.

Extended absorption is particularly suited for high-pressure absorption. In medium-pressure plants in general two absorption columns are needed to achieve an acceptable level of NO$_x$ emissions. Extended absorption is successfully used in retrofit applications.

8.4.2.3   *Compression/Expansion*

For many reasons the compression and expansion operations may be considered the heart of the nitric acid process. Although the compressors and expander in Figures 8.9 and 8.10 are depicted as independent units, in fact they are part of one machine. Figure 8.11 shows the compressor/expander combination of dual-pressure plants.

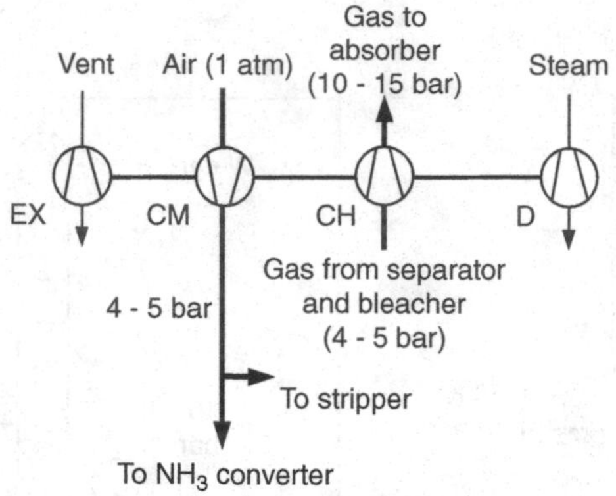

Figure 8.11   *Compression/expansion section of dual-pressure nitric acid plants. EX = expander; CM = medium-level compression; CH = high-level compression; D = makeup driver (steam turbine).*

As can be seen from Figure 8.11, the energy required for compression is partly delivered by expansion of the tail gas from the absorber. The make-up driver, which usually is a steam turbine, or occasionally an electric motor, supplies the remainder of the required energy. Compressors, expander, and make-up driver all share a common shaft.

8.4.3   NO$_x$ Abatement

Environmental regulations necessitate the removal of NO$_x$ from gases to be vented to the atmosphere, because of their harmful impact on the environment. Examples of NO$_x$-containing gases from stationary sources are flue gas from power plants and exhaust gas from gas turbines and nitric acid plants.

The concentration of nitric oxides in the gas leaving a conventional absorber in a nitric acid plant is about 2000 ppmv, while regulations require maximum concentrations of 200 ppmv or even lower depending on local legislation. The most commonly employed methods for NO$_x$ abatement in existing plants are (a combination of) selective catalytic reduction (SCR), non-selective catalytic reduction (NSCR), and extended absorption. For new nitric acid plants, NO$_x$ emissions can be well controlled by using advanced engineering processes, such as strong acid processes. The emission of N$_2$O (a very strong and stable greenhouse gas) is also a point of concern. Since 2000, processes for the reduction of N$_2$O emissions [10] have been successful implemented.

8.4.3.1   *Selective Catalytic Reduction (SCR)*

In SCR, NO$_x$ is converted into nitrogen by reducing it with ammonia:

$$4\,NO + 4\,NH_3 + O_2 \rightarrow 4\,N_2 + 6\,H_2O \qquad \Delta_r H_{298} = -408 \text{ kJ/mol NO} \qquad (8.13)$$

$$2\,NO_2 + 4\,NH_3 + O_2 \rightarrow 3\,N_2 + 6\,H_2O \qquad \Delta_r H_{298} = -669 \text{ kJ/mol NO}_2 \qquad (8.14)$$

This process is called "selective" because the excess oxygen is not consumed. The SCR process, developed in Japan in the 1970s, was a breakthrough because of this selectivity. The primary variable affecting NO$_x$ reduction is temperature. At too low a temperature, the rate is too low, resulting in ammonia slip. Moreover, solid deposits can be formed. At too high a temperature, the NO$_x$ exit concentration increases due to oxidation of ammonia to NO$_x$. Furthermore, excessive temperatures may damage the catalyst. Typically, the optimum temperature range is between 420 and 670 K [11].

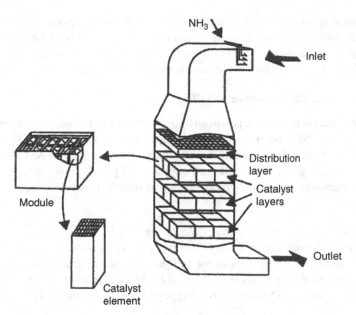

Figure 8.12 *Structure of an SCR monolith reactor. Reprinted with permission from [12] Copyright (2002) John Wiley & Sons Inc.*

The original SCR catalysts were pellets or spheres and were used in fixed catalyst beds, but current catalysts are usually shaped as honeycombs (monolithic catalyst) or parallel plates. The main advantages of these shaped catalysts for SCR are their low pressure drop, large surface area, and good resistance against dust. Figure 8.12 shows the structure of an SCR monolith reactor.

QUESTIONS:

> *You are in charge of an R&D team developing low-temperature processes for SCR. A coworker suggests the development of a catalyst that is active at very low temperature, in the range of 380 to 400 K. Would you agree with this idea? Explain. Take into consideration that the gas phase might contain SO₂. (Hint: the SCR catalyst might catalyze the oxidation of SO₂.)*

Selective catalytic reduction is used in many nitric acid plants in Europe and Japan, and also in some plants in the USA. Several advantages of SCR make it an attractive control technique. Since the SCR process can operate at any pressure, it is a viable control alternative for retrofitting existing low-pressure acid plants as well as for new plants. Another technical advantage is that because the temperature rise through the SCR reactor bed is small, energy recovery equipment (e.g., waste heat boilers and high-temperature turboexpanders) is not required, as is the case with the NSCR system, discussed in the next section.

The main disadvantages of the SCR system are the high costs involved. The cost of the catalyst can amount to 40−60% of operating costs and it is necessary to replace the catalyst every 2−3 years.

QUESTIONS:

> *SCR has the fundamental disadvantage that a pollutant is removed by a stoichiometric reaction with a valuable chemical. Can you think of a process that is fundamentally much more elegant? Why is that process not applied in practice?*
> *Estimate the cost of ammonia consumption. Is this significant?*

> *Instead of ammonia, urea can be used for SCR (the reactions occurring are nearly equal). SCR processes in nitric acid production plants use NH$_3$ whereas SCR units for diesel engines in trucks use urea (Section 10.6). Why? If you were involved in designing an SCR unit for a power plant, would you choose NH$_3$ or urea?*

8.4.3.2 Non-Selective Catalytic Reduction (NSCR)

Besides SCR with ammonia (or urea) there is also an option to use a simple reducing agent such as hydrogen or methane. In that case, besides NO$_x$ the oxygen present in the exhaust gas is converted. Therefore, this technique is called non-selective catalytic reduction (NSCR). The reaction temperature of NSCR is significantly higher than that of SCR. In exhaust gases from lean-burn engines and power plants, a large excess of oxygen is present, and therefore reduction with simple reducing agents usually is not economic.

QUESTIONS:

> *What are the reactions taking place during NSCR?*
> *Why is the temperature in the NSCR process higher than that in the SCR process?*
> *Typical O$_2$ concentrations in the tail gas from nitric acid plants are in the range of 2 to 3%. Assess the attractiveness of NSCR for nitric acid plants, diesel engines, and lean-burn gasoline engines (Section 10.6), and power plants (Section 5.3).*

In nitric acid plants, NSCR can be used effectively. The high reaction temperature required is realized by heat exchange with the effluent gas from the ammonia converter and by adding a fuel (usually natural gas). The gas/fuel mixture then passes through the catalytic reduction unit where the fuel reacts in the presence of a catalyst with NO$_x$ and oxygen to form elemental nitrogen and water (and carbon dioxide if carbon-containing fuels are used). Despite the associated fuel costs, NSCR offers advantages that continue to make it a viable option for new and retrofit applications in nitric acid plants. Flexibility is one advantage, especially for retrofit considerations. Additionally, heat generated by operating an NSCR unit can be recovered in a waste heat boiler and a tailgas expander to supply the energy for process compression with additional steam for export.

8.4.3.3 Comparison of SCR and NSCR

Table 8.5 compares a number of characteristics of the SCR and NSCR processes. Future legislation will probably require a higher removal efficiency than can be obtained with current SCR and NSCR technologies. In that case there will be a need for the development of SCR at high temperature (1070–1170 K) for removal of NOx from stationary sources, including nitric acid plants and power plants (5.3.7).

8.5 Chlorine

Originally, chlorine was produced from hydrogen chloride. In the first production process in the beginning of the nineteenth century hydrogen chloride was oxidized by manganese oxide (MnO$_2$):

$$2\,HCl + MnO_2 \rightarrow MnO + H_2O + Cl_2 \tag{8.15}$$

Table 8.5 *Characteristics of SCR and NSCR.*

Characteristic	SCR	NSCR
Temperature (K)	420–670	970–1120
Reducing agent	Ammonia/urea	H$_2$, CH$_4$, naphtha
Reaction of O$_2$	no	yes
Removal efficiency (%)	85–95	~90

The yield was low, typically 30% of the theoretically attainable value. In the second half of the nineteenth century, Deacon developed a catalytic process based on oxidation with air:

$$4\,HCl + O_2 \rightarrow 2\,H_2O + Cl_2 \qquad \Delta_r H_{298} = -114.6 \text{ kJ/mol} \qquad (8.16)$$

In this process, a yield of 75% could be obtained.

QUESTION:

> *The discovery of the Deacon process was a big step forward. What would be the reason for its attractiveness, except for the higher yield?*

At the time of the development of the chemical processes, it was already known that electrochemical processes were possible. However, suitable generators and stable anodes were not available yet. Only at the end of the nineteenth century were electrochemical processes introduced. These processes are based on electrolysis of an aqueous solution of sodium chloride (NaCl). By applying a direct current chlorine, hydrogen, and sodium hydroxide are produced:

$$2\,NaCl + 2\,H_2O \rightarrow 2\,NaOH + H_2 + Cl_2 \qquad (8.17)$$

This process is called the chlor-alkali process.

Electrolysis technology is not limited to sodium chloride. Occasionally, potassium chloride is used as the raw material, resulting in potassium hydroxide as one of the side products.

8.5.1 Reactions for the Electrolysis of NaCl

In practice, three major technologies are in use for the production of chlorine by electrolysis of sodium chloride:

- mercury cell electrolysis,
- diaphragm cell electrolysis, and, more recently,
- membrane cell electrolysis.

These technologies are discussed in more detail in Section 8.5.3.

All technologies have in common that an aqueous solution of NaCl ("brine") is fed to an electrolysis cell and that at the anode chlorine is formed:

$$2\,Cl^- \rightarrow Cl_2 + 2\,e^- \qquad (8.17a)$$

Except for the mercury cell process, at the cathode hydrogen and hydroxyl (OH^-) ions are formed:

$$2\,H_2O + 2\,e^- \rightarrow H_2 + 2\,OH^- \qquad (8.17b)$$

For this reaction to occur, a suitable cathode material must be selected, allowing the production of hydrogen. Many materials are known to be active for this reaction. Platinum is used most often.

In practice, the selectivity of an electrochemical cell is never 100%. There is a driving force for transport of the OH^- ions to the (positive) anode. When these ions reach the anode, they are oxidized, resulting in consumption of power and contamination of the chlorine produced by oxygen. In addition, other undesired reactions can occur in which hypochlorite (ClO^-) and chlorate (ClO_3^-) are formed. Thus, in the electrolysis processes it is crucial to keep the reaction products at the anode separated from those at the cathode in order to realize large product yields.

QUESTIONS:

How are hypochlorite and chlorate formed? Where are they leaving the cell? How can the formation of these by-products be reduced?

List the reactions in the electrolysis of water. Compare the products of water electrolysis with those formed in a NaCl solution.

Pt often is used as cathode. How would you design this cathode? (Hint: Pt is very expensive.)

For the economics of the electrolysis process the cost of the brine and the electricity consumed should be competitive.

8.5.2   Technologies for the Electrolysis of NaCl

How can the aforementioned side reactions be avoided? In the history of the development of chlorine production processes by electrolysis of sodium chloride two routes have been followed. The first is the total suppression of the formation of negative mobile anions that are attracted to the anode electrode (the mercury cell process). The second is to divide the electrochemical cell into an anode and cathode compartment, separated by a semi-permeable layer (diaphragm and membrane cell processes).

8.5.2.1   Mercury Cell Process

The mercury cell process is the oldest process for the production of chlorine by the electrolysis of sodium hydroxide on an industrial scale. In this process, mercury serves as a liquid cathode.

Mercury has remarkable properties. It might be expected that hydrogen would be formed at the cathode. However, at the mercury surface this does not happen; hydrogen is not produced. In the world of electrochemistry this is referred to as a high "overvoltage".

QUESTION:

Could you come up with a theory why Hg does not show high activity in the production of H_2? (Hint: reaction 8.17b is a complex reaction composed of several elementary steps. Could there be a rate-determining step that is very slow at the Hg surface?) Compare Hg with Pt. Explain the term "overvoltage".

Instead, Na^+ ions are reduced and a compound (an "amalgam") is formed with the mercury:

$$Na^+ + n\,Hg + e^- \rightarrow NaHg_n \tag{8.18}$$

This reaction equation shows a stoichiometric reaction, consuming the electrode material. Recycling of the mercury obviously is required. The amalgam formed can be regenerated by reaction with water:

$$NaHg_n + H_2O \rightarrow n\,Hg + NaOH \tag{8.19}$$

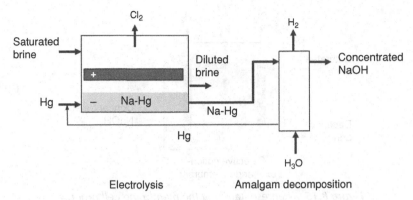

Figure 8.13 *Representation of the mercury cell process.*

This reaction enables recycling of the mercury. The remarkable fact that mercury is a liquid allows a convenient technology. In industrial applications, the amalgam flows from the electrochemical cell to the regenerator, where the amalgam decomposes, and the mercury formed flows back to the cell (Figure 8.13).

The sodium hydroxide produced is extremely pure and highly concentrated. However, a practical limitation is that the viscosity of mercury strongly increases with increasing content of sodium in the amalgam. This leads to a severe constraint of the allowable concentration of sodium in the amalgam. The consequence of the low maximum concentration is that the recycle stream is large. In addition, the high toxicity of mercury is a disadvantage. In particular because of the latter aspect, in Japan alternative mercury-free electrochemical cells have successfully been developed. New mercury cell processes are not built anymore and worldwide mercury-cell-based processes are being phased out.

QUESTION:

> *The regenerator ("secondary cell") is placed in parallel to the primary cell. How would you design a practical configuration (different heights?).*

8.5.2.2 *Diaphragm Cell Process*

The first cell involving separate cathode and anode compartments was the diaphragm cell (Figure 8.14). In the diaphragm cell, the anode and cathode compartments are separated by a microporous diaphragm. The function of the diaphragm is to inhibit the transport of the OH^- ions to the anode, while allowing a convective

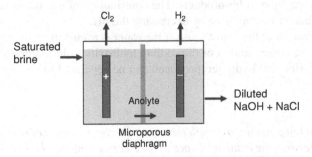

Figure 8.14 *Representation of the diaphragm cell process.*

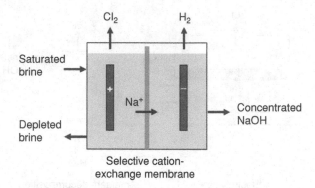

Figure 8.15   *Representation of the membrane cell process.*

flow of the anolyte (the portion of the electrolyte in the immediate vicinity of the anode) to pass. Initially, the diaphragm used to be made from asbestos materials. In later versions other materials have been selected. The conditions are such that the convective anolyte flow is slightly larger than that of the OH⁻ ion transport in the opposite direction due to the electric field. In contrast with the sodium hydroxide produced in the mercury cell process, the sodium hydroxide produced in this process is contaminated with 1–2% sodium chloride. This sodium chloride has to be removed, for instance by evaporation and crystallization.

QUESTIONS:

> *Explain why the NaOH product stream for the diaphragm process is contaminated with NaCl. What are major impurities in the Cl₂ product stream?*

8.5.2.3   Membrane Cell Process

The development of efficient cationic membranes has led to a breakthrough in sodium chloride electrolysis. These membranes allowed the development of the modern state-of-the-art membrane cell process (Figure 8.15). It has been an enormous challenge to find membrane materials that are able to withstand the severe process conditions: at one side the chlorine-containing anolyte is present and at the other side strongly caustic conditions prevail. Ion-exchange materials containing fluorosulfonate groups ($-SO_3F$) appeared to combine good rates, selectivity, and stability. They inhibit convective flow and are highly selective for cationic transport.

In practice, a small OH⁻ flow from the cathode to the anode compartment is unavoidable, resulting in some reaction in the anode compartment. The oxygen concentration in the chlorine product stream is a good measure of the extent of formation of by-products. The contribution of the side reactions can be reduced by subtle changes of the surface of the anode or by decreasing the pH.

During the diffusive transport of the cations also some water is transported, diluting the sodium hydroxide solution produced. Thus, the concentration of the sodium hydroxide solution produced is lower than that of the mercury cell. The chlorine and hydrogen produced can be separated in the compartments or outside of the cell.

QUESTIONS:

> *Why is it not possible to completely eliminate the transport of water through the membrane? Before feeding the brine feedstock to the electrochemical cell it is purified in order to reduce the amounts of, among others, calcium and magnesium. Why?*

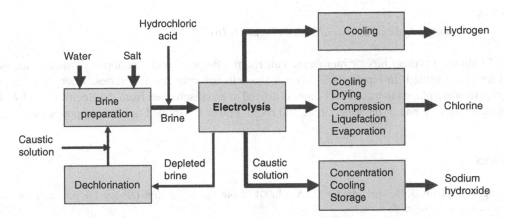

Figure 8.16 *Simplified flow diagram of the membrane cell process.*

Figure 8.16 shows a simplified flow diagram of the membrane cell process. The life of the (expensive) membrane depends on the purity of the brine. Therefore, after purification by precipitation and filtration, the brine is also purified with an ion exchanger. Saturated purified brine is introduced in the electrolysis cell. The depleted brine is dechlorinated and recirculated.

The caustic solution (30 – 36 wt%) leaving the cell must be concentrated. The chlorine gas must be purified to remove oxygen.

Figure 8.17 *Bipolar cell room by Krupp Uhde. Reprinted with permission from [13] Copyright (2000) Wiley-VCH.*

QUESTION:

Why is hydrochloric acid added (Figure 8.16)?

Figure 8.17 shows a typical bipolar membrane cell room. "Bipolar" and "monopolar" refer to the way the electrolyzer is assembled. In bipolar electrolyzers the cells are connected in series, whereas in monopolar electrolyzer they are placed in parallel. The media are fed to and discharged from the electrolyzers by a header system arranged along one of the walls of the cell room. Power is supplied from the opposite side.

References

[1] Hwang, S.C., Lein, R.D., and Morgan, D.A. (2000) Noble Gases, in *Kirk-Othmer Encyclopedia of Chemical Technology*. John Wiley & Sons, Inc.

[2] McNutt, M.K. (2011) Mineral Commodity Summaries 2011. http://minerals.usgs.gov/minerals/pubs/mcs/2011/mcs2011.pdf.

[3] Euro Chlor (2012) Chlorine Tree. http://www.eurochlor.org/media/29613/chlorine_tree_1000x720.swf (last accessed 17 December 2012).

[4] Davies, P., Donald, R.T., and Harbord, N.H. (1996) Catalytic Oxidations, in *Catalyst Handbook*, 2nd edn (ed. M.V. Twigg). Manson Publishing Ltd, London, pp. 490–502.

[5] Muller, T.L. (1997) Sulfuric acid and sulfur trioxide, in *Kirk-Othmer Encyclopedia of Chemical Technology*, 4th edn, vol. **23** (eds J.I. Kroschwitz and M. Howe-Grant). John Wiley & Sons, Inc., New York, pp. 363–408.

[6] Nehr, W. and Vydra, K. (1994) Sulfur, in *Ullman's Encyclopedia of Industrial Chemistry*, 5th edn, vol. **A25** (ed. W. Gerhartz). VCH, Weinheim, pp. 507–567.

[7] Piéplu, A., Saur, O., Lavalley, J.C., Legendre, O., and Nédez, C. (1998) Claus catalysis and H_2S selective oxidation. *Catalysis Reviews – Science and Engineering*, 40, 409–450.

[8] Van den Berg, P.J. and de Jong, W.A. (1980) *Introduction to Chemical Process Technology*. Delft University Press, Delft, The Netherlands.

[9] Thiemann, M., Scheibler, E., and Wiegand, K.W. (2000) Nitric acid, nitrous acid, and nitrogen oxides, in *Ullmann's Encyclopedia of Industrial Chemistry*. Wiley-VCH Velag GmbH, Weinheim.

[10] Kapteijn, F., Rodriguez-Mirasol, J., and Moulijn, J.A. (1996) Heterogeneous catalytic decomposition of nitrous oxide. *Applied Catalysis B*, 9, 25–64.

[11] Beretta, A., Tronconi, E., Groppi, G., and Forzatti, P. (1997) Monolithic catalysts for the selective reduction of NO_x with NH_3 from stationary sources, in *Structured Catalysts and Reactors* (eds A. Cybulski and J.A. Moulijn). Marcel Dekker, New York, pp. 121–148.

[12] Forzatti, P., Lietti, L., and Tronconi, E. (2002) Nitrogen Oxides Removal – Industrial, in *Encyclopedia of Catalysis* (ed. I.T. Horváth). John Wiley & Sons, Inc.

[13] Schmittinger, P., Florkiewicz, T., Curlin, L.C., Lüke, B., Scannell, R., Navin, T., Zelfel, E., and Bartsch, R. (2000) Chlorine, in *Ullmann's Encyclopedia of Industrial Chemistry*. Wiley-VCH Verlag GmbH, Weinheim.

9

Homogeneous Transition Metal Catalysis in the Production of Bulk Chemicals

9.1 Introduction

In homogeneous catalysis soluble catalysts are used, in contrast to heterogeneous catalysis, where solid catalysts are employed. One example of homogeneous catalysis has been discussed in Chapter 3, namely alkylation of isobutane with alkenes with an acid such as HF or H_2SO_4 as catalyst. In this chapter, we shall confine ourselves to homogeneous catalysts based on transition metals. Homogeneous catalysis is encountered in essentially all sectors of the chemical process industry, in particular in polymerization processes, in the synthesis of bulk chemicals (solvents, detergents, plasticizers), and in the production of fine chemicals.

It is useful to compare heterogeneous and homogeneous transition-metal-based catalysis (Table 9.1). In homogeneous catalysis the reaction mixture contains the catalyst complex in solution. This means that all metal atoms are exposed to the reaction mixture. In heterogeneous catalysis on the other hand, the metal is typically applied on a carrier material or as a porous metal sponge type material and only the surface atoms are active (Figure 9.1).

Thus, in terms of activity per metal center homogeneous catalysts are often far more active. The high dispersion of homogeneous catalysts also minimizes the effect of catalyst poisons; in homogeneous catalysis one poison molecule only deactivates one metal complex, whereas in heterogeneous catalysis a poison molecule can block one active site or even a pore containing many active sites (pore plugging). Homogeneous catalysts are also capable of being much more selective because with homogeneous catalysts there is usually only "one" type of active site, whereas heterogeneous catalysts often contain many different types of active sites, some of which could even catalyze undesired reactions. The discrete metal complexes in homogeneous systems provide a well-defined catalyst system, which is easy to study by conventional techniques (infrared spectroscopy, NMR, UV, etc.). Hence, another advantage of homogeneous catalysis arises, namely the possibility to follow the effect of changes in ligands and/or reaction conditions accurately. Thus, the selectivity for a specific product can be controlled by modification of the ligands. The behavior of heterogeneous catalysts, with their complex surfaces, is much more difficult to understand, so effective modification is fairly difficult. Because homogeneous catalysis is usually carried out in the liquid phase, temperature control is relatively easy.

Chemical Process Technology, Second Edition. Jacob A. Moulijn, Michiel Makkee, and Annelies E. van Diepen.
© 2013 John Wiley & Sons, Ltd. Published 2013 by John Wiley & Sons, Ltd.

Table 9.1   *Comparison of heterogeneous and homogeneous catalysis.*

	Heterogeneous	Homogeneous
Catalyst form	solid, often metal or metal oxide	metal complex
Activity	variable	high
Selectivity	variable	high, also chemo-, regio-, and stereo-
Stability	stable to high temperature	often decompose <373 K
Reaction conditions	often harsh	mild
Sensitivity against poisons	high	low
Diffusion/mass transfer limitations	important	unimportant
Mechanistic understanding	complex	relatively easy
Solvent	usually not required	usually required, can be product or by-product
Catalyst separation and recycle	not necessary	difficult

Why then, with all these advantages, is homogeneous transition metal catalysis not applied in every catalytic process? First of all, its high specificity is not always called for. Especially in oil refinery processes the complex feeds and products do not justify the use of a highly specific catalyst. Secondly, homogeneous catalysis also has disadvantages. The main problem is the separation of catalyst and product; this is often only feasible if the reaction products have a low molecular weight compared to the molecular weight of the homogeneous catalyst. Furthermore, the use of solvents in the reactor adds an additional separation step. Thirdly, the available temperature window is limited, in particular because catalyst complexes are often not resistant to high temperature. In homogeneous catalysis often precious metals are used. Furthermore, the ligands are also (often extremely) expensive. Therefore, the catalyst productivity indeed should be high, and catalyst losses should be minimized, requiring nearly complete recovery of metal and ligands. Fouling of the reactor and other equipment by the catalyst is another problem specific for homogeneous catalysis.

Homogeneous catalysis is a highly innovative area [1]. It is a good example of an area where progress is accelerated when chemical engineers and chemists cooperate closely from the start. Novel developments in homogeneous catalysis are for a large part related to new insights in organometallic chemistry. Table 9.2

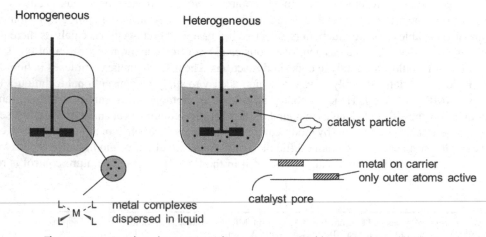

Figure 9.1   *Catalyst dispersion in homogeneous and heterogeneous catalysis.*

Table 9.2 *Reactions of industrial importance catalyzed by transition metals.*

Reaction	$\Delta_r H_{298}$ (kJ/mol)	Reaction number
Acetic acid by methanol carbonylation		
	− 138	(9.1)
Hydroformylation (oxo reaction)		
	− 115	(9.2)
	− 126	
Oligomerization		
	− 95^a	(9.3)
Synthesis of dimethyl terephthalate		
	− 1305	(9.4)
Synthesis of terephthalic acid		
	− 1360	(9.5)

^{a}per mol C_2H_4 added.

shows examples of industrially important reactions that are catalyzed by transition metals, which will be dealt with in this chapter.

QUESTIONS:

Why would separation of catalyst and product often only be feasible for products having a low molecular weight?

It might be assumed that homogeneous catalysis is limited to liquid phase reactions. However, homogeneous catalysis in the gas phase is possible (and important in practice!). Give examples. (Hints: (i) The atmosphere can be considered one big reactor. Does catalysis play a role here? (ii) Consult the chapter on inorganic bulk chemicals (Chapter 8).)

It is often thought that in bulk chemicals production only heterogeneous catalysis is applied. All examples in Table 9.2, however, belong to the class of bulk chemicals, but it is true that the majority of homogeneously catalyzed reactions are in the fine chemicals area (Chapter 12).

Table 9.3 *Capacity and process conditions of some modern homogeneously catalyzed processes.*

Process	World capacity (million t/a)[a]	Catalyst metal	Temperature (K)	Pressure (bar)
Acetic acid[b]	16	Rh or Ir	425–475	20–60
Oxo-alcohols[c]	9	Rh	370–390	20–50
1-Alkenes	7	Ni	350–390	70–140
Dimethyl terephthalate	2	Co	415–445	4–8
Terephthalic acid	50	Co	450–505	15–30

[a]2010 capacities; data from Tecnon Orbichem (www.tecnon.co.uk), SRI Consulting (www.sri.com).
[b]Process conditions are for methanol carbonylation.
[c]Process conditions are for hydroformylation of propene to butyraldehydes; products are mainly *n*-butanol, iso-butanol, and 2-ethylhexanol.

The mild reaction conditions are an advantage of homogeneously catalyzed reactions. Table 9.3 shows these conditions for the reactions in Table 9.2. As can be seen from Table 9.3, all processes take place at moderate temperatures. For comparison, Table 9.4 shows some data from some important heterogeneously catalyzed processes. It is clear that the conditions are more severe.

Homogeneous catalysts are not suitable for strongly endothermic reactions, such as cracking of C—C bonds in fluid catalytic cracking (FCC), because of their limited stability at high temperatures. In most cases the operating pressure is low-to-moderate but there are exceptions, for example the hydroformylation reaction with an unmodified cobalt catalysts, which is performed at a pressure of at least 200 bar (Section 9.3).

9.2 Acetic Acid Production

9.2.1 Background Information

Apart from food applications, acetic acid is mostly used for the production of vinyl acetate and acetic anhydride and as a solvent, for instance in the production of purified terephthalic acid (Section 9.5).

Acetic acid production has a larger number of feedstock and process options than any other bulk chemical, but commercially nearly all non-food acetic acid is produced by one of two homogeneously catalyzed processes.

Until late in the nineteenth century, acetic acid was manufactured by the age-old process of sugar fermentation to ethanol (Section 13.4) and subsequent oxidation to acetic acid, which is still the major route for

Table 9.4 *Capacity and process conditions of some modern heterogeneously catalyzed processes.*

Process	World capacity (million t/a)[a]	Catalyst	Temperature (K)	Pressure (bar)
Methanol	70	Cu	530	50–100
Ammonia	205	Fe	675	100–200
Sulfuric acid	240	V	700	1
FCC	>650	zeolite-Y	775	1
Ethene oxide	24	Ag	550	20

[a]2010 capacities; data from Fertecon (www.Fertecon.com, *Ammonia Outlook*, Issue 2010–3, January 2011), Jim Jordan & Associates (www.jordan-associates.com, SRI Consulting (www.sri.com), and Hydrocarbon Publishing Company (www.hydrocarbonpublishing.com).

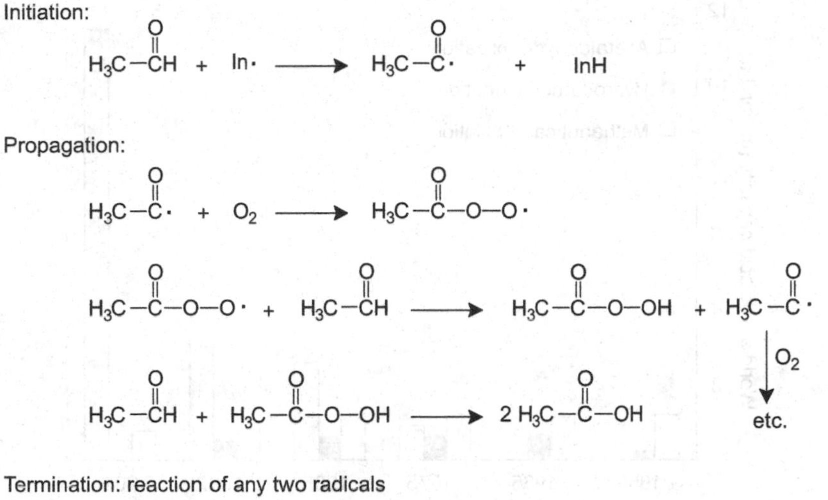

Initiation:

$$H_3C-\overset{\overset{\displaystyle O}{\|}}{C}H + In\cdot \longrightarrow H_3C-\overset{\overset{\displaystyle O}{\|}}{C}\cdot + InH$$

Propagation:

$$H_3C-\overset{\overset{\displaystyle O}{\|}}{C}\cdot + O_2 \longrightarrow H_3C-\overset{\overset{\displaystyle O}{\|}}{C}-O-O\cdot$$

$$H_3C-\overset{\overset{\displaystyle O}{\|}}{C}-O-O\cdot + H_3C-\overset{\overset{\displaystyle O}{\|}}{C}H \longrightarrow H_3C-\overset{\overset{\displaystyle O}{\|}}{C}-O-OH + H_3C-\overset{\overset{\displaystyle O}{\|}}{C}\cdot$$

$$\Big\downarrow O_2$$

$$H_3C-\overset{\overset{\displaystyle O}{\|}}{C}H + H_3C-\overset{\overset{\displaystyle O}{\|}}{C}-O-OH \longrightarrow 2\,H_3C-\overset{\overset{\displaystyle O}{\|}}{C}-OH \qquad \text{etc.}$$

Termination: reaction of any two radicals

Scheme 9.1 *Radical mechanism for acetaldehyde oxidation. In· is an initiator.*

the production of vinegar In 1916, the first industrial process for the production of synthetic acetic acid was commercialized. This process was based on the liquid phase oxidation of acetaldehyde:

$$CH_3CHO + 1/2\,O_2 \rightarrow CH_3COOH \qquad \Delta_r H_{298} = -292\ kJ/mol \qquad (9.6)$$

which basically proceeds through radical reactions (Scheme 9.1).

The acetaldehyde process is still widely used but since the development of the Monsanto methanol carbonylation process it has progressively been losing ground.

Direct liquid phase oxidation of naphtha or *n*-butane:

$$CH_3CH_2CH_2CH_3 + 5/2\,O_2 \rightarrow 2\,CH_3COOH + H_2O \qquad \Delta_r H_{298} = -986\ kJ/mol \qquad (9.7)$$

was once the preferred route to acetic acid because of the low cost of these hydrocarbons. The mechanism for this reaction is also based on radicals. It is not surprising that a major drawback of this process is that up to 50% of the feed goes to by-products (e.g., formic acid, higher acids, and aldehydes), many of which have very limited markets. Furthermore, the process requires a very complex purification train, adding to the investment and operating costs. Nowadays, only a small portion of acetic acid is manufactured by this process.

The manufacture of acetic acid by carbonylation of methanol:

$$CH_3OH + CO \rightleftarrows CH_3COOH \qquad \Delta_r H_{298} = -135\ kJ/mol \qquad (9.1)$$

at high temperature and pressure was described as early as 1913 [2], but due to the lack of suitable construction materials that could contain the corrosive reaction mixture at the high pressures needed, it took a long time before the carbonylation route was commercialized. The first commercial methanol carbonylation plant did not come on stream until 1963, when a new cobalt/iodide catalyst system was developed by BASF. In 1968, Monsanto introduced a carbonylation process using a novel highly active and selective rhodium-based catalyst that could operate at much lower pressure; this was commercialized successfully only two years later. For

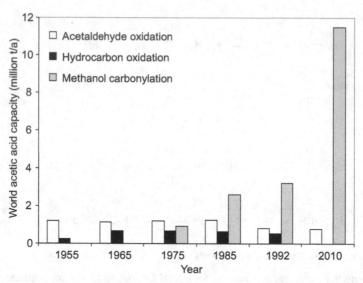

Figure 9.2 *Breakdown of the three most important technologies for acetic acid production; data from [6–8] and Tecnon Orbichem (tecnon.co.uk).*

years, this process has been the most attractive route for acetic acid manufacture. Figure 9.2 shows that the Monsanto methanol carbonylation technology, now owned by British Petroleum (BP), has indeed emerged surprisingly quickly. Currently, carbonylation accounts for over 75% of the global acetic acid production [3].

Despite the success of the Monsanto process, the search for new catalysts has continued. The strategies employed have involved either modifications to the rhodium-based system or the replacement of rhodium by another metal, in particular iridium [4]. An important driver for this is the high rhodium price (Box 9.1). The most recent significant development was the introduction of the iridium-based Cativa process by BP in 1996 and its commercialization in 2000 [5]. Very early on, researchers at Monsanto found that iridium is active in the carbonylation of methanol, but less active than rhodium. Promoters, specifically ruthenium, are key to the success of the Cativa process, which has several advantages over the Monsanto process (Section 9.2.3).

Table 9.5 shows a comparison of the process conditions and yields of the various processes, which are very different. The carbonylation technologies are closely related. In the Monsanto and Cativa processes,

Table 9.5 *Comparison of processes for acetic acid production.*

Process	Yield (mol %)	T (K)	p (bar)	Catalyst
Acetaldehyde oxidation	93–96	335–355	3–10	Mn/Co[b]
n-butane/naphtha oxidation	40–80	395–475[a]	45–55	Mn/Co[b]
Methanol carbonylation				
BASF	90 (CH_3OH), 70 (CO)	455–515	>500	Co/HI
Monsanto	99 (CH_3OH), 85–90 (CO)	425–475	30–60	Rh/HI
BP-Cativa	99 (CH_3OH), > 94 (CO)	455–465	20–40	Ir/HI + Ru

[a]The lower temperatures are used for naphtha, the higher for *n*-butane.
[b]Salts (usually acetate salts) of these metals are employed.

however, the pressure is much lower than in the now obsolete BASF process, resulting in less severe, lower cost operating conditions and higher acetic acid yields.

Box 9.1   Rhodium and Iridium Prices

Rhodium is one of the most expensive metals in the world, with a price of about 45 $/g (April 2012). April 2012 prices of iridium and platinum were about 35 $/g and 52 $/g, respectively. Until the early 1990s, an important incentive for replacing rhodium with iridium was the large price difference between the two noble metals; in 1992, rhodium was about 15 times as expensive as iridium. Between 1992 and 2012, the price ratio has been fluctuating between 1.3 and 16. The highest rhodium price and Rh/Ir price ratio occurred in July 2008 (Figure B.9.1.1).

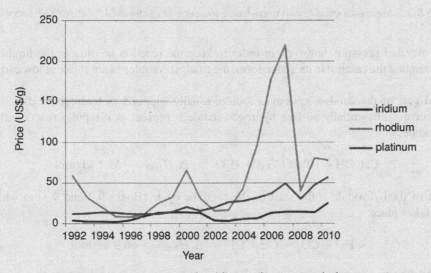

Figure B.9.1.1   *Historical noble metal prices (end-of-year).*

QUESTIONS:

> *What is the most important use of rhodium? Explain the sharp increase of the rhodium price from 2003 to July 2008 and the very sharp decline during the second half of 2008.*

9.2.2   Methanol Carbonylation – Reactions, Thermodynamics, and Catalysis

Methanol carbonylation is an exothermic reaction and thus is thermodynamically favored by low temperature. Figure 9.3 shows the effect of temperature and pressure on the equilibrium conversion of methanol to acetic acid. The feed contains equal molar amounts of methanol and carbon monoxide and it is assumed that acetic acid is the only product formed.

The methanol equilibrium conversion is almost complete at the reaction conditions employed. Indeed, at the temperatures employed the conversion is even nearly complete at atmospheric pressure. The reaction is

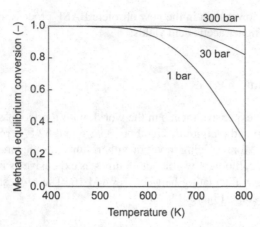

Figure 9.3   *Methanol equilibrium conversion to acetic acid (feed CH₃OH/CO = 1 mol/mol).*

carried out at elevated pressure, however, in order to keep the reaction mixture in the liquid phase and to generate and maintain the catalyst in its active form; the catalyst complex is not stable at low carbon monoxide pressure.

An essential part of the catalyst system is iodide, usually supplied as hydrogen iodide, which is very corrosive. Fortunately, essentially no free hydrogen iodide is present as it rapidly reacts with methanol to form methyl iodide:

$$CH_3OH + HI \rightleftarrows CH_3I + H_2O \qquad \Delta_r H_{298} = -53.1 \text{ kJ/mol} \qquad (9.8)$$

The formation of methyl iodide is the start of the catalytic cycle (Boxes 9.2 and 9.3) in which the actual carbonylation takes place:

$$CH_3I + CO \rightarrow CH_3COI \qquad \Delta_r H_{298} = -66.9 \text{ kJ/mol} \qquad (9.9)$$

The formed CH_3COI is then converted to acetic acid:

$$CH_3COI + H_2O \rightleftarrows CH_3COOH + HI \qquad \Delta_r H_{298} = -15.6 \text{ kJ/mol} \qquad (9.10)$$

The reaction rate in the rhodium-catalyzed Monsanto process shows a first-order dependence on the methyl iodide and rhodium concentrations:

$$r_{CH_3COOH} = k \cdot [\text{Rh}] \cdot [CH_3I] \qquad (9.11)$$

Remarkably, the reaction rate is independent on the concentrations of the two reactants (methanol and carbon monoxide). The consequence of this zero-order dependence is that at any conversion level the production rate is the same. Therefore, high conversions can be obtained even in a continuous stirred-tank reactor (CSTR) of relatively small volume.

QUESTIONS:

Why would HI be so corrosive?
Recall the design equation for a CSTR and show that only in the case of zero-order reactions complete conversion can be attained in a reactor of finite volume.

Box 9.2   Catalytic Cycle for the Monsanto Acetic Acid Process

Scheme B.9.2.1 shows the catalytic cycle for the Monsanto acetic acid process. Following the formation of methyl iodide, an oxidative addition reaction occurs between methyl iodide and $[Rh(CO)_2I_2]^-$. This has been found to be the rate-determining step. Methyl migration results in a penta-coordinated acyl complex (acyl: H_3C—CO—). After addition of carbon monoxide, acetyl iodide splits to yield the original rhodium complex. Subsequently, acetyl iodide reacts with water to form acetic acid and hydrogen iodide, which closes the catalytic cycle.

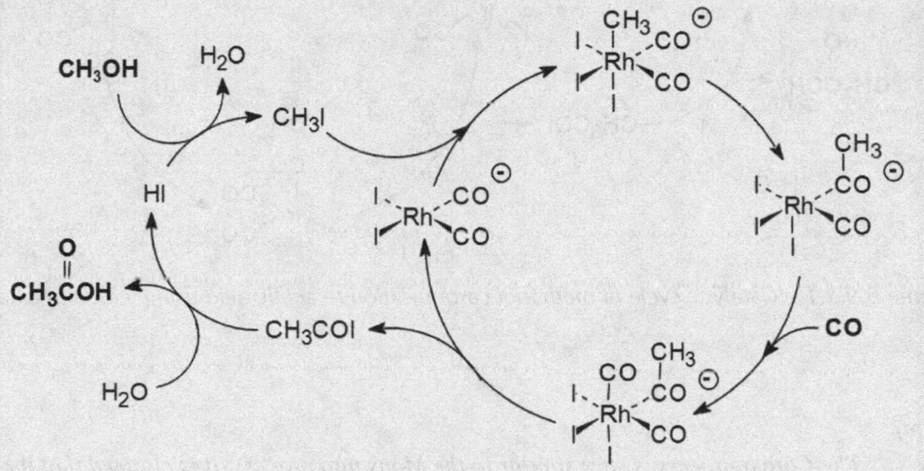

Scheme B.9.2.1   *Catalytic cycle of methanol carbonylation to acetic acid using a rhodium catalyst.*

Box 9.3   Catalytic Cycle for the Cativa Acetic Acid Process

The iridium cycle of the Cativa process (Scheme B.9.3.1) is similar to the rhodium cycle but has some key differences. The oxidative addition of methyl iodide to the iridium center is much faster than the equivalent reaction with rhodium and is no longer the rate-determining step [5]. The slowest step is the subsequent methyl migratory insertion to form the acyl complex, which is relatively slow with iridium [9]. This step involves substitution of I^- with carbon monoxide. Thus, the reaction rate dependence is totally different from that for the rhodium-catalyzed process:

$$r_{CH_3COOH} = k \frac{[Ir] \cdot [CO]}{[I^-]}$$

The inverse dependence on the ionic iodide concentration implies that removing it increases the reaction rate. Here promoters come into play. Carbonyl iodide complexes such as $[Ru(CO)_4I_2]$ are believed to reduce the concentration of I^- so that the reaction rate is increased [5]. Another key role of promoters seems to be the prevention of the build-up of "inactive" forms of the catalyst, such as $[Ir(CO)_2I_4]^-$ and $[Ir(CO)_3I_3]$ [5].

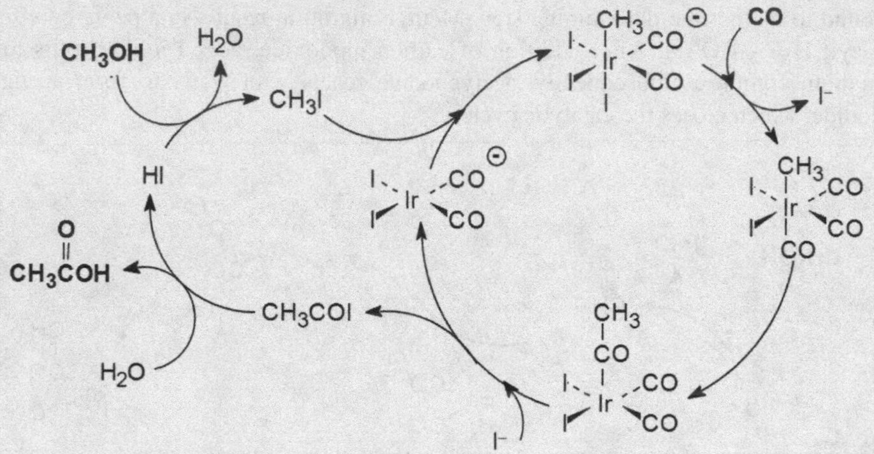

Scheme B.9.3.1   *Catalytic cycle of methanol carbonylation to acetic acid using a iridium catalyst.*

QUESTION:

The Cativa process is very similar to the Monsanto process. It is claimed that the catalyst in the Cativa process is more active and selective. What would be the advantages of Cativa in process design/operation?

9.2.3   Methanol Carbonylation – Processes

Figure 9.4 shows a simplified flow scheme of the Monsanto acetic acid process. Carbon monoxide and methanol are fed to a sparged CSTR containing the catalyst. Reaction takes place in the liquid phase under relatively mild conditions. Non-condensable by-products, mainly carbon dioxide and hydrogen, which are formed by the water–gas shift reaction, are vented from the reactor. This vent gas is combined with off-gas from the light ends column and sent to a scrubber to recover the volatile and toxic methyl iodide.

Liquid is removed from the reactor through a pressure reduction valve and enters a flash vessel, resulting in a gas and a liquid phase. The liquid phase, which contains the dissolved catalyst complex, is recycled to the reactor. The gas phase, containing most of the acetic acid product, water, methyl acetate, and methyl iodide, is sent to a distillation train for purification.

In the light ends column, methyl acetate, methyl iodide and part of the water are removed and recycled to the reactor. Wet acetic acid is taken as a side stream from this column and fed to the drying column, where dry acetic acid is removed as bottoms product. The overheads of the drying column, containing a mixture of acetic acid ($\approx35\%$) and water, are recycled to the reactor. Thus, a fixed amount of acetic acid and water is continuously circulating through the plant. The dry acetic acid is fed to the product column from which heavy by-products, mainly propionic acid, are removed as bottoms.

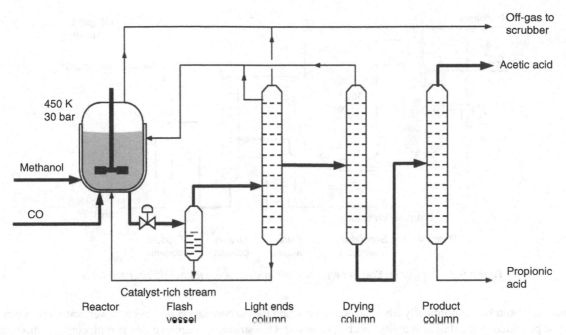

Figure 9.4 *Simplified flow scheme of the Monsanto process for acetic acid production.*

QUESTIONS:

> *How is the heat of the exothermic reaction removed?*
> *Why is such a large amount of acetic acid present in the overhead stream from the drying column?*
> *Why is the acetic acid/water mixture not separated further so that only water would have to be recycled to the reactor?*
> *What would be the origin of the propionic acid by-product?*
> *What are the advantages and disadvantages of the Monsanto process?*

The carbon monoxide pressure in the flash vessel is low, which may lead to loss of CO ligands and precipitation of insoluble rhodium species, for example, RhI_3. Therefore, a relatively large amount of water (10–15 wt%) must be retained in the process to keep the rhodium catalyst stable and in solution. In addition, excess water prevents the formation of methyl acetate and dimethyl ether:

$$CH_3OH + CH_3COOH \rightleftarrows CH_3COOCH_3 + H_2O \qquad \Delta_r H_{298} = -15.3 \text{ kJ/mol} \qquad (9.12)$$

$$2\,CH_3OH \rightleftarrows H_3COCH_3 + H_2O \qquad \Delta_r H_{298} = -23.6 \text{ kJ/mol} \qquad (9.13)$$

The disadvantages of adding water are the more costly separation train and the loss of carbon monoxide due to interference of the water–gas shift reaction:

$$CO + H_2O \rightleftarrows CO_2 + H_2 \qquad \Delta_r H_{298} = -41 \text{ kJ/mol} \qquad (9.14)$$

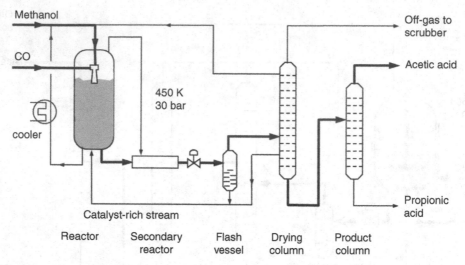

Figure 9.5   *Simplified flow scheme of the Cativa process for acetic acid production.*

Thus, it would be economically advantageous to operate the carbonylation process at lower water concentrations, provided that catalyst stability can be preserved. One strategy to achieve this is replacing the rhodium catalyst by a more robust iridium catalyst.

Figure 9.5 shows a typical flow scheme of the iridium-based Cativa process [5]. The reactor is not mechanically stirred; a jet loop reactor is used, in which the reactor contents are mixed by the jet mixing effect of the cooling loop. A (finishing) plug-flow reactor is installed after the first reactor for increased carbon monoxide conversion.

Because the iridium catalyst system is stable at low carbon monoxide pressures, a much lower water concentration (<8 wt%) can be used. As a result, one less distillation column is required and the light ends column and drying column can be combined into a single column. Furthermore, propionic acid is produced in much smaller amounts than in the Monsanto process, allowing the size of the product column to be reduced.

In the development of the carbonylation process, corrosion by hydrogen iodide has always been a point of concern. The key to the success of the Monsanto and Cativa processes is their very active catalyst system, which allows operation at much less severe conditions than those in the BASF process. Still, the need for expensive corrosion-resistant construction materials like Hastelloy C, titanium or even more exotic materials remains. Titanium is most often used.

QUESTION:

> *In the Cativa process it makes sense to install a finishing reactor in order to achieve a high conversion of CO. Is this also the case for the Monsanto process? (Hint: compare the kinetics of the two processes.)*

9.3   Hydroformylation

9.3.1   Background Information

Hydroformylation, often referred to as oxo synthesis, was discovered by Otto Roelen in Germany before World War II in one of the many programs aimed at the use of coal-derived synthesis gas (CO/H_2 mixtures). Roelen discovered that alkenes react with syngas, provided an appropriate catalyst is present.

Formally, a formyl group (CHO) and a hydrogen atom are added to the double bond. Therefore, the reaction has been called "hydroformylation", analogous to "hydrogenation" being the addition of hydrogen. The most important products are in the range C_4–C_{19}. With a share of about 75%, the most significant is the hydroformylation of propene [10] yielding two isomers:

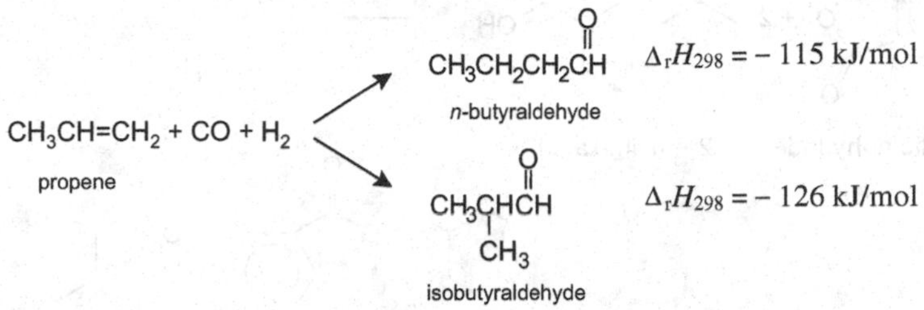

$$CH_3CH=CH_2 + CO + H_2$$

propene

$$CH_3CH_2CH_2\overset{\displaystyle O}{\overset{\displaystyle \|}{C}}H \qquad \Delta_r H_{298} = -115 \text{ kJ/mol}$$

n-butyraldehyde

$$CH_3\overset{\displaystyle O}{\underset{\displaystyle CH_3}{\overset{\displaystyle \|}{C}}}HC\overset{\displaystyle }{\overset{}{}}H \qquad \Delta_r H_{298} = -126 \text{ kJ/mol}$$

isobutyraldehyde

(9.3)

The products of hydroformylation usually are intermediates for the production of several types of alcohols, which are commonly known as oxo-alcohols. In most cases the normal aldehyde is the preferred product because it enjoys a much larger market than the branched aldehyde. For instance, normal butanol, produced by the direct hydrogenation of normal butyraldehyde, is used in the production of a wide variety of chemical intermediates, while iso-butanol is predominantly used as a solvent. Furthermore, only normal butyraldehyde can be used to produce 2-ethylhexanol, which is the most widely used plasticizer alcohol [11] (Box 9.4). The oxo-alcohols in the C_{11}–C_{17} range are utilized in detergents (Box 9.5).

Box 9.4   Synthesis and Application of Oxo-Alcohols in Plasticizers

Plasticizers for plastics are additives with low vapor pressures, most commonly phthalates, that give hard and rigid plastics like PVC the desired flexibility and durability. The vapor pressure should not be too low, which makes oxo-alcohols in the C_7–C_{11} range the most suitable precursors. At present some 300 plasticizers are manufactured, of which at least 100 are of commercial importance [12].

Di-2-ethylhexyl phthalate, also known as dioctyl phthalate (DOP), was the most common plasticizer, but now is suspected of causing cancer and already has been banned in a number of applications. Its precursor is 2-ethylhexanol, which is formed from *n*-butyraldehyde by the reaction sequence shown in Scheme B.9.4.1.

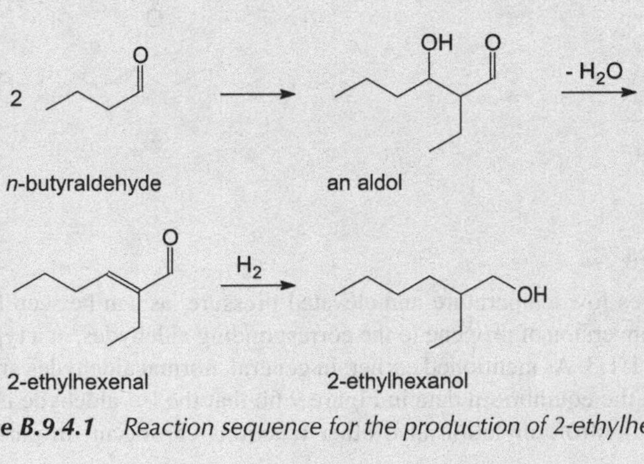

n-butyraldehyde   an aldol

2-ethylhexenal   2-ethylhexanol

Scheme B.9.4.1   *Reaction sequence for the production of 2-ethylhexanol.*

DOP is then produced by reaction of phthalic anhydride with 2-ethylhexanol:

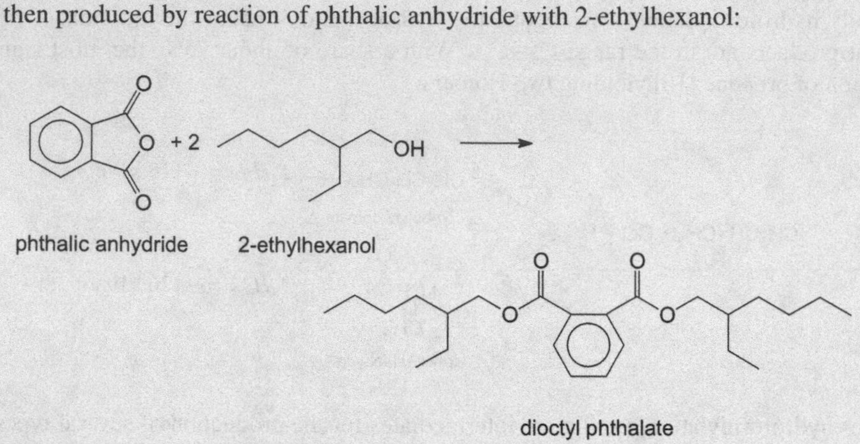

phthalic anhydride   2-ethylhexanol

dioctyl phthalate

Box 9.5 Oxo-Alcohols in Detergents

Detergents are surfactants (surface active agents), that is, compounds that lower the surface tension of a liquid or the interfacial tension between two liquids or a liquid and a solid. They are usually organic compounds with large hydrophobic hydrocarbon tails and small hydrophylic heads that are either anionic, cationic, nonionic, or amphoteric [13]. Commonly encountered surfactants of the anionic-head type are alkyl sulfates, which are used in a variety of detergents.

Alkyl sulfates are produced by sulfatation of C_{11}–C_{17} oxo-alcohols with sulfur trioxide followed by neutralization with sodium hydroxide:

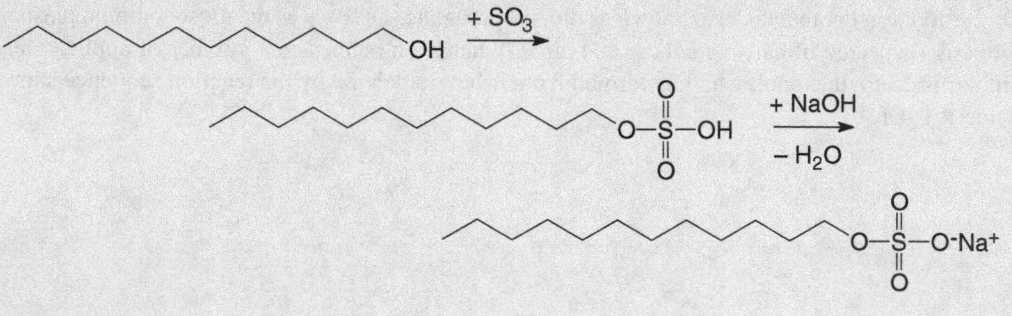

9.3.2 Thermodynamics

Hydroformylation requires low temperature and elevated pressure, as can be seen from Figure 9.6a, which shows the equilibrium conversion of propene to the corresponding aldehydes, at a typical molar inlet composition $H_2/CO/propene = 1:1:3$. As mentioned earlier, in general, normal aldehydes are the preferred products. However, it is clear from the equilibrium data in Figure 9.6b that the iso-aldehyde is the thermodynamically favored product. During hydroformylation also other reactions can occur, in particular hydrogenation to

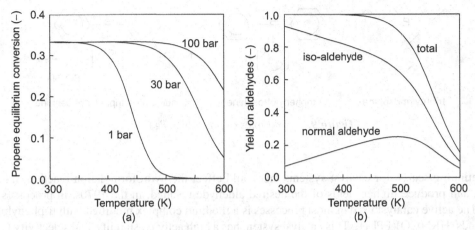

Figure 9.6   *Hydroformylation of propene: (a) propene equilibrium conversion; (b) normal/iso distribution of butyraldehyde (p = 30 bar; propene/syngas = 3 mol/mol).*

propane. Thermodynamically, the latter reaction is the preferential reaction, so a catalyst with high selectivity is required.

QUESTIONS:

> *The thermodynamically favored product of propene hydroformylation is not a butyraldehyde but propane. Show that this statement is true. Nonetheless, the products of propene hydroformylation are indeed butyraldehydes. Explain.*

9.3.3   Catalyst Development

The most important catalysts for hydroformylation are based on rhodium and cobalt, which are introduced as carbonyls. Rhodium and cobalt are not the only metals showing catalytic activity but they are the most active ones. The relative metal activity for hydroformylation is shown in Table 9.6. It is not surprising that in practice rhodium and cobalt are used.

A wide variety of ligands, in particular phosphines (Figure 9.7), have been shown to influence activity, stability, and selectivity of the catalyst complexes. The first catalyst used for the hydroformylation reaction, in the 1940s, was cobalt hydridocarbonyl, $HCo(CO)_4$, which requires operation at very high pressure (200–450 bar) to ensure catalyst stability [11]. In the 1960s, a more stable tributylphosphine-substituted cobalt carbonyl catalyst, $HCo(CO)_3P(n\text{-}C_4H_9)_3$, was developed, which also shows a much higher normal-to-iso ratio in the product distribution. However, the activity of the catalyst is lower and more by-products are formed [14].

Table 9.6   *Relative activities of metals for hydroformylation.*

Rh	Co	Ru	Mn	Fe	Cr, Mo, W, Ni
10^3–10^4	1	10^{-2}	10^{-4}	10^{-6}	0

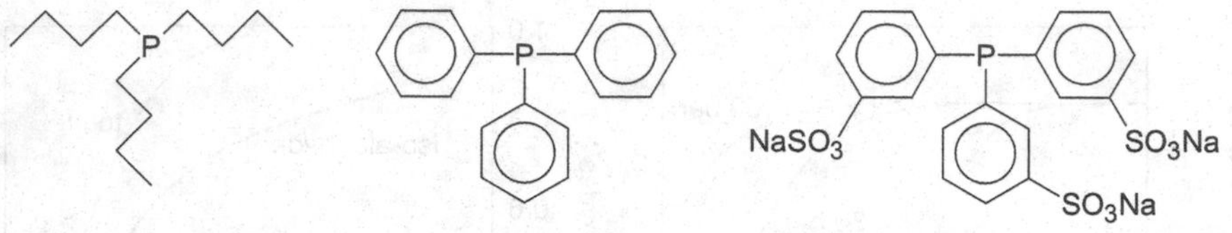

tributylphosphine   triphenylphosphine   sulfonated triphenylphosphine

Figure 9.7   *Examples of phosphine ligands.*

The continuing search for catalyst systems that could effect the hydroformylation reaction under milder conditions and produce higher yields of the desired aldehyde resulted, in the 1970s, in processes utilizing rhodium. The active catalyst used in most processes is a rhodium complex modified with triphenylphosphine (TPP) ligands, $HRh(CO)(PPh_3)_3$. This catalyst system has a high activity, stability, and selectivity [15]. Since the 1980s, a rhodium catalyst with a water-soluble ligand, triphenylphosphine-*m*-trisulfonic acid trisodium salt (TPPTS) has also been used. The normal-to-iso ratio of the product is very high, but the activity of this catalyst is significantly lower [14].

Kinetic studies have been performed for the hydroformylation reaction. For a simple cobalt system, $(HCo(CO)_4)$, the following rate expression has been found [16]:

$$r_{\text{hydroformylation}} = k\frac{[\text{Co}]\,[C_3H_6]\,[H_2]}{[\text{CO}]} \tag{9.15}$$

For a rhodium system containing TPP ligands the following rate equation has been reported [17]:

$$r_{\text{hydroformylation}} = k\frac{[\text{Rh}]\,[C_3H_6]^{0.6}\,[H_2]^{0.05}}{[PPh_3]^{0.7}\,[\text{CO}]^{0.1}} \tag{9.16}$$

Box 9.6 illustrates the catalytic cycle for rhodium-based hydroformylation.

QUESTIONS:

Despite the high price of the ligand, in practice the TPP/Rh ratio is high, in the order of 100 mol/mol. What is the reason?

Draw a catalytic cycle for the hydroformylation of propene with a HCo(CO)₄ catalyst. Interpret the rate equations based on the catalytic cycles. What are the consequences for reactor design?

During hydroformylation, side reactions occur, for example, double bond isomerization (e.g., 1-hexene to 2-hexene) and hydrogenation of the alkenes to the corresponding alkanes. The extent of these side reactions depends on the metal and the ligand added.

The selectivity to normal aldehydes of rhodium catalysts is much better than that of cobalt catalysts. In addition, rhodium catalysts show little hydrogenation or double bond isomerization activity. Therefore, processes for hydroformylation of propene, the predominant feedstock, are currently almost exclusively based on this type of catalyst.

Box 9.6 Catalytic Cycle for Rhodium-Based Hydroformylation

Scheme B.9.6.1 shows the catalytic cycle for hydroformylation of propene by $HRh(CO)L_2$ [14].

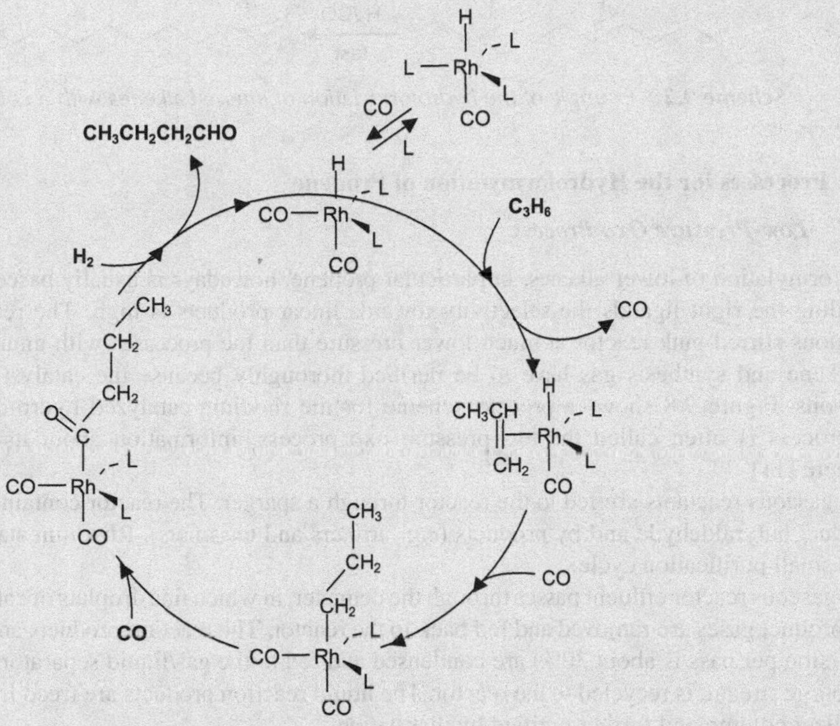

Scheme B.9.6.1 *Catalytic cycle in hydroformylation of propene; L = triphenylphosphine ligand.*

Unlike rhodium catalysts, cobalt catalysts strongly enhance double bond isomerization. This is why cobalt-based processes are quite flexible in that they can be used for the hydroformylation of a wide variety of alkenes, including internal alkenes, which would yield a large proportion of the undesired nonlinear products with a rhodium catalyst. Scheme 9.2 illustrates how double bond isomerization increases the selectivity to linear aldehydes.

QUESTIONS:

Hydroformylation in principle can be used for all types of alkenes. Why would 1-alkenes (with a terminal double bond) react faster than internal alkenes?

The major side reactions are hydrogenation of the alkenes to form alkanes and double bond isomerization. Are these desired or undesired reactions? Explain.

For mixtures of n- and iso-alkenes, Co catalysts produce a more favorable product distribution than Rh catalysts. Explain.

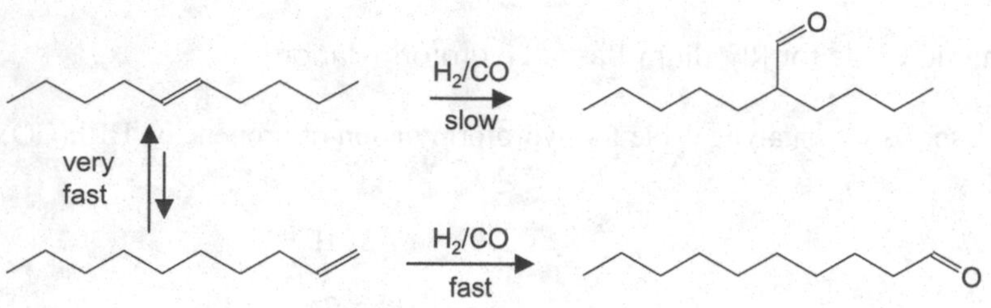

Scheme 9.2   *Example of the hydroformylation of internal alkenes with a cobalt catalyst.*

9.3.4   Processes for the Hydroformylation of Propene

9.3.4.1   *Low-Pressure Oxo Process*

Hydroformylation of lower alkenes, in particular propene, nowadays is usually based on rhodium catalysts. By adding the right ligands the selectivity towards linear products is high. The reaction takes place in a continuous stirred-tank reactor at much lower pressure than the processes with unmodified cobalt catalyst. The alkene and synthesis gas have to be purified thoroughly because the catalyst is extremely sensitive to poisons. Figure 9.8 shows a process scheme for the rhodium-catalyzed hydroformylation of propene. This process is often called the low-pressure oxo process; information about its development is given elsewhere [11].

The gaseous reactants are fed to the reactor through a sparger. The reactor contains the catalyst dissolved in product butyraldehyde and by-products (e.g., trimers and tetramers). Rhodium stays in the reactor, apart from a small purification cycle.

The gaseous reactor effluent passes through the demister, in which fine droplets of catalyst that are contained in the product gases are removed and fed back to the reactor. The reaction products and unconverted propene (conversion per pass is about 30%) are condensed and fed to the gas/liquid separator. Propene, apart from a small purge stream, is recycled to the reactor. The liquid reaction products are freed from residual propene in a stabilizer column and further purified by distillation.

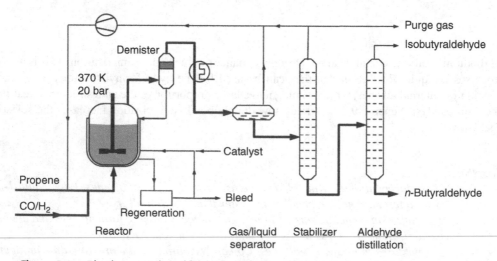

Figure 9.8   *Rhodium-catalyzed hydroformylation of propene (low-pressure oxo process).*

A portion of the catalyst solution passes through a purification cycle for reactivation of inactive rhodium complexes, and for removal of heavy ends and ligand decomposition products.

The demisting device is an essential part of the plant. It ensures that all rhodium remains in the reactor. In view of the high cost of rhodium, recovery should be as high as possible, see Ref. [15] for an enlightening calculation. In practice, the rhodium losses in the process are very small.

QUESTIONS:

> *What would be the main impurity in propene? Why is a purge needed?*
> *A disadvantage of the process scheme of Figure 9.8 is that by-products remain in the catalyst phase. In modern plants, the products are removed in the liquid phase together with the catalyst [1]. Draw a process scheme for such a process.*

9.3.4.2 Process with a Biphasic Catalyst System

A novel industrial process is based on a *water-soluble* rhodium catalyst (Ruhrchemie/Rhône-Poulenc, now Celanese) [18]. It contains polar ligands (sulfonated triphenylphosphines) which are highly soluble in water. The catalyst is insoluble in the organic phase formed, so a biphasic reaction medium results. Thus, separation is greatly simplified and, as a consequence, the rhodium losses are minimal. In fact, the catalyst system could be considered to be heterogeneous. Figure 9.9 shows a simplified flow scheme of the process.

In this process, the reaction also takes place in a continuous stirred-tank reactor containing the catalyst solution. Before entering the reactor, the syngas is first passed through a stripping column to recover the unreacted propene. The liquid reactor effluent is fed to a phase separator where dissolved gases are removed and the butyraldehydes are separated from the aqueous catalyst solution. The catalyst solution is returned to the reactor via a heat exchanger in which steam is generated.

QUESTIONS:

> *What would be the composition of the off-gas. Why is a demister not needed in this case?*
> *Why has water been selected as the catalyst solvent instead of an organic solvent?*
> *Why are none of the Rh-based processes suitable for the hydroformylation of higher alkenes (either terminal or internal)? (Hint: consider relative volatility and solubility.)*

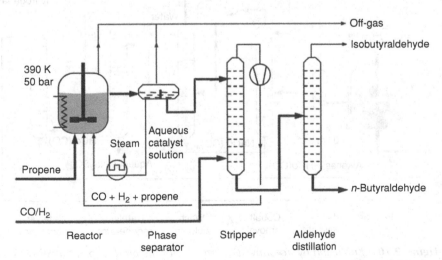

Figure 9.9 *Ruhrchemie/Rhône-Poulenc hydroformylation of propene using a water-soluble rhodium catalyst.*

9.3.5   Processes for the Hydroformylation of Higher Alkenes

9.3.5.1   *Conventional Cobalt-Based Process*

The first industrial oxo process was based on an unmodified cobalt catalyst: $HCo(CO)_4$. This catalyst system requires high pressures and temperatures to ensure the stability of the catalyst. Hydroformylation of higher alkenes to produce plasticizer and detergent-range alcohols still uses this type of catalyst.

To minimize catalyst consumption and to avoid problems in downstream processing (fouling of equipment, etc.), it is very important that cobalt is recovered. A process that offers an elegant solution to the problem of catalyst separation is the Kuhlmann (now Exxon Mobil) process. This hydroformylation process is based on the following cobalt cycle [14]:

$$2\,HCo(CO)_4 + Na_2CO_3 \rightarrow 2\,NaCo(CO)_4 + H_2O + CO_2 \tag{9.17}$$

The sodium salt is soluble in water. This enables recovery of the cobalt catalyst by scrubbing with water. The aqueous solution containing cobalt as the sodium salt is transformed to the active catalyst by reaction with sulfuric acid:

$$2\,NaCo(CO)_4 + H_2SO_4 \rightarrow 2\,HCo(CO)_4 + Na_2SO_4 \tag{9.18}$$

This complex is soluble in alkenes and, as a consequence, it can be returned to the reactor dissolved in the reactant alkene flow.

Figure 9.10 shows a flow scheme of the Kuhlmann hydroformylation process for the production of aldehydes. The alkene, with recycled and make-up catalyst, is fed to the hydroformylation reactor (usually

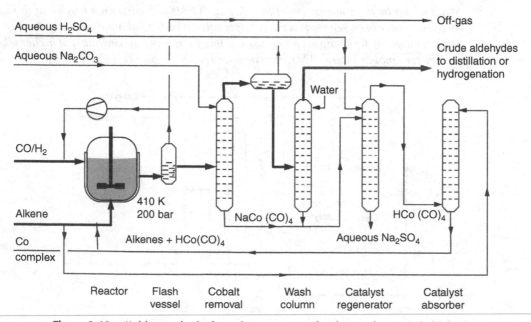

Figure 9.10   *Kuhlmann hydroformylation process for the production of aldehydes.*

a stirred tank or a loop reactor) together with the synthesis gas. The crude product, after passing through the flash vessel, in which the gas is separated, is treated in countercurrent flow with an aqueous sodium carbonate (Na_2CO_3) solution. The cobalt complex forms a water-soluble sodium salt. After scrubbing with water in the wash column, a virtually cobalt-free organic phase (crude aldehydes) and an aqueous phase containing the sodium salt remain. The crude aldehydes are sent to a distillation section (not shown). Here the excess alkenes are recovered and recycled to the reactor. The sodium salt is transformed to the original cobalt complex, $HCo(CO)_4$, by adding sulfuric acid in the regenerator. This complex is extracted by the alkene feedstock and recycled to the reactor.

9.3.5.2 *Low-Pressure Cobalt-Based Process for the Direct Production of Alcohols*

Shell has commercialized a process based on a trialkylphosphine-cobalt catalyst that is much more stable than the unmodified catalyst, allowing (and requiring) operation at lower pressure [19]. This process produces the alcohol rather than the aldehyde, which is an added advantage since alcohols are usually the desired end products of hydroformylation. A disadvantage is that the high hydrogenation activity also causes the formation of alkanes (up to 15% versus 2% for the non-modified catalyst). Therefore, it may be advantageous to use two reaction stages, with a lower hydrogen partial pressure in the first reactor to promote hydroformylation of the alkene rather than hydrogenation. Figure 9.11 shows a simplified flow scheme of this version of the process.

The alkene feed is added to the first reactor together with make-up and recycle catalyst complex. Here, conversion takes place with a hydrogen-poor synthesis gas to limit alkane formation. In the second reactor, hydrogen-rich gas is added to ensure the hydrogenation of the aldehyde to the alcohol. The separation takes place by depressurization in a flash vessel and subsequent distillation. In the first column, unconverted alkenes are removed and recycled to the reactor. A purge stream is added in order to prevent the build-up of alkanes. The second column separates the catalyst complex and heavy by-products from the crude alcohols.

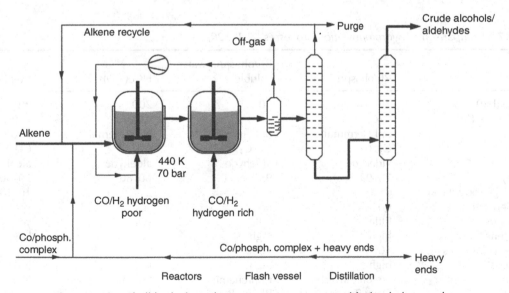

Figure 9.11   *Shell hydroformylation process using a modified cobalt complex.*

By-products are formed by dimerization of aldehydes (Section 9.3.1). Build-up of these aldols is prevented by purging part of the recycle stream.

Although this process looks much simpler than the Kuhlmann hydroformylation process, it has two distinct disadvantages:

- thorough purification of the synthesis gas is very important because the ligand is very sensitive to oxidation;
- the activity of the modified cobalt catalyst is much lower than that of the unmodified catalyst and, therefore, for the same productivity the reactors need to be five to six times larger than in the conventional process.

Advantages, apart from the ones already mentioned, are:

- higher linearity of the alcohol product;
- the possibility to separate the catalyst from the product by conventional distillation.

9.3.6   Comparison of Hydroformylation Processes

Table 9.7 summarizes characteristics of the major hydroformylation processes. The rhodium catalyst is attractive for the conversion of propene to the linear aldehyde. The selectivity may be as high as 95%. The branched product also has a market, although a much smaller one than the linear product, and its production is, therefore, desired in these small quantities. Hydroformylation of higher alkenes with internal double bonds (e.g., produced by oligomerization of ethene, Section 9.4) requires isomerization activity and, as a consequence, in that case cobalt catalysts are preferred. In all processes, except one, aldehydes are the primary products. These can be hydrogenated to alcohols in a separate reactor. The modified cobalt catalyst is used for the direct production of alcohols rather than aldehydes because this catalyst promotes hydrogenation.

Table 9.7   *Comparison of hydroformylation processes [7, 19, 20].*

Catalyst	Rh/phosph.	Rh/phosph., water soluble	HCo(CO)$_4$	Co/phosph.
Pressure (bar)	20	50	200	70
Temperature (K)	370	390	410	440
Alkene	only terminal	only terminal	also internal	also internal
	C$_3$, C$_4$	C$_3$, C$_4$	C$_{10+}$	C$_{10+}$
Product	aldehyde	aldehyde	aldehyde	alcohol
Linearity (%)	70–95	95	60–80	70–90
Alkane by-product (%)	0	0	2	10–15
Metal deposition	no	no	yes	yes
Heavy ends	little	little	yes	yes
Poison sensitivity	high	high	low	low
Relative metal costs	1000	1000	1	1
Ligand costs	high	high	low	high
Company	Union Carbide	Ruhrchemie/	Kuhlmann (now	Shell
	Davy Powergas	Rhône-Poulenc	Exxon Mobil)	
	Johnson Matthey			

Table 9.8 *Uses of alkenes produced by oligomerization of ethene.*

Alkene	Use
1-butene	polymer production
1-hexene	Co-monomer in polyethene production
1-octene	Co-monomer in polyethene production
C_6–C_{10} alkenes	Plasticizer production (after hydroformylation)
C_{11}–C_{18} alkenes	Production of detergent-range alcohols (by hydroformylation)

QUESTIONS:

Compare the four processes in the Table 9.7 with respect to the phases present. Is mass transfer expected to be limiting? Compare the four processes in this respect and elaborate on the reactor choice.

9.4 Ethene Oligomerization and More

9.4.1 Background Information

Linear 1-alkenes are important feedstocks for the chemical industry. They initially were produced by thermal cracking of waxes (high-molecular-weight alkanes) or by dehydration of 1-alkanols, but both routes have been abandoned, although dehydration has come into focus again. Since the mid-1970s, C_4–C_{18} alkenes have been produced by oligomerization of ethene [21]. Table 9.8 summarizes the uses of these alkenes.

A number of companies have developed ethene oligomerization processes, most of them based on Ziegler-type catalysts [22–24] (Chapter 11). Discussed here is the Shell Higher Olefins Process (SHOP), which is different from the other processes in a number of ways. Firstly, it uses a different catalyst (Box 9.7) and, secondly, it is a unique combination of three reactions (four if the hydroformylation reaction is counted as well), namely ethene oligomerization, double bond isomerization, and metathesis.

Table 9.9 compares the product qualities of the C_6–C_{18} 1-alkenes from various processes. The yield and linearity of the oligomerization stage of the SHOP process are exceptionally high.

The demand for 1-alkenes in the C_4–C_{10} range is growing faster than that in the C_{12+} range, which has made the selective formation of light 1-alkenes, such as 1-butene, 1-hexene, and 1-octene, an important topic of research during the last twenty years [24]. A number of such "on purpose" processes have been commercialized [21, 24].

Table 9.9 *Comparison of typical qualities of C_6–C_{18} 1-alkenes (wt%) from various processes [25].*

	Wax cracking	Chevron	Ethyl	Shell SHOP
1-alkenes	83–89	93–97	63–98	96–98
Branched alkenes	{3–12	{2–8	2–29	1–3
Internal alkenes			1–8	1–3
Alkanes	1–2	1	0–1	<0.1
Dienes	3–6	—	—	—
Total monoalkenes	92–95	99	<99	99.9

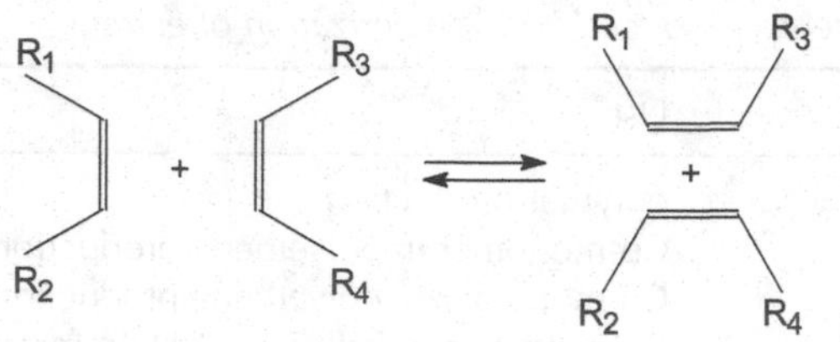

Scheme 9.3   *Metathesis of alkenes. R_1, R_2, R_3, R_4 = H, CH_3, CH_2CH_3, etc.*

9.4.2   Reactions of the SHOP Process

The *oligomerization* reaction is a catalytic chain-growth reaction that is very similar to polymerization, but the chains formed are much shorter: The key is a homogeneous catalyst with an unusual ligand (Box 9.7). This ligand allows oligomerization but because of its size inhibits chain growth to the polymer stage. For oligomerization of ethene the overall reaction is represented by:

$$(n+2)\ CH_2{=}CH_2 \longrightarrow$$

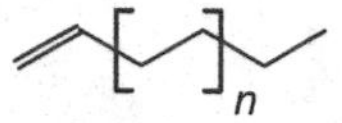

$$0 < n < 18$$

$$\Delta_r H_{298} \approx -95\ \text{kJ/mol}\ C_2H_4 \tag{9.3}$$

Oligomerization of ethene produces linear 1-alkenes with chain lengths ranging from C_4 to C_{40} with an even number of carbon atoms and with a statistical distribution (Anderson–Flory–Schulz distribution, Section 6.3.1) in which the alkenes $<C_{20}$ are favored. A typical general distribution is as follows: C_4–C_8 40%, C_{10}–C_{18} 40%, and C_{20+} 20%.

This is not the desired carbon number distribution. The C_{10}–C_{18} fraction contains valuable intermediates, but the C_4–C_8 fraction has limited commercial value, and the C_{20+} fraction cannot be sold as such. It would thus be useful to combine these latter fractions and produce more C_{10}–C_{18} alkenes. This can be done by *double bond isomerization* followed by *metathesis*. Although these are both solid-catalyzed reactions, they will be discussed here briefly to get a more complete picture of the SHOP process. Double bond isomerization yields an equilibrium mixture of internal alkenes. Metathesis involves the conversion of alkenes to produce alkenes of different size (also see Section 4.5.2). Formally, double bonds are broken with the simultaneous formation of new ones (Scheme 9.3).

Box 9.7   Catalytic Cycle for the Oligomerization of Ethene

The mechanism of the oligomerization of ethene has been studied in detail [26]. Based on these results, the catalytic cycle shown in B.9.7.1 has been postulated. The oligomerization reaction takes place by successive insertions of ethene into the nickel-hydride bond. Elimination yields the oligomerized products.

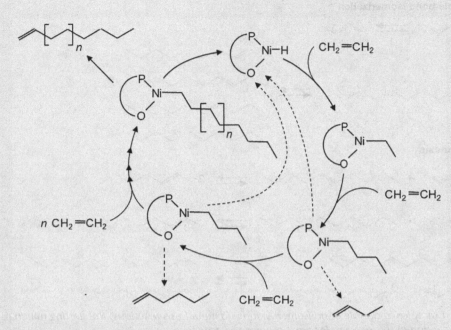

Scheme B.9.7.1 *Catalytic cycle in oligomerization of ethene. P = phosphine.*

QUESTION:

> *Draw (or write out) the catalytic cycle for the formation of 1-hexene.*

The mechanism of this reaction has been subject to a lot debate. Scheme 9.3 suggests a pairwise mechanism, but it is generally agreed that in reality the mechanism is a stepwise one [1, 27].

Scheme 9.4 shows examples of typical isomerization and metathesis reactions occurring in the SHOP process. Clearly, the product distribution shifts to the desired range.

QUESTIONS:

> *According to present insights the key intermediate in metathesis is a metallocyclobutane complex, formed by the reaction of an alkene and a carbene complex. Draw the catalytic cycle.*
>
> *A small amount of branched alkenes is formed during oligomerization. What would be the cause of this?*
>
> *What are the products of metathesis of propene and of a mixture of 1-butene and 2-butene?*

9.4.3 The SHOP Process

The oligomerization reaction is carried out in a series of reactors (Figure 9.12, only one shown) in a polar solvent with the rate of reaction being controlled by the rate of catalyst addition [22, 23]. The heat of reaction is removed by water-cooled heat exchangers between the reactors.

Double-bond isomerization

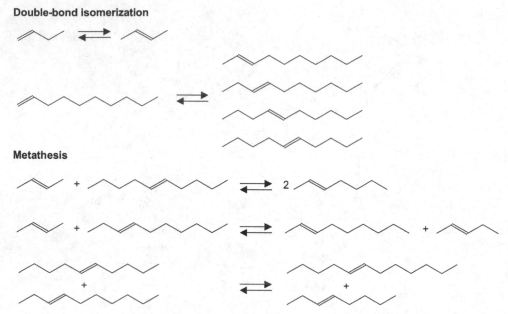

Metathesis

Scheme 9.4   *Examples of double bond isomerization and metathesis reactions; the starting materials for isomerization in these examples are 1-butene and 1-decene.*

Unconverted ethene is separated from the two liquid phases in a high-pressure separator and recycled. The catalyst solution is separated from the 1-alkene products and fed back into the oligomerization reactor. The 1-alkenes are separated into the desired product fractions in two distillation columns. Firstly, the C_4–C_{10} 1-alkenes are stripped off. Then, the C_{20+} fraction is removed from the desired C_{12}–C_{18} 1-alkenes, which can be separated into the desired fractions or individual 1-alkenes.

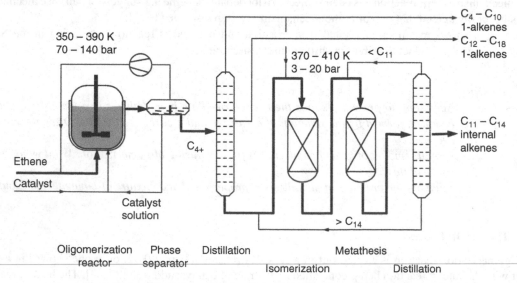

Figure 9.12   *Simplified flow scheme of the Shell Higher Olefins Process (SHOP).*

Part of the C_4–C_{10} fraction is combined with the C_{20+} fraction to be isomerized and subjected to metathesis. About 90% of the 1-alkenes are converted to internal alkenes by isomerization [22]. Subsequent metathesis of the lower and higher internal alkenes gives a mixture with a different carbon number distribution, which contains only about 11–15 wt% per pass of the desired C_{11}–C_{14} linear internal alkenes [22, 23]. These are separated by distillation and can then be converted to detergent alcohols by hydroformylation (Section 9.3). The lighter alkenes are recycled to the metathesis reactor, while the higher alkenes are again subjected to isomerization. These are recycled to extinction.

The SHOP process, like the Ruhrchemie/Rhône-Poulenc hydroformylation of propene, is an example of biphasic catalysis: the chelate ligand is dissolved in a polar solvent (1,4-butanediol) in which the non-polar products are almost insoluble. This use of a biphasic organic liquid/liquid system is one of the key features of the process [22]. Another key feature is the high flexibility of the process as a result of the combination of three different reactions, which allows tailoring of the carbon number distribution and the amounts of 1-alkenes produced.

QUESTIONS:

> *Would the molar flow rates of the light and heavy feeds to the isomerization reactor be the same? If not, which flow rate is higher and why?*
> *Why would it be a bad idea to skip the isomerization reactor and directly feed the product mix from the oligomerization process to metathesis?*

9.5   Oxidation of *p*-Xylene: Dimethyl Terephthalate and Terephthalic Acid Production

9.5.1   Background Information

Both terephthalic acid (TPA) and dimethyl terephthalate (DMT) are monomers for the production of polyethene terephthalate (PET) by polycondensation with ethene glycol (Scheme 9.5). In the production of the polymer, the acid has obvious advantages over the ester because the yield per kilogram of starting material is higher and no methanol recovery step is needed.

In most applications, and especially in the field of fibers, monomer purity is an essential factor in polymer quality. Monomers for polyester production are among the purest high-volume chemicals [28]. This is why in the past often the ester was preferred over the acid. The latter is difficult to purify; it is not volatile, does not readily dissolve in normal solvents, and does not melt but sublimes.

TPA and DMT are major outlets for *p*-xylene. Oxidation of *p*-xylene with air to produce *p*-toluic acid (C_6H_4-(COOH)(CH$_3$)) is relatively easy. Further oxidation to TPA is very slow, however, due to the high stability of *p*-toluic acid. Therefore, the earliest process for TPA production based on *p*-xylene involved oxidation by nitric acid [29], which is a very strong oxidant:

$$\text{CH}_3\text{-C}_6\text{H}_4\text{-CH}_3 + 4\,\text{HNO}_3 \longrightarrow \text{HOOC-C}_6\text{H}_4\text{-COOH} + 4\,\text{H}_2\text{O} + 4\,\text{NO} \qquad \Delta_r H_{298} = -750 \text{ kJ/mol}$$

$$(9.19)$$

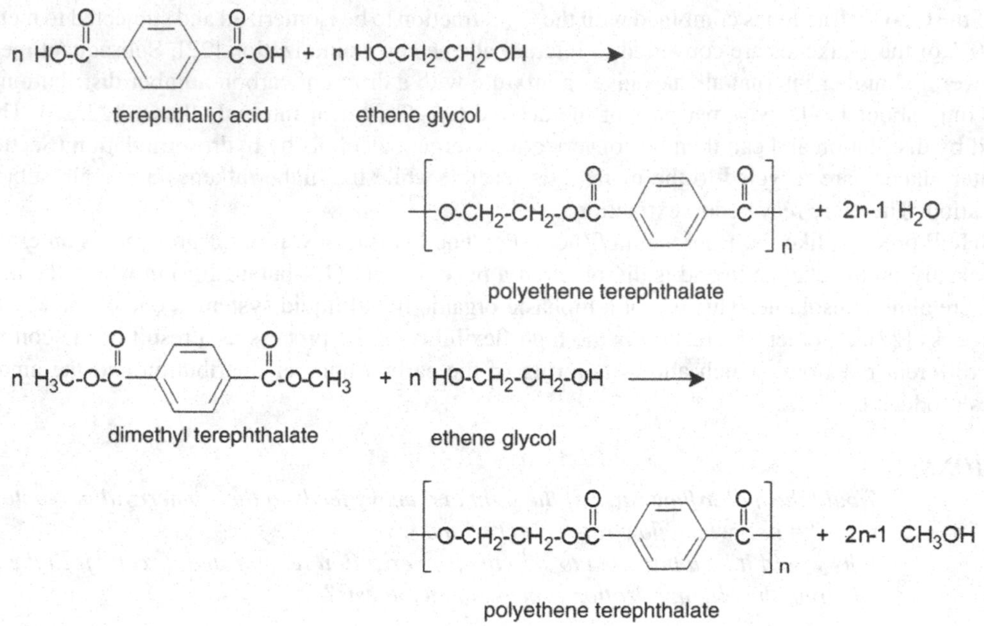

Scheme 9.5   *Reactions for the production of polyethene terephthalate (PET).*

Not surprisingly, the product obtained contains substantial amounts of impurities. The solution found was the conversion of the acid into the corresponding methyl ester, namely DMT, which was then purified by vacuum distillation and/or recrystallization.

9.5.2   Conversion of *p*-Toluic Acid Intermediate

The oxidation of *p*-xylene takes place by a radical mechanism with cobalt acetate or cobalt napthenate as the catalyst. As stated above, the oxidation stops because of the high stability of *p*-toluic acid. Two methods are available to continue.

QUESTION:

> *Is the statement that the oxidation is a radical process in disagreement with the use of a catalyst?*

One method is to esterify *p*-toluic acid with methanol to *p*-methyl toluate, which is readily oxidized. Esterification proceeds in the liquid phase without a catalyst. During the 1950s and 1960s a number of processes combining oxidation in air and esterification were developed to produce DMT, of which the most important is the Witten process. In this process, *p*-xylene is converted in four alternate oxidation and esterification steps as shown in Scheme 9.6. The overall reaction is given by equation 9.4 in Table 9.2.

The second solution is the addition of extremely reactive radicals, such as bromine atoms. These are formed by reaction of sodium bromide (NaBr) with acetic acid and oxygen. The bromine radicals then attack *p*-toluic

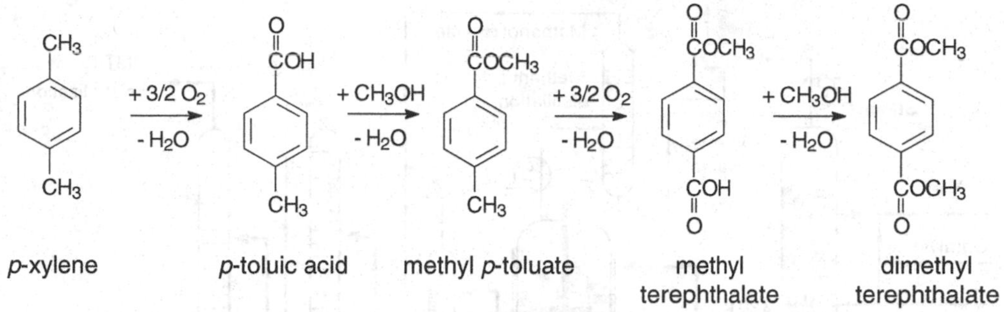

Scheme 9.6 *Reaction sequence for the production of dimethyl terephthalate (DMT).*

acid. Once the *p*-toluic acid radical has been formed, the reactions can proceed similarly to the oxidation of *p*-xylene, finally resulting in the formation of TPA. The overall reaction is as follows:

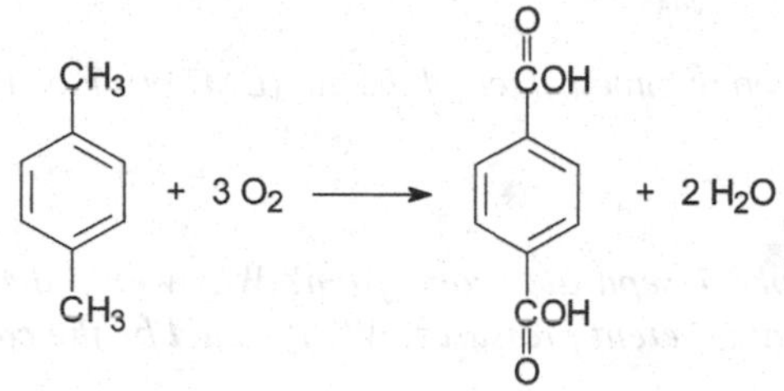

$$\Delta_r H_{298} = -1360 \text{ kJ/mol}$$

(9.5)

During the 1960s, a commercial process was developed based on this approach. The TPA produced is of a high purity. Since the discovery of this process, the Amoco process, there has been a trend towards the use of TPA rather than DMT in polymerization processes. Indeed, currently over 95% of the combined production of TPA and DMT is production of TPA.

9.5.3 Processes

Figure 9.13 shows a simplified flow scheme of the Witten process for the production of DMT. At the heart of the process are two reactors, one for oxidation and one for esterification. Fresh and recycled *p*-xylene, recycled *p*-methyl toluate, and catalyst (cobalt naphthenate) are fed to the oxidation reactor, in which mixing takes place by the upward movement of air. In this reactor, both *p*-xylene and *p*-methyl toluate are oxidized. The temperature in the reactor is controlled by evaporation and condensation of the formed water.

The product from the oxidation reactor is heated and fed to the esterification reactor, into which excess evaporated methanol is sparged. *p*-Toluic acid and methyl terephthalate are transformed to their corresponding esters non-catalytically. Overhead vapors are condensed, after which methanol is separated from the water and recycled. The liquid product is flashed to remove any remaining methanol and water.

The esterification products are sent to a distillation column. *p*-Xylene and *p*-methyl toluate are collected overhead and returned to the oxidation reactor. The crude DMT, after removal of heavy by-products and catalyst metal, is purified by crystallization. Although the cobalt catalyst used in this process is relatively cheap, its recovery is necessary since heavy metals cannot simply be discarded. The residue can be mixed with water from the oxidation reactor, in which the catalyst dissolves. After centrifugation the catalyst solution can be recycled.

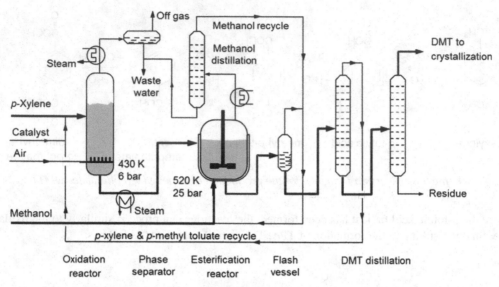

Figure 9.13   *Simplified flow scheme for the production of dimethyl terephthalate (DMT) (Witten process).*

QUESTIONS:

Where does the water leaving the phase separator come from? Why would the oxidation and esterification reactors operate at different pressures? What would be the composition of the off-gas?

The Amoco process produces polymer grade TPA directly by the oxidation of *p*-xylene. Figure 9.14 shows a simplified flow scheme. *p*-Xylene, acetic acid, and catalyst (cobalt acetate, promoted by bromine) are

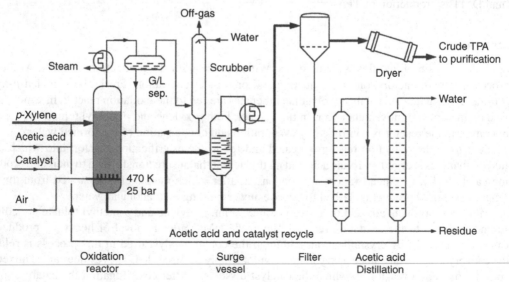

Figure 9.14   *Simplified flow scheme for the production of pure terephthalic acid (TPA) (Amoco process).*

introduced in the oxidation reactor. The reaction medium is agitated by introducing air at the bottom. Air is added in excess to maximize *p*-xylene conversion. Reaction takes place with acetic acid as the solvent for the catalyst. The heat generated is removed by vaporization of part of the solvent, and by condensation and reflux to the reactor. The generated steam can be used in the distillation part of the plant. A scrubber recovers acetic acid from the off-gases (nitrogen, unused oxygen, and carbon dioxide resulting from overoxidation).

Because of its limited solubility in acetic acid (0.13 g/kg at room temperature, 15 g/kg at 470 K [29]) most of the TPA crystallizes while it is being formed. A slurry forms, which is removed from the base of the reactor and sent to a surge vessel[1], which operates at lower pressure than the reactor. More TPA crystallizes. The suspension is filtered and the crude TPA is dried and sent to purification.

Due to the presence of acetic acid and bromine, highly corrosive process conditions exist. Therefore, the use of special materials such as Hastelloy C or titanium is necessary in the reactor and some other parts of the process.

The purity of the "crude" TPA is already over 99 wt% (technical grade TPA) [29, 30]. The most troublesome impurity is 4-formylbenzoic acid ($COOH-C_6H_4-COH$), commonly known as 4-carboxybenzaldehyde (4-CBA), which is an intermediate in the oxidation reactions and cocrystallizes with the TPA. 4-CBA causes termination of the esterification chain in the production of PET, which necessitates a purification step. An additional reason for removing 4-CBA is that it causes coloring of polymers.

The 4-CBA content may be reduced by selective catalytic hydrogenation in a fixed bed reactor. The crude TPA is slurried with water and heated until it dissolves entirely. 4-CBA is converted to *p*-toluic acid over a palladium catalyst. *p*-Toluic acid has a higher solubility than 4-CBA and does not cocrystallize with TPA [29, 30].

QUESTIONS:

> *What is the function of the heat exchanger at the top of the surge vessel? Why does complete oxidation to CO_2 and water not occur? Why is excess p-xylene used in the production of DMT, while excess air is used in the production of TPA?*

9.5.4 Process Comparison

The Amoco technology for the production of polymer-grade TPA accounts for nearly all new plants built. The reasons for this are its very high yield (>95%) accomplished in a single reactor pass and the low solvent loss. Furthermore, as mentioned previously, polymerization from TPA has several advantages over polymerization from DMT. The only important advantage of the Witten process for the production of DMT is that no expensive construction materials are required because no very strong oxidants such as bromine radicals are used. Figure 9.15 shows the capacities for TPA and DMT since 1980, illustrating the increasing popularity of the Amoco process.

9.6 Review of Reactors Used in Homogeneous Catalysis

As has been seen in previous sections, in homogeneous catalysis usually a reaction takes place between a gaseous and a liquid reactant in the presence of a catalyst that is dissolved in the liquid phase. The reaction

---

[1] Surge vessel: Vessel in which liquid is retained at a set level in order to prevent downstream or upstream equipment damage due to flow fluctuations.

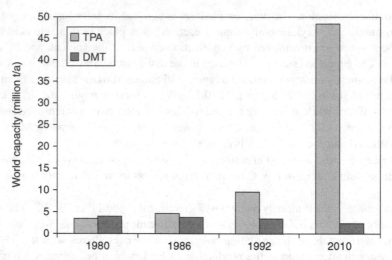

Figure 9.15   *World capacity for TPA and DMT production from 1980 to 2010. (Data from [28] and SRI Consulting.)*

generally takes place in the liquid phase. The contacting pattern between gas and liquid can be one of the following:

- Gas bubbles dispersed in continuous liquid phase:
  ○ Bubble column (with gas distribution device)
  ○ Sparged stirred tank (mechanical gas dispersion).
- Liquid droplets dispersed in continuous gas phase:
  ○ Spray reactor
- Thin flowing liquid film in contact with gas:
  ○ Packed column
  ○ Wetted-wall column.

9.6.1   Choice of Reactor

The choice of reactor depends on factors such as the desired volume ratio of gas to liquid, the rate of the reaction (fast or slow) in relation to the mass transfer characteristics, the kinetics (positive, negative, zero order), the ease of heat removal and temperature control, and so on. Figure 9.16 shows examples of the reactor types mentioned above, while Table 9.10 shows some important parameters. The numbers in the table should be used with care because they depend on the physical and chemical properties of the reacting system and the detailed design of the reactor. In literature most data have been obtained for model systems (often air/water) and, therefore, large discrepancies are possible. The data are meant to serve as a rough guide and to help in understanding the basics of reactor design.

QUESTION:

> *Reactors for reactions in the liquid phase are chosen based on the optimal usage of the total reactor volume. The reactors in Figure 9.16 are selected for increasingly faster reactions, that is, the sparged stirred tank and bubble column are suitable for slow reactions, while spray columns and packed columns are used for fast reactions. Explain this based on the data given in Table 9.10.*

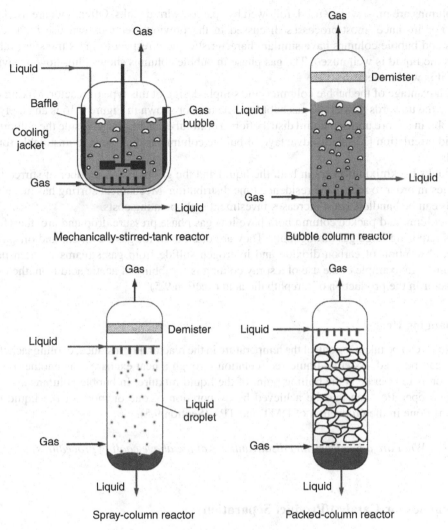

Figure 9.16 *Reactors for gas/liquid reactions.*

Table 9.10 *Mass transfer parameters of various reactors [31, 32].*

Reactor type	Liquid holdup (%)	Gas holdup (%)	$k_l \, a_l \, (s^{-1})^a$	$\beta \, (m^3{}_{liquid}/m^3{}_{film})^b$
Stirred tank	>70	2–30	0.15–0.5	10^3–10^4
Bubble column	>70	2–30	0.06–0.3	10^3–10^4
Spray column	<30	>70	0.05–0.1	10–40
Packed column	5–15	50–80	0.05–0.1	10–40

$^a k_l$ = mass transfer coefficient ($m^3{}_{liquid}/(m^2{}_{interface} \cdot s)$), value: 5×10^{-5}–3×10^{-4} for packed column; 10^{-4}–5×10^{-4} for other reactors; a_l = interfacial area ($m^2{}_{interface}/m^3{}_{liquid}$).
$^b \beta$ = Liquid phase volume/volume of diffusion layer within the liquid phase.

Bubble columns are most widely used, followed by sparged stirred tanks. Often both are used for a certain application. For instance, most processes discussed in the previous sections can use both reactor types. Stirred tanks and bubble columns have similar characteristics with respect to mass transfer (Table 9.10). In both reactors the liquid is well mixed. The gas phase in bubble columns shows plug-flow behavior, while in stirred tanks it is well mixed.

The major advantage of the bubble column is the simple design of this type of reactor. Mixing of the liquid is achieved by the upwards flow of the gas bubbles. The reactor shown in Figure 9.16 is an empty vessel. If it is desired to obtain a more uniform liquid distribution, a draft tube can be placed inside the column to enhance internal liquid circulation [33]. A disadvantage of bubble columns is that they are unsuitable for processing viscous fluids.

In stirred tanks backmixing occurs in both the liquid and the gas phase. A number of stirred tanks can be placed in series in order to reduce the residence time distribution. Mechanical stirring has the advantage that viscous fluids can be handled, but it increases investment and operating costs.

The spray column and packed column both have low gas phase pressure drop and are, therefore, suitable for processes requiring large gas throughputs. They are often used for gas treatment and off-gas scrubbing. For instance, absorption of carbon dioxide and hydrogen sulfide from gas streams is often performed in packed columns. An example of the use of a spray column is scrubbing of acetic acid from the off-gas of the oxidation reactor in the production of terephthalic acid (Section 9.5).

9.6.2   Exchanging Heat

Several measures can be taken to control the temperature in the reactor. For instance, cooling jackets or internal cooling coils can be used. External liquid recirculation through a heat exchanger is another possibility. An elegant solution is to operate at the boiling point of the liquid mixture. In bubble columns and stirred tanks near isothermal operation can thus be achieved by evaporation of one or more of the liquid components present. This is done in the production of DMT and TPA (Section 9.5).

QUESTION: *What are advantages and disadvantages of the above cooling procedures?*

9.7   Approaches for Catalyst/Product Separation

In homogeneous catalysis, complete catalyst recovery is important for several reasons:

- catalyst metal is expensive (especially rhodium);
- ligands are expensive (e.g., phosphines);
- catalyst metal or cocatalyst is hazardous to the environment (e.g., cobalt, MeI);
- metals act as oxidation catalysts;
- catalyst components are usually not allowed in the product;
- metals are deposited in downstream equipment.

Product purification not only includes recovery of the catalyst. It also consists of removal of cocatalysts, decomposition products of the ligands, unconverted reactants, and by-products. The latter two, of course, are not specific for homogeneous catalysis. An example of a process that uses a cocatalyst is the production of acetic acid. Methyl iodide acts as the cocatalyst. This substance is extremely toxic and, therefore, must stay in the system.

Despite the large number of attractive properties of homogeneous catalysis, especially the high activity and selectivity, many promising homogeneous catalytic systems cannot be commercialized because of one very important problem, namely the difficulties associated with separating the catalyst from the product and recycling of the catalyst. This problem arises because the most commonly used separation method – distillation – requires elevated temperatures unless the product is very volatile, while most homogeneous catalysts decompose at temperatures above 420 K [34]. Most of the commercially important homogeneously catalyzed reactions either involve volatile (low molecular weight) reactants or catalysts that do not contain unstable organic ligands.

QUESTIONS:

Is the above statement true for the processes described in this chapter? Describe how the catalyst/product separation is accomplished for each process and whether the catalyst is recycled or not.

Strategies to overcome the separation problems can broadly be grouped into two types: (i) designing the catalyst in such a way that it dissolves in a solvent that, under some conditions, is immiscible with the reaction product (biphasic catalysis) and (ii) immobilizing the catalyst by anchoring it to some kind of support (either insoluble or soluble) or entrapping it inside the pores of solid particles.

9.7.1 Biphasic Catalyst Systems

The use of biphasic reaction systems for the recovery and recycle of homogenous catalysts, in which the catalyst is in one liquid phase and the product is in the other, is one of the few successful developments in heterogenizing homogeneous catalysts.

An aqueous/organic biphasic system is applied successfully in the Ruhrchemie/Rhône-Poulenc propene hydroformylation process (Section 9.3). A few other aqueous/organic biphasic systems are also used on a commercial but much smaller scale [35].

So far, the SHOP process (Section 9.4) is the only commercial organic/organic biphasic catalytic process. The selection of pairs of non-miscible organic liquids is often difficult; non-conventional solvents or pairs of solvents, such as fluorous, ionic, and supercritical liquids, have been reported [35].

QUESTION:

Besides propene, butene can also be hydroformylated in the aqueous/organic catalyst system discussed in Section 9.3. Why is this process not suitable for hydroformylation of higher alkenes?

9.7.2 Immobilized Catalyst Systems

It is appealing to combine the advantages of homogeneous and heterogeneous catalysis by immobilization of the homogeneous catalyst to obtain a tailor-made catalyst that can be used in slurry reactors or fixed bed reactors, thus facilitating separation or eliminating the separation problem altogether. Although this approach is probably the most widely studied [36] and there are many patents on immobilized catalysts, so far there is only one commercial process, the Acetica process for the production of acetic acid by carbonylation of methanol [34, 37].

Immobilization of a homogeneous catalyst can be achieved by attaching it to dendritic, polymeric organic, inorganic or hybrid supports. Here the ligand is functionalized with a group that enables anchoring to such a support.

The two main difficulties preventing the successful use of immobilized catalysts are (i) metal leaching from the support and (ii) reduced activity and selectivity upon immobilization. Metal leaching causes the metal and also the modified ligand to slowly but steadily detach from the support and dissolve in the liquid phase, thus delaying the problem of homogeneous catalysis, not solving it [35].

In the Acetica process, the $[Rh(CO)_2 I_2]^-$ complex is electrostatically bound to an ion exchange resin [37]. In this case leaching cannot be avoided and a guard bed with fresh ion exchange resin is used to re-adsorb any dissolved $[Rh(CO)_2I_2]^-$. After some time, the guard bed, now loaded with the catalyst complex, becomes the catalyst bed [34].

An interesting idea is to anchor the homogeneous catalyst to monolith blocks that are arranged in a stirrer-like fashion [38] (Figure 9.17). Using this configuration a reasonably large surface area of the immobilized catalyst is retained, while many problems of handling a suspension of finely divided supported catalyst species are eliminated.

A different way of immobilizing catalyst complexes is to entrap them in the pores of porous solid particles, for example, by synthesizing the complexes inside the pores of a zeolite, from which they cannot escape

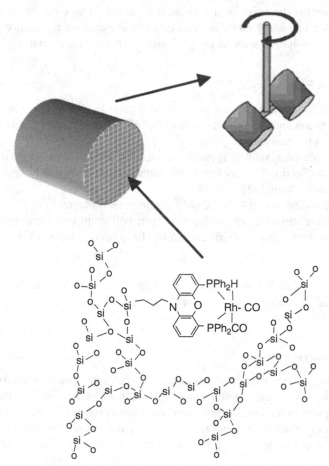

Figure 9.17 *Schematic representation of a homogeneous complex immobilized on silicon monoliths that are implemented in the blades of a mechanical stirrer. Adapted with permission from [38] Copyright (2001) Wiley-VCH.*

("ship in a bottle") or by entrapping the complexes in porous materials and depositing a membrane at the outer surface. This way of immobilization does not involve chemical linkage between the catalyst and the support; the fixation is a result of steric hindrance.

References

[1] Van Leeuwen, P.W.N.M. (2004) *Homogeneous Catalysis. Understanding the Art*. Kluwer Academic Publishers, Dordrecht, The Netherlands, p. 406.

[2] Cheung, H., Tanke, R.S., and Torrence, G.P. (2000) *Acetic Acid*. Wiley-VCH Verlag GmbH, Weinheim.

[3] Bertleff, W., Roeper, M., and Sava, X. (2007) Carbonylation, in *Ullmann's Encyclopedia of Industrial Chemistry*. Wiley-VCH Verlag GmbH, Weinheim. doi: 10.1002/14356007.a05_217.pub2

[4] Haynes, A. (2006) Acetic acid synthesis by catalytic carbonylation of methanol, in *Catalytic Carbonylation Reactions*, vol. 18 (ed. M. Beller). Springer, Berlin/eidelberg, pp. 179–205.

[5] Jones, J.H. (2000) The Cativa^{TM} process for the manufacture of acetic acid. *Platinum Metals Review*, **44**, 94–105.

[6] Hagemeyer, H.J. (1991) Acetaldehyde, in *Kirk-Othmer Encyclopedia of Chemical Technology*, 4th edn, vol. **1** (eds J.I. Kroschwitz and M. Howe-Grant). John Wiley & Sons, Inc., New York, pp. 94–109.

[7] Chauvel, A. and Lefebvre, G. (1989) Petrochemical Processes 1. Synthesis Gas Derivatives and Major Hydrocarbons. Technip, Paris.

[8] Agreda, V.H. (1993) *Acetic Acid and Its Derivatives*. Marcel Dekker, New York.

[9] Howard, M.J., Jones, M.D., Roberts, M.S., and Taylor, S.A. (1993) C1 to acetyls. catalysis and process. *Catalysis Today*, **18**, 325–354.

[10] Bahrmann, H. and Bach, H. (2000) Oxo synthesis, in *Ullmann's Encyclopedia of Industrial Chemistry*. Wiley-VCH Verlag GmbH, Weinheim. doi: 10.1002/14356007.a18_321

[11] Tudor, R. and Ashley, M. (2007) Enhancement of industrial hydroformylation processes by the adoption of rhodium-based catalyst" Part I. *Platinum Metals Review*, **51**, 116–126.

[12] Cadogan, D.F. and Howick, C.J. (2000) Plasticizers, in *Ullmann's Encyclopedia of Industrial Chemistry*. Wiley-VCH Verlag GmbH, Weinheim. doi: 10.1002/14356007.a20_439

[13] Smulders, E., von Rybinski, W., Sung, E., Rähse, W., Steber, J., Wiebel, F., and Nordskog, A. (2000) Laundry detergents, in *Ullmann's Encyclopedia of Industrial Chemistry*. Wiley-VCH Verlag GmbH, Weinheim. doi: 10.1002/14356007.a08_315.pub2

[14] Billig, E. and Bryant, D.R. (2000) Oxo process, in *Kirk-Othmer Encyclopedia of Chemical Technology*. John Wiley & Sons, Inc. doi: 10.1002/0471238961.15241502091212.a01

[15] Arnoldy, P. (2002) Process aspects of rhodium-catalyzed hydroformylation, in *Rhodium Catalyzed Hydroformylation*, vol. 22 (eds P.W.N.M. Leeuwen and C. Claver). Springer, The Netherlands, pp. 203–231.

[16] Chaudhari, R.V., Seayad, A., and Jayasree, S. (2001) Kinetic modeling of homogeneous catalytic processes. *Catalysis Today*, **66**, 371–380.

[17] van Koten, G. and Van Leeuwen, P.W.N.M. (1999) Homogeneous catalysis with transition metal complexes, in *Catalysis: An Integrated Approach to Homogeneous, Heterogeneous and Industrial Catalysis*, 2nd edn (eds R.A. Van Santen, P.W.N.M. Van Leeuwen, J.A. Moulijn, and B.A. Averill). Elsevier, Amsterdam, The Netherlands, pp. 307.

[18] Kohlpaintner, C.W., Fischer, R.W., and Cornils, B. (2001) Aqueous biphasic catalysis: Ruhrchemie/Rhône-Poulenc oxo process. *Applied Catalysis A*, **221**, 219–225.

[19] Bahrmann, H. and Back, H. (1991) Oxo synthesis, in *Ullmann's Encyclopedia of Industrial Chemistry*, 5th edn, vol. **A18** (ed. W. Gerhartz). VCH, Weinheim, pp. 321–327.

[20] Falbe, J. (1980) *New Syntheses With Carbon Monoxide*. Springer Verlag, New York.

[21] van Leeuwen, P.W.N.M., Clément, N.D., and Tschan, M.J.L. (2011) New processes for the selective production of 1-octene. *Coordination Chemistry Reviews*, **255**, 1499–1517.

[22] Vogt, D. (2004) Nonaqueous organic/organic separation (SHOP process), in *Aqueous-Phase Organometallic Catalysis*, 2nd edn (eds B. Cornils and W.A. Herrmann). Wiley-VCH Verlag GmbH, Weinheim, pp. 639–645.

[23] Al-Jarallah, A.M., Anabtawi, J.A., Siddiqui, M.A.B., Aitani, A.M., and Al-Sa'doun, A.W. (1992) Ethylene dimerization and oligomerization to butene-1 and linear α-olefins: a review of catalytic systems and processes. *Catalysis Today*, **14**, 1–121.

[24] Forestière, A., Olivier-Bourbigou, H., and Saussine, L. (2009) Oligomerization of monoolefins by homogeneous catalysts. *Oil & Gas Science and Technology (Revue de l'Institut Francais du Petrole)*, **64**, 649–667.

[25] Turner, A. (1983) Purity aspects of higher alpha olefins. *Journal of the American Oil Chemists Society*, **60**, 623–627.

[26] Keim, W. (1990) Nickel: an element with wide application in industrial homogeneous catalysis. *Angewandte Chemie (International Edition in English)*, **29**, 235–244.

[27] Hérisson, J.L. and Chauvin, Y. (1971) Catalyse de transformation des oléfines par les complexes du tungsténe. II. Télomérisation des oléfines cycliques en présence d'oléfines acycliques. *Makromolekulare Chemie (Macromolecular Chemistry and Physics)*, **141**, 161–176.

[28] Sheehan, R.J. (1995) Terephthalic acid, dimethyl terephthalate and isophtalic acid, in *Ullmann's Encyclopedia of Industrial Chemistry*, 5th edn, vol. **A26** (ed. W. Gerhartz). VCH, Weinheim, pp. 193–204.

[29] Park, C.M. and Sheehan, R.J. (2000) Phthalic acids and other benzenepolycarboxylic acids, in *Kirk-Othmer Encyclopedia of Chemical Technology*. John Wiley & Sons, Inc. doi: 10.1002/0471238961.1608200816011811.a01

[30] Sheehan, J. (2000) Terephthalic acid, dimethyl terephthalate, and isophthalic acid, in *Ullmann's Encyclopedia of Industrial Chemistry*. Wiley-VCH Verlag GmbH, Weinheim. doi: 10.1002/14356007.a26_193

[31] Trambouze, P., Van Landeghem, H., and Wauquier, J.P. (1988) *Chemical Reactors, Design/Engineering/Operation*. Gulf Publishing Company, Houston, TX.

[32] Krishna, R. and Sie, S.T. (1994) Strategies for multiphase reactor selection. *Chemical Engineering Science*, **49**, 4029–4065.

[33] Zehner, P. and Kraume, M. (1992) Bubble columns, in *Ullmann's Encyclopedia of Industrial Chemistry*, 5th edn, vol. **B4** (ed. W. Gerhartz). VCH, Weinheim, pp. 275–330.

[34] Cole-Hamilton, D.J. (2003) Homogeneous catalysis – new approaches to catalyst separation, recovery, and recycling. *Science*, **299**, 1702–1706.

[35] Wiebus, E. and Cornils, B. (2006) Biphasic systems: water–organic, in *Catalyst Separation, Recovery and Recycling*, vol. **30** (eds D.J. Cole-Hamilton and R.P. Tooze). Springer, The Netherlands, pp. 105–143.

[36] Reek, J.N.H., Leeuwen, P.W.N. M., Ham, A.G.J., and Haan, A.B. (2006) Supported catalysts, in *Catalyst Separation, Recovery and Recycling*, vol. **30** (eds D.J. Cole-Hamilton and R.P. Tooze). Springer, The Netherlands, pp. 39–72.

[37] Yoneda, N., Kusano, S., Yasui, M., Pujado, P. and Wilcher, S. (2001) Recent advances in processes and catalysts for the production of acetic acid. *Applied Catalysis A*, **221**, 253–265.

[38] Sandee, A.J., Ubale, R.S., Makkee, M., Reek, J.N.H., Kamer, P.C.J., Moulijn, J.A., and Van Leeuwen, P.W.N.M. (2001) ROTACAT: a rotating device containing a designed catalyst for highly selective hydroformylation. *Advanced Synthesis & Catalysis*, **343**, 201–206.

10
Heterogeneous Catalysis – Concepts and Examples

10.1 Introduction

In previous chapters, many processes based on heterogeneous catalysis have been discussed. This is no coincidence. Heterogeneous catalysis is a crucial technology. It is the workhorse in chemical reaction engineering. In this light, homogeneous transition metal catalysis, as discussed in Chapter 9, can be considered the show horse. Usually applications of homogeneous catalysis are on a smaller scale and more specific than applications of heterogeneous catalysis.

In industrial catalysis, it is generally preferable from an engineering perspective to use heterogeneous catalysts in continuous processes where possible. The main advantages of heterogeneous catalysis compared to homogeneous catalysis are:

- easy catalyst separation;
- flexibility in catalyst regeneration;
- lower cost.

This chapter surveys the reactors commonly applied in heterogeneous catalysis, while attention is also paid to novel reactor types and special catalyst applications. The selection of a reactor configuration for heterogeneous catalysis is determined by many, often conflicting, factors. Firstly, there are the process needs, such as safety and environmental requirements, and feasibility of scale-up to economic size. Secondly, there are the process "wants", which usually include maximum conversion, maximum selectivity, high throughput, easy scale-up, and minimum overall process costs.

Some reactors are intrinsically safer than others. On one hand, in fixed bed reactors hot spots can occur, possibly leading to runaways, whereas in fluidized bed reactors temperature control is relatively easy and hot spots will not occur. On the other hand, fluidized bed reactors are much more difficult to scale-up and catalyst particles have to be able to withstand the high mechanical stress of a fluidized bed.

Often, with increasing conversion the selectivity decreases so that it is not possible to have both maximum conversion and maximum selectivity. Both conversion and selectivity can be dependent on the degree of mixing

Chemical Process Technology, Second Edition. Jacob A. Moulijn, Michiel Makkee, and Annelies E. van Diepen.
© 2013 John Wiley & Sons, Ltd. Published 2013 by John Wiley & Sons, Ltd.

of the phases present (from plug-flow to completely mixed), the contacting pattern (cocurrent, countercurrent, crosscurrent), the particle size, the temperature (distribution), and so on.

A systematic approach to reactor selection has been presented [1] and several excellent textbooks on design aspects of chemical reactors are available [2,3]. Reactor selection is intimately linked with catalyst properties such as size, activity, and so on. Discussed firstly are the influence of the various catalyst properties. Subsequent sections deal with conventional reactor types and some unconventional techniques and applications.

10.2   Catalyst Design

The choice of catalyst morphology (size, shape, porous texture, activity distribution, etc.) depends on intrinsic reaction kinetics as well as on diffusion rates of reactants and products. The catalyst size and morphology cannot be chosen independently of the reactor type, because different reactor types place different demands on the catalyst. For instance, fixed bed reactors require relatively large particles to minimize the pressure drop, while in fluidized bed reactors relatively small particles must be used. However, an optimal choice is possible within the limits set by the reactor type.

10.2.1   Catalyst Size and Shape

The catalyst particle size influences the observed rate of reaction: the smaller the particle, the less time required for the reactants to move to the active catalyst sites and for the products to diffuse out of the particle. Furthermore, with fast reactions in large particles the reactants may never reach the interior of the particle, thus decreasing catalyst utilization. Catalyst utilization is expressed as the internal effectiveness factor η_i, which is defined as follows:

$$\eta_i = \frac{\text{observed reaction rate}}{\text{rate without internal gradients}} = \frac{r_{obs}(c, T)}{r_{v,chem}(c_s, T_s)} \qquad (10.1)$$

where

$r_{obs}(c, T)$ = observed reaction rate, an average of the rates inside the particle at different
          concentrations c and temperatures T (mol/(s·m^{3}particle)),
$r_{v,chem}(c_s, T_s)$ = reaction rate at surface concentration c_s and temperature T_s (mol/(s·m^{3}particle)).

The internal effectiveness factor is a function of the generalized Thiele modulus [1, 3]. For an nth order reaction:

$$\varphi_{gen} = \frac{V}{SA} \sqrt{\frac{k_v}{D_{eff}} \cdot \frac{n+1}{2} \cdot c_s^{n-1}} \qquad (10.2)$$

in which V/SA is the volume/external surface area ratio of the catalyst (for a sphere: $r_p/3$, in which r_p is the particle radius), k_v is the intrinsic reaction rate coefficient (per unit volume), D_{eff} is the effective diffusion rate of the molecule, and c_s the concentration at the catalysts surface.

QUESTIONS:

> *According to Figure 10.1 $\eta_i \leq 1$. Is this always the case? (Hint: what happens when an exothermic reaction is performed?)*
> *What are the dimensions of k_v?*

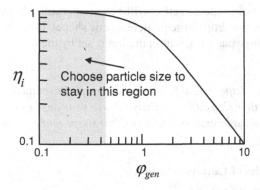

Figure 10.1 *Internal effectiveness factor as a function of the generalized Thiele modulus for a first-order reaction.*

In order for diffusional limitations to be negligible, the effectiveness factor should be close to 1, that is, nearly complete catalyst utilization, which requires that the Thiele modulus is sufficiently small (typically <0.5) (Figure 10.1). The physical meaning of the Thiele modulus is the ratio between intrinsic catalytic activity and internal mass transfer. There is no doubt that in the future, for existing processes, the catalyst activity will be improved by R&D efforts, while mass transfer is less accessible for fundamental improvements. Therefore, it is to be expected that mass transfer will become an increasingly important issue. The surface to volume (SA/V) ratio should be as large as possible (particle size as small as possible) from a diffusion (and heat transfer) point of view. There are many different catalyst shapes that have different SA/V ratios for a given size. Figure 10.2 surveys the most common catalyst shapes.

Of course, the relative surface area is not the only factor influencing the performance of the catalyst. Smaller particles result in larger pressure drop over the reactor and in fixed bed systems this is the main limiting factor. In fixed bed and moving bed reactors the catalyst strength is also an important factor. The particles must be able to withstand the forces exerted by the bed above and by the pressure drop. In general, compact particles have a higher crushing strength than hollow ones, and the crushing strength increases with particle size. For the process economics it is also important that the manufacturing costs are within acceptable limits. The more exotic the catalyst shape, the more expensive is its production.

With increasing particle size, the SA/V ratio and, hence, the reaction rate decreases, which is unfavorable, while the pressure drop, and the manufacturing costs also decrease, which is favorable.

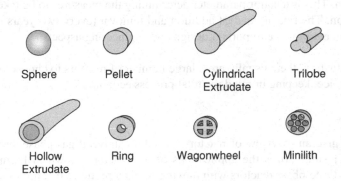

Figure 10.2 *Various catalyst shapes.*

The definitive catalyst size and shape selection will be a compromise between high reaction rate (small particle, exotic shape), low pressure drop (large particle, exotic shape), large crushing strength (low porosity), and low manufacturing cost (large particle), within the limits set by the reactor type.

QUESTIONS:

What particle shape is most favorable with respect to catalytic activity? Is there a connection between particle size and selectivity? Prove your answer for a first-order reaction. Does a relationship exist between catalyst shape and reactor type?

10.2.2   Mechanical Properties of Catalyst Particles

An important aspect, which is crucial for reactor selection, has only been mentioned briefly. The mechanical transport properties required for a chosen reactor type play a decisive role in catalyst design. For instance, fluidized bed and entrained flow reactors require spherical particles in the range of 30 to 150 μm (microspheres) with a certain size distribution in order to obtain good fluidization behavior. Such microspheres can be produced by incorporating the catalyst material in a matrix of, for example, silica (Section 3.4.2 and Section 10.5.2.3). The advantage of the microspheroidal shape over other catalyst shapes is the mechanical attrition resistance. This is very important in entrained flow reactors and in fluidized bed reactors where catalyst particles are recirculated rapidly between the reactor and a regenerator (as in FCC (Section 3.4.2)).

Attrition resistance is also important for use of catalysts in moving bed reactors. Usually, the particle size distribution should be narrow (2–6 mm) and no fines must be present. These fines can result from attrition and can clog the catalyst bed because they accumulate in the spaces between the larger particles.

10.3   Reactor Types and Their Characteristics

The most important factors for the choice and design of a reactor for heterogeneous catalysis are:

- The number and type of phases involved (G/S, G/L, G/L/S, L/L/S). Freedom exists in choosing the volume fractions of the various phases. Measures must be taken for appropriate mass and heat transport between the phases, for example, determined by degree of mixing and contacting patterns (cocurrent, countercurrent, crosscurrent).
- The kinetics of the main and side reactions. These determine the desired temperature and concentration profile, the residence time distribution (degree of mixing), and so on.
- The heat of reaction. This is a major parameter determining the measures to be taken for heat transfer.
- Catalyst deactivation. The rate of catalyst addition and removal (every two years or continuously, etc.), and the need for regeneration are important design and engineering aspects.

For any particular situation, there usually are a large number of reactors to choose from. The challenge is to make the optimal choice, keeping in mind the total process economics.

10.3.1   Reactor Types

Figures 10.3 and 10.4 give an overview of reactors for solid-catalyzed gas phase and gas/liquid reactions, respectively. Table 10.1 summarizes the applications of the reactors shown in Figures 10.3 and 10.4. In Chapter 12 some other three-phase reactors with moving catalyst particles that are used in the production of fine chemicals are discussed.

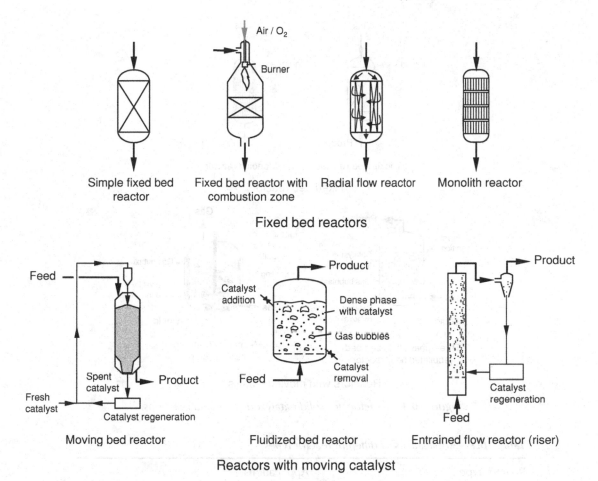

Fixed bed reactors

Reactors with moving catalyst

Figure 10.3 *Reactors for solid-catalyzed gas phase reactions.*

Numerous variations exist on each reactor type, such as the use of many small diameter tubes in fixed bed reactors, a different gas/liquid contacting pattern, and the manner of temperature control.

Occasionally, both gas phase operation and liquid phase operation are possible for reactions occurring in a single fluid phase. In that case, flexibility exists and the question arises what is the best option. The liquid phase has the advantage of better heat transfer properties and a high flexibility in the range of concentrations that can be used (by using a solvent low concentrations are possible, enabling better temperature control); a disadvantage is the low diffusion rate.

In industrial practice the adiabatic fixed bed reactor is the most commonly used reactor type. It is essentially a vessel, generally a vertical cylinder, in which the catalyst particles are randomly packed. It is the simplest reactor for use with solid catalysts and it offers the highest catalyst loading per unit volume of reactor.

In principle, with one-phase flow the fluid can flow upward, downward or radially through the bed. Downward flow is most common, while radial flow is sometimes applied to decrease the pressure drop over the bed. Upward flow is hardly ever applied, because flow rates are often high and undesired movement of the catalyst particles might be the result. Similarly, with two-phase flow, operation can be cocurrent downward, cocurrent upward, or countercurrent. Again, the first configuration, also called trickle flow, is used most frequently. The

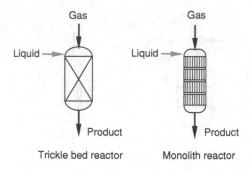

Fixed bed reactors

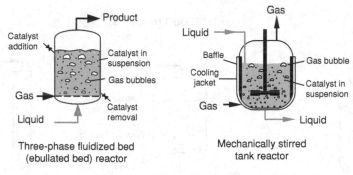

Reactors with moving catalyst

Figure 10.4   *Reactors for solid-catalyzed gas/liquid reactions.*

Table 10.1   *Applications of different reactor types.*

Reactor type	Applications
Gas phase reactions	
Simple fixed bed reactor	HDS of naphtha
	Catalytic reforming (semi-regenerative)
	Steam reforming
	Water–gas shift
	Methanation
	Ammonia synthesis
	Methanol synthesis
Monolith reactor	Exhaust gas cleaning
Fixed bed with combustion zone	Autothermic reforming
Radial flow reactor	Catalytic reforming
	Methanol synthesis
Moving bed reactor	Catalytic reforming (continuous regenerative)
Fluidized bed reactor	Classical FCC process
Entrained flow reactor	Modern FCC process
Gas/liquid phase reactions	
Trickle bed reactor	HDS of heavy oil fractions
Moving bed reactor	Hydrogenation of residues (metal removal)
Three-phase fluidized bed reactor	Hydrogenation of residues

main example is hydrodesulfurization (HDS) of heavy oil fractions. Countercurrent operation is not employed industrially except in special cases (notably catalytic distillation, Chapter 14).

The moving bed reactor is similar to the fixed bed reactor. However, the catalyst bed moves downward due to gravitational forces. The fluid phase may flow from top to bottom or horizontally (axial flow). A complication is the need to circulate the catalyst, but this offers the possibility of continuous catalyst regeneration. Practical applications of this technology are limited. Examples are the continuously regenerative reforming process (Section 3.4.3) and the "bunker" reactor in the Shell HYCON process for catalytic hydrogenation of residual oil fractions (Section 3.4.5).

Monolith reactors are so-called structured reactors: they are ceramic or metallic blocks consisting of parallel channels covered with a catalyst coating. They are especially suited for gas phase reactions requiring low pressure drop.

In fluidized bed reactors the catalyst particles move randomly due to the upward flow of gas, liquid, or gas and liquid (also called ebullated bed reactor). The advantages of this type of reactor are the excellent heat transfer properties and the ease of catalyst addition and removal. Catalytic cracking (classical process), catalyst regeneration in catalytic cracking (a non-catalytic process, Section 3.4.2), and the LC-fining process for the catalytic hydrogenation of residues (Section 3.4.5) are examples of processes employing fluidized beds.

When in a gas-fluidized-bed reactor the gas velocity is increased, at one point the catalyst particles become entrained by the gas flow. Then the term entrained flow reactor or riser reactor applies. Such reactors are suitable for the use of very active catalysts that deactivate fast. The most important example is the FCC process using a zeolite catalyst (Section 3.4.2).

Mechanically-stirred-tank reactors are not very common for solid-catalyzed bulk chemicals production. They are, however, the main reactors in the production of fine chemicals (Chapter 12).

QUESTIONS:

> *For each process in Table 10.1 explain why this type of reactor has been chosen. Could alternatives also be employed? Why would countercurrent flow be preferable in certain gas/liquid reactions, provided that there are no practical constraints? Give some practical examples.*

10.3.2 Exchanging Heat

Fixed bed reactors exhibit fairly poor heat transfer characteristics. The particles are large (1–5 mm), resulting in a small surface/volume ratio and, hence, a small heat transfer area. Moreover, the catalyst material is usually a thermal insulator. Temperature control in fixed bed reactors, therefore, is difficult and it is the rule rather than the exception that a non-isothermal temperature profile in both axial and radial direction results with the accompanying danger of local hot spots (dependent on the system and the conditions). Clearly, the phase (liquid or gas) also plays a role. In the liquid phase, reactor temperature gradients are much smaller than those in gas phase processes.

The adiabatic fixed bed reactor is only suitable for reactions with a small heat effect and for reactions in which the temperature change due to reaction does not influence the selectivity. If in an adiabatic fixed bed reactor the adiabatic temperature rise/fall becomes too large (e.g., with large heats of reaction), a number of smaller adiabatic fixed beds with intermediate cooling or heating can be placed in series. Figure 10.5 shows some possible configurations.

If the process imposes special requirements on temperature control, for example, in the case of very exothermic or endothermic reactions, heat transfer must take place throughout the reactor. Multitubular reactors are very suitable for such reactions. Typically, hundreds to thousands of tubes surrounded by the heat transfer medium are used. For highly exothermic reactions, the tubes are placed in a vessel with the heat

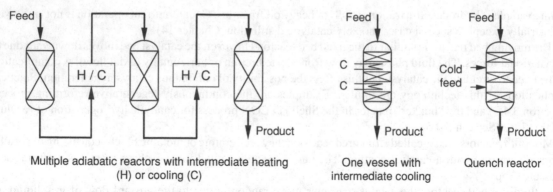

Multiple adiabatic reactors with intermediate heating
(H) or cooling (C)

One vessel with
intermediate cooling

Quench reactor

Figure 10.5   *Adiabatic fixed bed reactors with inter-stage heating/cooling.*

transfer medium (usually boiling liquids) flowing in the space surrounding the tubes (Figure 10.6). The flow is directed back and forth across the tubes by baffles to provide good temperature control. Using this method near-isothermal operation can be achieved. An example is the production of ethene oxide by partial oxidation of ethene (Section 10.5.2.2).

Endothermic reactions requiring a large heat input at high temperature level can be carried out in tubes placed in a furnace (Figure 10.6). The main example of this type of operation is steam reforming of methane or naphtha (Section 5.2).

Moving bed reactors exhibit similar heat transfer characteristics as fixed bed reactors but temperature gradients can be held within limits by the circulation of the solid catalyst.

Slurry reactors exhibit excellent heat transfer properties. Due to the relatively small particle size of the catalyst, the heat transfer area is relatively large and high heat transfer rates can be obtained. Backmixing of the catalyst results in a uniform temperature throughout the reactor. Additional heat transfer area can be installed in the form of internal cooling/heating coils.

The heat transfer characteristics of entrained flow reactors are intermediate between those of fluidized bed and moving bed reactors.

In multiphase operation, excellent temperature control is possible by allowing for heat exchange by evaporation. This can be the reason to choose multiphase operation rather than, for instance, a single gas phase reaction.

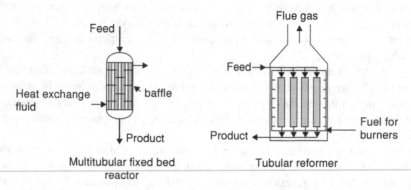

Multitubular fixed bed
reactor

Tubular reformer

Figure 10.6   *Fixed bed reactors with heating/cooling throughout reactor.*

10.3.3 Role of Catalyst Deactivation

Catalyst deactivation, and in particular the timescale of deactivation, is an important issue in the choice of reactor type. Deactivation is the loss of catalyst activity with time, and can be either chemical (fouling or poisoning) or thermal (loss of surface area due to sintering) [4].

10.3.3.1 Fouling – Coke Formation

The most common cause of catalyst deactivation is fouling due to the occurrence of undesired reactions of reactants, intermediates, or products on the catalyst surface. The main example of this type of deactivation is coke formation. It occurs by condensation reactions (cycloaddition, dehydrogenation) of aromatic compounds on the catalyst surface to form large structures of low hydrogen content. Coke is often simply indicated as carbon (C). Optimal conditions for coke formation are a reducing environment, high temperature, and low hydrogen pressure. Therefore, it is logical that in previous chapters the problems associated with coke deposition were most pronounced in endothermic reactions proceeding at high temperature (>750 K), namely in FCC (Section 3.4.2), catalytic reforming (Section 3.4.3), and steam reforming (Section 5.2). Coke is also formed in thermal (non-catalytic) processes such as steam cracking of ethane and naphtha (Chapter 4).

Coke formation can be limited by addition of hydrogen, either as H_2 or as H_2O:

$$C + 2\,H_2 \rightleftarrows CH_4 \tag{10.3}$$

$$C + 2\,H_2O \rightleftarrows CO_2 + 2\,H_2 \tag{10.4}$$

An alternative is regeneration of the catalyst by burning off the coke.

Table 10.2 illustrates the effect of the time scale of catalyst deactivation on the reactor choice. In steam reforming, the use of excess steam not only limits coke formation but it also enhances the conversion of methane (or naphtha) to synthesis gas. In contrast, in the case of catalytic reforming operation takes place at conditions which are not optimal for the reforming reactions from a thermodynamic point of view (during reforming hydrogen is produced, so addition of hydrogen decreases the feed conversion). A compromise is necessary between conversion and catalyst deactivation. If it is chosen to operate at high hydrogen pressure, the equilibrium conversion is relatively low, but simple adiabatic fixed bed reactors can be used because catalyst deactivation is slow. It is also possible to opt for a lower hydrogen pressure, but this adds to the investment costs due to the requirement of an additional reactor (swing reactor) or the use of a moving bed reactor, which is more cumbersome from the viewpoint of construction and operation. The advantages are a higher conversion and lower costs for the recycle of hydrogen.

Table 10.2 *Effect of catalyst deactivation by coke formation on choice of reactor.*

Process	Measure to limit coke formation	Deactivation time	Reactor type
Steam reforming	Excess steam addition	2 years	Tubular fixed bed reactor
Catalytic reforming			
Semi-regenerative	Large H_2/HC ratio	0.5–1.5 years	Multiple fixed bed reactors
Fully regenerative	Moderate H_2/HC ratio	days–weeks	Multiple fixed bed reactors with swing reactor
Continuously regenerative	Moderate H_2/HC ratio	days–weeks	Moving bed reactor
FCC	None (regeneration)	seconds	Entrained flow reactor

HC = hydrocarbon.

10.3.3.2   Sintering

Sintering can occur when the temperature becomes too high. This can happen, for instance, as a result of maldistribution of the fluid phases and incomplete wetting of the catalyst particles in trickle bed reactors (Section 3.4.5). Locally, the heat is not removed efficiently, leading to hot spots. Water can accelerate the sintering process.

10.3.3.3   Poisoning

Poisoning of the catalyst occurs by the adsorption of impurities in the feed on specific catalytic sites, thus rendering them inactive for reaction. Common poisons are sulfur (e.g., in catalytic reforming (platinum catalyst, Section 3.4.3), steam reforming (nickel catalyst, Section 5.2)), carbon monoxide and carbon dioxide (in ammonia synthesis (iron catalyst, Section 6.1) ammonia, and ethyne.

Poisoning can be limited by purification of the feed. For example, hydrodesulfurization (HDS) is applied to remove sulfur (Section 3.4.5). The removal of carbon monoxide and carbon dioxide from synthesis gas for ammonia production is achieved by the water–gas shift reaction, subsequent absorption of carbon dioxide, and a final methanation step (Section 5.4).

QUESTIONS:

> *Sulfur is a poison for Pt catalysts used in catalytic reforming and for Ni catalysts used in steam reforming. In ammonia synthesis, the Fe catalyst is poisoned by CO and CO_2. What would be the mechanisms of these poisoning processes?*
> *In Table 10.2, several different processes are listed in which coke deposition takes place. In which cases is "C" a good description of the deposited material and in which cases is this an unrealistic oversimplification?*

10.3.4   Other Issues

10.3.4.1   *Mixing of Reactants*

As has become clear from the previous section, reactors with a high degree of backmixing, such as fluidized bed reactors, have a more isothermal temperature profile and are easier to control than other reactors (e.g., fixed bed reactors). On the other hand, extensive backmixing is usually disadvantageous from the viewpoint of reaction rate, the level of conversion, and the potentially poorer selectivity.

For first- and higher-order reactions, the reaction rate increases with increasing reactant concentration. Hence, in plug-flow reactors the reaction rate is high in the entrance zone of the reactor and decreases along the length of the reactor. A well-mixed reactor (e.g., CSTR) always operates at the lowest reactant concentration (which is the same throughout the reactor) and thus at the lowest reaction rate. Therefore, the maximum obtainable conversion in a plug-flow reactor is higher than that in a CSTR of the same size.

Consecutive reactions such as A → P → Q, in which P is the desired product, pose another problem. In order to obtain maximum selectivity towards P, plug-flow conditions should be aimed for. However, for highly exothermic reactions, from the point of view of avoiding large temperature gradients, a thermally well-mixed system is preferred. Provided they are very fast, consecutive reactions of the type

$$A \xrightarrow{\ B\ } P \xrightarrow{\ B\ } Q \tag{10.5}$$

occasionally can be best performed in multiphase operation: the reaction then is limited to a thin diffusion layer around the gas bubbles and the consecutive reaction does not occur.

A solution to obtain a well-mixed temperature together with plug-flow concentration conditions can be to circulate the catalyst and remove the heat externally (Section 10.5.2).

QUESTIONS:
> *Why, in reaction 10.5, is plug flow preferable from a concentration point of view?*
> *Explain why consecutive reactions such as reaction 10.5 can be carried out at relatively high selectivity in multiphase operation? How would you realize this idea in practice?*
> *Give advantages and disadvantages of temperature control by external heat exchange?*

10.3.4.2 Safety

Important hazards are high temperature excursions and runaways. Highly exothermic reactions can lead to such events when cooling is not sufficient. For instance, in a trickle bed reactor the even distribution of the liquid over the catalyst bed is of decisive importance. In exothermic reactions, liquid maldistribution can lead to local hot spots, possibly leading to a temperature rise in the entire reactor, and, a total runaway.

10.3.4.3 Scale-up

The ease of scale-up of a chemical process, and with that the time and cost involved in going from the laboratory to a working industrial plant, depends on several factors. Some are intrinsically related to the reactor and process, for instance, the reactor hydrodynamics (gas, liquid, solid flow). Others depend, more or less, on external factors, such as maximum size of equipment, proven versus new technology, experience of the company with the particular technology, and so on (see also Chapter 15).

The scale-up of a fixed bed reactor from laboratory scale to the full size plant is much easier than the scale-up of a fluidized bed reactor. The reason for the difficulty of scaling up fluidized beds is that the fluidization characteristics of the bed not only depend on the local properties but also on the reactor size. For illustration, various fixed bed reactors used in industrial processes have been scaled-up in one step from pilot reactors of 13–44 mm diameter to full-scale reactors of 3–9 m. In contrast, the scale-up to fluidized bed reactors with diameters of 7.5–17 m requires several steps, with each step only one or two orders of magnitude larger than the previous step (Chapter 15).

QUESTION:
> *Compare the various reactor types in terms of attainable conversion, relative reactor volume, safety, ease of operation and scale-up, and investment costs.*

10.4 Shape Selectivity — Zeolites

Most solid catalysts are only "class selective". For example, a classical hydrogenation catalyst will hydrogenate double bonds in all types of reactant molecules irrespective of their size or of the position of the double bond [5]. Tremendous progress has been made in developing catalysts that show selectivity with respect to the position of the unsaturated bond (regioselectivity). Stereoselectivity is also often possible, leading to the desired enantiomeric compound (Chapter 12). A breakthrough has been the synthesis of solid catalysts that show shape selectivity as a result of spatial constraints. They can be tuned to selectively convert only particular molecules or produce only a certain product. The most important examples are enzymes and molecular sieves, of which zeolites are a special group. The recently discovered classes of metal-organic frameworks (MOFs)

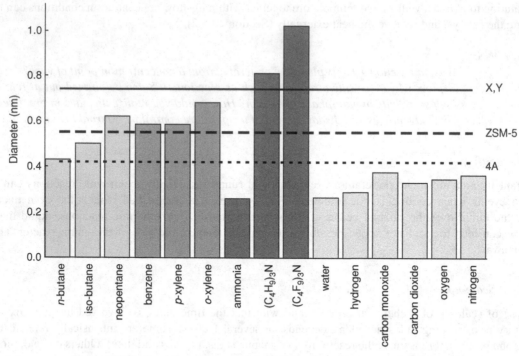

Figure 10.7   *Pore sizes of zeolites and dimensions of some molecules [6].*

and covalent-organic frameworks (COFs), which have a homogeneous channel and/or cage structure, bear the potential of having a similar impact.

A remarkable property of zeolites is their porosity (up to 0.4 ml/g), although they are crystalline compounds. The pore size distribution is well defined and the pores are in the nanometer range, corresponding to molecular dimensions. The pore sizes of most zeolites are between 0.3 and 0.9 nm (Figure 10.7). For comparison, the dimensions of some molecules that are interesting regarding chemical process technology are also shown in Figure 10.7. The fact that the dimensions of the pores are in the range of molecular sizes enables shape selectivity. Often three types of shape selectivity are distinguished: reactant, transition state, and product selectivity (Figure 10.8).

Reactant selectivity occurs if, in a mixture of feed molecules, only those with dimensions smaller than the pores can enter and react. An example is hydrocracking of a mixture of linear and branched heptanes: the linear molecules are cracked whereas the branched alkanes cannot enter the pores and, consequently, are not transformed.

In transition-state selectivity only pathways involving certain transition geometries will be operational. An example is the disproportionation of m-xylene (m-xylene $\rightarrow$ trimethylbenzene + benzene), where only one of the three trimethylbenzene isomers is formed, because the others require a coupling geometry that is sterically hindered.

In product selectivity, only products that are small enough to diffuse from the interior of the zeolite to the outside can be formed. In the example of Figure 10.8 xylene is formed from toluene. Of the possible products (o-, m-, and p-xylene) only the para isomer, which is the preferred product, is found in large quantities.

Shape selectivity provides a means to selectively convert one type of reactant or selectively produce the desired product, where otherwise the thermodynamically-determined product distribution would result. Of course, in zeolites, as in other catalysts, intrinsic reaction kinetics also play a role with respect to selectivity;

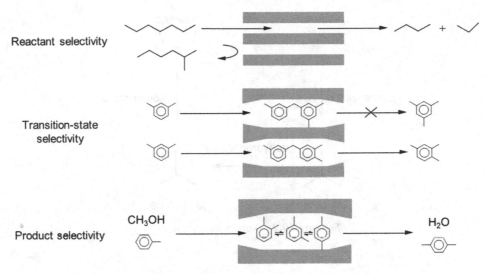

Figure 10.8   *Three types of shape selectivity.*

if the desired reaction is fast compared with the undesired reaction, a high selectivity can also be obtained, even if no shape selectivity occurs.

There are also examples where shape selectivity is a disadvantage. An example is processing of heavier feeds in the FCC process discussed in Section 3.4.2. Due to the decreasing supplies of light crude oils, there is a trend towards processing of heavier FCC feeds. In FCC zeolites are applied because of their high activity for cracking. However, the shape selectivity of the catalyst is an additional property. In the processing of heavy feeds, this shape selectivity will prevent the larger molecules from entering the pores and hence prevent their conversion.

QUESTION:

> *Suggest a solution for the challenge of processing heavy feeds in FCC.*

10.4.1   Production of Isobutene

The capacity of plants for the production of oxygenates such as MTBE and ETBE (Scheme 10.1) from isobutene and methanol or ethanol has increased significantly during the 1980s and 1990s as a result of

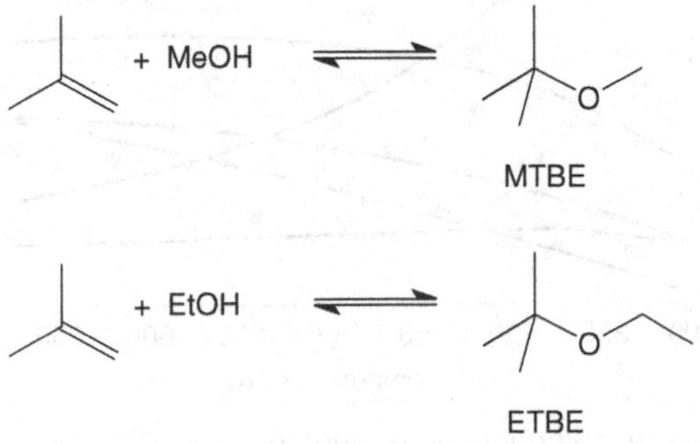

Scheme 10.1   *Routes to MTBE and ETBE.*

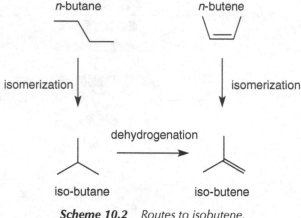

n-butane

n-butene

isomerization

isomerization

dehydrogenation

iso-butane

iso-butene

Scheme 10.2   *Routes to isobutene.*

government regulations regarding oxygen content in gasoline (see also Chapter 6 for MTBE production). This has led to a shortage of isobutene.

C$_4$ hydrocarbons are produced by steam crackers (Chapter 4) and catalytic crackers. Isobutene can be produced from these streams by isomerization of *n*-butene or from *n*-butane when a dehydrogenation step is added (Scheme 10.2). However, until recently the seemingly simple butene-to-isobutene isomerization was not very attractive due to the lack of a selective catalyst. The reaction was hampered by the dimerization and oligomerization of butenes, which led to large amounts of C$_{5+}$ hydrocarbons and thus poor yields of isobutene. In addition, rapid coking of the catalyst took place.

The exothermic dimerization reactions can be suppressed by operating at high temperature, but this also leads to low isobutene yields as a result of the unfavorable thermodynamic equilibrium (compare Figure 10.9, which shows the equilibrium composition for hexanes). Furthermore, under these conditions, coke formation

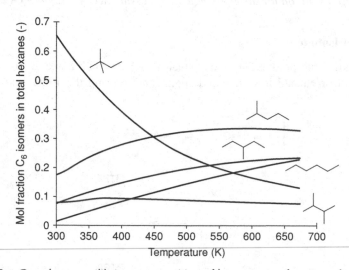

Figure 10.9   *Gas phase equilibrium composition of hexanes as a function of temperature.*

results in rapid catalyst deactivation. An alternative is the application of shape-selective skeletal isomerization catalysts [7–9] (Box 10.1).

Box 10.1 Butene to Isobutene Isomerization Using Ferrierite

The isomerization reaction on the zeolite Ferrierite (a zeolite of medium pore size) proceeds via a dimerization step to form a C_8 intermediate, which then undergoes skeletal isomerization followed by selective cracking to produce two isobutene molecules. The Ferrierite catalyst was found to produce isobutene at much higher selectivity than other catalysts due to its shape selectivity: The C_8 isomer responsible for isobutene formation cannot leave the zeolite pores, while less branched C_8 isomers can (Scheme B10.1.1) [7, 8]. At first sight it might have been expected that single-branched C_8 would be the major product. Fortunately, this is not the case. As tertiary carbenium ions are more stable than secondary or primary carbenium ions, formation of the highly branched C_8 isomer is favored. Therefore, isobutene is the major product. In addition to the higher selectivity, Ferrierite has a much greater stability than most other catalysts for this reaction [9].

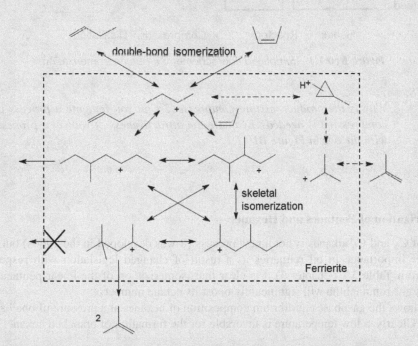

Scheme B10.1.1 *Proposed mechanism for butene isomerization in Ferrierite [8].*

QUESTIONS:

What type of shape selectivity is this? Is direct isomerization of n-butene to isobutene as shown on the right of Scheme B10.1.1 be likely?

Figure B10.1.1 shows a simplified flow scheme of the butene isomerization process. A C_4 hydrocarbon feed containing normal butenes (and butanes) is preheated and fed to a fixed bed reactor, where conversion to isobutene takes place (typical product distribution: 35 wt% isobutene, 40 wt% *n*-butene, 5 wt% C_1–C_3, 20 wt% C_{5+}) [8]. The reaction takes place in the vapor phase. The reactor effluent is cooled in the feed–effluent heat exchanger prior to compression. The C_{5+} material, which is suitable for blending into the gasoline pool, is then separated from the isomerate by distillation.

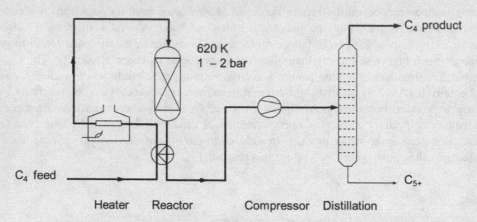

Figure B10.1.1   *Simplified flow scheme of n-butene isomerization.*

QUESTIONS:

Why is the product mixture compressed? Can you imagine a process in which no compression is needed? Evaluate the attractiveness of your new process compared with the one in Figure B10.1.1.

10.4.2   Isomerization of Pentanes and Hexanes

Isomerization of C_5 and C_6 alkanes is not a new process (it was developed in the 1960s) but it has become of much greater importance in oil refineries as a result of changed legislation with respect to gasoline composition. From Table 3.2 (Chapter 3) it is clear that isomerization of the linear pentanes and hexanes present in a straight-run naphtha will significantly boost its octane number.

Figure 10.9 shows the gasphase equilibrium composition of hexanes at a pressure of one bar as a function of temperature. Clearly, a low temperature is favorable for the formation of branched hexanes.

QUESTION:

How does the equilibrium composition depend on pressure?

Many isomerization processes have been developed [10, 11]. An example is described in Box 10.2. Two types of catalysts are used, a $Pt/Cl/Al_2O_3$ catalyst and platinum on the zeolite H-Mordenite. The former catalyst is more active but the latter has a much longer lifetime and is less susceptible to deactivation by, for instance, water. Therefore, water does not need to be removed. In addition, problems as a result of the presence of chlorine are avoided, which makes zeolite-based processes much simpler. For the isomerization of C_7 and C_8 alkanes, which are currently more of interest, more robust acid catalysts based on sulfated zirconia have recently been developed and are currently being used.

QUESTIONS:

> *Pt/Cl/Al₂O₃ is a so-called bi-functional catalyst. What are its two functions? (Hint: describe the role of Cl and of Pt in the isomerization process.) Why is Pt/Cl/Al₂O₃ sensitive to water? Sulfated zirconia is a solid acid catalyst. Guess the structure of the active sites.*

Box 10.2   The Total Isomerization Package Process

The Total Isomerization Package (TIP) process is a combination of the Shell Hysomer process and the Isosiv process of Union Carbide. This process is interesting because it combines catalytic isomerization of linear C_5 and C_6 alkanes over a zeolitic catalyst (Pt/H-Mordenite) with separation by means of shape selectivity in another type of zeolite (molecular sieve Linde 5A) in one process.

In the Hysomer process linear alkanes are isomerized in the presence of hydrogen. The reactor is an adiabatic fixed bed reactor, usually divided into two catalyst beds with intermediate cooling by hydrogen injection. Heat production is predominantly caused by hydrogenation of aromatics that are present and by some hydrocracking (the isomerization reaction itself is nearly thermoneutral). The process flow scheme is very similar to that of the catalytic reforming process (Section 3.4.3). Only about 40–50% of the normal alkanes in the feed are converted as a result of thermodynamic limitations at the relatively high temperature used.

In the Isosiv process, the unconverted normal alkanes are separated from the branched alkanes by means of adsorption: the normal alkanes are selectively adsorbed since iso-alkanes have too large a molecular diameter to enter the pores. Usually, the Isosiv process operates by a pressure difference between adsorption and desorption, that is, pressure swing adsorption (Section 6.1.6). However, when the Isosiv process is integrated with the Hysomer process, desorption is performed by purging with hydrogen instead of by pressure reduction [11]. Close integration of both processes is possible as they operate at similar temperatures and at the same pressure.

As a result of recycle of the normal alkane fraction, the TIP process yields a product with a much higher octane number than the Hysomer process only. Figure B10.2.1 shows this for a feed consisting of 60% pentanes, 30% hexanes, and 10% cyclic compounds.

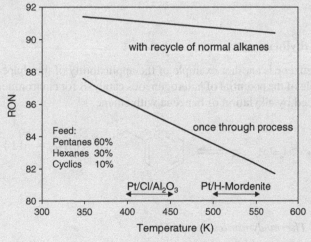

Figure B10.2.1   *Effect of temperature on product octane number in isomerization without and with recycle of normal alkanes.*

Depending on the percentage of normal alkanes in the feedstock, the Hysomer reactor is situated upstream or downstream of the Isosiv adsorbers (Figure B10.2.2).

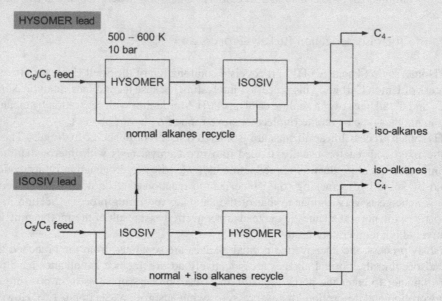

Figure B10.2.2   *Alternative configurations of the TIP process.*

QUESTIONS:

In the Hysomer process H$_2$ is present. Why?
Why does the RON decrease with temperature?
When would you use the Hysomer lead configuration and when the Isosiv lead configuration?

10.4.3   Production of Ethylbenzene

The production of ethylbenzene is another example of the applicability of the shape selectivity of zeolites. It also is an excellent example of the potential of heterogeneous catalysis for environmentally friendly processes.
   Ethylbenzene is produced by alkylation of benzene with ethene:

$$\Delta_r H_{298} = -114 \text{ kJ/mol} \qquad (10.6)$$

10.4.3.1   *Reactions and Thermodynamics*

The production of ethylbenzene by alkylation of benzene with ethene is accompanied by various side reactions of which consecutive alkylations, leading to the formation of polyethylbenzenes, are the most important

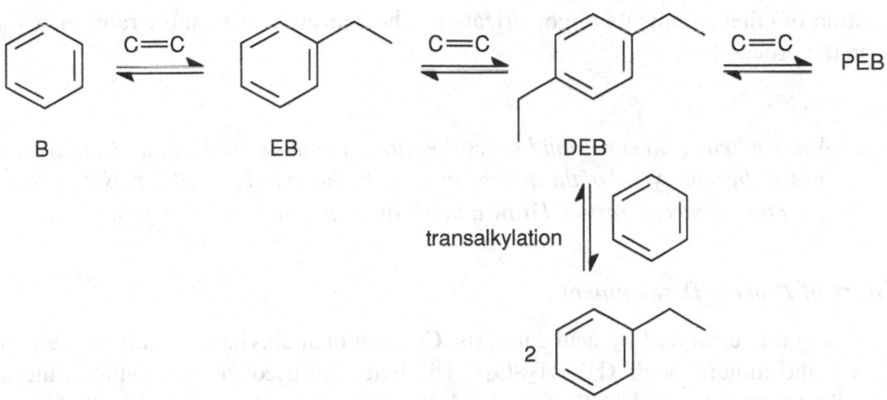

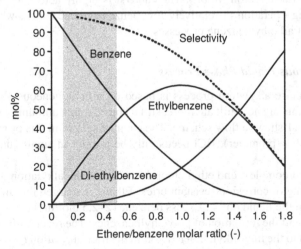

Scheme 10.3   *Main reactions during alkylation of benzene with ethene. B = benzene; EB = ethylbenzene; DEB = diethylbenzenes; PEB = polyethylbenzenes.*

(Scheme 10.3). These can be recycled and converted to ethylbenzene by transalkylation, and therefore are not by-products in the normal sense. The alkylation reactions are exothermic and thermodynamically favored by low temperature and elevated pressure. Transalkylation is virtually independent on temperature and pressure.

De-alkylation and isomerization also occur, as well as oligomerization of ethene (combination of two ethene molecules). The latter reaction is the first step in the formation of most of the heavy by-products (e.g., butylbenzenes, diphenyl compounds, and heavy aromatics and non-aromatics), which either end up as impurities in the ethylbenzene product or as fuel. Control of ethene oligomerization has been a major issue in catalyst development for ethylbenzene production.

Figure 10.10 shows the equilibrium composition and selectivity as a function of the ethene/benzene mole ratio at 500 K and 1 bar. The conversion of benzene increases with increasing ethene/benzene ratio (= decreasing excess benzene), while the selectivity towards ethylbenzene (based on benzene) decreases. In actual practice, excess benzene is added to minimize the formation of polyalkylated compounds. Furthermore,

Figure 10.10   *Effect of ethene:benzene mole ratio on the alkylation of benzene to ethylbenzene (T = 500 K; p = 1 bar; the E/B mole ratio of industrial processes is indicated).*

a low concentration of ethene limits its oligomerization. The benzene/ethene mole ratio may vary from 2 to 16 depending on the process.

QUESTIONS:

> *A low ethene concentration lowers the rates of both desired and undesired reactions. Show that at the same yield of the desired product higher catalyst stability is expected at relatively low ethene concentration. Draw a block diagram for a possible process.*

10.4.3.2   *History of Process Development*

The alkylation reaction is catalyzed by acid catalysts. Conventional alkylation catalysts are metal chlorides (BF_3, $AlCl_3$, etc.) and mineral acids (HF, H_2SO_4). The latter are used in, for example, the alkylation of isobutane with alkenes as discussed in Section 3.4.4. Traditional processes for the alkylation of benzene with ethene are based on an aluminum chloride catalyst. These processes take place in the liquid phase with the catalyst either in a separate fluid phase (e.g., Union Carbide–Badger [12]) or in homogeneous form (Monsanto–Lummus) as discussed below.

Although the conventional catalysts are effective for alkylation, they suffer some important disadvantages. Firstly, they are corrosive, which means that the reactor section of the process must be constructed of special materials (glass lined, brick lined or Hastelloy). Secondly, the use of such catalysts results in waste disposal problems.

As a result of environmental pressure and the costs associated with waste disposal, several new processes have been developed during the 1970s and 1980s, all based on zeolite catalysts. Zeolites have the obvious advantages of being noncorrosive and harmless to the environment.

The first process to become commercially successful was a vapor phase process with a ZSM-5 catalyst (1970s, Mobil–Badger). Nowadays, most plants built after 1980 use this process. Although the Mobil–Badger process has the advantages mentioned above, it has suffered the disadvantage of rather rapid catalyst deactivation (weeks) due to coke deposition. This problem has been overcome in later versions of the process.

Other zeolite-based processes have been developed since, among them one employing catalytic distillation [12]. In 1995, a new liquid phase alkylation process, based on a very selective zeolite called MCM-22, was commercialized by Mobil (now Exxon Mobil). The catalyst is highly active for alkylation but inactive for oligomerization, permitting operation at relatively low benzene/ethene ratios, while achieving the highest yield and product purity of all ethylbenzene processes.

10.4.3.3   *The Homogeneous Liquid Phase Process*

The conventional liquid phase alkylation process is based on a homogeneous aluminum chloride ($AlCl_3$) catalyst. Although new plants are not built anymore, still a considerable amount of ethylbenzene is produced by this process. Figure 10.11 shows a flow scheme. The catalyst system in this process consists of $AlCl_3$ and HCl (which acts as a catalyst promoter). HCl needs only be present in small quantities and is supplied as ethyl chloride.

Dry benzene, the catalyst complex, and ethyl chloride are fed to the alkylation reactor, which is an empty vessel with an internal lining of corrosion-resistant brick. Ethene is sparged into the liquid in the reactor.

Alkylation takes place in the liquid phase. Excess benzene (2–2.5 mol/mol ethene) is used in order to limit oligomerization of ethene and the formation of polyethylbenzenes. The reactor effluent is mixed with recycle polyethylbenzenes and fed to the transalkylation reactor. In the transalkylation reactor the polyethylbenzenes, mainly di- and tri- ethylbenzene, are converted to ethylbenzene. The soluble catalyst complex is recycled to the alkylation reactor. The alkylate is treated with caustic to neutralize any HCl still present, remove residual

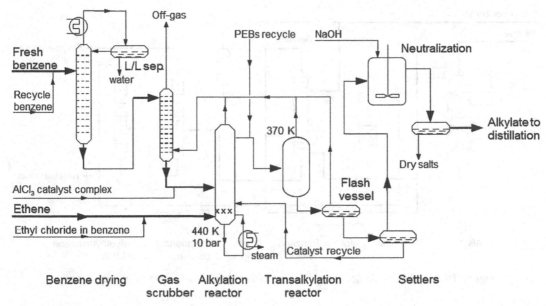

Figure 10.11   *Monsanto–Lummus liquid phase alkylation process for the production of ethylbenzene.*

$AlCl_3$, and to ensure that the alkylate is free of chlorinated compounds. The crude ethylbenzene product, still containing benzene and polyethylbenzenes, is then separated by distillation.

QUESTIONS:

How is the optimal ethene/benzene ratio in the reactor realized? Why is ethyl chloride added instead of HCl? How is the heat of reaction removed? What is the composition of the off-gas? Suggest a distillation sequence and explain.

Apart from the obvious disadvantages of corrosion, the need to remove the catalyst from the product, and the production of polluting waste, the benzene feed must be virtually free of water in order to prevent destruction of the catalyst complex, which requires an additional drying step. Furthermore, the catalyst complex must be prepared separately (not shown in Figure 10.11). Both steps add to investment and operation costs.

10.4.3.4   The Zeolite-Catalyzed Gas Phase Process

The gas phase alkylation process (Figure 10.12) is based on the zeolite ZSM-5. This catalyst contains pores of molecular dimensions, enabling shape selectivity. The ethylbenzene molecules can diffuse freely through the pores, whereas diffusion of the polyethylbenzene molecules is strongly inhibited (product selectivity).

The catalyst deactivates due to coke deposition, which is favored by high temperatures. Regeneration of the catalyst takes place by burning off coke in oxygen-poor air or suitable N_2/O_2 mixtures. This operation takes about 24 hours.

Benzene drying is not required since this catalyst is very resistant to water. A mixture of make-up and recycle benzene is preheated and fed to a multistage fixed bed reactor. Part of the ethene required is added at this point. The benzene/ethene ratio at the entrance of the reactor is higher than in the $AlCl_3$-catalyzed liquid phase process, namely 8–15 mol/mol. Cooling is performed by quenching with cold ethene after each catalyst bed in order to control the temperature rise resulting from the exothermic alkylation reaction.

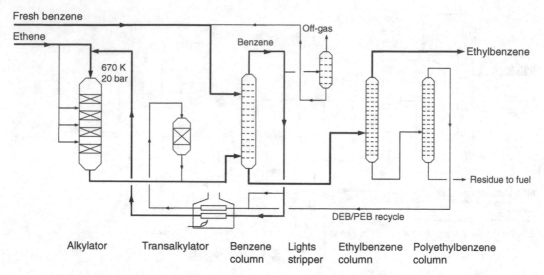

Figure 10.12   *Mobil–Badger gas phase process for the production of ethylbenzene.*

Separation of the product stream takes place by distillation. Although the ZSM-5 catalyst is very selective for ethylbenzene, the formation of polyethylbenzenes cannot be prevented completely. The polyethylbenzene stream from the distillation section is mixed with benzene and fed to a transalkylator where additional ethylbenzene is formed.

Heat is recovered at various points in the process (not shown) by generation of steam for use in a styrene plant (Section 10.5).

QUESTIONS:

Sketch the temperature profile along the reactor. The benzene/ethene ratio in the zeolite-based process is higher than that in the AlCl₃-catalyzed process. Is this logical in view of the shape selectivity of the zeolite?

Recently, catalysts with higher selectivity have been developed. What do you think is the consequence for process design?

The major advantages of the zeolite-based gas phase process are the absence of aqueous waste streams and the fact that no corrosive substances are used, so that high-alloy materials and brick linings are not required. Furthermore, equipment for catalyst recovery and waste treatment is eliminated. On the other hand, due to the larger benzene/ethene ratio, the costs for benzene recovery and recirculation are higher.

10.5   Some Challenges and (Unconventional) Solutions

10.5.1   Adiabatic Reactor with Periodic Flow Reversal

10.5.1.1   *Thermodynamics versus Kinetics*

For exothermic equilibrium reactions a dilemma is faced. In order to reach a favorable equilibrium between reactant and product the temperature must be as low as possible, while a sufficiently high temperature is

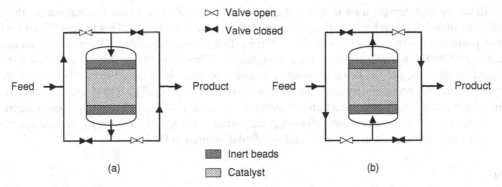

Figure 10.13 Principle of the RPFR: (a) first half of the cycle; (b) second half of the cycle.

required in order for the reaction to proceed at an acceptable rate. Furthermore, the temperature increases towards the exit of the catalyst bed due to the exothermicity of the reaction, lowering the obtainable conversion. Thus, it could be said that the temperature profile is exactly the wrong way around. The consequence for process operation is that the feed must be preheated, while the product stream must be cooled. This is usually done in feed–effluent exchangers. To prevent the reactor temperature from increasing too much, heat must be removed between reaction stages (e.g., by employing a quench reactor). This obviously increases investment and operation costs.

An alternative, which makes use of the heat capacity of the catalyst bed, has been conceived by Boreskov *et al.* [13] and refined by Matros [14, 15]: the adiabatic fixed bed reactor with periodic flow reversal (RPFR). Figure 10.13 shows the principle of the RPFR. The main idea is to utilize the heat of reaction within the catalyst bed itself.

Feeding a hot catalyst bed with relatively cold gas will cool the inlet side of the bed; on the other hand, the temperature at the exit of the bed will increase due to the heat produced by the reaction. By reversing the direction of flow the heat contained in the catalyst bed will bring the cold inlet stream to reaction temperature. The part of the bed that has now become the outlet zone is relatively cold, which is favorable for the reaction equilibrium. After some time the inlet has cooled again, while the outlet has become warmer. Then the flow is reversed again and a new cycle begins. After a sufficient number of flow reversals, an oscillating, but on the average stationary state, is attained. By reversing the flow at the right time heat can be kept in the reactor, while the temperature in the middle zone will remain above the reaction ignition temperature. Once the process has been started up, the heat of reaction is sufficient to keep the process going. Figure 10.14 shows a typical temperature profile over the reactor as a function of time.

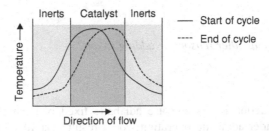

Figure 10.14 Temperature profile in RPFR at the start and end of a cycle.

As a sufficiently high temperature is maintained only in the middle part of the reactor, part of the bed can consist of inert material (with high heat capacity and large particle diameter). This lowers the cost while the conversion remains the same. An additional advantage is that the pressure drop is over the bed reduced.

Currently, the RPFR has three commercial applications. These are the oxidation of sulfur dioxide for sulfuric acid production, the oxidation of volatile organic compounds (VOCs) for purification of industrial exhaust gases, and NO_x reduction by ammonia in industrial exhaust gases. The RPFR seems to be particularly interesting for weakly exothermic reactions to avoid the necessity of feed-effluent heat exchangers. Other possible future applications are in steam reforming and partial oxidation of methane to produce synthesis gas, production of methanol and ammonia, and catalytic dehydrogenation [16].

QUESTIONS:

Reverse-flow operation has an obvious disadvantage. What is this disadvantage and how could it be solved?

Why have the above three processes been chosen for the application of the RPFR?

10.5.1.2 An Endothermic Reaction – the Production of Styrene

Styrene is one of the most important large volume chemicals. It is mainly used as a monomer for polystyrene, which is used in the manufacture of all sorts of packaging materials. Other major applications are in ABS (acrylonitrile-butadiene-styrene) and SBR (styrene-butadiene rubber) and other copolymers. Styrene is produced from ethylbenzene by dehydrogenation:

$$\text{(structure)} \rightleftharpoons \text{(structure)} \quad + \text{ } H_2 \qquad \Delta_r H_{298} = 125 \text{ kJ/mol} \qquad (10.8)$$

The reaction is closely related to the primary dehydrogenation reactions occurring in steam cracking (Chapter 4) but needs a catalyst to prevent side reactions, such as isomerization and cracking with associated coke formation.

10.5.1.2.1 Thermodynamics

The benzene dehydrogenation reaction is highly endothermic and requires a high temperature, as shown in Figure 10.15, which also shows the effect of pressure. Low ethylbenzene partial pressures are clearly preferred. These two observations lead to the use of steam as a means of both supplying heat and lowering the partial pressure of ethylbenzene.

QUESTION:

What could be another reason for adding steam?

10.5.1.2.2 Processes

Most processes for the production of styrene use adiabatic fixed bed reactors [17]. Complete conversion of pure ethylbenzene under adiabatic operation would result in a very large temperature drop (Figure 10.16), resulting in thermodynamic as well as kinetic limitations. Hence, steam is added to the feed for adiabatic fixed bed reactors and usually two or more reactors are placed in series, as shown

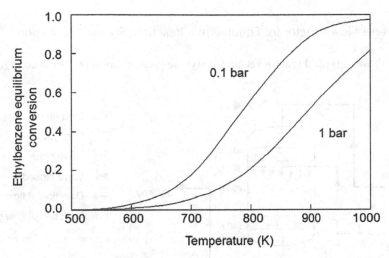

Figure 10.15   *Equilibrium ethylbenzene conversion versus temperature and pressure.*

In Figure 10.16. The steam/ethylbenzene ratio ranges from about 12 to 17 mol/mol in order to supply enough heat for the endothermic reaction, leading to high flow rates with associated large pressure drop [17].

QUESTION:
> *What can be done to limit the pressure drop in a fixed bed reactor?*

The RPFR could be a good alternative for endothermic reactions such as the production of styrene (Box 10.3).

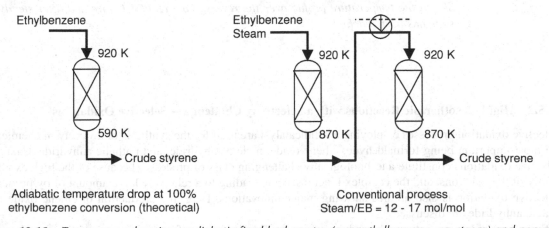

Figure 10.16   *Temperature drop in an adiabatic fixed bed reactor (pure ethylbenzene, no steam) and conventional reactor configuration for styrene manufacture.*

Box 10.3   Reverse-Flow Reactor for Endothermic Reaction: Styrene Production

Figure B10.3.1 shows a typical reactor set-up for styrene production in an RPFR configuration [18].

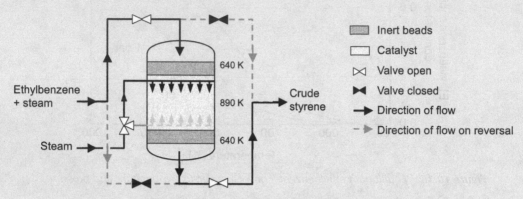

Figure B10.3.1   *RPFR for styrene production.*

Ethylbenzene and steam are introduced at one reactor end, while additional steam is fed cocurrently at one or more downstream locations (one shown in Figure B.10.3.1). The flow direction of the streams is periodically reversed between the reactor ends. The inert beads serve as the heat exchange medium and also prevent the occurrence of the reverse reaction, which would have taken place at the cooler reactor ends if catalyst material had been present.

The RFPR operates at intermediate steam/ethylbenzene ratio (8–10 mol/mol). Near-isothermal operation at the desired temperature (circa 900 K) can be obtained if steam is added at different points along the length of the reactor. Furthermore, the inlet and outlet streams are much cooler than in conventional operation (circa 640 K) due to heat exchange with the inert material. Therefore, energy is used much more efficiently.

QUESTION:

> *Sketch the temperature profile over the reverse-flow reactor before and after steady state has been reached.*

10.5.2   Highly Exothermic Reactions with a Selectivity Challenge – Selective Oxidations

Selective oxidation processes employing a solid catalyst are used for the synthesis of a variety of chemicals, the most important being formaldehyde, ethene oxide, maleic anhydride, and phthalic anhydride. Catalytic selective oxidations constitute a technologically challenging class of processes because of the high exothermicity of the reactions, and the complex kinetic scheme leading to a relatively large amount of by-products. Selective oxidation is also an area showing major innovations. Here, the production of ethene oxide and maleic anhydride are discussed.

Selective oxidation of ethene to produce ethene oxide is interesting for several reasons. Silver is the only metal known to catalyze this reaction with sufficient selectivity and this epoxidation reaction is unique to ethene [19]. No satisfactory catalyst has been found yet for the analogous reactions for the selective oxidation of propene and butene.

Maleic anhydride production is interesting from the point of view of both catalyst design and reactor technology.

10.5.2.1   Production of Ethene Oxide

Ethene oxide, because of its reactivity, is an important raw material for the production of a wide range of intermediates and consumer products. The main outlet for ethene oxide is the manufacture of ethene glycol, accounting for about 70%. Other uses are in the production of surfactants, ethanolamines, and so on. After polyethene, ethene oxide is the second largest consumer of ethene.

10.5.2.1.1   Background

Direct oxidation of ethene has largely replaced a more complex route to ethene oxide. Originally, ethene oxide was produced by indirect oxidation of ethene via ethene chlorohydrin:

$$2\,Cl_2 + 2\,H_2O \;\rightleftharpoons\; 2\,ClOH + 2\,HCl \qquad \Delta_r H_{298} = 141.9\ \text{kJ/mol} \tag{10.9}$$

$$2\,CH_2{=}CH_2 + 2\,ClOH \;\rightleftharpoons\; 2\,HOCH_2{-}CH_2Cl \qquad \Delta_r H_{298} = -538.1\ \text{kJ/mol} \tag{10.10}$$

$$2\,HOCH_2{-}CH_2Cl + Ca(OH)_2 \;\rightleftharpoons\; 2\,\underset{\displaystyle O}{CH_2{-}CH_2} + CaCl_2 + 2\,H_2O \quad \Delta_r H_{298} = -7\ \text{kJ/mol} \tag{10.11}$$

Although this process provided good ethene oxide yields, most chlorine was lost to useless calcium chloride ($CaCl_2$) and unwanted chlorine-containing by-products were generated. This not only was inefficient but also caused major pollution problems, resulting in the replacement of this route by direct oxidation of ethene with either oxygen or air.

The chlorohydrin route is still used in a number of plants for the manufacture of propene oxide from propene, mainly because attempts to achieve direct oxidation of propene to propene oxide failed. Many of these propene oxide plants are converted chlorohydrin-based ethene oxide plants.

10.5.2.1.2   Reactions and Kinetics

The basic reaction involved in the direct oxidation of ethene to ethene oxide is the following:

$$CH_2{=}CH_2 + 1/2\,O_2 \;\longrightarrow\; \underset{\displaystyle O}{CH_2{-}CH_2} \qquad \Delta_r H_{298} = -105\ \text{kJ/mol} \tag{10.12}$$

The only significant by-products are carbon dioxide and water, which are either formed by complete combustion of ethene or by combustion of ethene oxide:

$$CH_2{=}CH_2 + 3\,O_2 \;\longrightarrow\; 2\,CO_2 + 2\,H_2O \qquad \Delta_r H_{298} = -1324\ \text{kJ/mol} \tag{10.13}$$

$$\underset{\displaystyle O}{CH_2{-}CH_2} + 5/2\,O_2 \;\longrightarrow\; 2\,CO_2 + 2\,H_2O \qquad \Delta_r H_{298} = -1220\ \text{kJ/mol} \tag{10.14}$$

All three reactions, and especially the latter two, are highly exothermic and thermodynamically complete at the operating conditions of ethene oxide synthesis. Moreover, the activation energies of the undesired reactions are higher than that of the desired reaction, which implies that even small changes in temperature can have a large effect on the selectivity towards ethene oxide. This temperature sensitivity demands very good temperature control.

QUESTIONS:

> *An important process parameter is the ethene/oxygen ratio in the reactor. What is the optimum ratio (qualitatively)? Explain.*

A catalyst is needed that promotes the partial oxidation of ethene. All commercial processes use a supported silver catalyst. A promoter is added to increase the selectivity towards ethene oxide by reducing the total-combustion reactions. Box 10.4 shows a typical rate expression for the partial oxidation of ethene.

Box 10.4   Kinetics of Ethene Oxidation

One of the numerous rate expressions available for the selective oxidation of ethene is the following Langmuir–Hinshelwood equation [20,21]:

$$r_{EO} = \frac{k K_E p_E K_O p_O}{(1 + K_E p_E + K_O p_O)^2}$$

in which r_{EO} is the rate of formation of ethene oxide (mol/g/h), k is the reaction rate coefficient for reaction 10.12 (15×10^{-3} mol/g/h), K_E and K_O are the adsorption coefficients (10 bar^{-1} and 0.6 bar^{-1}), and p_E and p_O the partial pressures (bar) for ethene and molecular oxygen, respectively.

QUESTIONS:

> *Based on the reactions occurring and the rate expression, what would be the best reactor type, a plug-flow reactor or a CSTR. How can the reactants best be fed (e.g., together at the reactor entrance, one at entrance and the other distributed, etc.)?*

10.5.2.1.3 Process

The need for effective temperature and selectivity control dominates reactor and process design for the selective oxidation of ethene to ethene oxide. Firstly, the ethene conversion per pass must be limited to about 7–15% in order to suppress the consecutive oxidation of ethene oxide. Secondly, a multitubular reactor is preferred over other reactors because it allows better temperature control. The current large plant capacities (up to >300 000 t/a) require large diameter reactors. However, the need to minimize the radial temperature gradients and the risk of hot spots calls for small diameter tubes. Therefore, ethene oxide reactors contain several thousand tubes (length: 6–12 m, internal diameter: 20–50 mm [22]) in parallel. The heat liberated by the reaction is removed by a high boiling hydrocarbon or water, which flows around the reactor tubes. The process shown in Figure 10.17 employs a hydrocarbon-cooled reactor. Condensation of the vapors of this coolant in an external boiler is used to recover heat by steam generation.

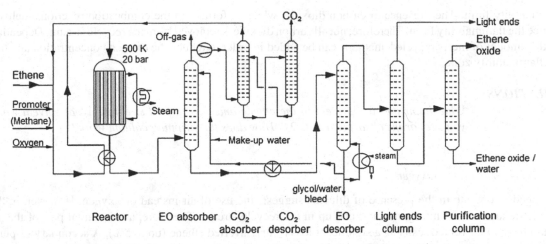

Figure 10.17   *Production of ethene oxide by oxidation of ethene with oxygen.*

In principle, a fluidized bed reactor, possibly with internal heat transfer area, offers better temperature control than the multitubular reactor. However, the consequence of choosing fluidized bed technology is that a catalyst that is resistant against attrition is needed [23].

The temperature rise due to the exothermic reactions is further controlled by adding inert diluents to the reactor feed. This suggests air as the source of oxygen rather than pure oxygen. However, as will become clear below, the use of air has clear disadvantages. Nearly all modern plants use oxygen as the oxidizing agent. Figure 10.17 shows a flow scheme of the oxygen-based ethene oxide process employing a multitubular reactor.

In this scheme, methane is used as the diluent but other gases can also be used. The choice is based on the thermal properties of the gas. Methane has a higher heat capacity and thermal conductivity than nitrogen, which results in better isothermicity.

The preheated feed mixture consisting of ethene, oxygen, and inerts (impurities in the reactants, added methane, and recycled carbon dioxide) is fed to the reactor. A large excess of ethene is used to keep its conversion low and the selectivity towards ethene oxide high. Although the pressure has been shown not to influence the conversion much at typical reaction temperatures, the reaction is conducted at elevated pressure to facilitate the subsequent absorption of ethene oxide in water. The gases leaving the reactor are cooled by heat exchange with the feed gas and sent to a column where ethene oxide (only present in a concentration of about 1–2 mol%) is absorbed in water. The aqueous solution rich in ethene oxide that leaves the bottom of the ethene oxide absorber passes through a stripping column (EO desorber), separating ethene oxide at the top. The bottom stream, containing water and some ethene glycol ($HO-CH_2-CH_2-OH$), which is formed when ethene oxide comes into contact with water, is returned to the ethene oxide absorber. Two distillation columns separate ethene oxide from light ends (carbon dioxide, acetaldehyde, other hydrocarbon traces) and residual water.

A small part of the gas leaving the top of the ethene oxide absorber is purged to prevent the build-up of inerts (mainly carbon dioxide, argon, and methane). After compression, a side stream of the recycle gas goes to an absorber where carbon dioxide is removed, before being recycled to the reactor. This is necessary to keep the carbon dioxide concentration at an acceptable level.

Ethene and oxygen in certain proportions form explosive mixtures. The feed to the reactor contains 20–40 mol% ethene and about 7 mol% oxygen to produce a reaction mixture that is always above the upper flammability limit. The presence of carbon dioxide, which is formed in the combustion reactions, helps to reduce the flammability limit. Therefore, not all carbon dioxide is removed from the recycle stream. Depending on the amount of inerts recycled, methane can be added in order to reduce the oxygen concentration and thus the flammability zone.

QUESTIONS:

> *What impurities are present in the ethene and oxygen feed streams? What solvent would you use for the absorption of CO_2? How does the stripping column work?*

10.5.2.1.4   Air versus Oxygen

The need to operate in the presence of diluents suggests the use of air instead of oxygen. However, in this case large amounts of nitrogen would end up in the recycle stream. Therefore, a substantial part of the gas would have to be vented, causing a significant loss of unconverted ethene (up to 5%). A secondary or purge reactor is needed with associated ethene oxide absorption system in order to improve the ethene conversion before venting. Therefore, additional investment is required. However, this is partially offset by the need for a carbon dioxide removal system in the oxygen process.

Also, the operating costs for the two processes differ. The price of oxygen, especially, has a great bearing on the production costs. Furthermore, in the oxygen process considerable steam input is required in the carbon dioxide removal unit. The higher production rate per unit volume of catalyst and the less costly recycle of ethene when using oxygen may be a decisive advantage.

Table 10.3 compares the operating parameters in oxygen- and air-based ethene oxide plants.

QUESTIONS:

> *Why is a CO_2 removal system not required in the air-based process? Estimate the residence time of the gas in the reactor.*

10.5.2.2   Production of Maleic Anhydride

Maleic anhydride is an important intermediate in the chemical industry. It is used in copolymerization and addition reactions to produce polyesters, resins, plasticizers, and so on. Maleic anhydride is mainly produced

Table 10.3   *Comparison of oxygen- and air-based ethene oxide process [22].*

Process	Oxygen based	Air based
Concentrations (vol.%)		
C_2H_4	15–40	2–10
O_2	5–9	4–8
CO_2	5–15	5–10
inert	rest (mainly CH_4 and Ar)	rest (mainly N_2)
Temperature (K)	570–630	570–630
Pressure (bar)	10–20	10–30
GHSV (m^3_{gas})/((m^3_{cat}).h)[a]	2000–4000	2000–4500
C_2H_4 conversion (%)	7–15	20–65
Selectivity (%)	75–80	70–75

[a]GHSV = gas hourly space velocity.

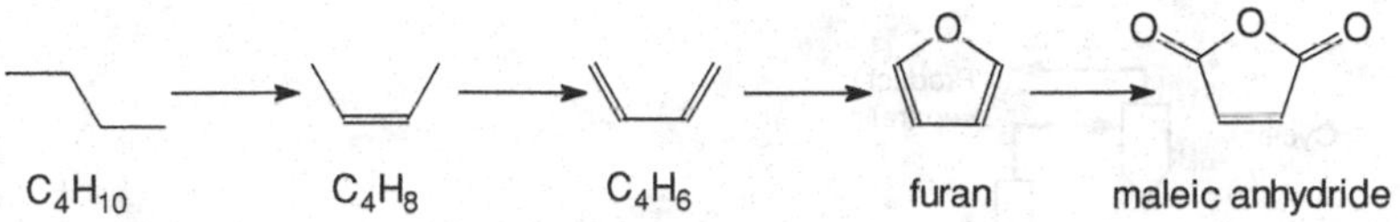

C$_4$H$_{10}$ C$_4$H$_8$ C$_4$H$_6$ furan   maleic anhydride

Scheme 10.4   *Simplified reaction pathway for maleic anhydride production.*

by the selective oxidation of *n*-butane, according to the overall reaction:

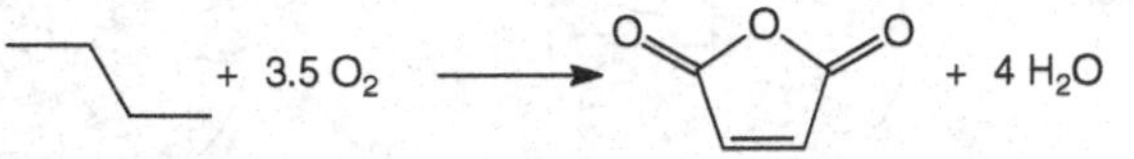

+ 3.5 O$_2$ $\longrightarrow$ + 4 H$_2$O $\Delta_r H_{298} = -1260$ kJ/mol   (10.15)

The commercial processes are catalytic and based on fluidized beds or multitubular fixed bed reactors, as used in the production of ethene oxide. In principle, a fixed bed reactor is not the optimal choice: the intermediates (Scheme 10.4) are more reactive than the starting material and the reactions are highly exothermic. Hot spots may easily occur leading to total combustion.

Note that the exothermicity of the desired reaction is much larger than that of the partial oxidation of ethene (reaction 10.12) in the production of ethene oxide. This results in much lower single-reactor capacities for the production of maleic anhydride in fixed bed reactors (up to 40 000 t/a [24]).

Scheme 10.4 shows the simplified reaction network for the production of maleic anhydride. Just as in the selective oxidation of ethene to ethene oxide, by-products (CO$_2$, H$_2$O, and CO) are formed by complete or partial oxidation of *n*-butane and maleic anhydride.

Fluidized bed reactors suggest themselves as an alternative to circumvent the problems encountered with fixed bed reactors. They allow better temperature control and higher feed concentrations. However, catalyst attrition is a major problem. Box 10.5 describes an innovative process based on a configuration consisting of a riser and a fluidized bed reactor.

QUESTIONS:

> *In the past, benzene and other aromatics were the predominant feedstocks for the production of maleic anhydride. At present, n-butane is a more important feedstock. What would be the reason?*
>
> *Summarize advantages and disadvantages of fluidized bed reactor technology for the production of maleic anhydride.*

Box 10.5   Production of Maleic Anhydride in a Riser Reactor [24–26]

An interesting option for the production of maleic anhydride has been developed by Monsanto and DuPont. It involves the use of a riser reactor analogous to the reactor applied in the FCC process (Section 3.4.2). The catalyst played an important role in this development. A catalyst with the required high attrition resistance can be produced by incorporating the active ingredients in a matrix, such as in the FCC process, but this decreases the activity and might lead to lower selectivity. In a new technology, developed by DuPont, the active material is encapsulated in a porous silica shell. The pore openings of the shell should permit unhindered diffusion of the reactants and products, without affecting conversion or selectivity. Figure B10.5.1 shows the reactor configuration and a schematic of the conventional catalyst and the novel catalyst proposed by DuPont.

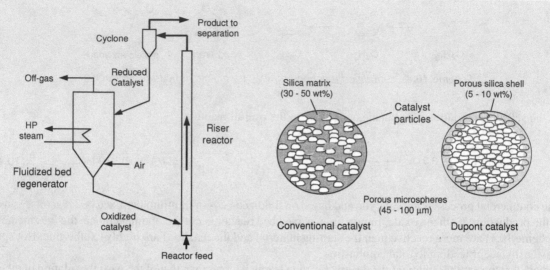

Figure B10.5.1   *Riser reactor and catalyst for catalytic oxidation of n-butane.*

In the riser reactor, the oxidized catalyst transfers oxygen to butane to produce maleic anhydride. The catalyst is separated from the product in a cyclone. Subsequently, the catalyst is reoxidized in a conventional fluidized bed reactor.

Separation of the two steps enables fine tuning of the reaction conditions:

- the oxidation is performed without molecular oxygen, so the reaction mixture is never within the explosion limits;
- the residence times in the reactor and the regenerator can be chosen independently and, as a consequence, optimal conditions for both reactors can be chosen.

QUESTIONS:

> *Why is the lower activity due to incorporation in a matrix not a problem in FCC? Give the pros and cons of this process concept compared to the conventional processes. What would have been the critical steps in the development of this process?*
>
> *In practice, the catalyst stability under actual process conditions appeared to be a point of concern. What would be the cause of this? (Hint: The reaction temperature is about 770 K. Consider the properties of silica and the products formed in overall reaction.) What would be the effect on the flow properties of the catalyst?*

10.6   Monolith Reactors – Automotive Emission Control

Monolith reactors are conceptually the simplest structured reactors and they are most widespread. Monoliths are continuous structures consisting of narrow parallel channels (typically 1–3 mm diameter). A ceramic or metallic support (produced in the form of blocks by extrusion) is coated with a layer of material in which

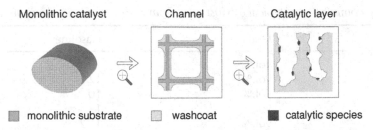

Monolithic catalyst   Channel   Catalytic layer

■ monolithic substrate   □ washcoat   ■ catalytic species

Figure 10.18   *Example of a monolith with square channels [27].*

active ingredients are dispersed (the washcoat) (Figure 10.18). The shape of the channels (or capillaries) differs widely: circular, triangular, square, hexagonal, and so on. The shape of the entire monolithic block can be adapted to fit in the reaction chamber.

QUESTION:

> *Extrusion is a common method for producing shaped solid materials. A well-known example is the production of pasta. Ceramic monoliths are usually also produced by extrusion. How would this be done in practice?*

Monolithic reactors are very suitable for applications requiring low pressure drop, such as the conversion of harmful components in car exhausts; the pressure drop in the exhaust line of the engine has a great effect on the efficiency of the engine (fuel consumption). The low pressure drop compared to conventional fixed bed reactors is due to the flow through straight channels rather than through the tortuous paths in catalyst particles. The application of monoliths in the cleaning of exhaust gases from cars can be considered one of the greatest successes of the last decades in the areas of chemical engineering and catalysis.

The automobile is one of the major sources of environmental pollution and photochemical smog, as illustrated in Table 10.4. Thus, legislation to control harmful emissions from car exhaust gases is becoming increasingly strict.

Gasoline and diesel engines have to be considered separately, as their operation and exhaust gas composition (Table 10.5) differ profoundly. Until recently, the success of catalytic converters was limited to the treatment of exhaust gases from gasoline engines. In these converters, introduced in the early 1990s, a so-called three-way catalyst is used for the simultaneous abatement of hydrocarbons, carbon monoxide, and NO_x. Since then every country has introduced their own emission standards which are periodically updated for the different applications.

Table 10.4   *Contribution (%) to various emissions from mobile and stationary sources in the USA (2008) [28].*

Component emitted	Mobile	Stationary fuel combustion	Other stationary sources
SO_2	5	84	11
NO_x	57	33	10
HC	37	9	54
CO	83	8	9

NO_x = NO, NO_2; HC = hydrocarbons.

Table 10.5   *Average composition of typical exhaust gases from gasoline and diesel engines in passenger cars.*

Compound	Unit	Gasoline	Diesel
N_2	vol.%	74	67
CO_2	vol.%	12	13
H_2O	vol.%	10	11
O_2	vol.%	1	10
CO	vol.%	1–2	0.05
HC	vol.% "C_1"	0.25	0.05
NO_x	vol.%	0.25	0.15
SO_x	ppmv	<10[a]	10–150
particulate matter	mg/m^3	1–10	20–200

[a]Based on 10 ppm S in fuel.

For diesel engines, which emit relatively large amounts of particulate matter (PM), traps were introduced from 2000 on. Around 2005, NO_x-abatement systems for heavy-duty trucks based on the SCR (selective catalytic reduction) technology became available.

10.6.1   Exhaust Gas Composition

Table 10.5 shows the average composition of exhaust gases from gasoline and diesel engines that have not undergone treatment.

Carbon dioxide and water are the products of complete combustion. Carbon monoxide and hydrocarbons are present in the exhaust gas as a result of incomplete combustion. The presence of hydrocarbons originates mainly from regions that are not reached by the flame (e.g., layers near the combustion chamber walls). Nitrogen oxides are formed by reaction of nitrogen and oxygen from the air when sufficient oxygen is present and the temperature is high. The sulfur compounds present in gasoline or diesel are oxidized largely to sulfur dioxide.

The exhaust gas composition for gasoline engines depends on many parameters, the most important one being the air/fuel intake ratio, which is usually expressed using λ. $\lambda > 1$ implies a lean or oxidizing environment and $\lambda < 1$ a rich or reducing atmosphere. λ is defined as:

$$\lambda = (\text{air/fuel})_{\text{actual}}/(\text{air/fuel})_{\text{stoich}} \qquad (10.16)$$

Figure 10.19 shows the influence of λ on the exhaust gas composition of a gasoline engine. The nitric oxide production curve goes through a maximum. At rich conditions ($\lambda < 1$) the amount of air is low and as a result the emission of nitric oxide is low. At lean conditions ($\lambda > 1$) the temperature drops, which causes a decrease in nitric oxide formation. The maximum amount of nitric oxide is formed near stoichiometric conditions.

With increasing λ the carbon monoxide and hydrocarbon emissions decrease because of better combustion when more oxygen is present. The slight increase of hydrocarbons at high λ is surprising at first sight. The explanation for this increase is that the temperature drop at lean conditions results in a less complete combustion.

Carbon monoxide and hydrocarbon emissions from diesel engines are much lower than those from gasoline engines, because diesel engines always operate at lean conditions. NO_x emissions are also lower. However, the exhaust gas from diesel engines contains a relatively large amount of particulate matter. This consists of dry carbon (soot), a soluble organic fraction (SOF), and sulfuric acid. SOF contains unburned hydrocarbons, oxygenates (ketones, aldehydes, etc.), and polycyclic aromatic hydrocarbons.

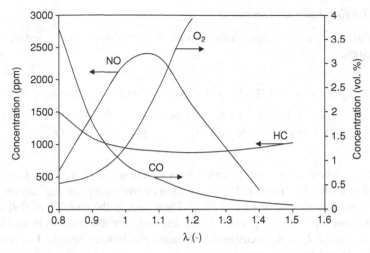

Figure 10.19   *The concentration of CO, NO, HC (hydrocarbons), and O₂ emitted by a gasoline engine as a function of λ [29].*

10.6.2   Reduction of Exhaust Gas Emissions

10.6.2.1   *Control Strategies*

Legislation requires the reduction of the carbon monoxide, NO_x, hydrocarbon, and particulate emissions of cars and trucks. Basically two categories of emission reducing techniques exist: primary and secondary measures. Primary measures include:

Speed limitations. The emission of NO_x significantly increases with the driving speed. This increase is due to both a higher temperature and the operation in a relatively lean mixture at constant high speed.

Fuel reformulation. Hydrotreating of fossil fuels has a favorable influence on sulfur dioxide emissions but only a small influence on NO_x emissions, since NO_x formation is mainly due to radical processes involving nitrogen from the air.

Fischer–Tropsch diesel fuel and biodiesel (hydrogenated vegetable oils and fats) could have more favorable properties than diesel fuel from conventional processes.

Engine modifications. A gasoline combustion engine operating at for instance $\lambda > 1.3$ would result in a major reduction of carbon monoxide and NO_x emissions (up to 50%). The limiting factor at high λ, due to the dilution of the fuel, is the ignition. In principle, this problem can be avoided by using a higher compression ratio. "High-compression lean-burn engines" are currently used commercially. An additional advantage of this strategy is a decrease in fuel consumption (estimated to be 15%). Nevertheless, with current emission standards, end-of-pipe NO_x abatement technology still has to be used.

In diesel engines, NO_x emissions can be reduced by engine modifications but usually at the cost of increased particulate emissions or vice versa. For example, exhaust gas recirculation (EGR) reduces NO_x emissions, but increases particulate emissions. On the other hand, high-pressure injection suppresses particulate emissions, while increasing NO_x emissions. In the development of diesel engines enormous progress has been made. Nevertheless, because of increasingly severe emission standards both PM- and NO_x-abatement technologies are required.

Secondary measures (end-of-pipe solutions) are based on the treatment of exhaust gas, often by catalytic processes.

10.6.2.2   *Gasoline Engine Exhaust Treatment*

Secondary measures to reduce exhaust emissions from gasoline engines all are based on catalytic conversion of the harmful emissions:

$$C_mH_n + (m + n/2)\,O_2 \rightarrow m\,CO_2 + n/2\,H_2O \qquad \Delta_r H_{298} < 0 \tag{10.17}$$

$$CO + 1/2\,O_2 \rightarrow CO_2 \qquad \Delta_r H_{298} = -288\ kJ/mol \tag{10.18}$$

$$CO + NO \rightarrow 1/2\,N_2 + CO_2 \qquad \Delta_r H_{298} = -373\ kJ/mol \tag{10.19}$$

Figure 10.20 shows the evolvement of control technologies with time. Obviously, one reactor in which all three species are converted is preferable. However, the oxidation of hydrocarbons and carbon monoxide requires other reaction conditions than the conversion of NO_x. Therefore, in the past (1976–1979) NO_x control was not attempted (regulations were not very strict then), and only carbon monoxide and hydrocarbons were converted by an oxidation catalyst. Recirculation of exhaust gas in later designs has helped to reduce NO_x emissions, due to a lower combustion temperature.

In later years (1979–1986), a dual catalyst system was applied. In the first reactor NO_x (and part of the hydrocarbons and carbon monoxide) was converted, while in the second reactor the remaining hydrocarbons and carbon monoxide were oxidized with a secondary supply of oxygen. A consequence of this arrangement was that the engine had to operate at rich conditions because at that time NO_x could only be converted in a reducing environment.

In 1981, emission standards became more stringent and the solution was found in the application of the so-called "three-way catalyst" (TWC). The name stems from the fact that the three components (NO_x, CO and hydrocarbons) are removed simultaneously in one reactor. Improved catalysts and fuel/air control techniques have enabled this development. Figure 10.21 shows the position of the three-way catalytic converter in the exhaust pipe.

In parallel with studies related to the active catalyst components (activity, stability), a number of engineering problems had to be tackled. Firstly, the pressure drop over the catalytic converter should be low and the

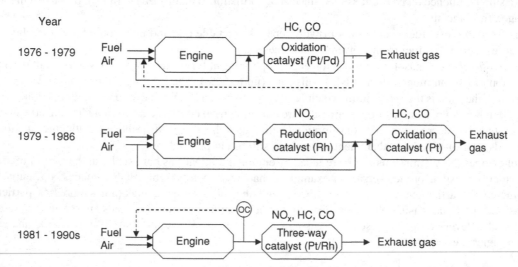

Figure 10.20   *Development of emission control technologies for gasoline engines. OC = oxygen control.*

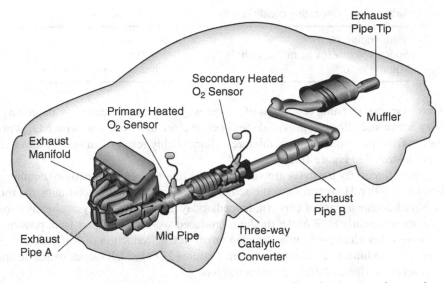

Figure 10.21 *Position of the three-way catalytic converter in the exhaust pipe of a gasoline engine.*

converter should be as light as possible in view of fuel economy. Secondly, the catalyst should be resistant to the extreme temperatures and corrosive environment in the exhaust pipe.

Not surprisingly, the first converters were reactors packed with catalyst particles, that is, conventional fixed bed reactors. The main concern with these reactors was attrition of the catalyst particles as a result of vibrational and mechanical stresses. Today the majority of catalytic converters are monolith reactors. These reactors combine a lower pressure drop with smaller size and weight, and hence provide better fuel economy than the conventional fixed bed reactors.

The catalyst is mounted in a stainless steel container (the most difficult part of the reactor production) with a packing wrapped around it to ensure resistance to vibration (Figure 10.22).

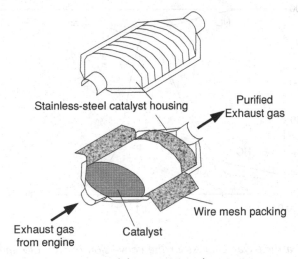

Figure 10.22 *Monolithic reactor and mounting system.*

Table 10.6 *Operating conditions in a three-way catalytic converter.*

Temperature (K)	570–1170 K
Space velocity ($m^3_{gas} \cdot m^{-3}_{reactor} \cdot h^{-1}$)	$1–2 \times 10^5$
volume ratio	Catalyst/cylinder = 0.8–1.5

Table 10.6 shows typical operating conditions of a three-way catalyst. The catalysts, based on platinum and rhodium, are very active and effective, provided they operate close to $\lambda = 1$, otherwise NO_x is not reduced to nitrogen. Deviation from this stoichiometric mixture will inevitably result in an overall lower conversion of the pollutants, as illustrated in Figure 10.23.

The presently used three-way catalysts are very effective due to their high activity combined with their outstanding thermal stability. However, since the active components are mainly platinum and rhodium, costs and availability have become a point of concern. Indeed, today automotive catalysis already consumes about 40% of the platinum and nearly 80% of the rhodium produced worldwide. Apart from possible availability problems, there were other challenges. In a reducing environment, reduction of sulfur dioxide to hydrogen sulfide was observed. In addition, at intermediate temperatures N_2O (which causes ozone destruction in the stratosphere) was formed. These problem have been solved.

QUESTIONS:

Explain how H_2S can be formed. Compare the space velocity of the three-way catalytic converter with "normal" chemical reactors. The current emission limits for gasoline-fueled cars are so low that most of the emission occurs during start-up of the catalytic system. How could this start-up-related emission be reduced?

10.6.2.3 *Diesel Engine Exhaust Treatment*

Carbon monoxide and hydrocarbon emissions from diesel engines can be treated relatively easily with a noble metal oxidation catalyst, such as used for the treatment of exhaust gas from gasoline engines. Particulate

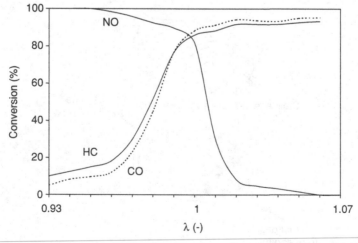

Figure 10.23 *Efficiency of a three-way catalyst for the conversion of NO, CO and hydrocarbons (HC) as a function of λ [30].*

control and NO_x abatement are more problematic; particulates exhibit a low reactivity in oxidation and NO_x reduction is easiest in a reducing environment, while oxidizing conditions prevail in diesel exhaust gas.

10.6.2.3.1 Removal of Particulates

Particulates, which for a large part consist of soot, can be trapped in a so-called absolute filter with periodic thermal regeneration. In practice the procedure is as follows. The pressure drop over the filter is continuously measured and when the pressure drop exceeds a certain value, the engine management system starts a regeneration procedure: the temperature of the filter is raised and the trapped diesel soot is burned by the oxygen in the exhaust gas.

QUESTION:

> *How would you accomplish raising the temperature of the particulate filter for periodic*
> *regeneration?*

Besides filter systems with periodic regeneration, systems based on catalytic regeneration that operate more continuously have also been developed. The first periodic catalytic regeneration system, which was introduced by PSA Peugeot Citroën in 2000, uses catalytic fuel additives that are active in the oxidation of soot in combination with a particulate filter (Figure 10.24).

A catalyst ("additive") is added to the diesel fuel. This additive ends up in the soot deposited in the particulate filter, making the soot more reactive towards oxidation. Analogous to the filter systems with periodic thermal regeneration, the pressure drop is monitored. When filter clogging is observed, the engine management computer initiates regeneration of the filter. A complete regeneration requires only a few minutes and is performed every 500–800 km without the driver noticing.

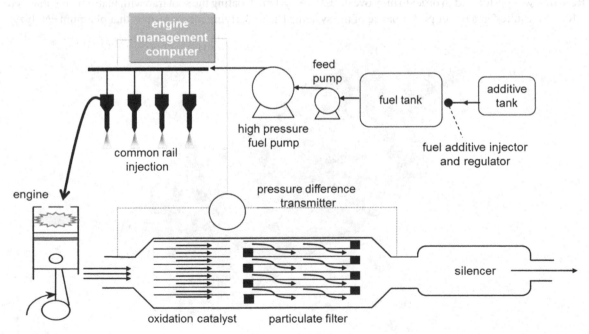

Figure 10.24 *Schematic representation of the PSA Peugeot Citroën system for the control of particulate emissions [31].*

Later, filters with catalytic coatings were successfully introduced, but these are less active than the additives and they have to be operated at higher temperature.

QUESTIONS:

The oxidation catalyst (Figure 10.24) is a monolith of the flow-through type, while the particulate filter is a so-called wall-flow monolith. What is the difference?

Is the robustness of monoliths with a catalytic coating a point of concern?

A catalytic coating needs a higher operating temperature than a catalytic additive. Why would this be the case?

If you had to choose between a fuel additive and a catalytic coating technology, which would you choose? Explain.

An original idea for a soot removal device came from researchers at Johnson Matthey, who observed that NO_2 is much more reactive for soot oxidation than O_2. They developed a two-stage particulate filter unit – a Continuously Regenerating Trap (CRT®) – consisting of an oxidation monolith and a "classical" soot trap (a wall-flow monolith) (Figure 10.25). The oxidation monolith oxidizes CO and hydrocarbons to CO_2 and part of the NO to NO_2:

$$NO + \tfrac{1}{2}O_2 \rightarrow NO_2 \qquad \Delta_r H_{298} = -56.5 \text{ kJ/mol} \tag{10.20}$$

The NO_2 then reacts with the soot that has been trapped in the soot trap:

$$C + NO_2 \rightarrow NO + CO \qquad \Delta_r H_{298} = 166.5 \text{ kJ/mol} \tag{10.21}$$

Research was performed in order to improve the CRT® system. Coating the soot trap with platinum ("catalyzed CRT®") resulted in a better performance of the system. The underlying mechanism is that platinum catalyzes

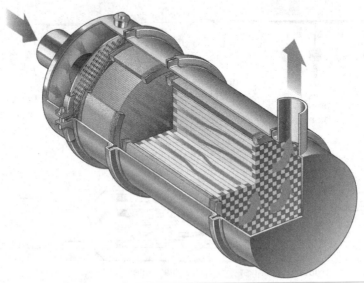

Figure 10.25   *Illustration of the Continuously Regenerating Trap (CRT®) system. Image provided courtesy of Johnson Matthey. Copyright (2007) Johnson Matthey*

re-oxidation of NO produced by reaction 10.21 to NO_2. As is well-known, platinum is poisoned by sulfur. Thus, the catalyzed CRT® technology requires low-sulfur diesel fuel. It should be noted that NO_x is not eliminated in this process.

10.6.2.3.2 NO_x Removal

The removal of NO_x from diesel exhaust gases is a real challenge: a reduction reaction must be performed in an oxidizing environment. Not surprisingly, the catalytic converters used in gasoline engines do not work for diesel exhaust gas. In this respect, diesel engines are similar to power plants, nitric acid plants, and so on. For the abatement of NO_x emissions from these stationary sources, "Selective Catalytic Reduction (SCR)" is a standard process (Section 8.4.3).

For NO_x abatement in exhaust gases from the heavy-duty engines in trucks, SCR with ammonia is now also commercially applied in order to meet the current emission standards set by legislation. The source of ammonia usually is urea, which upon heating decomposes according to the reaction:

$$\underset{H_2N}{\overset{O}{\underset{}{\|}}}\underset{NH_2}{C} + H_2O \rightleftharpoons 2\,NH_3 + CO_2 \qquad \Delta_r H_{298} = 101\ \text{kJ/mol} \qquad (10.22)$$

The main SCR reactions are:

$$4\,NO + 4\,NH_3 + O_2 \rightarrow 4\,N_2 + 6\,H_2O \qquad \Delta_r H_{298} = -408\ \text{kJ/mol NO} \qquad (10.23)$$
$$2\,NO_2 + 4\,NH_3 + O_2 \rightarrow 3\,N_2 + 6\,H_2O \qquad \Delta_r H_{298} = -669\ \text{kJ/mol } NO_2 \qquad (10.24)$$

Usually, these reactions take place in one monolithic reactor, with the upstream part being used for hydrolysis of urea and the downstream part for SCR.

QUESTION:

Why is urea used and not ammonia (standard in the reduction of NO_x emissions from stationary sources)?

For diesel-fueled passenger cars, SCR units would be too large. Currently, for these cars "NO_x Storage and Reduction (NSR)" systems using so-called "Lean NO_x Traps" (LNTs) are being developed. These systems rely on a three-way catalyst like the catalyst used for the treatment of emissions from gasoline engines, but modified with an oxide which forms a nitrate with NO_x that is stable under oxidizing conditions and unstable under reducing conditions. Many oxides exhibit this behavior and, among others, barium oxide appears to be a satisfactory oxide. During normal lean operation, NO is oxidized and the product is trapped by the barium oxide:

$$NO + \tfrac{1}{2}O_2 \rightarrow NO_2 \qquad (10.20)$$
$$BaO + 2\,NO_2 + \tfrac{1}{2}O_2 \rightarrow Ba(NO_3)_2 \qquad (10.25)$$

$Ba(NO_3)_2$ is stable under normal driving conditions, but when rich conditions are created in the reactor, the nitrate decomposes and a complex network of reactions leads to the formation of a mixture of nitrogen, carbon dioxide and water:

$$Ba(NO_3)_2 + HC \rightarrow BaO,\ N_2,\ CO_2,\ H_2O \qquad (10.26)$$

In practice, when the trap is saturated (after 50–60 seconds), extra fuel is injected over a very short time frame (2–5 seconds), resulting in a reducing environment and, as a consequence, the NO_x trap is regenerated. After regeneration, the engine is returned to lean burn operation.

This technology requires fuel with a very low sulfur content, because sulfur can be stored in the trap as sulfate, leading to deactivation. Despite the fact that currently the sulfur level in diesel fuel is extremely low (<10 ppm S), improvement of the sulfur tolerance and life time of these NO_x traps is still a considerable challenge.

QUESTION:

> *The LNT is a combination of a three-way catalyst (containing Pt and Rh) and an oxide (often BaO) trapping NO_x. Describe the function of the three-way catalyst during normal operation and during regeneration.*

10.6.2.4   Effect of Future Legislation

The Euro 6 and Euro VI emission standards for light- and heavy-duty vehicles, respectively, coming in force around 2014–2016, require that the exhaust gases for both gasoline and diesel engines are the same. This has consequences both for diesel- and for gasoline-powered cars. Not surprisingly, for diesel-powered cars besides the particulate filter a NO_x-abatement system also has to be in operation [32]. It should be noted that these stricter regulations also affect emission control for gasoline engines, because the increasingly used lean burn direct-injection gasoline engines emit significant amounts of nano- to micro-sized particulate matter. These engines are more efficient in fuel consumption than classical gasoline engines and approach the efficiency of advanced diesel engines, but they share with diesel engines the undesired particulate emission.

QUESTION:

> *Explain why lean burn gasoline engines, like diesel engines, emit particulates.*

References

[1] Krishna, R. and Sie, S.T. (1994) Strategies for multiphase reactor selection. *Chemical Engineering Science*, **49**, 4029–4065.

[2] Trambouze, P., Van Landeghem, H., and Wauquier, J.P. (1988) *Chemical Reactors, Design/Engineering/Operation.* Gulf Publishing Company, Houston, TX.

[3] Fogler, S.H. (1986) *Elements of Chemical Reaction Engineering.* Prentice-Hall, London.

[4] Moulijn, J.A., Van Diepen, A.E., and Kapteijn, F. (2001) Catalyst deactivation: is it predictable?: What to do? *Applied Catalysis A*, **212**, 3–16.

[5] Weisz, P.B. (1980) Molecular shape selective catalysis. *Pure and Applied Chemistry*, **52**, 2091–2103.

[6] van de Graaf, J.M. (1999) Permeation and Separation Properties of Supported Silicalite-1 Membranes – a Modeling Approach. PhD Thesis, Delft University of Technology, The Netherlands.

[7] Maxwell, I.E., Naber, J.E., and de Jong, K.P. (1994). The pivotal role of catalysis in energy related environmental technology. *Applied Catalysis A*, **113**, 153–173.

[8] Mooiweer, H.H., de Jong, K.P., Kraushaar-Czarnetski, B., Stork, W.H.J., and Krutzen, B.C.H. (1994) Skeletal isomerization of olefins with the zeolite ferrierite as catalyst, in *Zeolites and Related Microporous Materials: State of the Art 1994*, vol. **84** (eds J. Weitkamp, H.G. Karge, H. Pfeifer, and W. Hölderick). Elsevier, Amsterdam, The Netherlands, pp. 2327–2334.

[9] Houzvicka, J. and Ponec, V. (1997) Skeletal isomerization of n-butene. *Catalysis Reviews, Science And Engineering*, **39**, 319–344.

[10] Meyers, R.A. (2004) *Handbook of Petroleum Refining Processes*, 3rd edn. McGraw-Hill.

[11] Sie, S.T. (1997) Isomerization reactions, in *Handbook of Heterogeneous Catalysis* (eds G. Ertl, H. Knözinger, and J. Weitkamp). VVH, Weinheim, pp. 1998–2017.

[12] Chen, S.S. (1997) Styrene, in *Kirk-Othmer Encyclopedia of Chemical Technology*, 4th edn, vol. **22** (eds J.I. Kroschwitz and M. Howe-Grant). John Wiley & Sons, Inc., New York, pp. 956–994.

[13] Boreskov, G.K., Matros, Yu.Sh., and Kselev, O.V. (1979) Catalytic processes carried out under nonstationary conditions. *Kinetika i Kataliz*, **20**, 773–780 (in Russian).

[14] Matros, Yu.Sh. (1989) *Catalytic Processes Under Unsteady-State Conditions*. Elsevier, Amsterdam, The Netherlands.

[15] Matros, Yu.Sh. (1985) *Unsteady-State Processes in Catalytic Reactors*. Elsevier, Amsterdam, The Netherlands.

[16] Matros, Y.S. and Bunimovich, G.A. (1996) Reverse-flow operation in fixed bed catalytic reactors. *Catalysis Reviews, Science And Engineering*, **38**, 1-68.

[17] James, J.W. and Castor, W.M. (1994) Styrene, in *Ullmann's Encyclopedia of Industrial Chemistry*, 5th edn, Vol. **A25** (ed. W. Gerhartz). VCH, Weinheim, pp. 329–344.

[18] Snyder, J.D. and Subramaniam, B. (1994) A novel reverse flow strategy for ethylbenzene dehydrogenation in a packed-bed reactor. *Chemical Engineering Science*, **49**, 5585-5601.

[19] Satterfield, C.N. (1991) *Heterogeneous Catalysis in Industrial Practice*, 2nd edn. McGraw-Hill, New York, 279–285.

[20] Farrauto, R. and Bartholomew, H. (1997) *Fundamentals of Industrial Catalytic Processes*. Blackie, London.

[21] Park, D.W. and Gau, G. (1987) Ethylene epoxidation on a silver catalyst: unsteady and steady state kinetics. *Journal of Catalysis*, **105**, 81–94.

[22] Rebsdat, S. and Mayer, D. (1994) Ethylene oxide, in *Ullmann's Encyclopedia of Industrial Chemistry*, 5th edn, **A10** (ed. W. Gerhartz). VCH, Weinheim, pp. 117–135.

[23] Rebsdat, S. and Mayer, D. (2000) Ethylene oxide, in *Ullmann's Encyclopedia of Industrial Chemistry*. Wiley-VCH Verlag GmbH, http://dx.doi.org/10.1002/14356007.a10_117 (last accessed 19 December 2012).

[24] Lerou, J.J. and Mills, P.L. (1993) Du Pont butane oxidation process, in *Precision Process Technology; Perspectives for Pollution Prevention* (eds M.P.C. Weijnen and A.A.H. Drinkenburg). Kluwer, Dordrecht, The Netherlands, pp. 175–195.

[25] Contractor, R.M. (1999) Dupont's CFB technology for maleic anhydride. *Chemical Engineering Science*, **54**, 5627–5632.

[26] Lintz, H.G. and Reitzmann, A. (2007) Alternative reaction engineering concepts in partial oxidations on oxidic catalysts. *Catalysis Reviews-Science And Engineering*, **49**, 1–32.

[27] Van Diepen, A.E., Van de Riet, A.C.J.M., and Moulijn, J.A. (1996) Catalysis in fine chemicals production. *Revista Portuguesa de Quimica*, **3**, 23–33.

[28] Shaw, D.J. and Andersen, B.S. (2011) Cleaner Cars, Cleaner Fuel, Cleaner Air. National Association of Clean Air Agencies (NACAA), www.4cleanair.org (last accessed 19 December 2012).

[29] Van Diepen, A.E., Makkee, M., and Moulijn, J.A. (1999) Emission control from mobile sources: otto and diesel engines, in *Environmental Catalysis* (eds F.J.J.G. Janssen and R.A. Van Santen). Imperial College, London, pp. 257–291.

[30] Lepperhoff, G., Pischinger, F., and Koberstein, E. (1985) In *Ullmann's Encyclopedia of Industrial Chemistry*, 5th edn, vol. **A3** (ed. W. Gerhartz). VCH, Weinheim, pp. 189–200.

[31] Setiabudi, A. (2004) Oxidation of Diesel Soot: Towards an Optimal Catalysed Diesel Particulate Filter. PhD Thesis, Delft University of Technology, The Netherlands, p. 16.

[32] (2012) www.dieselnet.com (last accessed 19 December 2012).

General Literature

Chauvel, A. and Lefebvre, G. (1989) Petrochemical Processes 1. Synthesis Gas Derivatives and Major Hydrocarbons. Technip, Paris.

Chauvel, A. and Lefebvre, G. (1989) Petrochemical Processes 2. Major Oxygenated, Chlorinated and Nitrated derivatives. Technip, Paris.

Ertl, G., Knözinger, H., and Weitkamp, J. (eds) (1997) *Handbook of Heterogeneous Catalysis*, vol. **1–5**. VCH, Weinheim.

Henkel, K.D. (1992) 'Reactor types and their industrial applications', in *Ullmann's Encyclopedia of Industrial Chemistry*, 5th edn, vol. **B4** (ed. W. Gerhartz *et al.*). VCH, Weinheim, pp. 87–120.

Thoenes, D. (1994) *Chemical Reactor Development; from Laboratory Synthesis to Industrial Production.* Kluwer, Dordrecht, The Netherlands.

Salmi, T.O., Mikkola, J.-P., and Wärnå, J.P. (2011) Chemical Reaction Engineering and Reactor Design, CRC Press, Taylor & Francis Group.

Van Santen, R.A., Van Leeuwen, P.W.N.M., Moulijn, J.A., and Averill, B.A. (eds) (1999) *Catalysis: An Integrated Approach to Homogeneous, Heterogeneous and Industrial Catalysis.* 2nd edn, Elsevier, Amsterdam, The Netherlands.

11

Production of Polymers — Polyethene

11.1 Introduction

Natural polymeric materials, such as wood and horn, have been used since prehistoric times. During the last century modified forms of natural polymers have been produced (e.g., the vulcanization of rubber, modified forms of cellulose such as cellulose nitrate and, later, cellulose acetate). The first fully synthetic polymers were the phenol-formaldehyde resins, which were developed in the beginning of the twentieth century. Since then, the synthetic polymers industry has seen a spectacular growth, as shown in Figure 11.1 for the production of plastics (or thermoplastics), the largest class of polymers by far (approximately 85% of synthetic polymers).

In 2010, approximately 265 million metric tons of plastics were produced worldwide. Figure 11.2 indicates the applications of the various types of plastics and their demand in Europe. World figures show a similar trend.

Plastics become flexible solids above a particular temperature, and become rigid again upon cooling below this temperature. This flexible/rigid cycle can be repeated on reheating and cooling again. Plastics, when flexible, can be molded into shapes that are preserved in the rigid state after cooling. Thermoplastic fibers, such as PET and the polyamides Nylon and Kevlar, can be drawn (pulled out) into strands with considerable tensile strength and durability. The non-fibers (PE, PP, PVC, PS, PUR) cannot undergo such processing.

Two other types of synthetic polymers are thermoset resins, such as phenol-formaldehyde, and synthetic rubbers (e.g., styrene-butadiene rubber, SBR). Thermoset resins (network copolymers) do not become flexible at all until the temperature becomes so high that thermal decomposition occurs. Synthetic rubbers, or elastomers, can be deformed quite severely by a small stress but regain their original shape on removal of the stress.

11.2 Polymerization Reactions

Polymerization can proceed according to two different mechanisms, referred to as step growth and chain growth polymerization.

Chemical Process Technology, Second Edition. Jacob A. Moulijn, Michiel Makkee, and Annelies E. van Diepen.
© 2013 John Wiley & Sons, Ltd. Published 2013 by John Wiley & Sons, Ltd.

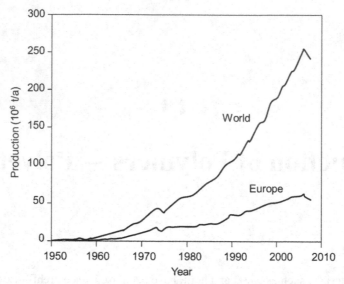

Figure 11.1   *World and European production of plastics [1].*

11.2.1   Step growth Polymerization

Step growth polymerization, generally producing polyesters and polyethers, involves the reaction between the functional groups (HO–, HOOC–, etc.) of any two molecules (either monomers or polymers). By repeated reaction long chains are gradually produced. Most commonly, the reactions are condensation reactions resulting in the elimination of a small molecule like water or methanol. An example is the production of polyethene terephthalate (PET) by reaction of ethene glycol with either terephthalic acid or dimethyl terephthalate (Scheme 11.1)

In both cases, the two different molecules involved in the polymerization reactions each contain two identical functional groups (that can be indicated as AA and BB). Scheme 11.2 shows typical reactions for this type of polymerization.

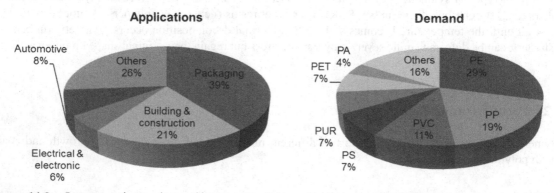

Figure 11.2   *European plastics demand by segment (left) and by polymer type (right) in 2011 (total 47 million t) [2]. PE = polyethene, PP = polypropene, PVC = poly(vinyl chloride), PS = styrene, PUR = polyurethane, PET = polyethene terephthalate, PA = polyamides.*

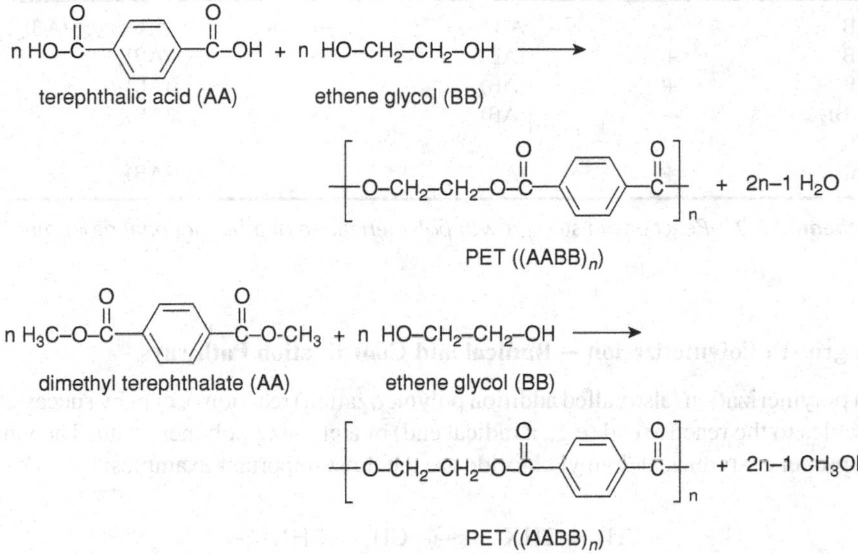

Scheme 11.1   *Reactions for the production of poly(ethene terephthalate) (PET).*

An example of a different type of condensation polymerization is the production of the polyester nylon 6 from caprolactam:

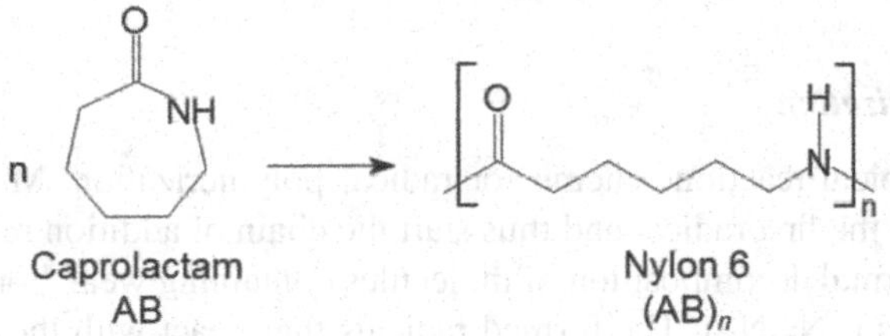

$$(11.1)$$

In this case, two different functional groups are present on the same molecule (indicated as AB). Scheme 11.3 shows the reactions taking place in this type of polymerization.

QUESTION:

According to Scheme 11.3, the product mixture is expected to be of a complex composition in many cases. Illustrate this for the production of PET.

AA	+	BB	→	AABB
AABB	+	AA	→	AABBAA
AABBAA	+	BB	→	AABBAABB
AABB	+	AABB	→	AABBAABB
etc.				

Scheme 11.2   *Reactions in step growth polymerization of two monomers with different functional groups.*

AB	+	AB	→	ABAB (or (AB)$_2$)
AB	+	(AB)$_2$	→	(AB)$_3$
AB	+	(AB)$_3$	→	(AB)$_4$
(AB)$_2$	+	(AB)$_2$	→	(AB)$_4$
...				
(AB)$_r$	+	(AB)$_s$	→	(AB)$_{r+s}$

Scheme 11.3   *Reactions in step growth polymerization of a bifunctional monomer.*

11.2.2   Chain growth Polymerization – Radical and Coordination Pathways

In chain growth polymerization (also called addition polymerization) reaction occurs by successive addition of monomer molecules to the reactive end (e.g., a radical end) of a growing polymer chain. The vinyl monomers such as ethene, propene, styrene, and vinyl chloride are the most important examples:

$$n\,CH_2 = CHX \rightarrow -(-CH_2 - CHX)_n - \qquad (11.2)$$

in which $X = H$, CH_3, C_5H_6, or Cl, respectively.

An initiator or a catalyst is usually required to start the chain growth reaction.

Chain growth polymerization can be classified (in order of commercial importance) as radical, coordination, anionic, and cationic polymerization, depending on the type of initiation. The next two sections briefly discuss radical and coordination polymerization, respectively.

11.2.2.1   *Radical Polymerization*

Scheme 11.4 outlines a typical reaction scheme for radical polymerization. Most radical polymerizations need an initiator to produce the first radical and thus start the chain of addition reactions. The most common initiation reaction is the thermal decomposition of molecules containing weak bonds, for example, peroxides (–O–O–) or azo compounds (–N=N–). The formed radicals then react with the monomers. Once initiated, a chain will grow by repeated additions of monomer molecules with simultaneous creation of a new radical

Initiation:	R-R	→	2 R$^\bullet$	$k_i \approx 10^{-4}$–10^{-6}/s (300–350 K)
	R$^\bullet$ + M	→	RM$_1$$^\bullet$	
Propagation:	RM$_1$$^\bullet$ + M	→	RM$_2$$^\bullet$	$k_p \approx 10^2$–10^4 m^3/kmol/s (300–350 K)
	RM$_2$$^\bullet$ + M	→	RM$_3$$^\bullet$	
	Etc.			
	RM$_{n-1}$$^\bullet$ + M	→	RM$_n$$^\bullet$	
Termination	RM$_m$$^\bullet$ + RM$_n$$^\bullet$	→	RM$_m$$^=$ + RM$_n$	disproportionation
	RM$_m$$^\bullet$ + RM$_n$$^\bullet$	→	RM$_m$-M$_n$R	recombination
Chain transfer:	RM$_n$$^\bullet$ + S	→	RM$_n$ + S$^\bullet$	to solvent
	RM$_n$$^\bullet$ + M	→	RM$_n$ + M$^\bullet$	to monomer

Scheme 11.4   *Steps in radical polymerization.*

M$_1$, M$_n$, number of monomers in chain.

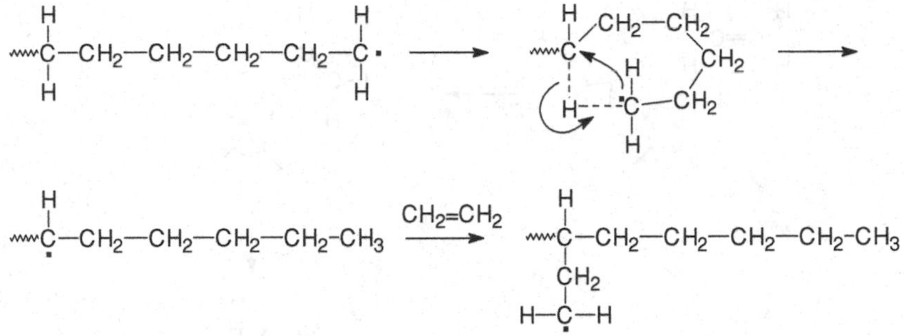

Figure 11.3 *Backbiting in the synthesis of LDPE by radical polymerization.*

site. This propagation is very fast, so very long polymer chains will form already in the earliest stage of the reaction.

Termination can occur by disproportionation or recombination. In the case of disproportionation, the final polymers have, on average, the same length as the growing chains. Termination occurs by transfer of a hydrogen atom from one of the radicals to the other, leading to unsaturation at one chain end. Recombination results in polymers with on average double the length of the growing chains.

Chain transfer is common in many radical polymerization processes. It involves the transfer of the radical end of a growing polymer chain to another species, for instance a monomer or a solvent molecule. Chain transfer reduces the average polymer size and molecular weight. Thus, by adding a chain transfer agent, a modifier, the average degree of polymerization can be controlled [3]. When polymerization is carried out in solution the solvent can act as chain transfer agent and, as a consequence, when the solvent is not inert the average chain length will be small.

QUESTION:

> *In the early production of rubber from styrene and butadiene (styrene-butadiene rubber, SBR), the product was a material with a very high molecular weight that required heat treatment to make it processable. Researchers at Standard Oil Development Company and the US Rubber company discovered that addition of a mercaptan modifier during polymerization not only produced a lower molecular weight and more flexible rubber, but also increased the rate of polymerization. Use of a mercaptan modifier became standard in the production process [4]. Explain the beneficial influence of the addition of mercaptan.*

In the production of low-density polyethene (LDPE)[1], side chains are generated also by internal chain transfer, in which the end of the chain abstracts a hydrogen atom from an internal $-CH_2-$ group, a process termed "backbiting". Figure 11.3 illustrates this process. Backbiting has a significant influence on the structure and, hence, on the properties of the final polymer.

Chain growth polymerization generally is fast, irreversible, and moderately-to-highly exothermic. In contrast, step growth polymerization is usually slow, equilibrium limited, and thermoneutral to slightly exothermic.

---

[1] The IUPAC naming terminology has been used but some traditional names for polymers are retained in common usage, for example, polyethylene (polyethene) and polypropylene (polypropene).

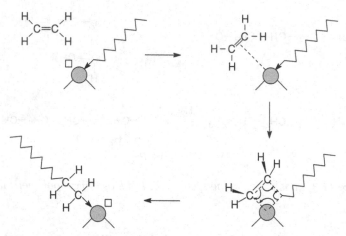

Figure 11.4   *General mechanism of propagation in coordination polymerization of ethene (⬤ represents a metal atom and □ indicates a vacancy for two electrons).*

QUESTIONS:

What are the reaction equations describing the radical polymerization of ethene?
Explain the differences mentioned between chain and step growth polymerization.

11.2.2.2   *Coordination Polymerization*

In coordination polymerization, transition metal catalysts are usually involved. The most important catalysts are the so-called Ziegler–Natta catalysts and "Phillips" catalysts, both discovered to be effective for alkene polymerizations in the 1950s. Figure 11.4 indicates the main features of chain propagation in the coordination polymerization of ethene. A growing polymer chain is coordinatively bound to a metal atom that has another coordinative vacancy (partially empty d-orbitals). A new ethene molecule is inserted by the creation of bonds between one of its carbon atoms and the metal and between the other carbon atom and the innermost carbon atom of the existing chain. Branching does not occur through this mechanism. High-density polyethene (HDPE) is produced by this type of polymerization.

QUESTION:

Why does coordination polymerization result in high-density polyethene? The pressure applied in coordination polymerization is low compared to that in radical polymerization. Explain.

When propene is used as polymerization feedstock, polymers with very different properties can be produced, depending on the microstructure of the polymer chain (Figure 11.5). In isotactic polypropene all methyl groups point into the same direction when the backbone (the carbon atoms in the chain) is stretched. In contrast, in syndiotactic polypropene the methyl groups alternate along the chain, while the atactic polymer (not shown) lacks regularity. The microstructure of the propene polymer determines the material properties to a large degree. Isotactic polypropene is used for the production of car bumpers, while atactic polypropene is a rubber-like material used in chewing gum and carpets.

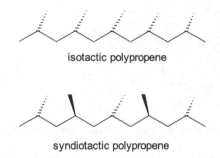

isotactic polypropene

syndiotactic polypropene

Figure 11.5   *Microstructures of polypropene.*

On Ziegler–Natta catalysts, the polymer chains would grow very long. Therefore, a chain-transfer agent is added to the system. Most commonly H_2 is used, which donates a hydrogen atom to terminate and detach the chain from the metal atom [3].

It is not surprising that polymerization catalysis draws a lot of attention and many new systems have been discovered. Even plastic materials that only melt at temperatures of over 770 K (!) can be produced.

11.3   Polyethenes – Background Information

11.3.1   Catalyst Development

The chance discovery (in 1935 by ICI) that ethene can be polymerized at very high pressure (over 2000 bar) to produce a high molecular weight semicrystalline material, was the starting point for the manufacture of polyethene. After the discovery of polyethene materials with higher density, it was called low-density polyethene (LDPE). LDPE was first produced commercially in the United Kingdom in 1938.

In the 1950s, Ziegler discovered titanium- and vanadium-based catalysts capable of polymerizing ethene at lower pressures than those used in the production of LDPE (only a little later Natta used these catalysts for the production of polypropene). The polymer formed by this new process had a higher density (a property that could easily be measured) than LDPE, and thus was called high-density polyethene (HDPE). At about the same time, researchers at Phillips Petroleum Co. discovered that catalysts of chromium oxide suppported on silica were also capable of producing HDPE at moderate temperatures and pressures. Phillips catalysts are less active than Ziegler–Natta catalysts, and therefore are limited to the production of polyethenes. Commercial production of HDPE started in the late 1950s, using both types of catalyst systems. HDPE produced with these catalysts is linear without branching.

Since the introduction of the original Ziegler–Natta and Phillips catalysts a lot of development work has been performed (also see Box 11.1). These developments have made possible the control of physical properties and molecular weight. The addition of 1-alkenes during catalytic polymerization of ethene results in polymers with a density similar to LDPE or lower. When higher 1-alkenes are added, the resultant polymer chain will have short branches. These are all of the same length, since they are simply the rest of the 1-alkene molecule. Figure 11.6 shows the incorporation of 1-butene in a growing polyethene chain.

A small degree of branching of these polymers can be deliberately introduced and controlled by the type of comonomer added. Commercial production of such polymers started in 1968 by Phillips Petroleum Co. The most important of these catalytically produced ethene copolymers is linear low-density polyethene (LLDPE).

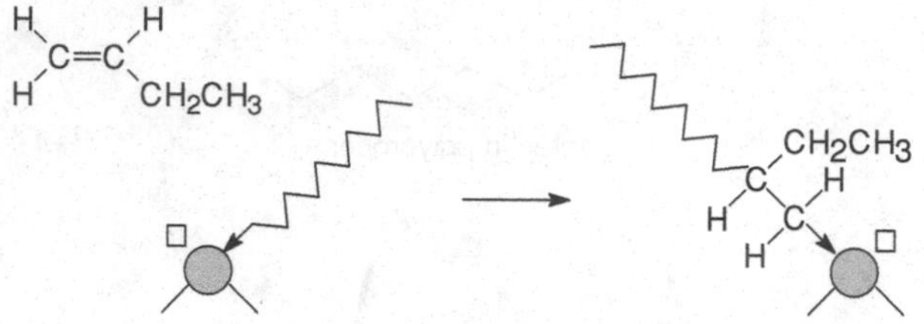

Figure 11.6   *Incorporation of 1-butene in polyethene to produce a short branch.*

Box 11.1   Kaminsky Catalysts

A new family of catalyst was discovered in the 1970s. These catalysts (Kaminsky catalysts) contain a metallocene complex, usually based on zirconium and methylaluminoxane (Figure B.11.1.1) [5]. Metallocene catalysts enabled the synthesis of ethene/1-alkene copolymers with highly uniform branching. Chain growth takes place at Zr.

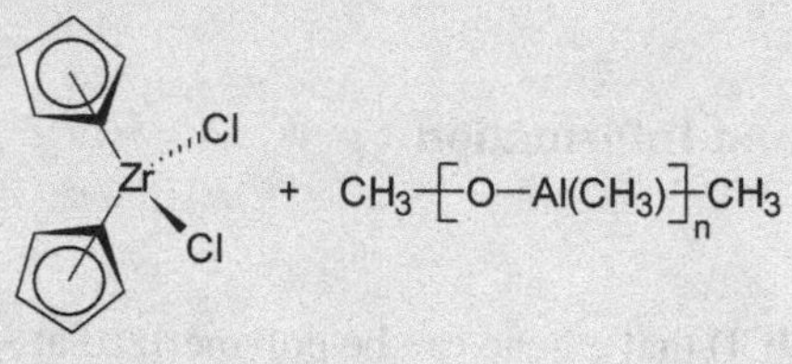

Figure B.11.1.1   *Metallocene catalyst (Kaminsky type).*

11.3.2   Classification and Properties

In the previous section, the three most important polyethene classes, namely LDPE, HDPE, and LLDPE, have already been encountered. Table 11.1 gives a commercial classification of the various polyethenes.

The polymer density (and crystallinity) depends on the degree of branching as well as on the molecular weight. Linear unbranched polymers have a more densely packed molecular structure, and hence contain larger portions of crystalline (ordered) material than branched polymers. This is due to the interference of

Table 11.1   *Commercial classification and properties of polyethenes [5].*

Acronym	Designation	Density (kg/m^3)
HDPE	High-density polyethene	>940
UHMWPE[a]	Ultrahigh-molecular-weight polyethene	930–935
MDPE	Medium-density polyethene	926–940
LLDPE	Linear low-density polyethene	915–925
LDPE	Low-density polyethene	910–940
VLDPE	Very-low-density polyethene	880–915

[a]Linear polymer with molecular weight of over 3×10^6 g/mol.

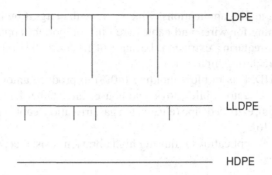

Figure 11.7   *Simple representation of the backbone structures of LDPE, HDPE and LLDPE.*

the side branches with the alignment of the polymer chains. Figure 11.7 summarizes the chain structure of the main polyethenes. HDPE has no branches, LLDPE has short and uniform branches while LDPE has branches of more random length. The differences are a result of the different operating conditions, catalysts, and comonomers employed in the production of these polymers.

In the case of LDPE production, the polymer density is controlled by the operating temperature. A higher temperature results in more side reactions causing branching and hence a lower density.

In the low-pressure catalytic processes to produce HDPE and LLDPE (Section 11.4.3), the density is controlled by the type and amount of comonomer added to the polymerization reactor. Common comonomers are 1-alkenes, ranging from 1-butene to 1-octene. The amount of comonomer present varies from 0 mol% in the highest density HDPEs to about 3 mol% in LLDPE. VLDPE contains over 4 mol% comonomers of different types. Due to the presence of side chains of variable length, this polymer has an even lower density than LDPE.

With increasing density, certain polymer properties increase, for example, yield strength, stiffness, and impermeability to gases and liquids, while others decrease, for example, transparency and low-temperature brittleness resistance. Therefore, the density correlates with the possible applications of a specific polymer to a large extent.

QUESTION:

> *The polymerization of ethene is an example of chain growth polymerization. Earlier the reaction pathways have been classified as radical, coordination, anionic, and cationic polymerization, depending on the type of initiation. For the various classes of PE specify to which category they belong.*

11.3.3   Applications

In 2007, the world production of polyethenes was approximately 80 million metric tons, divided mainly among LDPE, LLDPE and MDPE, and HDPE in roughly equal shares.

LDPE has a wide variety of applications of which those requiring processing into thin film take up the largest segment (60–70%). Polyethene film applications include food and nonfood packaging, carry-out bags, and trash can liners. Other applications of LDPE include extrusion coating (e.g., for packaging of milk), injection molding for the production of toys, and sheathing for wires and cables [5].

LLDPE is replacing LDPE in film applications, which is now the largest outlet for LLDPE (about 70%). Injection molding and sheathing for wires and cables are other important applications [6]. LLDPE is less likely to replace applications requiring extrusion, because of its lower flexibility and lubricity. Neither is LLDPE used in applications requiring high clarity.

The major application of HDPE is in blow molding (40%) to produce among other things milk bottles, containers, drums, fuel tanks for automobiles, toys, and housewares. Other large outlets are films and sheet (e.g., for wrapping, carrier bags), extruded pipe (e.g., water, gas, irrigation, cable insulation), injection molding (e.g., crates, containers, toys) [6].

UHMWPE is used in special applications requiring high abrasion resistance, such as bearings, sprockets, and gaskets.

11.4   Processes for the Production of Polyethenes

11.4.1   Monomer Production and Purification

Most ethene for the production of polyethene is produced by cracking of ethane or naphtha in the presence of steam (Chapter 4). Besides ethene, the cracked gas also contains ethane and other compounds in smaller quantities (e.g., methane, hydrogen, ethyne). Table 11.2 gives typical specifications for polymer-grade ethene. Note that the purity is specified on a ppm level.

Firstly, sour gases (CO_2, H_2S) and water are removed by absorption and drying, respectively. Then ethene is further separated and purified by cryogenic distillation. Ethyne is one of the most undesired compounds, and its presence must be limited to below 1 ppm [7]. This is achieved by selective hydrogenation of the triple bond.

QUESTION:

> *Discuss how the specifications for the various components listed in Table 11.3 can be achieved.*

Table 11.2   *Specifications for polymer-grade ethene.*

Component	Value
Ethene (mol% minimum)	99.95
Methane + ethane	Balance
Other impurities(mol ppm max)	
Ethyne	1.0
Oxygen	1.0
Carbon monoxide	1.0
Carbon dioxide	1.0
Water	2.0
Sulfur	2.0
Hydrogen	5.0
Propene	10.0
C4 +	10.0

Table 11.3 *Typical reaction enthalpies of chain growth polymerization [8].*

Polymer	$\Delta_r H_{298}$ (kJ/mol monomer)	$\Delta T_{adiabatic}$ (K)[a]
Polyethene	−95.0	1610
PVC	−95.8	730
Polystyrene	−69.9	320
Polymethyl methacrylate	−56.5	270

[a]Complete monomer conversion and heat capacity of 2 kJ/kg K assumed (and no dilution).

11.4.2 Polymerization – Exothermicity

The main difference between polymerization processes and processes for the manufacture of smaller molecules lies in the fact that polymer systems are highly viscous. In chain growth polymerization an extra challenge is the often high exothermicity of the reaction (Table 11.3).

As can be seen from Table 11.3, the reaction enthalpy for polymerization of ethene has one of the largest negative values, with an accompanying excessive adiabatic temperature rise of circa 1600 K!! (ΔT_{ad} is the temperature rise if no cooling is applied). In addition, at temperatures above about 570 K ethene decomposition into carbon, hydrogen and methane poses a serious threat. This reaction is highly exothermic ($\Delta_r H_{298} = -120$ kJ/mol) and once started is difficult to control, leading to a thermal runaway. Hence, very special attention has to be paid to temperature control in the polymerization of ethene, both to ensure polymer quality and safe operation of the process.

QUESTION:

According to a rule of thumb hydrogenation reactions are exothermic. Does this agree with the exothermicity of ethene decomposition?

11.4.3 Production of Polyethenes

The three classes of polyethene, LDPE, HDPE and LLDPE, are produced according to quite different processes. Table 11.4 shows important characteristics.

11.4.3.1 Production of LDPE

The production of LDPE is very interesting because of the combination of a very high pressure and a high propensity to runaway of the reacting system. In a sense, nowadays it is an eye opener that such a challenging process has been successfully developed. Because of its high learning curve value and the fact that this technology will be around for a long time the process is treated here in some detail. Moreover, safety is

Table 11.4 *Polyethene production processes [9].*

Reactant	Polymer	Process	Phase	Catalyst	T (K)	p (bar)
Ethene	LDPE	Solution or bulk, PFR or CSTR	Liquid	None, H_2O_2 as initiator	470–570	1500–3500
Ethene	HDPE	Suspension, fluidized bed	Liquid	$TiCl_4$ or Cr- oxide	60–80	20–35
Ethene and 1-butene	LLDPE	Solution, fluidized bed	Vapor	$TiCl_4$ (or Cr oxide)	100	7–20

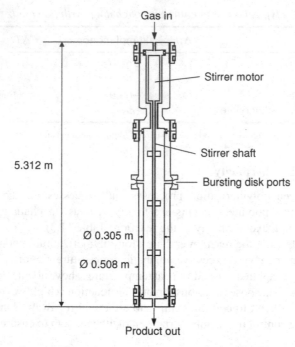

Figure 11.8   *General arrangement of 0.25 m³ autoclave [10, 11].*

crucial for the survival of a chemical company and LDPE production plants are very instructive in this respect.

The formation of LDPE proceeds through a radical mechanism (Scheme 11.4) without a catalyst (though with initiators). Therefore, a relatively high temperature and very high pressure are required. At the high pressures used (above the critical pressure of ethene), ethene behaves like a liquid and is a solvent for the polymer. LDPE is produced in either a stirred autoclave (CSTR) or a tubular reactor (PFR).

11.4.3.1.1 Reactors and Process

The stirred autoclave (Figure 11.8) consists of a cylindrical vessel with a length/diameter ratio of 4:1 to 18:1. Reactors for larger plants have volumes of about 1 m³ and a residence time of 30–60 s. The reactor wall thickness, typically 0.1 m, which is required to contain the high pressure of up to 2100 bar, severely limits the removal of heat.

In fact, the reactor can be considered to operate adiabatically; the fresh cold ethene entering the reactor is the only way to remove the heat released by the reaction. Accordingly, the conversion must be limited (to about 20%) in order to limit the temperature increase. Another reason to limit the conversion is the increased viscosity with higher polymer content, which makes stirring difficult and can lead to local hot spots. Most reactors have two or more zones with increasing temperatures, the first zone typically operating at 490 K and the final at 570 K. Initiators can be injected at several points.

Bursting disks are mounted directly into the reactor walls to provide unrestricted passage of the reactor contents in the event of a pressure rise due to decomposition of ethene [10].

The tubular reactor is essentially a long double-pipe heat exchanger. Industrial reactors generally have a length of between 200 and 1000 m, with inside diameters of 2.5–7 cm. This reactors is built by joining

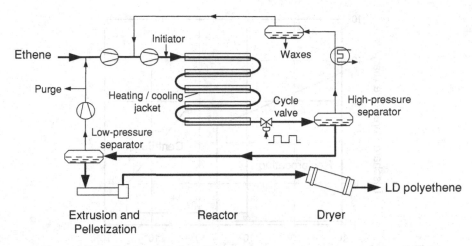

Figure 11.9   *Production of LDPE in a tubular reactor (shown schematically).*

straight lengths of 10–20 m in series, connected by U-bends (Figure 11.9). The tubes are surrounded with jackets in which a heat transfer fluid circulates. In the first part of the reactor ethene is heated to a temperature between 370 and 470 K. The heat of polymerization raises the temperature to 520–570 K. In the final part of the reactor the temperature of the reaction mixture decreases. Optionally, initiator (oxygen or a peroxide) can be added at several points along the reactor. This increases the conversion but also increases the reactor length. Typically, the ethene conversion ranges from 15 to 20% with single initiator injection, and can be up to 35% with multiple injection.

To achieve the high velocity required for effective heat transfer to the cooling jacket, the reactor pressure is controlled by a cycle valve that opens periodically to reduce the pressure from about 3000 to 2000 bar. An additional advantage of this type of pressure control is that the periodically increased flow velocity through the reactor reduces contamination of the walls with polymer. Pressure cycling also has disadvantages, because the tubes have to endure both large pressure changes and temperature gradients through the walls. Especially fatigue plays an important role.

QUESTION:
> *What problems could result from contamination of the reactor walls with polymer material?*

Due to the low ethene conversion, a recycle system is required; this is similar for both the autoclave and the tubular reactor. The high-pressure separator separates most of the polymer from the unreacted monomer. The separator overhead stream, containing ethene and some low molecular weight polymer (waxes) is cooled, after which the waxes are removed. The resulting ethene stream is recycled to the reactor.

The polymer stream leaving the high-pressure separator is fed to a low-pressure separator, where most of the remaining ethene monomer is removed. This ethene is also recycled to the reactor. Some ethene is purged at this point to prevent accumulation of feed impurities. The molten polymer from the low-pressure separator is fed to an extruder, which forces the product through a die with multiple holes. Polymer strands are cut under water into pellets of about 3 mm diameter and then dried in a centrifugal drier.

QUESTIONS:
> *What kind of impurities are present in the ethene feed?*
> *Which are the most important contributors to the polyethene production costs?*

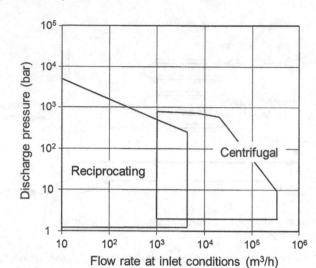

Figure 11.10   *Operating ranges of reciprocating and centrifugal compressors.*

11.4.3.1.2   Ethene Compressors – High-Pressure Technology

High pressure was first applied at the end of the nineteenth century for the liquefaction of the so-called permanent gases such as oxygen and nitrogen. The use of high pressures in the chemical industry started at about the same time; it was triggered by the discovery of the dyestuff mauveine (Chapter 2) and the development of the dye industry. Some dye intermediates were produced in batch autoclaves at pressures of 50–80 bar. At the beginning of the twentieth century, ammonia and methanol (Chapter 6) were two of the first chemicals produced in continuous high-pressure plants. Nowadays, ammonia synthesis takes place at pressures ranging from 150 to 350 bar, although an early process operated at 1000 bar.

Ammonia and methanol plants currently use centrifugal compressors. These have several advantages over the earlier used reciprocating compressors; for example, low investment and maintenance costs, high reliability, low space requirement, and high efficiency at high flow rates. However, for the production of LDPE the maximum attainable discharge pressure of centrifugal compressors (about 1000 bar at the lowest flow rates) is not nearly enough (Table 11.4 and Figure 11.10).

With reciprocating compressors higher pressures can be achieved than with centrifugal compressors. However, the early reciprocating compressors had throughputs that were too low for a high-pressure polyethene process to be economical. Improved techniques in high-pressure engineering enabled the increase in throughput of reciprocating compressors from about 5 t/h at 1250 bar in the early 1950s to 120 t/h at 3500 bar nowadays.

In LDPE plants a primary, centrifugal, compressor is used to raise the pressure of ethene to typically 300 bar and a secondary, reciprocating, compressor, often referred to as a hypercompressor, is used to increase the pressure to 1500–3000 bar. The development of reciprocating compressors for these high operating pressures and high throughputs was not an easy task. Several problems had to be solved [12]:

- ethene leakage past the plunger;
- breakage of the plunger due to misalignment;
- fatigue failure of high-pressure components;

- formation of low molecular weight polymer in the compressor, leading to obstruction of the gas ducts inside the compressor cylinder;
- decomposition of ethene into carbon, hydrogen, and methane due to overheating, for instance in dead zones where the gas is stagnant.

Ethene leakage not only means a loss of ethene but also represents an explosion danger. Plunger misalignment is the most common cause of failure of the compressors. Plunger breakage may lead to serious fires due to a large escape of ethene. Fatigue failures in frame or cylinder parts of the compressor have led to serious damage [12]. In cases where fragments of material fell inside the cylinder, thus blocking the delivery duct, excessive overpressure resulted, leading to explosion of the compressor. Obstruction of gas ducts by low molecular weight polymer and decomposition of ethene can have a similar effect.

QUESTION:

Where does ethene compression in an LDPE plant fit in Figure 11.10?

11.4.3.1.3 Safety of High-Pressure Polymerization Reactors

High-pressure polyethylene reactors operate on the edge of stability, sometimes leading to sudden runaways. A runaway reaction is initiated by local hot spots developing in the reactor. In these hotter reactor parts the temperature becomes higher than the temperature in the surrounding fluid and decomposition of ethene to carbon, methane, and hydrogen occurs. Figure 11.11 shows some of the incidents leading to hot spots and decomposition reactions in an autoclave reactor.

A runaway reaction occurs less frequently in tubular processes than in autoclave processes due to the smaller reactor content and larger cooling surface area. However, fouling of the reactor walls by polymer deposits in a tubular reactor is an additional possible cause of runaways. Since the reactor operates at its stability limit, a small decrease in heat transfer, and hence a small rise in (local) temperature, could trigger decomposition reactions.

The decomposition reactions are highly exothermic. Especially in the adiabatic autoclave reactors, once initiated these decompositions cause a runaway in a matter of seconds [13]. As an example, Figure 11.12 shows the response of an autoclave reactor to a disturbance in the amount of ethyne in the feed (simulation results [13]). This is not an academic exercise; small amounts of ethyne are present in the ethene feed; ethyne can decompose into free radicals and induce runaway reactions. The figure explains why the specification with respect to the maximum ethyne content is severe (Table 11.2).

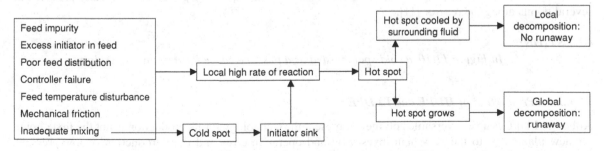

Figure 11.11 *Possible causes of runaway reactions (autoclave process).*

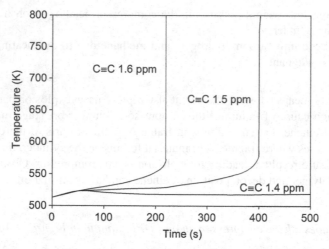

Figure 11.12   *Response of autoclave temperature to different ethyne levels in the feed [13].*

QUESTION:

> *Why is it to be expected that the ethene feed is contaminated with ethyne?*

Similar behavior resulted from a small disturbance in the reactor feed temperature and a disturbance in the initiator concentration (from 7.3 to 8.0 ppm) [13].

In both autoclave reactors and tubular reactors the danger of a runaway reaction is usually held in check by adjusting the initiator concentration in the ethene feed. Even very small temperature changes (as measured at various points along the reactor) should be counteracted immediately. Therefore, a high-temperature alarm for a polymerization reactor is set at only a few degrees above the normal operating temperature, whereas in other reactor types a margin of 10–20 K is common [14].

Decomposition reactions not only lead to a temperature excursion but also to a pressure peak. To protect the autoclave from excessive overpressure, without having to resort to a large vent area, it is fitted with pressure relief valves (or bursting disks) that are actuated at an overpressure as small as practical. The hot decomposition products released (at sonic velocity) would be subject to severe reaction with air (decarbonization, cracking) and, therefore, they are rapidly quenched before venting to the atmosphere [11].

In tubular reactors, the problem of excessive overpressure is not as serious as in autoclave reactors because these reactors contain less ethene due to the smaller volume, and hence a possible pressure rise is lower. The first measure is fitting of one or more so-called dump valves to the reactor. These dump valves are controlled externally. In addition, bursting disks, which are set to burst at a specified pressure, are usually provided at several points along the reactor [11].

QUESTION:

> *In Figure 11.10, a cold spot is listed as a possible cause for a runaway. Explain.*

11.4.3.2   *Production of HDPE and LLDPE*

Although LDPE is a very versatile polymer, at present the technology for its production is not the first choice for new plants due to the very high investment and operation costs and the introduction of low-pressure processes for the production of linear low-density polyethene (LLDPE), which can be used in many of the traditional LDPE outlets. In addition, HDPE can also be produced in low-pressure plants, making them more flexible than high-pressure plants.

Many of the common catalytic polymerization processes are suitable for the production of both HDPE and LLDPE. Such processes include bulk polymerization (polymer dissolves in monomer), solution polymerization (polymer and monomer dissolve in a hydrocarbon solvent), slurry polymerization (catalyst–polymer particles suspended in a hydrocarbon diluent), and fluidized bed polymerization (catalyst–polymer particles fluidized in gaseous monomer) [10].

LLDPE can also be produced in high-pressure processes such as described in Section 11.4.3.1, in the presence of a Ziegler-Natta-type catalyst and with the right comonomer. Although this route is not seen as suitable for a new investment, several manufacturers have modified existing LDPE plants for the production of LLDPE.

This section focuses on the fluidized bed polymerization process (also referred to as gas phase polymerization) because this is the most flexible process; it is capable of producing both HDPE and LLDPE, and can use a wide range of solid and supported catalysts. It also is the most economic process because it does not require a diluent or solvent, and hence no equipment or energy for thier recovery and recycle. Many fluidized bed polymerization plants have been built as dual-purpose plants ("swing plants") with the ability to produce HDPE or LLDPE according to demand [10].

11.4.3.2.1   Fluidized Bed Polymerization Process

Figure 11.13 shows a flow scheme of a fluidized bed process. A wide variety of heterogeneous catalysts can be used. Ethene, hydrogen, and comonomer (for the production of LLDPE, mainly 1-butene and 1-hexene) are fed to the reactor through a grid, which provides an even distribution of the gas. The recirculating ethene gas fluidizes the catalyst (in the shape of spherical particles of about 50 μm) on which the polymer grows and enables efficient mass and heat transfer. The characteristic shape of the reactor (about 25 m high, length-to-diameter ratio around 7 [15]), with a cylindrical reaction section and a larger diameter section where the

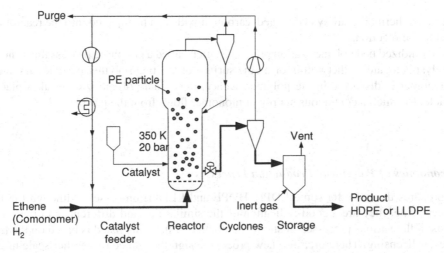

Figure 11.13   *Union Carbide Unipol process for the production of HDPE and LLDPE.*

gas velocity is reduced, allows entrained particles to fall back into the bed. Remaining polymer particles are separated from gaseous monomer in a cyclone and returned to the reactor.

QUESTIONS:

> *What is the purpose of adding hydrogen?*
> *Why is a purge stream required?*

The residence time of the PE-catalyst particles in the reactor ranges from 3 to 5 hours, during which they reach an average size of about 500 μm [15]. The powder is taken off at regular intervals near the bottom of the reactor through valves. The polymer contains an appreciable amount of ethene, which is mostly separated in a cyclone operating at near atmospheric pressure and recycled to the reactor. The remaining ethene is purged from the polymer with nitrogen and stored.

Because of the high activity of current polymerization catalysts, over 1500 kg of polymer is produced per gram of metal (Cr, Ti, etc., depending on the catalyst used). Therefore, the catalyst remains in the product and no residual catalyst removal unit is required. Although this impurity level is low it still limits the number of applications of the product and there is an incentive to develop systems with higher turnover. A catalyst has been developed that can yield 6000 kg/g metal (!).

Temperature control is very important to keep the polymer particles below their melting point (398 K and 373 K for HDPE and LLDPE, respectively [15]) and to prevent a thermal runaway reaction. When using cooling water from a cooling tower, the available temperature difference for cooling the gas should be at least 40–50 K. From this value it can be calculated that the single-pass conversion of ethene is limited to about 2% (somewhat higher if higher alkene comonomers are used, as in LLDPE production).

QUESTIONS:

> *What would be the consequences of melting of the polymer particles?*
> *How is the temperature regulated?*
> *Verify that the maximum ethene conversion is 2% as stated above assuming a constant heat capacity of ethene of 40 J/mol/K and a heat of reaction of 100 kJ/mol ethene. Also calculate the recirculation ratio (mol ethene reactor out/mol feed ethene in). Neglect the purge stream in both calculations.*

In the case that a thermal runaway is detected, carbon dioxide can be injected into the reactor to poison the catalyst and stop the reaction.

Although the fluidized bed polymerization process is often called a gas phase process this is not actually the case. The catalyst is located either within or on the surface of existing polymer particles and an appreciable amount of monomer is dissolved in the polymer. Hence, the actual polymerization takes place within the polymer particle, to which a continuous supply of monomer flows from the gas phase [10].

11.4.3.3   Economics of Polyethene Production Processes

Table 11.5 compares the costs for some LDPE, HDPE and LLDPE processes. Although the figures given are not of recent date, they are a good indication of the similarities and differences in the investment and operating costs of the various processes. It should be noted that the scale listed is based on the largest stream size available for licensing. Therefore, in a new process design the choice of another scale might influence the outcome of a process comparison.

Table 11.5 *Production costs for polyethene processes in $/t (1996 US Gulf Coast prices) [16].*

Product	LDPE		LLDEP		HDPE		
Process	Autoclave	Tubular	Fluidized bed	Solution	Fluidized bed	Ziegler	Phillips
Capacity (10^3 t/a)	117	200	225	200	200	200	200
Capital cost (10^6 $)	85	116	98	138	90	135	105
Monomer	447	443	450[a]	452[a]	449	456	445
Catalysts, chemicals	20	18	29	31	26	22	20
Electricity	31	33	15	9	15	16	16
Other utilities	5	2	5	17	5	10	11
Labor	10	6	6	6	6	6	6
Maintenance	15	13	9	15	9	15	13
Overheads	35	29	22	31	22	26	29
Production costs	**565**	**544**	**534**	**561**	**553**	**553**	**539**
Depreciation	71	59	44	69	45	68	53
Total costs	**636**	**603**	**578**	**630**	**577**	**620**	**592**

[a]Includes cost of butene monomer at a unit price equal to 1.05 times that of ethane; other locations or the use of other comonomers could lead to a higher monomer cost.

Clearly, the monomer costs (mainly ethene but also comonomers in the case of LLDPE) contribute most to the total cost (approximately 70%). The monomer and catalyst and chemicals costs per metric ton of polymer produced are similar for all processes.

Large differences are observed in capital costs. For instance, in general LDPE processes are not competitive with LLDPE processes due to the large investment costs involved in high-pressure equipment and the large energy requirements for compression. Nevertheless, the LDPE processes are not obsolete, because for some applications LDPE cannot be replaced by LLDPE yet. Another example of the importance of the capital cost factor is the enormous difference between the solution process and the fluidized bed process for LLDPE production. Even at much lower capacity the capital costs are much higher for the solution process. It should be noted, however, that the costs for the solution process are based on the DuPont process [10], which at that time included a catalyst residue removal stage. Therefore, the comparison might be unfair regarding the solution process. A similar observation can be made when comparing the two slurry processes for HDPE production. The data for the Ziegler process are based on an early process, which also included a facility for the removal of catalyst residue. Since the late 1960s it has been possible to eliminate this step [10], resulting in appreciably lower cost.

References

[1] Oceaneye (2012) Plastic. http://oceaneye.eu/en/le_plastique (last accessed 23 December 2012).
[2] PlasticsEurope (2012) Plastics – The Facts 2012, http://www.plasticseurope.org/Document/plastics-the-facts-2012.aspx?Page=DOCUMENT&FolID=2 (last accessed 23 December 2012).
[3] Campbell, I.M. (1994) *Introduction to Synthetic Polymers*. Oxford University Press, Oxford.
[4] Anon. (2011) Chain Transfer, http://en.wikipedia.org/wiki/Chain_transfer (last accessed 19 December 2012).
[5] Pebsworth, L.W. and Kissin, Y.V. (1996) Olefin polymers (polyethylene), in *Kirk-Othmer Encyclopedia of Chemical Technology*, 4th edn, vol. **17** (eds J.I. Kroschwitz and M. Howe-Grant). John Wiley & Sons, Inc., New York, pp. 702–784.
[6] Wells, G.M. (1991) *Handbook of Petrochemicals and Processes*. Gower Publishing Company Ltd, Hampshire.

[7] Sundaram, K.M., Shreehan, M.M., and Olszewski, E.F. (1994) Ethylene, in *Kirk-Othmer Encyclopedia of Chemical Technology*, 4th edn, vol. **9** (eds J.I. Kroschwitz and M. Howe-Grant). John Wiley & Sons, Inc., New York, pp. 877–915.

[8] Nauman, E. (1994) Polymerization reactor design, in *Polymer Reactor Engineering* (ed. C. McGreavy). Blackie, London, pp. 125–147.

[9] Chaudhuri, U.R. (2010) *Fundamentals of Petroleum and Petrochemcial Engineering*. CRC Press (Francis & Taylor Group), pp. 385.

[10] Whiteley, K.S., Heggs, T.G., Koch, H., *et al.* (1992) Polyolefins, in *Ullmann's Encyclopedia of Industrial Chemistry*, 5th edn, vol. **A21** (ed. W. Gerhartz). VCH, Weinheim, pp. 487–530.

[11] Bett, K.E. (1995) High pressure technology, in *Kirk-Othmer Encyclopedia of Chemical Technology*, 4th edn, vol. **13** (eds J.I. Kroschwitz and M. Howe-Grant). John Wiley & Sons, Inc., New York, pp. 169–236.

[12] Traversari, A. and Beni, P. (1974) Design of a safe secondary compressor. Symposium on Safety in High pressure Polyethylene Plants, Tulsa, OK, pp. 8–11.

[13] Zhang, S.X., Read, N.K., and Ray, H.W. (1996) Runaway phenomena in low-density polyethylene autoclave reactors. *AIChE Journal*, 42, 2911–2925.

[14] Prugh, R.W. (1988) Safety, in *Encyclopedia of Polymer Science and Engineering*, vol. **14** (ed. J.I. Kroschwitz). John Wiley & Sons, Inc., New York, pp. 805–827.

[15] Brockmeier, N.F. (1987). Gas-phase polymerization, in *Encyclopedia of Polymer Science and Engineering*, vol. **7** (ed. J.I. Kroschwitz). John Wiley & Sons, Inc., New York, pp. 480–488.

[16] Whiteley, K.S. (2000) Polyethylene, in *Ullmann's Encyclopedia of Industrial Chemistry*. Wiley-VCH Verlag GmbH. doi: 10.1002/14356007.a21_487.pub2

12

Production of Fine Chemicals

Traditionally, chemical engineers have been mainly involved in bulk chemicals production. Numerous novel processes have been introduced and the chemical engineers have evidently been successful. Furthermore, the contribution of the chemical engineering community in the field of environmental catalysis is noteworthy, with, as a prime example, the catalytic converter used in the cleaning of car exhaust gases. In the past, less attention has been devoted to fields where the production volumes are smaller and the chemical complexity is higher, for example, in the production of pharmaceuticals and agrochemicals. Fortunately, also in these fields the impact of chemical reaction engineering is becoming increasingly visible.

12.1 Introduction

In the chemical industry, usually a distinction is made between bulk chemicals (or commodities), fine chemicals, and specialties. Fine chemicals include advanced intermediates, bulk drugs, vitamins, and pesticides, flavor and fragrance chemicals, and active pharmaceutical ingredients (API) or additives in formulations (Figure 12.1). Just as bulk chemicals, fine chemicals are identified based on specifications (what they are). In contrast, specialties are identified based on performance (what they can do). Some examples of specialty chemicals are adhesives, diagnostics, disinfectants, pesticides, pharmaceuticals, dyestuffs, perfumes, and specialty polymers.

There is a thin line between fine chemicals and specialties. For example, ibuprofen is a fine chemical as long it is sold based on its specifications, but once it is formulated, tableted and marketed as, for instance, Nurofen or Advil, it becomes a specialty.

Fine chemicals differ from bulk chemicals in many respects, as shown in Table 12.1. The price and volume limits are slightly arbitrary; a precise distinction between bulk and fine chemicals cannot be made. Box 12.1 illustrates the position of fine chemicals in the chemical industry.

One of the most important features of fine chemicals production is the great variety of products, with new products continually emerging. Often, significant fluctuations in the demand exist. In combination with the small volumes (often <100 t/a), plants dedicated to the production of a certain chemical would be much too costly. Most fine chemicals are produced in multiproduct or multipurposes plants (MPPs), in which a number

Chemical Process Technology, Second Edition. Jacob A. Moulijn, Michiel Makkee, and Annelies E. van Diepen.
© 2013 John Wiley & Sons, Ltd. Published 2013 by John Wiley & Sons, Ltd.

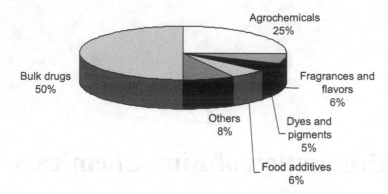

Figure 12.1 *Fine chemicals market ($ base) [1].*

of different chemicals are produced using the same pieces of equipment, usually in batch or semi-batch operation (Sections 12.4 and 12.5).

Specific for the production of fine chemicals and pharmaceuticals is the formation of unacceptably large amounts of (often hazardous) waste, as illustrated by Table 12.2. The *environmental* or *E factor* was introduced by Sheldon in the late 1980s [4, 5] and is defined as the mass ratio of waste to desired product, where the actual amount of waste produced in a process is defined as everything but the desired product. It includes

Table 12.1 *Fine versus bulk chemicals.*

Parameter	Fine chemicals	Bulk chemicals
Price ($/kg)	>5	<5
Production (t/a)	<10 000	>10 000
Product variety	high (>10 000)	low (~300)
Chemical complexity	high	low
Added value	high	low
Applications	limited number (often one)	many
Synthesis	multistep	few steps
	various routes	one or few routes
Catalysis	sometimes	often
Reactant and product phase	liquid and often solid	usually gas or liquid
Raw materials consumption (kg/kg)	high	low
Energy consumption (kJ/kg)	high	low
By-products/waste formation (kg/kg)	high	low
Toxic compounds	often (e.g., phosgene, HCN)	exception
Plants	often MPP[a]	dedicated
Operation	usually batch	usually continuous
Investment ($)	low	high
($/kg)	high	low
Labor	high	low
Market fluctuations	high	low
Producers	limited number	many

[a]MPP = multiproduct plant or multipurpose plant.

Box 12.1   Atorvastatin [2, 3]

Table B12.1.1 illustrates the big differences in size and prices in the bulk chemicals, fine chemicals, and specialties sectors. Atorvastatin is the active ingredient of Lipitor (trademark of Pfizer), a drug used for lowering blood cholesterol and the best-selling drug in the history of pharmaceuticals. The other compounds mentioned in Table B12.1.1 play a role in its synthesis, either as base chemicals or as intermediates.

Table B12.1.1   *Example of the position of fine chemicals in the chemical industry: Atorvastatin (molecular structures of (I), (II) and (III) are shown in Figure B.12.1.1).*

	Bulk chemicals		Fine chemicals			Specialties
			(I)		(III)	
Parameter	Methanol	Acetic acid	Hydroxynitrile	(II)	Atorvastatin	Lipitor
Molecular formula	CH_3OH	CH_3COOH	$C_6H_{10}NO_3$	$C_{14}H_{30}NO_4$	$C_{33}H_{34}FN_2O_5$	–
Applications	>100	>50	10	1	1	–
Price ($/kg)[a]	0.3	0.7	Not available	250	2500	50 000
Production (t/a)[a]	40×10^6	15×10^6	100	400	500	Not available
Producers[a]	100	25	15	5	1	1
Plant type[b]	C, D	C, D	B,M	B, M	B, M	F
Synthesis steps	1	2	7	15	20	–

[a]Indication.
[b]B = batch; C = continuous; D = dedicated; M = multipurpose; F = formulation.

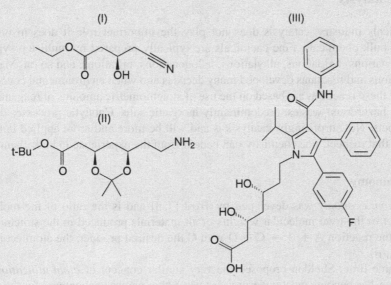

Figure B12.1.1   *Molecular structures of (I), (II), and (III).*

Table 12.2   *Production and waste formation in various sectors of industry [6].*

Industry segment	Production (metric ton/year)	Amount of waste produced (E factor) (kg waste/kg product)
Oil refining	10^6-10^8	circa 0.1
Bulk chemicals	10^4-10^6	<1–5
Fine chemicals	10^2-10^4	5–50
Pharmaceuticals	$10-10^3$	25–>100

reagents, solvent losses, all process auxiliaries, and, in principle, also fuel. It generally does not, however, include water.

Furthermore, in order to evaluate the environmental impact of alternative routes the nature of the waste must be taken into account. To this end, Sheldon [4] introduced the *environmental quotient* (EQ), which is E multiplied by an arbitrarily assigned unfriendliness quotient, Q. For example, sodium chloride could be assigned a Q value of 1 and a heavy metal salt, say, 100–1000, depending on its toxicity, ease of recycling, and so on. The magnitude of Q is obviously difficult to quantify but a quantitative assessment of the environmental impact of processes on this basis is, in principle, possible.

There are a number of options for reducing the amount of waste produced:

- Improved stoichiometry of the reaction(s). This can be achieved by choosing different, mostly catalytic, synthesis routes with a better "atom economy" (Section 12.2.1).
- Improved selectivity, either by choosing a different synthesis route or by using improved catalysts.
- Improved process design, for example, recycling of excess reactants and/or formed by-products.
- Reduced use of solvents, recycling of solvents, or use of more environmentally friendly solvents (water).

12.2   Role of Catalysis

In the fine chemicals industry, catalysis does not play the important role it does in oil refining and in the production of bulk chemicals. Fine chemicals are typically prepared by multistep synthesis reactions, involving hydrogenations, oxidations, alkylations, halogenations, nitrations, and so on. Many of these steps are based on reactions and reactants developed many decades ago when environmental concerns were absent. A large number of these reactions are based on the use of stoichiometric amounts of reagents producing large volumes of (often hazardous) wastes, predominantly inorganic salts. Catalytic processes do not produce so many waste products. Not surprisingly, catalysis is and will be more and more applied in the production of fine chemicals. In that respect, fine chemistry can benefit from the experience in bulk chemical processes.

12.2.1   Atom Economy

The concept of *atom economy* was developed by Trost [7, 8] and is the ratio of the molecular weight of the desired product to the total molecular weights of all materials produced in the stoichiometric equation. For example, for the reaction A + B → C + D with C the desired product, the atom economy is given by $M_{W,C}/(M_{W,C} + M_{W,D})$.

At about the same time, Sheldon proposed the very similar concept of *atom utilization* [9, 10], but the name *atom economy* has become most widespread in use. Other names encountered for the same concept are *atom efficiency* and *atom selectivity*. Figure 12.2 shows an example comparing two different reactions for the production of acetophenone.

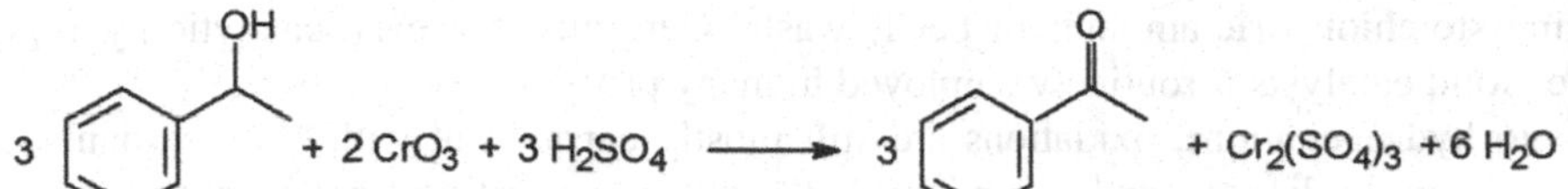

$$\%\ \text{atom economy} = \frac{3 \cdot 120}{3 \cdot 120 + 392 + 6 \cdot 18} \cdot 100\% = 42\%$$

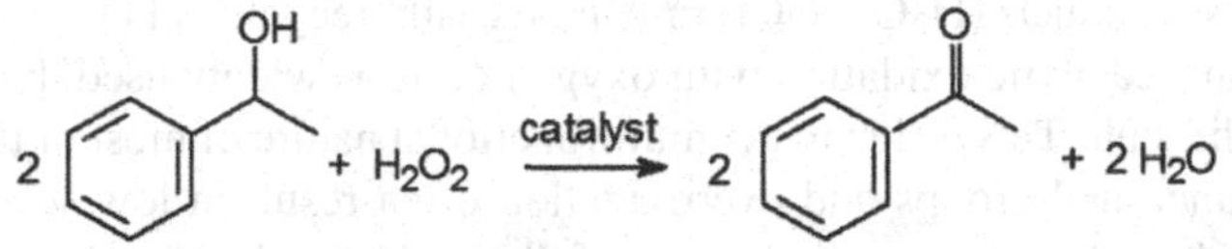

$$\%\ \text{atom economy} = \frac{2 \cdot 120}{2 \cdot 120 + 2 \cdot 18} \cdot 100\% = 87\%$$

Figure 12.2 *Atom economy in the oxidation of 1-phenylethanol to acetophenone (methyl phenyl ketone).*

Like the E factor, the atom economy is a useful measure of the environmental acceptability of a process. In contrast to the E factor, it is a theoretical number, that is, it assumes a chemical yield of 100%, exactly stoichiometric amounts of reactants and does not take into account substances not appearing in the stoichiometric equation, such as solvents used in the purification of the product. It does allow the minimum amount of waste expected to be predicted quickly.

In the production of bulk chemicals, examples of reactions with an atom economy of 100% are numerous. To name just a few, solid-catalyzed methanol production from hydrogen and carbon monoxide (Section 6.2) and the homogeneously catalyzed carbonylation of methanol to acetic acid (Section 9.2) are both processes with 100% atom economy.

In contrast, such processes (or even reactions) are rare in the production of fine chemicals and pharmaceuticals. Nevertheless, they do exist. For instance, with a few exceptions catalytic hydrogenations, which currently are routinely employed in many processes in the fine chemicals industry, generally have an atom economy of 100%.

The theoretical E factor for a process having an atom economy of 100% is zero, but the actual E factor is almost always higher, because the product yield is usually less than 100%. For reactions with an atom economy of less than 100%, the E factor is often further increased from its theoretical value as a result of the need to neutralize acids or bases that are formed as by-products. Note that in order to make a fair comparison of E factors and atom economies of reaction alternatives, the analysis should be based on the same starting materials.

QUESTIONS:

> *Give some more examples of reactions with an atom economy of 100%. Give examples of catalytic reactions with lower atom economy. What type of hydrogenation reactions exhibit <100% atom economy? What is the theoretical E factor for the first reaction of Figure 12.2? Also estimate the EQ.*

12.2.2 Alternative Reagents and Catalysts

A solution to the large amounts of waste generated in conventional processes for the production of fine chemicals is replacement of classical stoichiometric reagents with cleaner, catalytic alternatives. For example, many reduction reactions used to be performed with a mixture of a metal and an acid, such as iron and hydrochloric acid, producing stoichiometric amounts of FeCl$_2$ waste. Currently, however, catalytic hydrogenation with hydrogen over solid catalysts is routinely employed in many processes.

In contrast to hydrogenations, oxidations are still mostly carried out with toxic oxidants, such as permanganates, manganese dioxide, and chromium dichromates, generating large amounts of inorganic salts. Fortunately, in recent years, a number of "green" oxidants have proved to be efficient in catalytic oxidations. Some of the most attractive oxidants are oxygen (O$_2$), hydrogen peroxide (H$_2$O$_2$), sodium hypochlorite (NaOCl), or even organic peroxides, for example, *tert*-butyl-hydroperoxide ((CH$_3$)$_3$–C–O–OH). These oxidants are much more attractive from an environmental point of view, because they offer high atom economy and are either converted into harmless products (H$_2$O, NaCl) or can be easily recycled [11].

In the production of bulk chemicals, catalytic oxidation with oxygen or air is widely used, but application to fine chemicals is generally more difficult. This is due to the multifunctional nature of most of the molecules of interest; oxidation of "wrong" functional groups and overoxidation often result in low selectivity. Still, catalytic oxidation using oxygen/air has also been applied successfully in the production of fine chemicals. An illustrative example is the production of hydroquinone (Section 12.2.3).

QUESTIONS:

> *Why is the choice of oxidant in the production of bulk chemicals usually limited to O$_2$ or air, while it is wider in the production of fine chemicals?*
> *How can organic peroxides be recycled?*
> *Why are selective oxidations more difficult to perform reactions than selective hydrogenations?*

Another means of reducing the amounts of inorganic salts generated is to replace stoichiometric mineral acids (e.g., H$_2$SO$_4$ and HNO$_3$) and Lewis acids (e.g., AlCl$_3$ or ZnCl$_2$) with recyclable solid acids, such as zeolites, preferably in catalytic amounts. A prominent example is the Friedel–Crafts acylation of aromatic compounds, which is an important industrial process since it allows the formation of new C–C bonds onto

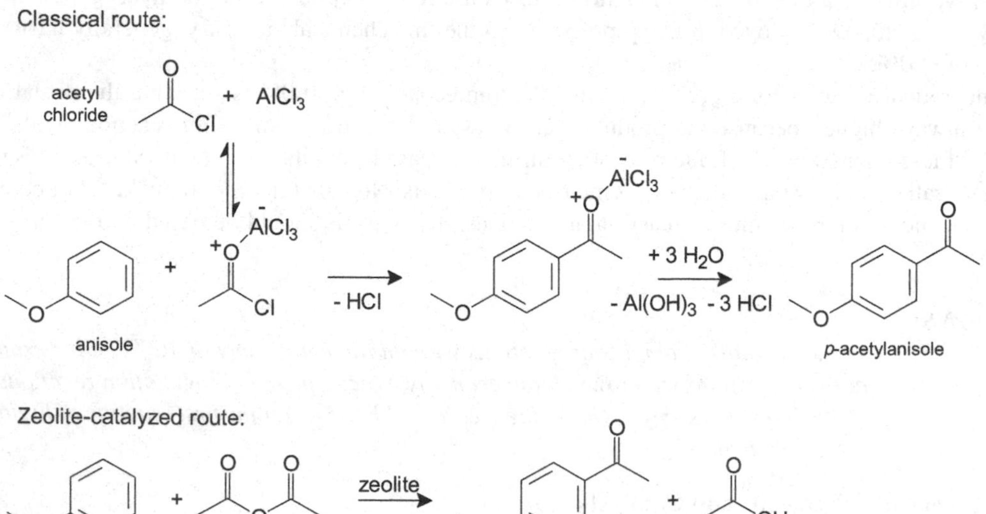

Classical route:

Zeolite-catalyzed route:

Scheme 12.1   *Mechanism of the classical Friedel–Crafts acylation of anisole to p-acetylanisole and overall reaction for the zeolite-catalyzed route.*

aromatic rings. The produced aromatic ketones are widely used for the production of pharmaceuticals, agrochemicals, and fragrances.

Friedel–Crafts acylations generally require more than a stoichiometric amount of, for instance, aluminum chloride ($AlCl_3$), due to the formation of a stable complex between the metal chloride and the product (Scheme 12.1). This requires subsequent hydrolysis, leading to the destruction of the "catalyst" and the production of large amounts of HCl and corrosive waste streams to be neutralized and disposed of. Furthermore, the production of acetyl chloride, which is used as the acylating agent, also generates HCl.

The commercialization of the first zeolite-catalyzed Friedel–Crafts acylation by Rhône-Poulenc (now Rhodia, a member of the Solvay group) for the acylation of anisole (methoxybenzene) with acetic anhydride (Scheme 12.1), may be considered a benchmark in the area of catalytic acylations. The reaction takes place in a fixed bed reactor with zeolite beta as the catalyst. This continuous process takes place in the liquid phase.

Figure 12.3 and Table 12.3 show the enormous simplification and improvement resulting from replacing the classical acylation technology by the new zeolite-catalyzed technology. Not only has the amount of waste

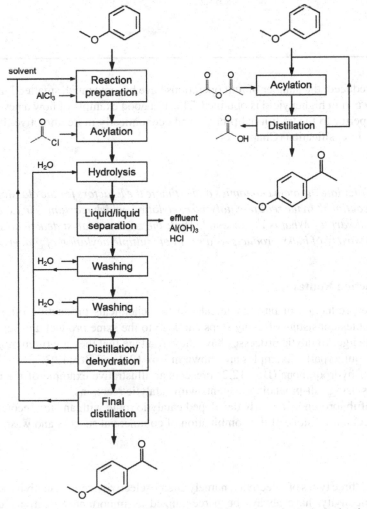

Figure 12.3 *Simplified flow schemes for the acylation of anisole to p-acetylanisole. Left: classical batch process. Right: new zeolite-catalyzed continuous process [13].*

Table 12.3   *Comparison of classical Friedel–Crafts and zeolite-catalyzed acylation [14, 15].*

	Classical Friedel–Crafts	Zeolite catalyzed
Type	homogeneous	heterogeneous
Process	batch	continuous
Reagent/catalyst	$AlCl_3$: 1.1 equivalent, not recoverable	zeolite H-beta catalyst, regenerable
Solvent	yes, 1,2-dichloroethane	no
Phase separation	yes	no
Distillation of organic phase	yes	yes
Solvent recycle	yes	no
Yield (%)	85–95	>95
Aqueous effluent (kg/kg product)	4.5	0.035
Composition of waste stream (wt%)		
H_2O	68.7	99
Al^{3+}	5	–
Cl^-	24	–
Organics	2.3	1

been considerably reduced, the number of unit operations have become much smaller. Furthermore, a product with a higher purity and in higher yield is obtained. This is a good example of how a new process can be both environmentally superior to the traditional process and economically more attractive. It also is an excellent example of Process Intensification (Chapter 14).

QUESTIONS:

Calculate the atom economies and estimate the E factors for the classical and new acylation reactions? In the zeolite-catalyzed acetylation, which reactant is in excess, anisole or acetic anhydride? What is the reason? (Hint: the zeolite catalyst deactivates by the formation of a variety of bulky products as a result of multiple acylation of p-acetylanisole [12].)

12.2.3   Novel Reaction Routes

In contrast with the production of bulk chemicals, in fine chemicals production often a large number of different reaction routes consisting of many steps can lead to the same product and selecting the best route may be quite a challenge. Catalytic processes have the advantage that they are often more direct and shorter than their classical counterparts. Examples are shown in Box 12.2 and Box 12.3.

The production of hydroquinone (Box 12.2) presents an illustrative example of the replacement of stoichiometric reactions using salt-producing reagents with catalytic reactions.

The production of ibuprofen by a newly developed catalytic route with an atom economy of nearly 100% (Box 12.3) is an excellent example of the combination of commercial success and waste minimization.

12.2.4   Selectivity

In organic syntheses, three types of selectivity, namely chemoselectivity, regioselectivity, and stereoselectivity, including enantioselectivity, have always been recognized as important. Selectivity can be achieved by choosing suitable starting materials, reagents, solvents, reaction conditions, and, most importantly, catalysts.

Box 12.2 Production of Hydroquinone [6]

Hydroquinone was traditionally manufactured by oxidation of aniline with stoichiometric amounts of manganese dioxide to produce benzoquinone, which was then reduced with iron and hydrochloric acid; this latter step was later replaced by a catalytic hydrogenation. The aniline was produced from benzene via nitration and reduction with Fe/HCl. The overall process generated over 10 kg of inorganic salts ($MnSO_4$, $FeCl_2$, Na_2SO_4, and NaCl) per kg hydroquinone. In contrast, a more modern route to hydroquinone, involving the oxidation of *p*-di-isopropylbenzene over a titanium-based catalyst, followed by acid-catalyzed rearrangement of the bishydroperoxide, produces less than 1 kg of inorganic salts per kg hydroquinone. In this process only acetone is formed as a (useful) by-product. Scheme B12.2.1 shows both routes.

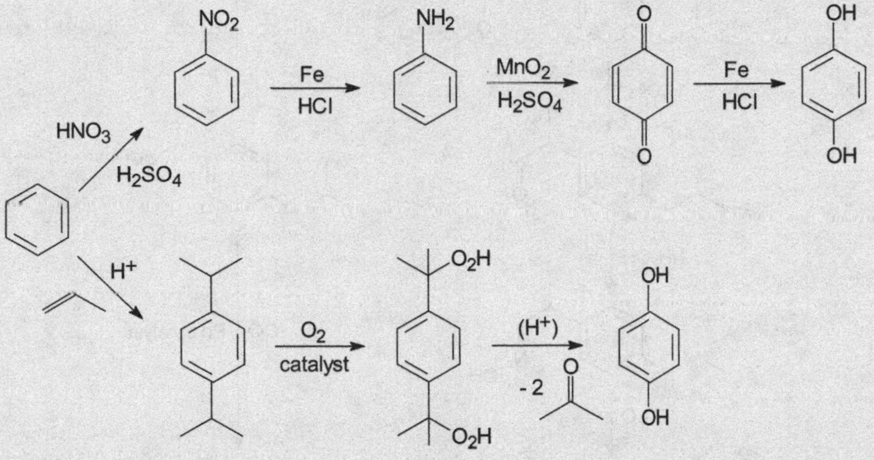

Scheme B12.2.1   *Two routes to hydroquinone.*

QUESTION:

Estimate the atom economy and E factor for both routes.

Box 12.3 Production of Ibuprofen [6]

Scheme B12.3.1 shows the classical route developed by Boots (the inventor of the drug) with an overall atom economy of 40%, and a new route developed by Boots/Hoechst-Celanese (BHC, now BASF).

Both routes start from 2-methylpropyl benzene and share the intermediate *p*-isobutylacetophenone but the classical route involves five additional reaction steps with substantial inorganic salt formation, while the alternative requires only two steps, one of which is a heterogeneously catalyzed hydrogenation, while the other involves homogeneously catalyzed carbonylation. Both of these catalytic steps have an atom economy of 100% and no waste is produced. The overall atom economy is 77%, but it can be >99%, when the acetic acid formed in the first step is recovered and recycled.

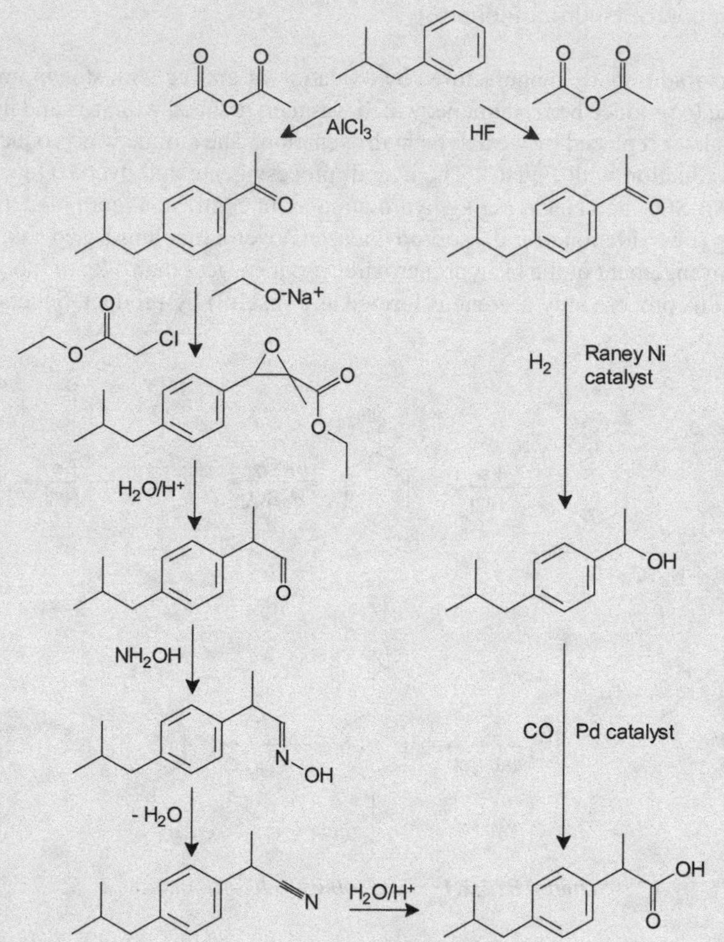

Scheme B12.3.1   *Two routes to ibuprofen: Boots (left) and BHC (right).*

QUESTIONS:

What type of reaction is the first step in both routes?
What is the advantage of using HF as catalyst (and solvent) instead of AlCl₃? What is the disadvantage?
Estimate the E factors of both routes.
How can the new process be improved further? (Hint: A high product purity is required. Which step(s) result(s) in contamination of the product?)

12.2.4.1   *Chemoselectivity*

Chemoselectivity is the preferential outcome of one reaction over a set of other possible reactions, starting from the same reactants. An example is the selective partial hydrogenation of an alkyne (C≡C) to the corresponding alkene (C=C), but not further to the alkane (C–C) [16 (Chapter 3)].

Table 12.4   *Preferred catalysts for some hydrogenation reactions.*

Group	Example reaction	Preferred catalysts
Aromatic nitro groups	$ArNO_2 \rightarrow ArNH_2$	Pt/C, Pd/C
Aliphatic nitro groups	$RNO_2 \rightarrow RNH_2$	Pd/C, Rh/C, Raney Ni
Benzyl derivatives	$ArCH_2X \rightarrow ArCH_3 + HX$	Pd/C
	$X = OR, NR_2$	
Alkenes	$-CH{-}CH- \rightarrow -CH_2{-}CH_2-$	Pd/C, Pt/C, PtO_2
Alkynes	$-C{\equiv}C- \rightarrow -CH_2{-}CH_2-$	Pd/C
	$-C{\equiv}C- \rightarrow -CH{=}CH-$	$Pd\text{-}Pb/CaCO_3$
Aliphatic C=O groups	$R_2C{=}O \rightarrow R_2CHOH$	Ru/C
Aromatic C=O groups	$Ar(C{=}O)R \rightarrow ArCH(OH)R$	Pd/C
Nitriles	$RC{\equiv}N \rightarrow RCH_2NH_2$	Raney Ni, Rh/Al_2O_3
Aromatic ring		Rh/Al_2O_3

In another definition, chemoselectivity refers to the selective reactivity of one functional group in the presence of others. This type of chemoselectivity has traditionally been obtained by protecting one or more groups, thus blocking their reactivity. However, this adds extra steps to the synthesis and since protecting and unprotecting reactions rarely give 100% yields, precious materials may be lost. It can be more efficient to use chemoselective reagents or catalysts. The chemoselectivity mainly depends on the relative reducibility of the different functional groups present and on the catalyst used. Table 12.4 illustrates this for hydrogenation reactions. The conventional and most used hydrogenation catalysts are the noble metals platinum, palladium, rhodium, and ruthenium supported on activated carbon, along with Raney nickel and a few supported nickel and copper catalysts. For specific hydrogenations the conventional catalysts may be modified and special catalysts have been developed for hydrogenations that are very difficult to achieve with high selectivity.

Figure 12.4 shows some guidelines for the chemoselective hydrogenation of different functional groups. Thus, the selective hydrogenation of aromatic nitro groups is easy in the presence of most other functional groups, whereas the selective hydrogenation of aromatic rings is often difficult.

An important reaction in the production of fine chemicals, and a moderately difficult one, is the selective hydrogenation of aldehydes in the presence of carbon–carbon double bonds. An example is the production of cinnamyl alcohol (Scheme 12.2), for which iridium and platinum catalysts seem to give the best results [16 (Chapter 3)]. Another example is the selective hydrogenation of aromatic nitro compounds (Box 12.4).

Box 12.4   Selective Hydrogenation of Aromatic Nitro Compounds

Standard solid catalysts are well suited for the hydrogenation of simple aromatic nitro compounds to the corresponding anilines (Table 12.4) but until the mid-1990s the selective hydrogenation of an aromatic nitro group in the presence of other easily reducible functional groups was only possible using H_2 from the stoichiometric reaction of Fe and HCl. This changed when scientists at Ciba-Geigy developed two new catalytic systems that allow highly selective hydrogenation of an aromatic nitro group in the presence of C=C, C=O, and C$\equiv$N groups, as well as chlorine or bromine substituents, and even C$\equiv$C groups [17, 18]. Scheme B12.4.1 shows an example. More recently, a promising supported gold catalyst has been developed for this type of reaction [18, 19].

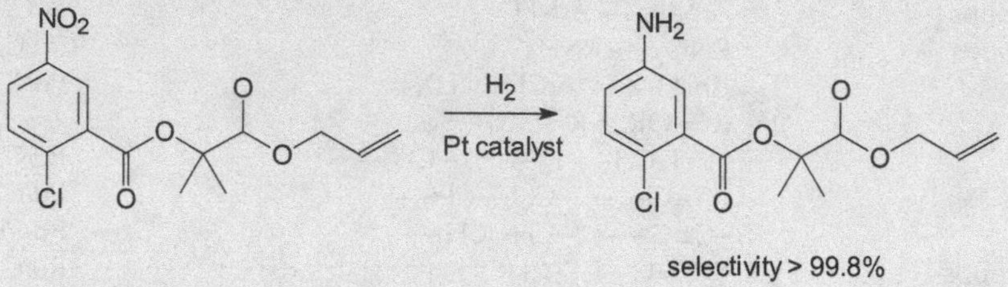

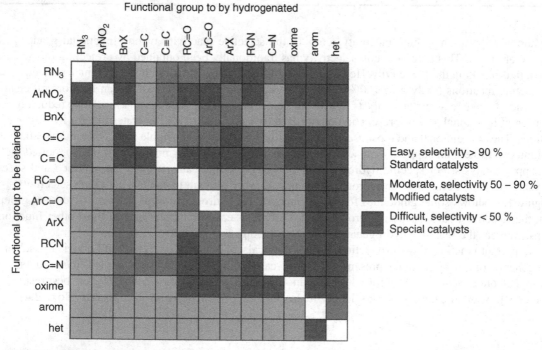

Scheme B12.4.1 *Chemoselective hydrogenation of a nitroallylester. (Catalyst: Pt-Pb/CaCO₃ or Pt/C with H₃PO₂-VO(acac)₂; acac is acetyl acetonate.)*

Figure 12.4 *Rules for chemoselective hydrogenation of functional groups. Ar = aryl; BnX = benzyl derivative with X = OR, NR₂; ArX = aryl halide with X = Cl, Br, I; arom is aromatic ring; het = heteroaromatic ring.*

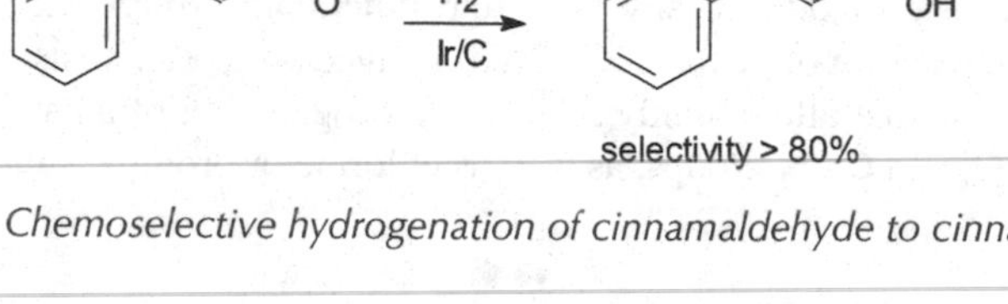

Scheme 12.2 *Chemoselective hydrogenation of cinnamaldehyde to cinnamyl alcohol.*

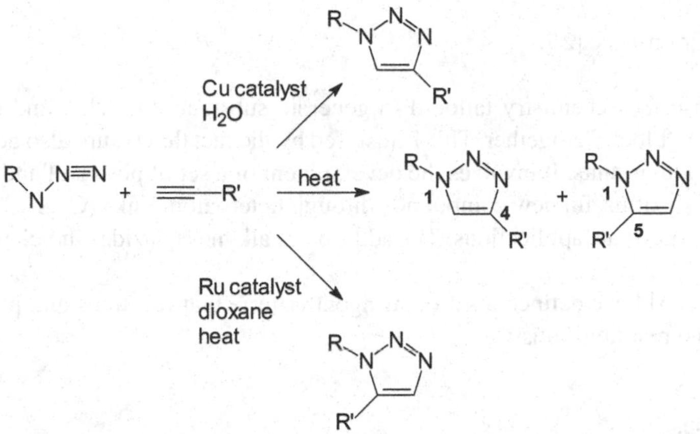

Scheme 12.3   *Regioselective formation of [1,2,3]-triazoles by cycloaddition of alkynes to azides.*

QUESTIONS:

What is the function of Pb in the catalysts for the hydrogenation of the alkyne (Table 12.4) and of the nitroallylester in Scheme B12.4.1?

Often a price has to be paid for achieving high chemoselectivity. Explain.

12.2.4.2   Regioselectivity

Regioselectivity is the preference of one direction of chemical bond making or breaking over all other directions. It applies to the formation of molecules that have the same molecular formula but differ in how their atoms are connected.

The cycloaddition of an alkyne to an azide shown in Scheme 12.3 is an example of a reaction in which regioselectivity plays a role. In the conventional, non-catalytic, approach a high temperature is required, which may lead to decomposition of the triazole product [20, 21]. Another disadvantage is that often a mixture of the two possible regio-isomers (1,4 and 1,5) in a ratio of about 1:1 is produced, resulting in low yields of the desired regio-isomer. Recently, methods have been described for the selective production of only one regio-isomer with high yield. A powdered copper catalyst affords the specific synthesis of the 1,4 regio-isomer with a yield of over 90% [22]. In contrast, with homogeneous ruthenium-based catalysts the 1,5 regio-isomer is favored (yield 40–>90%) [23, 24].

The cycloaddition pathway shown here has a remarkably broad scope and is what Sharpless called the "cream of crop" of *click chemistry*, a philosophy Sharpless introduced in 2001 [25] (Box 12.5).

12.2.4.3   Enantioselectivity

In the production of fine chemicals, and especially pharmaceuticals, there is increasing interest in the production of chirally pure compounds consisting of a single enantiomer (optical isomer). In pharmaceuticals it is often not safe to use mixtures of enantiomers; the desired effect almost always resides in only one of the enantiomers, while the other enantiomer is inactive or even has a damaging effect.

An example is the thalidomide (Figure 12.5) tragedy. Thalidomide (with over 50 trade names, including Softenon, Contergon, and Distoval) was introduced as a sedative drug in the late 1950s and was typically used by pregnant women as a treatment for morning sickness. However, the drug was found to cause severe birth defects, which was attributed to the (*S*)-enantiomer, and in 1961 thalidomide was withdrawn from the market.

Box 12.5   Click Chemistry [25]

Click chemistry describes chemistry tailored to generate substances quickly and reliably by joining small units, modular "blocks", together. This is inspired by the fact that nature also generates substances by joining small modular units. It involves the development of a set of powerful and selective reactions for the rapid synthesis of useful new compounds through heteroatom links (C–X–C) that work reliably in both small- and large-scale applications. The addition of alkynes to azides in Scheme 12.3 is a typical example.

Sharpless *et al.* [25] have defined a set of stringent criteria that reactions and processes must meet for this purpose. The reactions must:

- be modular,
- be wide in scope,
- give very high yields,
- only produce harmless by-products,
- be stereoselective (but not necessarily enantioselective), and
- have good atom economy.

The required process conditions include:

- simple reaction conditions (ideally, the process should be insensitive to oxygen and water),
- readily available starting materials and reagents,
- no use of solvents or use of a benign solvent (preferably water) or one that can be easily removed,
- simple product recovery, and
- purification, if required, by non-chromatographic methods (e.g., distillation or crystallization).

The purpose of click chemistry is to accelerate the discovery of substances with useful properties, with an emphasis on new medicines. Further details can be found elswhere [25].

QUESTION:
> *Explain the logic of the statement about purification.*

Although the effect of chirality on biological activity was already known in the 1960s, it took about thirty years before sufficient regulatory pressure stimulated the shift towards marketing drugs in enantiomerically pure form [9, 26]. This generated a need for economically feasible methods for the synthesis of the desired enantiomer only.

The classical method used in the preparation of a single enantiomer is to prepare a mixture of both enantiomers and then resolve this racemic mixture using various techniques, often involving the destruction of the undesired enantiomer. A cleaner and more economic route is the direct manufacture of the desired enantiomer through enantioselective catalysis.

The first commercial catalytic enantioselective synthesis[1] is the Monsanto process for the production of L-dopa, a chiral amino acid that is used in the treatment of Parkinson's disease [27, 28]. Reaction 12.1 shows

---

[1] Also known as asymmetric, chiral, or stereoselective synthesis.

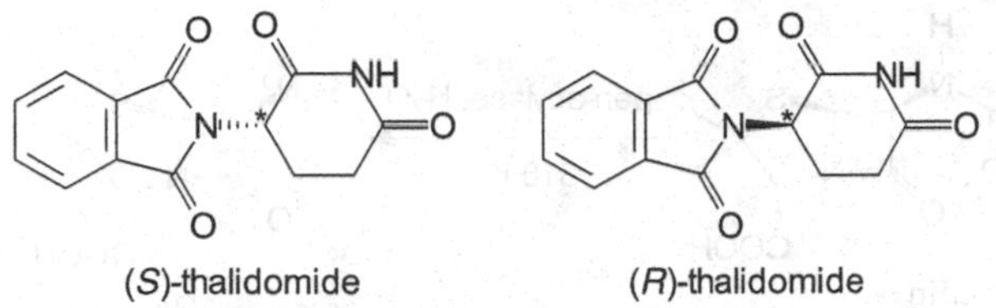

(*S*)-thalidomide   (*R*)-thalidomide

Figure 12.5 *The two enantiomers of thalidomide (* is the chiral center).*

the key step in this synthesis, namely an enantioselective hydrogenation, which was discovered by Knowles in 1968. The 2001 Nobel Prize in Chemistry was awarded to Knowles and Noyori for their work on chirally catalyzed hydrogenation reactions (and to Sharpless for his work on chirally-catalyzed oxidation reactions).

Reaction 12.1 takes place in the presence of a soluble rhodium catalyst [29]. The rhodium complex solution is mixed with a slurry of the solid reactant in aqueous ethanol. The solid product, essentially optically pure, is collected by filtration. The catalyst and the residual racemate remain in solution.

QUESTIONS:

Which separation principle is described above?
Ibuprofen (Box 12.3) also has a chiral center. On which carbon atom?

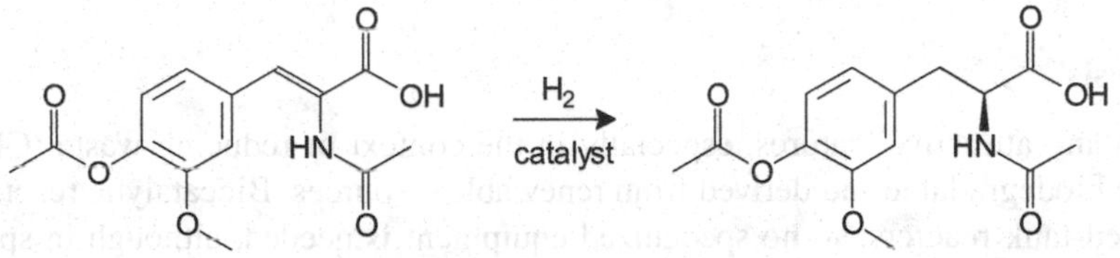

$$(12.1)$$

Most other examples of commercially applied enantioselective catalysis also can be found in the pharmaceutical industry and most of these are hydrogenations [1, 26, 30]. Some enantioselective isomerizations and epoxidations are also used commercially [1, 31].

An elegant example is the Takasago process for the manufacture of (−)-menthol, one of eight sterereoisomers and an important product in the fragrances and flavors industry. The key step is an enantioselective double bond isomerization that is catalyzed by a rhodium complex (reaction 12.2). The complete (−)-menthol synthesis is remarkable because three chiral centers are created, all of which are necessary to produce the characteristic menthol odor and local anesthetic action [32]. Menthol is one of the largest volume chiral chemicals with an estimated production of over 20 kt/a in 2009, of which about 35% is synthetic [31].

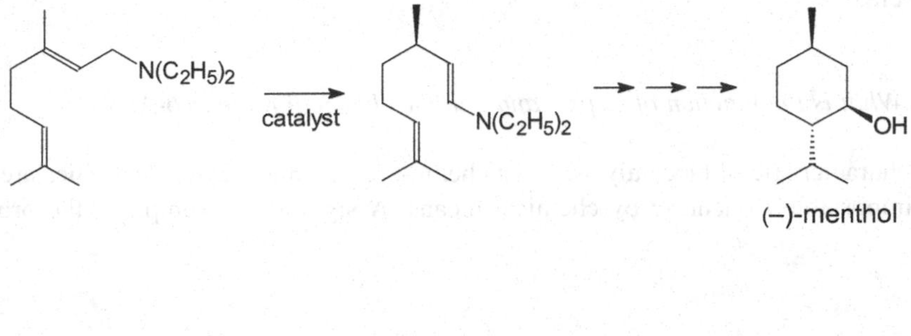

(−)-menthol

$$(12.2)$$

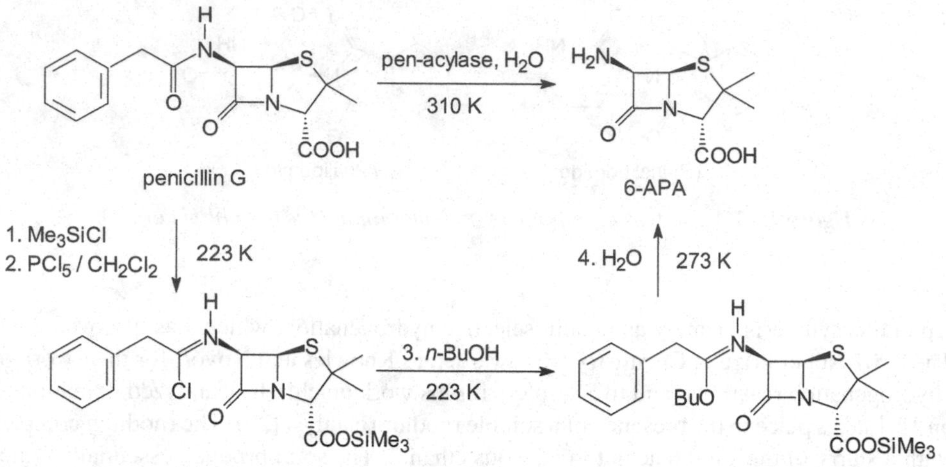

Scheme 12.4   *Comparison of the complex chemical and one-step enzymatic production of 6-APA. Me = methyl (CH₃), Bu = butyl (C₄H₉).*

The number of commercial applications of enantioselective catalysis with transition metal complexes is still limited. The main obstacle for wider application is that the ligands are expensive and only a few are commercially available [26]. Other aspects to be considered are catalyst stability, catalyst separation, and possible toxicity of the metals.

12.2.5 Biocatalysis

Biocatalysis has many attractive features, especially in the context of reducing waste (Chapter 13). Enzymatic catalysts are biodegradable and derived from renewable resources. Biocatalytic reactions are generally performed in stirred-tank reactors, so no specialized equipment is needed, although in specific cases other reactor types are preferred [70]. Usually, the reaction conditions are mild and an environmentally compatible solvent, water, is used. The use of enzymes often circumvents the need for functional group protection and deprotection steps.

An excellent example is the production of 6-amino penicillinic acid (6-APA) from penicillin G [33–35] (Scheme 12.4). The chemical route (steps 1, 2, 3, 4) was developed by the Nederlandsche Gist en Spiritusfabriek (now a DSM subsidiary) and is known as the *Delft Cleavage*. This chemical deacylation was difficult to execute as penicillin G contains both a secondary and tertiary amide functional group, the former being the desired cleavage site but the latter being more susceptible to chemical cleavage. Other disadvantages of this route are that it requires four chemical steps, a very low temperature (223 K), various stoichiometric reagents, including the highly reactive and hazardous phosphorus pentachloride (PCl₅), and the use of an undesirable solvent (CH₂Cl₂). In contrast, the enzymatic hydrolysis of penicillin G to 6-APA using immobilized penicillin acylase (pen-acylase) proceeds in one step at ambient temperature and with water as the solvent.

QUESTION:

> *What is the function of steps 1 and 2 in the chemical route to 6-APA?*

An attractive characteristic of biocatalysis is that chemo-, regio-, and stereoselectivities are attainable that are difficult or impossible to achieve by chemical means. A significant example is the production of the high intensity sweetener aspartame. The enzymatic process is completely regio- and enantioselective [36] (Box 12.6).

Box 12.6   Production of Aspartame

Scheme B12.6.1 shows the production of the high intensity sweetener aspartame by enzymatic coupling. The NH$_2$ group of aspartic acid is protected by a benzyloxycarbonyl (Z), which renders the nitrogen atom inactive so no aspartic acid molecules can react with each other.

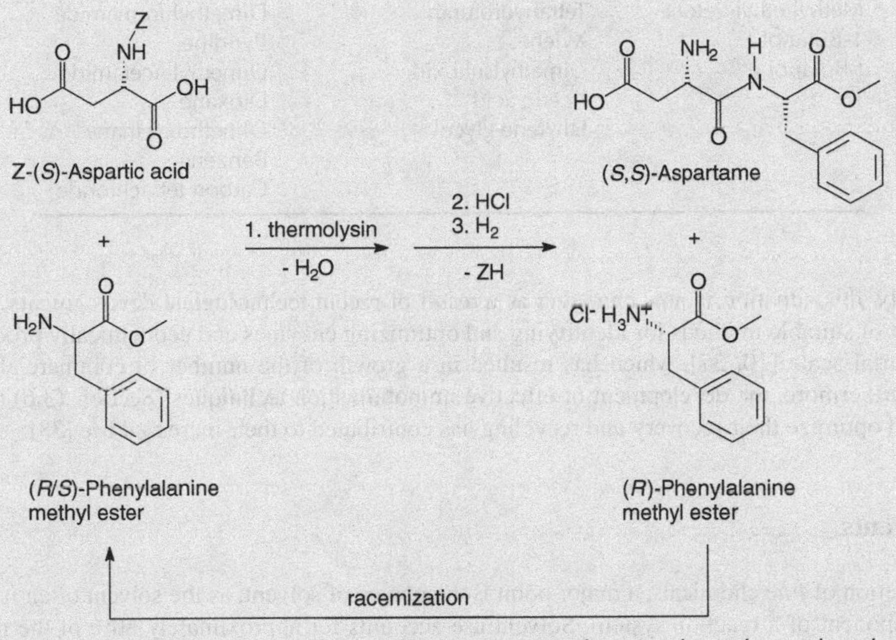

Scheme B12.6.1   *Aspartame through enzymatic coupling. Z = benzyloxycarbonyl.*

The enzyme thermolysin has been "programmed" in such a way that it only reacts with a carboxyl group that is attached to a carbon atom to which a nitrogen atom is also bound. Therefore, there is no need to protect the other carboxyl group. In addition, thermolysin is specific for the (*S*)-enantiomer of the phenylalanine methylester of which only the NH$_2$ group is reactive. Thus a racemic mixture of phenylalanine can be used as the starting material instead of the more expensive (*S*)-phenylalanine, because only this enantiomer takes part in the coupling reaction to form Z-(*S*,*S*)-aspartame. After removal of the protecting group Z by catalytic hydrogenation to yield aspartame, the unreacted (*R*)-enantiomer can be recovered, isomerized back to the racemic mixture, and recycled to the process.

Although the advantages of biocatalysis have been known for a long time, progress in using enzymes in the fine chemicals and pharmaceutical industries has been hampered by the lack of consistent production and formulation of enzymes, limited scope of substrates compared with classical chemical methods, limited stability, and difficult separation and reuse [37].

Table 12.5   *Pfizer solvent selection guide [10, 41].*

Preferred	Usable	Undesirable
Water	Cyclohexane	Pentane
Acetone	Heptane	Hexane(s)
Ethanol	Toluene	Di-isopropyl ether
2-Propanol	Methylcyclohexane	Diethyl ether
1-Propanol	*Tert*-butyl methyl ether	Dichloromethane
Ethyl acetate	Iso-octane	Dichloroethane
Isopropyl acetate	Acetonitrile	Chloroform
Methanol	2-Methyltetrahydrofuran	*N*-methylpyrrolidinone
Methyl ethyl ketone	Tetrahydrofuran	Dimethyl formamide
1-Butanol	Xylenes	Pyridine
t-Butanol	Dimethylsulfoxide	Dimethyl acetamide
	Acetic acid	Dioxane
	Ethylene glycol	Dimethoxyethane
		Benzene
		Carbon tetrachloride

Fortunately, this situation is now changing as a result of recent technological developments, such as the development of suitable methods for identifying and optimizing enzymes and economically producing them on an industrial scale [10, 37], which has resulted in a growth of the number of commercially available enzymes. Furthermore, the development of effective immobilization techniques (Section 13.6) to formulate enzymes and optimize their recovery and recycling has contributed to their increased use [38].

12.3   Solvents

In the production of fine chemicals, a major point is the choice of solvent, as the solvent often is the largest-volume component of a reaction system. Solvent use accounts for approximately 80% of the material and 75% of the energy usage for the production of active pharmaceutical ingredients (APIs) [39]. An additional problem is the (perceived) necessity for solvent switches between the various reaction steps. This makes recycling of solvents difficult owing to the occurrence of cross contamination [9]. In fact, often less than 50% of the used solvent is recycled [40]. Thus, solvent losses, either in effluents or as air emissions, account for a considerable portion of the waste produced.

12.3.1   Conventional Solvents

Many popular solvents, such as chlorinated hydrocarbons, have been banned or are likely to be banned in the near future. Therefore, alternatives need to be found. Table 12.5 shows a list reported by Pfizer for the selection of traditional solvents and Table 12.6 shows alternatives for undesirable solvents.

An illustrative example of a process in which the problem of using different solvents in different reaction steps has been solved is the manufacture of sertraline hydrochloride (Figure 12.6) by Pfizer. The redesign of the process, among other improvements, resulted in the restructuring of a three-step reaction sequence by employing ethanol as the only solvent. This eliminated several distillations and the need to recover four solvents (dichloromethane, tetrahydrofuran, toluene, and hexane) [42]. Moreover, it resulted in a reduction

Table 12.6 *Pfizer solvent replacement table [10, 41].*

Undesirable solvent	Alternatives
Pentane	Heptane
Hexane(s)	Heptane
Di-isopropyl ether or diethyl ether	2-Methyltetrahydrofuran or *tert*-butyl methyl ether
Dioxane, dimetoxyethane	2-Methyltetrahydrofuran or *tert*-butyl methyl ether
Dichloromethane	Ethyl acetate, *tert*-butyl methyl ether, toluene, 2-methyltetrahydrofuran
Dichloroethane, chloroform, carbon tetrachloride	Dichloromethane
N-methylpyrrolidinone, dimethyl formamide, dimethyl acetamide	Acetonitrile
Pyridine	Triethylamine (if pyridine used as base)
Benzene	Toluene

in solvent usage from 250 to 25 m^3/t of product [14]. Pfizer continued to improve the process and, today, the sertraline process stands as one of the greenest processes reported by the pharmaceutical industry, with an E factor of 8 [43]. This new process is a good example of Process Intensification (Chapter 14).

QUESTIONS:

What are the main reasons for banning some solvents?

Explain why pentane is referred to as an undesirable solvent and heptane as a better alternative (Table 12.6)? Answer the same question for dichloroethane and dichloromethane.

12.3.2 Alternative Solvents

The number one choice of solvent from a "green chemistry" point of view is water. It is non-toxic, non-flammable, abundantly available, and inexpensive [44]. However, many reactants and reagents are not compatible with water. A possible solution is biphasic operation, as is successfully applied in homogeneously catalyzed hydroformylation, with the additional advantage of easy separation of the transition metal catalyst from the product (Section 9.3).

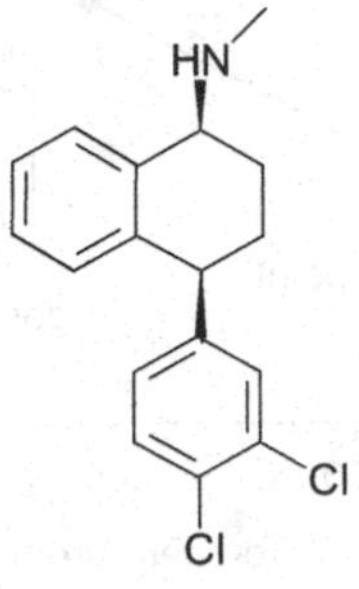

Figure 12.6 *Sertraline.*

Not surprisingly, aqueous biphasic operation is not always the answer. Therefore, alternative solvents, such as supercritical fluids (SCFs), especially carbon dioxide ("scCO$_2$") [45–47], and ionic liquids (IL) [42, 48–54], have also been extensively studied.

QUESTION:

> *When water is used in a biphasic system, occasionally environmental concerns can be related to the use of water. Explain.*

12.3.2.1   Supercritical Fluids

A supercritical fluid (SCF) [45–47] is a substance at a temperature and pressure above its critical point. The main distinctive characteristic of a supercritical fluid is that its density is very sensitive to small changes in temperature and pressure; this enables a phase change to be achieved by a small change of pressure. Beyond the critical point, no distinct liquid or vapor phase can exist, and the new supercritical phase exhibits properties of both states; for example, diffusion in SCFs is fast compared to diffusion in liquids, while solubilities are higher than in the gas phase. These unique properties of SCFs [45] make them interesting for various technological applications such as supercritical fluid extraction and as reaction media.

Carbon dioxide is the most used supercritical fluid due to its relatively low critical temperature and pressure (Figure 12.7), non-toxicity, low cost and non-flammability. Supercritical carbon dioxide allows applications at mild conditions, thus avoiding the degradation of temperature-sensitive compounds.

The use of scCO$_2$ offers the possibility of continuous instead of batch operation. The application of scCO$_2$ in continuous flow fixed bed reactors has drawn a lot of attention [47]. The world's first, and so far only, multipurpose, supercritical continuous flow fixed bed reactor went on stream in 2002 for the hydrogenation of isophorone to trimethylcyclohexanone (TMCH) (Box 12.7). Although the process was technically successful, the energy-intensive compression of carbon dioxide soon made the process uncompetitive when energy prices increased [55].

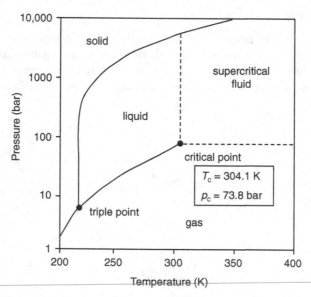

Figure 12.7   *Pressure–temperature phase diagram for CO$_2$.*

Box 12.7   Hydrogenation of Isophorone to Trimethylcyclohexanone (TMCH) in scCO$_2$

Conventional hydrogenation technologies easily lead to the overhydrogenated by-products shown in Scheme B12.7.1, while the required purity of TMCH is high. The hydrogenation and hydrogenolysis products and isophorone itself have similar boiling points, so the separation and purification of TMCH is difficult and greatly adds to both cost and environmental impact of the process. As a result of the high selectivity to TMCH, the scCO$_2$ technology eliminated the need for any downstream purification of the product.

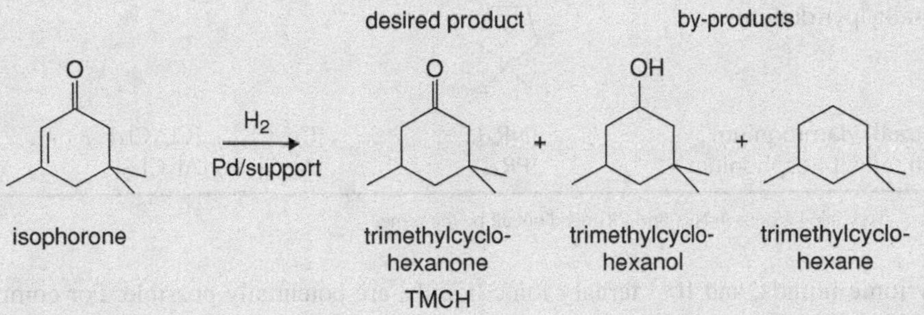

Scheme B12.7.1   *Reaction scheme illustrating the possible products obtained in the hydrogenation of isophorone.*

QUESTIONS:

Which type of selectivity/selectivities is/are involved in the hydrogenation of isophorone?

Which special properties of scCO$_2$ compared to liquid solvents are responsible for the high selectivity to TMCH?

However, there may be a completely new opportunity if current plans for carbon capture and storage (CCS, Section 5.3) are indeed implemented on a large scale. CCS is considered a promising route for achieving a large reduction in carbon dioxide emissions. CSS, if implemented, is expected to make available considerable quantities of pressurized liquid carbon dioxide [55]. Then the use of scCO$_2$ as an alternative solvent might become competitive again.

QUESTION:

Compare the use of supercritical water and supercritical CO$_2$.

12.3.2.2   Ionic Liquids

Ionic liquids (ILs) [42, 48–54] are salts that are liquid at low temperatures (<373 K) and are usually composed of a bulky organic cation and a smaller inorganic ion (Table 12.7). In contrast with volatile organic compounds (VOCs), ionic liquids have negligible vapor pressure and ideally can be recycled and reused repeatedly. One of the most interesting aspects of ionic liquids is that they can be tuned as regards to properties such as polarity, reactant solubility, and solvent miscibility by modification of the cation and the anion. At least a

Table 12.7   *Typical cation/anion combinations in ionic liquids.*

Cations	Structure[a]	Anions
1-alkyl-3-methylimidazolium		
1-alkyl-1-methylpyrolidinium		$[BF_4]^-$, $[PF_6]^-$, $[SbF_6]^-$, $[CF_3SO_3]^-$, $[CaCl_2]^-$, $[AlCl_4]^-$, $[AlBr_4]^-$, $[AlI_4]^-$, $[AlCl_3Et]^-$, $[NO_3]^-$, $[NO_2]^-$, $[SO_4]^{2-}$
N-alkylpyridinium		
tetraalkylammonium	$[NR_4]^+$	$[Cu_2Cl_3]^-$, $[Cu_3Cl_4]^-$,
tetraalkylphosphonium	$[PR_4]^+$	$[Al_2Cl_7]^-$, $[Al_3Cl_{10}]^-$

[a]R = alkyl; alkyl groups in NR_4 and PR_4 need not all be the same.

million binary ionic liquids, and 10^{18} ternary ionic liquids, are potentially possible. For comparison, about 600 conventional molecular solvents are commercially available [56].

In the last 15 years, tremendous progress has been made in the application of ionic liquids in catalytic and biocatalytic processes. The first commercial process employing ionic liquids is BASF's BASIL (Biphasic Acid Scavenging utilizing Ionic Liquids) process (Box 12.8). This process is a general solution for all kinds of reactions in which an acid (such as HCl) is formed that must be removed.

QUESTIONS:

> *Ionic liquids have several exciting advantages. Are there any disadvantages? If so, which?*

12.4   Production Plants

In the bulk chemicals industry, dedicated plants operating in continuous mode predominate. They are optimized to produce well-defined products in a precisely determined manner. In a sense, they are both highly efficient and highly inflexible. For example, redesigning a dedicated plant for producing a different type of product is generally not possible.

In the production of fine chemicals, reactions are usually carried out batchwise rather than continuously. Typical plant types are the so-called multiproduct plants and multipurpose plants (MPPs).

QUESTION:

> *Illustrate the inflexibility in bulk chemicals production regarding the incorporation of new findings (e.g., the development of a revolutionary catalyst for ammonia and methanol plants).*

12.4.1   Multiproduct and Multipurpose Plants (MMPs)

Multiproduct plants allow the production of a fixed set of products with similar synthesis steps, usually a product group of similar chemical structure. Each product goes through the same pieces of equipment (stages) and follows almost the same sequence of operations. Usually only one product is manufactured at a time.

Box 12.8 BASIL Process for the Synthesis of Dialkoxyphenylphosphine [54, 56, 57]

In the conventional process for the synthesis of dialkoxyphenylphosphines from dichlorophenylphosphine (Scheme B12.8.1) HCl is generated as a by-product that must be removed. This is normally done by adding a tertiary amine as acid scavenger, but this process results in a thick ammonium salt slurry that is difficult to separate from the reaction mixture.

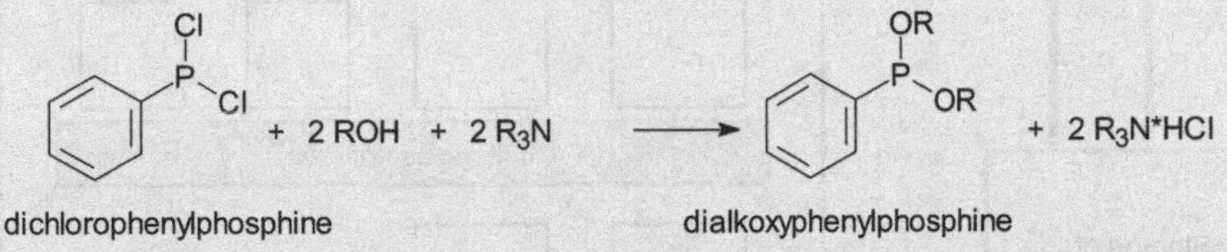

Scheme B12.8.1 *Classical synthesis of dialkoxyphenylphosphines.*

When using 1-methylimidazole as the acid scavenger (Scheme B12.8.2), an ionic liquid is formed (1-methyl-imidazolium chloride). After the reaction two phases are formed, which can easily be separated; the upper phase is the pure product and the lower phase is the ionic liquid. The latter is deprotonated, regenerating the imidazole for reuse.

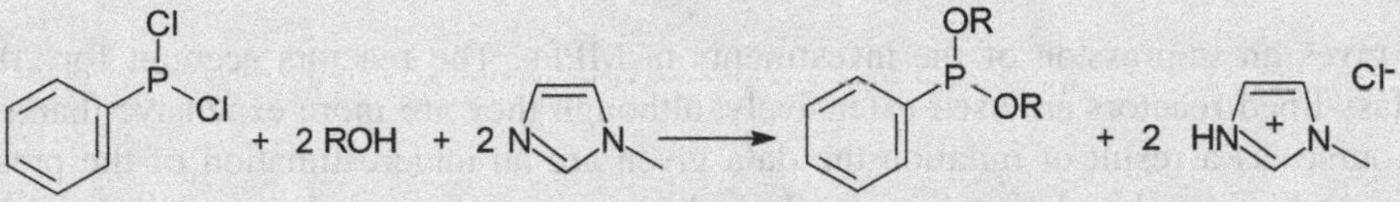

Scheme B12.8.2 *1-Methylimidazole as an acid scavenger.*

In addition to the excellent scavenging ability of 1-methylimidazole, the ionic liquid appeared to function as a (nucleophilic) catalyst. The combined effect is an enormous increase of the productivity of the process by a factor of 80 000 (!) to 690 000 $kg\ m^{-3}\ h^{-1}$. This enabled BASF to carry out the reaction, which previously needed a batch reactor with a volume of 20 m^3, in a continuously operated jet reactor the size of a thumb, a great example of Process Intensification! (Chapter 14).

Multipurpose plants allow the production of a broad range of products, which may vary considerably with respect to number and type of synthesis steps. Different equipment is required for different products and different products may follow different routes through the equipment. In multipurpose plants, the products and processes may not be known from the start and they may also change over time.

QUESTION:

   Which plant type is more flexible, a multiproduct plant or a multipurpose plant?

12.4.1.1 Equipment in MPPs

An MPP typically consists of a number of equipment items differing in functionality and size, and auxiliary facilities (Figure 12.8). These items are described in more detail elsewhere [1 (Chapter 7)].

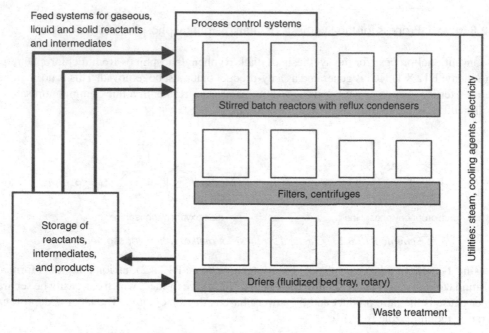

Figure 12.8   *Typical equipment in an MPP.*

Table 12.8 gives an impression of the investments in MPPs. The reactors account for 20–40% of the total costs. Glass-lined reactors are used extensively, although they are more expensive than stainless steel reactors. Of course, as a result of inflation the data given are an underestimation of the present costs. In addition, more sophisticated and expensive analytical instruments are used nowadays, and together with increased automation this leads to higher investment costs. On the other hand, operating costs are lower because less labor is required.

QUESTIONS:

Why are stainless steel reactors cheaper than glass-lined reactors?
The data in Table 12.8 are from 1982. Estimate the current investment costs per m³. Do you think the cost distribution has changed since 1982? How?

Table 12.8   *Impression of investments in MPPs; investment in 1000 US $/m³ (1982).*

	Glass lined		Stainless steel	
	1 m³	4 m³	1 m³	4 m³
Reactor system	70	32	50	25
Solids & liquid recovery	29	23	29	23
Drying, size reduction	16	14	16	14
Liquid storage	7	7	7	7
Analytical instruments	2	1	2	1
Utilities	6	6	6	6
Effluent treatment	22	22	22	22
Building & Civil	18	18	18	18
Total	170	125	150	118

MPPs need to be flexible in order to facilitate the multistep processes and multiproduct plant approach in fast-changing markets. As a result of this (and because they resemble laboratory flasks), batchwise operated stirred-tank reactors are almost universally used as the primary reactor technology in MPPs. Unlike in the production of bulk chemicals, the reactors are usually not designed for a specific process but purchased from a vendor (Box 12.9).

Box 12.9 Characteristics of Batch Reactors

Table B12.9.1 shows the main characteristics of commercially available standard reactors. For most equipment the volume–price dependency can be described by:

$$Cost = A \cdot Volume^B$$

For glass-lined reactors B = 0.32–0.42 and for stainless steel reactors B = 0.75.

Table B12.9.1 *Main characteristics of standard reactors available on the market [58].*

Capacity (m^3)	Diameter (m)	Weight (t)	Heat exchange area (m^2)	type[a]	Pressure (bar)	Material[b]
8	2.0	9.5	18.6	DJ	15	GS
14	2.5	14	26.6	DJ	15	GS
25	2.8	21	40.0	DJ	15	GS
32	3.1	24	45.6	DJ	15	GS
10	2.4	11	19.8	DJ	6	GS
20	2.8	18	33	DJ	6	GS
25	3.0	22	38.8	DJ	6	GS
1	1.2	1.8	4.45	DJ	6	SS
4	1.8	2.5	12.9	DJ	6	SS
10	2.4	5.2	11.2	EC	6	SS
25	3.0	7.9	19.5	EC	6	SS
0.7	0.9	0.8	3.4	DJ	1.5	G

[a]DJ = double jacket; EC = welded external coil.
[b]GS = glass-lined steel; SS = stainless steel; G = glass.

QUESTIONS:

Obviously, the larger the reactor, the lower the cost per unit volume. Why are the reactor volumes used nevertheless limited?

Why is the exponent in the cost estimation equation higher in the case of stainless steel reactors?

Why are most reactors made of glass-lined steel? Why is only the smallest made of glass?

If, based on design calculation, you would require a reactor with a volume of 12.5 m^3, what would you do?

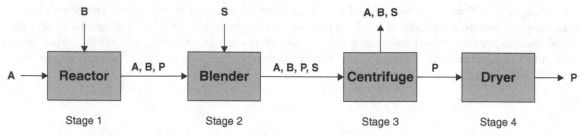

Figure 12.9 *Schematic of the stages for the production of product P from A and B.*

12.4.1.2   Scheduling of Batch Plants

12.4.1.2.1   Production Campaigns

The schedule by which each product is manufactured in an MPP depends on several factors, including demand and availability of equipment and product storage. In general, batch operations involve running *production campaigns*, in which either a single product or a set of products is produced. Campaigns typically last only a few days to a few weeks, so that in the same plant ten to thirty different chemical reactions may be performed in the course of a year.

A campaign consists of *batch sequences* in which one or more products are produced according to a certain schedule. Batch sequences consist of a number of *process stages*, that is, unit operations such as reaction, crystallization, drying, and so on. The process chemistry for each process stage must be translated into recipes for each product. The key elements of a recipe are the time required for each stage, the equipment used, and the raw materials and utilities required. Figure 12.9 shows an example for a single product batch process and Table 12.9 shows the accompanying recipe (excluding utilities).

Figure 12.10 shows the use of the equipment for the batch process described by Figure 12.9 and Table 12.9 in a so-called Gantt chart. The length of the rectangles is directly proportional to the processing time for each stage, which includes charging, chemical and/or physical processing of the batch, discharging, and possibly cleaning. Times for transfer between two stages are usually included in the processing time, as is the case in this example. One might wonder whether this procedure is efficient: the idle times for the various pieces of equipment are very different and the time needed for a complete campaign is 13.5 hour.

Table 12.9   *Possible recipe for batch process of Figure 12.9.*

Stage		Procedure	Processing time (hours)
1	Reaction	Charge A and B (0.5 h) Mix/heat for 1 h Hold at 250 K to form solid product P for 4 h Pump solution through cooler to blend tank (0.5 h)	6
2	Precipitation	Add solvent S (0.5 h) Mix with S at 300 K for 2 h Pump to centrifuge (2 h)	4.5
3	Centrifugation	Centrifuge solution for 1.5 h to separate product P	1.5
4	Drying	Load tray dryer (0,5 h) Dry in a tray dryer at 330 K for 1 h	1.5

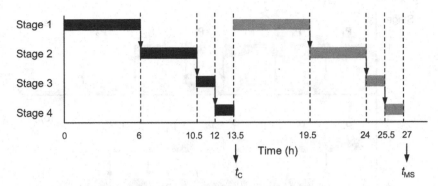

Figure 12.10 *Gantt chart for the process described by Figure 12.9 and Table 12.9 (non-overlapping mode). t_C = cycle time; t_{MS} = makespan.*

Figure 12.10 shows operation in non-overlapping mode; processing of each batch only starts after the preceding batch has been finished and no two batches are processed at the same time. In overlapping mode, as shown in Figure 12.11, processing of a new batch starts with the first stage before the second stage of the preceding batch has been completed.

The cycle time, t_C, is the time interval between the completion of two successive batches. In the case of non-overlapping operation it is the total processing time required for one batch, here 13.5 h. In overlapping operation, the cycle time is equal to the time required for the stage having the longest processing time (the *bottleneck* stage), here 6 h.

The makespan, t_{MS}, is the total time required to produce a given number of batches, and in this case is two. The makespans for two batches for non-overlapping operation and overlapping operation are 27 h and 19.5 h, respectively. Thus, overlapping operation is more efficient. Even so, only the bottleneck stage – reaction – is operated without interruption, that is, with zero idle time, while the utilization of the other equipment, especially that of the centrifuge and the dryer, is still poor (idle time is shown in Figures 12.10 and 12.11 for every stage as the horizontal spaces between rectangles).

QUESTIONS:

> *For efficient batch operation, idle times for all stages should be as small as possible. How could idle times be reduced further for improved equipment utilization? (Hint: in an MPP several units of equipment and equipment of different sizes are usually available.) Draw a Gantt chart for your solution.*

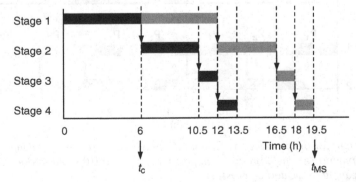

Figure 12.11 *Gantt chart for the process described by Figure 12.9 and Table 12.9 (overlapping mode).*

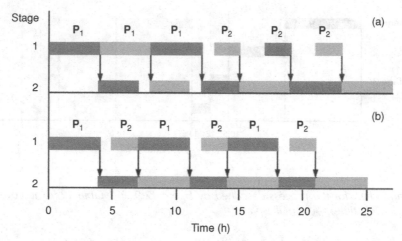

Figure 12.12   *Gantt chart for the production of P_1 and P_2 in two-stage processes: (a) single product campaign, $t_{MS} = 27$ h; (b) multiproduct campaign, $t_{MS} = 25$ h.*

For multiproduct processes, the scheduling can take place as separate single product campaigns following each other or as mixed product campaigns. Figure 12.12 shows the scheduling of a multiproduct plant producing two products P_1 and P_2 in two-stage processes. In a single product campaign (Figure 12.12a) first four batches of P_1 are produced, followed by the production of four batches of P_2. In a mixed product campaign (Figure 12.12b) the sequence is different. First a batch of P_1 is produced followed by a batch of P_2, and so on. The latter protocol results in reduced production time and improved equipment utilization. However, if cleaning of equipment is necessary when switching from one product to another, the situation may be different (Figure 12.13). For the successive arrangement, the makespan increases by one hour only, while for the mixed arrangement the makespan increases by five hours, making the mixed arrangement worse than the successive one.

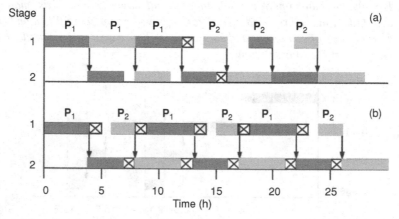

Figure 12.13   *Gantt chart for the production of P_1 and P_2 in two-stage processes taking into account cleaning of equipment: (a) single product campaign, $t_{MS} = 28$ h; (b) multiproduct campaign, $t_{MS} = 30$ h with cleaning (indicated by crosses).*

QUESTIONS:

What are the processing times for products P_1 and P_2 in stages 1 and 2?

Besides the increased makespan, what are other disadvantages of the necessity of frequent cleaning of equipment in mixed product campaigns?

12.4.1.2.2 Policies for Intermediate Storage

Operation and scheduling of a batch plant can be affected significantly by the need for storage of intermediates between process stages, for instance in case the next unit is not yet available. The decision on storage of intermediates is usually dictated by the properties of the intermediate materials under storage conditions. For example, unstable materials must be processed without any delay and cannot be stored. Different storage policies can be envisaged:

- unlimited intermediate storage (UIS),
- no intermediate storage (NIS), and
- finite intermediate storage (FIS).

Unlimited intermediate storage means that the intermediate products in a process can be stored without any problem. In contrast, in the case of no intermediate storage, the intermediates are unstable and storage should be avoided. In practice, the number and capacity of storage tanks is usually limited, and although intermediates are sufficiently stable for storage the situation is referred to as finite intermediate storage.

Figure 12.14 shows Gantt charts for a four product, three-stage plant with an unlimited storage option.

The examples discussed above are typical for multiproduct plants. For multipurpose plants, the variety of arrangements becomes even greater. Therefore, there is no simple way to determine the makespan. It is usually determined by an optimization procedure that takes into account processing sequences for different products and the possibility of splitting a production route into smaller steps. The number of possible processing sequences is generally very large. In practice, many technological constraints are imposed on the production routes, resulting in a limited number of options for the arrangement of the stages. In general, if cleaning and/or transfer times are large compared to processing times, single product campaigns are favored [1 (Chapter 7)].

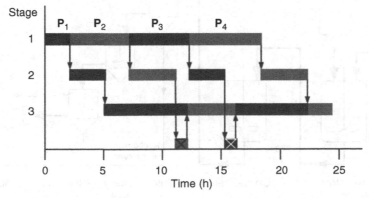

Figure 12.14 Gantt chart for the production of products P_1, P_2, P_3 and P_4 in three-stage processes with the option of unlimited storage.

QUESTIONS:

> *In a sense, a kitchen is a plant for the production of food. What type of MPP is it? Draw a few typical Gantt charts for the production process. Compare the kitchen at home with a kitchen of a catering company.*

12.4.2   Dedicated Continuous Plants

Some dedicated plants do exist in the fine chemicals industry. For example, aspirin (acetylsalicylic acid) is produced in such a dedicated plant (Figure 12.15) [59, 60]; in fact, the production volume in this case is tremendous (over 35 kt/a). It is questionable whether this process fits the category "fine chemicals production".

Salicylic acid and acetic anhydride react to produce acetylsalicylic acid and acetic acid according to reaction 12.3.

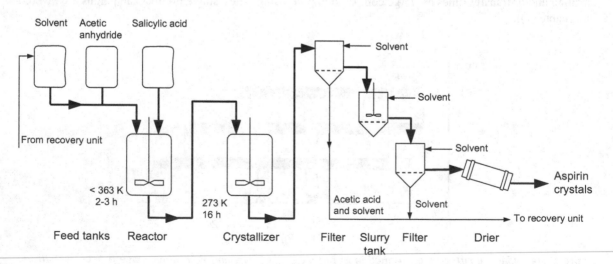

$$\Delta_r H_{298} = -84 \text{ kJ/mol} \tag{12.3}$$

The reaction takes place in a glass-lined or stainless steel batch reactor, in which the temperature must be kept below 363 K. At the end of the reaction period (2–3 h), the liquid product mixture is pumped into a crystallizer where it is cooled to 273 K. The aspirin crystallizes from the acetic acid/acetic anhydride mother liquor (16 h). The resulting suspension is transferred to a filter for removal of acetic acid and solvent (often also acetic acid, but hydrocarbons can be used instead). After washing with solvent, the aspirin crystals are slurried and filtered again. The aspirin crystals are then dried and sent to sifting, granulation, and tableting.

Figure 12.15 *Simplified scheme of a dedicated batch plant for aspirin production.*

Acetic acid, which is formed as a by-product, is recovered. Solvent and possibly unconverted anhydride are recycled to the reactor.

QUESTIONS:

> *What are the consequences of the different time scales for reaction and crystallization?*
> *Why is acetic anhydride used as reactant rather than acetic acid?*

Examples also exist where dedicated plants have been designed for purely technical reasons. An example is the dihydroxylation reaction represented by reaction 12.4.

$$(12.4)$$

The reaction is carried out at low temperature at a residence time of a few seconds. Longer residence times would lead to further oxidation. Accordingly, although the production of this intermediate is only a few kilograms per year [1], a continuous process using a tubular reactor is the only feasible option[2].

There is an increasing tendency to build dedicated plants for fine chemicals that used to be manufactured in MPPs. This is mainly due to rising requirements towards product purity; perfect cleaning of equipment before the next product is manufactured either is impossible or very expensive. Moreover, more catalytic processes are being implemented in fine chemicals manufacture because of environmental needs. Heterogeneous catalytic processes are less flexible and often require dedicated plants or at least a set of dedicated equipment items.

12.5 Batch Reactor Selection

In the fine chemicals industry, reaction systems are quite diverse. They can be classified based on the phases present as in the production of bulk chemicals, except that liquid phase reactions producing solid products dominate.

Currently, the majority of the reactors are mechanically-stirred-tank reactors used in the batch mode. The most important choices in the selection of batch reactors concern:

- reactor volume;
- selection of the mixer;
- speed of the agitator;
- geometry of the tank, including baffles;
- heat exchange area (internal and external).

Section 12.6 illustrates that the importance of heat transfer is easily underestimated.

---

[2] Attainable residence times for the three basic reactor types are approximately: 15 min to 20 h for a batch reactor, 10 min to 4 h for a single CSTR, and 0.5 s to 1 h for a tubular reactor (PFR) [61].

12.5.1 Reactors for Liquid and Gas–Liquid Systems

Figure 12.16 shows some of the batch reactor systems most often encountered in practice. Most are mechanically stirred. Double-jacket reactors are most easily built, but the heat exchange area is limited. Exchange area may also be installed inside the reactors but its presence hinders stirring and cleaning. Moreover, no glass-lined reactors equipped with internal cooling systems are available on the market. Therefore, when a large heat exchange area is needed, an external heat exchanger is usually required. A reflux condenser can be used when a reaction is carried out at boiling point. Alternatively, part of the liquid can be circulated through an external heat exchanger. In this case, obviously, the reactor is not mechanically stirred but mixing is provided by liquid circulation.

If one reactant is gaseous, it can be supplied through a sparger at the bottom of the reactor. This type of operation is termed semi-batch: one reactant, in this case a liquid, is loaded in the reactor, while the other, in this case a gas, is introduced continuously.

Other possible reactor types for gas–liquid reactions are bubble column reactors (empty, packed or with trays), spray columns, and so on [28, 61] (Figure 9.16). These reactors can also be used if, for whatever reason, mechanical stirring is not desired. When no gas is involved in the reaction, an inert gas, usually nitrogen, can be used to promote the agitation and mixing of two fluid phases [28].

QUESTIONS:

> *Give the advantages of a configuration where the reaction is carried out in a boiling solvent. How is the desired temperature realized?*

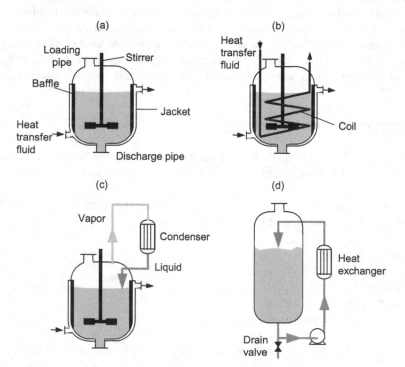

Figure 12.16 *Batch reactor systems: (a) with double jacket; (b) with double jacket and internal cooling coil; (c) with cooling by vapor phase condensation; (d) with external heat exchanger on liquid recirculation loop. (a), (b) and (c) are mechanically-stirred-tank reactors.*

12.5.2   Reactors for Gas–Liquid–Solid Systems

Most solid-catalyzed reactions for the production of bulk chemicals are gas phase reactions. In contrast, in the production of fine chemicals, gas–liquid–solid catalyst systems are common, for example, for catalytic hydrogenations, oxidations, hydrodesulfurizations, and reductive aminations. The reasons that a liquid phase is present are numerous. One reason can be low volatility of the generally bulky molecules, which together with their limited thermal stability (due to the often complex structure) prohibits operation in the gas phase, which would require relatively high temperatures. Furthermore, the selectivity might decrease with increasing temperature. It is also possible that it is desired to surround the catalyst particles with a liquid layer in order to:

- avoid deactivating deposits, and thus ensure higher catalyst effectiveness;
- achieve better temperature control due to the higher heat capacity of liquids;
- modify the active catalyst sites to promote or inhibit certain reactions.

In these instances, a liquid phase, often a solvent, is added on purpose. However, a disadvantage is that an extra barrier is introduced between the gaseous reactant (hydrogen, oxygen, etc.) and the catalyst. Then the mass transfer rate of the gaseous reactant through the liquid film may be the rate limiting step.

QUESTIONS:

> *Explain why in hydrogenation in multiphase reactors hydrogen transfer from the gas bubbles to the liquid often is the rate-determining step. Comparing water with an organic solvent: in which solvent do you expect the highest rate of hydrogenation?*

Gas–liquid–solid reactors are either fixed bed reactors (usually cocurrent trickle bed) or reactors in which the catalyst particles move. Trickle bed reactors have large-scale applications in oil refining processes such as hydrotreating and catalytic hydrogenation of residues (Chapter 3). Despite their continuous operation, trickle bed reactors have significant potential in fine chemicals production because they are well suited for high-pressure operations.

Figure 12.17 shows some examples of reactors with moving catalyst particles. The most commonly used reactors in fine chemicals production are still suspension (or slurry) reactors. The stirred-tank reactor, bubble column, and jet loop reactor are all suspension reactors, in which very fine catalyst particles (1–200 μm) are distributed throughout the volume of the liquid. Many variations exist for each reactor type. For instance, stirred tanks may have various types of agitators, cooling jackets or cooling coils or both. Bubble columns may be empty, packed, or fitted with trays.

In all reactor types presented in Figure 12.17, good mixing is important to aid in the transport of a gaseous reactant from the gas phase to the catalyst. The mechanically-stirred-tank reactor is most commonly used in batch processes. The catalyst particles are suspended in the liquid, which is almost perfectly mixed by a mechanical agitator. Cooling is usually accomplished by coils within the reactor or by a cooling jacket. Another option is circulation of the liquid/solid slurry over external cooling elements.

In bubble columns, agitation of the liquid phase, and hence suspension of the catalyst, is effected by the gas flow. Gas recycle causes more turbulence and thus better mixing. Often, circulation of the liquid is required to obtain a more uniform suspension. This can either be induced by the gas flow (airlift loop reactor) or by an external pump. In the latter case, it is possible to return the slurry to the reactor at high flow rate through an ejector (venturi tube). The local underpressure causes the gas to be drawn into the passing stream, thus providing very efficient mixing. This type of reactor is called a jet loop or venturi reactor. Jet loop reactors tend to replace stirred-tank reactors in hydrogenations. The external heat exchanger on the liquid circulation loop enables high heat removal capacity, which is a great advantage in highly exothermic reactions. A limitation

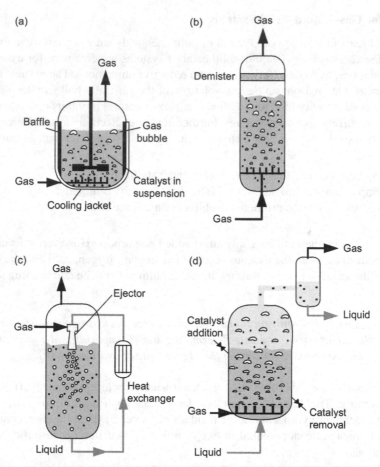

Figure 12.17   *Reactors with moving catalyst particles: (a) mechanically-stirred-tank reactor; (b) bubble column reactor; (c) jet loop reactor; (d) three-phase fluidized bed reactor.*

on the use of jet loop reactors is that the catalyst must be compatible with the pump, that is, possess low hardness and high attrition resistance.

The (not very commonly used) three-phase fluidized bed reactor (also called ebullated bed reactor) differs from the suspension reactors in the use of larger catalyst particles (0.1–3 mm) and the formation of a well-defined agitated catalyst bed. Whereas the suspension reactors can operate in both batch mode and continuously with respect to the liquid phase (and catalyst), the ebullated bed reactor (see also Section 3.4.5) only operates in the continuous mode, and hence is generally not the appropriate choice for fine chemicals.

QUESTIONS:

> *Compare fixed catalyst with moving catalyst reactors for the production of fine chemicals. What are advantages and disadvantages? Compare the moving catalyst reactors.*
> *In Table 12.10 typical values for mass transfer parameters are presented. How do the reactors compare with respect to mass transport in the interior of the catalyst particles?*

Table 12.10 compares characteristics of a number of three-phase reactors.

Table 12.10 Comparison of mechanically-stirred tank, bubble column and jet loop reactor for catalytic gas–liquid reactions [28, 58, 62–64].

Characteristic	Stirred tank	Bubble column	Jet loop	Trickle flow
Particle diameter (mm)	0.001–0.2	0.001–0.2	0.001–0.2	1.5–6
Fraction of catalyst (m_{cat}^3/m_r^3)	0.001–0.01	0.001–0.01	0.001–0.01	0.55–0,6
Liquid hold-up (m_l^3/m_r^3)	0.8–0.9	0.8–0.9	0.8–0.9	0.01–0.25
Mass transfer parameters				
Gas–liquid, $k_l a_l$ (s^{-1})	0.03–0.3	0.05–0.24	0.2–3.5	0.01–0.1
Liquid–solid, $k_s a_s$ (s^{-1})	0.1–0.5	≈0.25	0.1–1	0.06

12.6 Batch Reactor Scale-up Effects[3]

This section focuses on the scale-up of homogeneous stirred-tank batch reactors. In general, the temperature and concentration profiles and the mixing conditions in a full-scale reactor differ significantly from those in a laboratory reactor. Therefore, it is not surprising that selectivities usually change upon scale-up.

12.6.1 Temperature Control

Batch reactors are often operated as shown in Figure 12.18. At first, the temperature of the heat transfer medium (T_h) is high in order to heat the reaction mixture to reaction temperature (T_r). When an exothermic reaction starts, the temperature of the heat transfer medium is lowered, often to the lowest possible level. As soon as the reaction is under control (Figure 12.18a), the temperature may be raised to keep the reaction mixture at the optimal reaction temperature to achieve sufficiently high reaction rates. If a reaction is fast and/or highly exothermic, it might be impossible to keep the temperature under control even with maximum cooling (Figure 12.18b). Obviously, an event like this, a so-called runaway, is undesired. Batch reactors are unsuitable for such reactions, so other procedures are called for. Often, so-called semi-batch operation is used (Section 12.6.3).

12.6.2 Heat Transfer

Usually, the mechanically-stirred-tank reactors in the fine chemicals industry are of the jacketed type. As a consequence, the heat transfer area per unit volume decreases with increasing reactor volume. In principle, heat transfer can be improved by more intensive mixing. In the average commercial reactor, however, this does not result in drastically increased heat transfer rates. Heat transfer may also be improved by using a lower coolant temperature, but:

- use of cooling liquids at very low temperature is costly;
- components from the reaction mixture might precipitate at the reactor wall.

It is also possible to use internal or external heat exchange surfaces that allow large heat transfer rates at relatively mild cooling temperature. Moreover, if necessary, heat production rates can be reduced by using lower concentrations of reactants or catalysts. Another option is semi-batch operation with controlled dosing of (one of the) reactants.

---

[3] The late A. Cybulski has made a large contribution to this section.

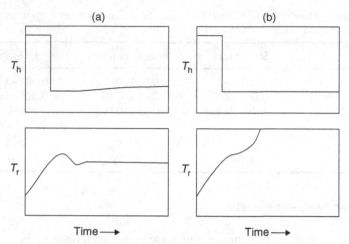

Figure 12.18   *Control of reactor temperature (T_r) by manipulating the temperature of the heat transfer medium (T_h): (a) reaction under control; (b) reaction out of control.*

QUESTIONS:

Give a simple expression for the amount of heat transferred, Q, in a batch stirred-tank reactor (double jacket).

Derive an equation for the overall heat transfer coefficient, U. If the internal heat transfer coefficient could be increased by 100%, what would be the effect on U?

12.6.3   Example of the Scale-up of a Batch and Semi-Batch Reactor

The influence of heat transfer on the selectivity and yield in scaling up batch and semi-batch reactors is illustrated here with the results of a simulation using a series reaction. This reaction is composed of two irreversible elementary steps, both exothermic and both with first-order kinetics:

$$A \xrightarrow{k_1} P \xrightarrow{k_2} S \tag{12.5}$$

A is the reactant, P the desired product, and S an unwanted by-product. k_1 and k_2 are the reaction rate coefficients (s^{-1}), whereby the pre-exponential factors and the activation energy in the Arrhenius equation for k_2 are much larger than those in that of k_1. Table 12.11 shows the required mole and energy balance equations with accompanying data, specific for this example.

Figure 12.19 shows the results of calculations for reactors of small and large volume, V. Clearly, the two reactors behave quite differently as a function of the time in the reactor. At large residence time, the selectivity of the reaction in a large reactor is significantly lower than that in a laboratory reactor. Imagine what the operators must have thought when in the laboratory it was found that after 5000 s the product concentration was 400 mol m^{-3} and in the commercial reactor at identical conditions the concentration was close to zero! Their first reaction would probably have been to ask for a new catalyst.

Nothing surprising has happened, however. The reason for the unsatisfactory yield is that the area-to-volume ratio available for heat transfer, A_h/V, of the large reactor is much smaller than that of the small reactor, which resulted in a temperature runaway. Hence, the rate of the second reaction increased, consequently resulting

Table 12.11   *Mole and energy balance equations for reaction 12.5 in a batch reactor.*

Mole balance equations:

$$\frac{dC_A}{dt} = r_1 = -k_1 C_A \tag{12.6}$$

$$\frac{dC_P}{dt} = r_2 - r_1 = k_1 C_A - k_2 C_P \tag{12.7}$$

Energy balance equation[a]:

$$\rho\, c_p \frac{dT}{dt} = (-r_1)(-\Delta H_1) + (-r_2)(-\Delta H_2) + U(T_h - T) A_h / V \tag{12.8}$$

Initial conditions: $t = 0$; $C_A = C_{A0}$; $C_P = C_{P0}$; $T = T_0$

Explanation of symbols and reaction data

$-r_1$ = rate of reaction of A (mol m^{-3} s^{-1})

$-r_2$ = rate of reaction of P (mol m^{-3} s^{-1})

$k_i = k_{i0} \exp(-E_i/RT)$ (s^{-1})

k_{i0} = pre-exponential factor (s^{-1})

E_i = activation energy for reaction i (J mol^{-1})

R = gas constant (8.314 J mol^{-1} K^{-1})

ΔH_i = reaction enthalpy (kJ mol^{-1})

c_p = heat capacity (kJ kg^{-1} K^{-1})

ρ = density (kg m^{-3})

U = overall heat transfer coefficient

C_i = concentration of i (mol m^{-3})

T_r = temperature of reaction mixture (K)

T_h = temperature of heat transfer medium (K)

A_h = area available for heat transfer (m^2)

V = reactor volume (m^3)

$k_{10} = 0.5$ s^{-1}

$k_{20} = 10^{11}$ s^{-1}

$E_1 = 20$ kJ mol^{-1}

$E_2 = 100$ kJ mol^{-1}

$\Delta H_1 = -300$ kJ mol^{-1}

$\Delta H_2 = -250$ kJ mol^{-1}

$c_p = 4$ kJ kg^{-1} K^{-1}

$\rho = 1000$ kg m^{-3}

$U = 500$ W m^{-2} K^{-1}

$C_{A0} = 1000$ mol m^{-3}, $C_{P0} = 0$ mol m^{-3}

$T_{0(r)} = 295$ K

$T_h = 345$ K $0 < t < 3.6$ ks

295 K $3.6 < t < 5.4$ ks

$A_{h, small} = 9.5$ m^{-1}, $A_{h,large} = 2.6$ m^{-1}

$V_{small} = 0.063$ m^3, $V_{large} = 6.3$ m^3

[a]The reactant and products are all dissolved in water. They are supposed not to influence the heat capacity and density of the mixture.

in decreased selectivity. Obviously, the procedure followed in this case was far from optimal; the heat supply should have been switched off after initiating the reaction and intensive cooling should have started. Then, after the temperature peak had passed, moderate cooling of the reaction mixture would have maintained the temperature at the desired level.

If a temperature rise is difficult to control, semi-batch operation is generally used. At the start, only a small amount of A is loaded into the reactor and, when the desired temperature is reached, more A is steadily added over time. By carefully dosing A, the rate of heat evolution is controlled and a runaway can be prevented.

In the simulation of the semi-batch mode, the process conditions are the same as in the batch reactor except for the dosing scheme of A. The initial concentration of A is lower (400 mol m^{-3}) and the remaining amount of A is dosed at a rate of 0.3 mol m^{-3} s^{-1} starting at time $t_d = 1800$ s. Figure 12.20 shows the results of calculations for a large semi-batch reactor.

Table 12.12 compares the results of the computations for both batch and semi-batch reactors.

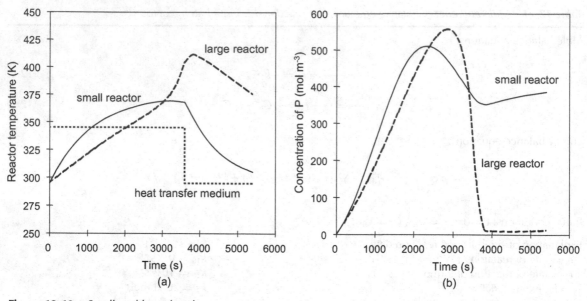

Figure 12.19   *Small and large batch reactors: (a) temperature and (b) concentration of product P versus time.*

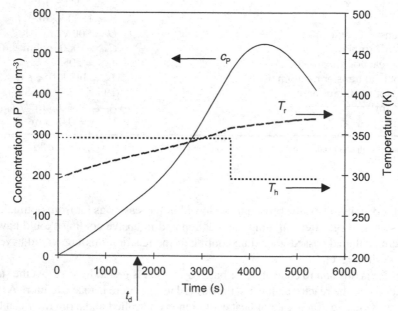

Figure 12.20   *Concentration of product P (left) and reactor temperature (right) versus time in semi-batch reactor; t_d is the time when dosing of reactant A starts.*

Table 12.12 *Batch and semi-batch reactor results (after 5400 s).*

Reactor	V (m³)	A_h/V (m⁻¹)	Conversion (% mol)	Selectivity (% mol)	Yield on P (% mol)
Batch	0.063	9.5	92.5	41.6	38.5
Batch	6.3	2.6	97.8	1.2	1.2
Semi-batch	6.3	2.6	85.6	47.2	40.2

The data presented clearly show that even in the case of an inappropriate cooling policy, the yield and selectivity in a large semi-batch reactor are better than those in a much smaller batch reactor. It is also clear that this type of modeling greatly helps in the design and operation of (semi-)batch processes.

QUESTION:

The simulations show that the difference between the two reactors is very large. Which parameters of a reaction system are critical in practice? (Hint: compare the parameter values in Table 12.12.)

Semi-batch operation is also often employed when two or more different reactants are present. Initially, one reactant and possible additives are loaded into the reactor and, when the desired temperature is reached, the other reactant(s) and additives are added with time. Figure 12.21 illustrates semi-batch operation for the exothermic reaction:

$$A + B \rightarrow P \qquad (12.9)$$

Initially, the reactor contains only A, possibly dissolved in an inert solvent. After a predetermined temperature has been reached, B is fed continuously or in portions while the temperature of the heat transfer medium is

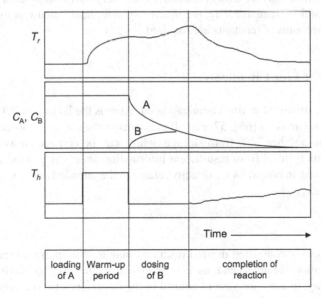

Figure 12.21 *Semi-batch operation of a batch stirred-tank reactor for the reaction $A + B \rightarrow P$ (T_r is the temperature of the reaction mixture; T_h is the temperature of the heat transfer medium; C_A and C_B are the concentrations of A and B).*

decreased. By appropriately adjusting the feed rate of B, the reactor stays far from its runaway temperature. After a certain time, when the reaction rate begins to fall below a certain value, the temperature of the heat transfer medium may be raised to keep the reaction mixture at the optimal temperature.

QUESTION:

> *In the case that a runaway is possible, semi-batch operation coupled with a low coolant temperature can lead to safe operation. However, it is possible that if the temperature is too low a dangerous situation develops, whereas this would not have been the case at higher temperature. Explain this.*

12.6.4   Summary of the Scale-up of Batch Reactors

A laboratory unit can be transposed to an industrial unit based on the reaction time alone, provided the temperatures and the concentration of all components involved are identical in both units. However, in practice this is often difficult to achieve. One reason is that for processes involving highly exothermic reactions maintaining the same temperature in a large unit as in the small laboratory unit is difficult. Furthermore, mixing and hydrodynamic conditions in a large tank are different from those in a small tank, resulting in concentration differences. Especially when the reaction networks are complex, unexpectedly low yields and/or selectivities may be the result.

Fast changes in market demand often require the quick development of a new process. A traditional approach towards reactor scale-up is then usually preferred: The process is tested in semi-technical and/or pilot reactors whose size is not less than 10% of the size of the industrial reactor. Ranges of safe operation and satisfactory yields are determined on the intermediate scales without knowledge of the reaction kinetics. This procedure may lead to a process that is suboptimal. Therefore, if there is sufficient time and budget, a chemical reaction engineering approach based on kinetic models, including mass and heat transfer phenomena, is recommended.

In practice, continuous-flow microreactors (Chapter 14) are very useful in process development. They can be constructed in a fast and inexpensive way, are inherently safer than batch reactors because of the small volume, and only small amounts of reactants are required.

12.7   Safety Aspects of Fine Chemicals

An illustration of the risks involved in fine chemicals production is the large number of incidents with batch processes. According to Rasmussen [65], 57% of all incidents in the chemical process industries originated from batch processes and another 24% from storage, which can be considered as a batch operation also. Frequently, these incidents resulted from insufficient information about the process and from design errors. Most of the incidents leading to runaways are directly related to the intended reactions and not to, for instance, unexpected side reactions.

12.7.1   Thermal Risks

Undesired reactions or poorly-controlled desired reactions may lead to thermal runaways. For example, in storage tanks heat dissipation is limited and, as a consequence, even slowly occurring exothermic reactions can cause thermal runaway. In a reactor, loss of control of the desired reactions can occur due to the fact that heat removal typically shows a linear and heat production an exponential relationship with temperature. It is, therefore, necessary to determine a range of stable reactor operation conditions. Furthermore, a runaway of the desired reactions can result in secondary reactions that cause the formation of undesired products.

Semi-batch operation is not always safe either. For instance, if the temperature in a semi-batch reactor is chosen too low, the reaction rates will be low and, consequently, it is possible that large amounts of intermediates or raw materials accumulate. If these participate in highly exothermic reactions, a runaway might occur. So, a runaway can be caused by too low an initial temperature or coolant temperature!

QUESTION:

> *Different heat transfer systems may be chosen: electrical heaters, steam heaters, or heating by a circulating liquid. Which operation is safest?*

12.7.2 Safety and Process Development

Safety is the major factor when working with hazardous materials and/or highly exothermic reactions. Usually, purely economic considerations are less important, although the safest possible process in the long run is also the most profitable. Three basic routes exist for protection against chemical process hazards [66,67]. These routes are schematically depicted in Figure 12.22.

Containment usually is an expensive means of reducing damage resulting from what has already happened. Installation of on-line systems for the detection of process deviations, associated with trip systems for corrective actions, adds to the initial investment but decreases the risk. The objective of trip systems is to stop the process driving to explosion. Trip systems are becoming cheaper thanks to progress in electronics. The third route, that is, the design of an inherently safe process, requires the deepest understanding of the process and, hence, the highest research expenditure. However, the risk of hazardous situations is much reduced. A process is inherently safe if no fluctuation, disturbance or failure (of equipment, control systems, or people) can cause an accident. Inherent safety is served by [66–68]:

- identifying all reactions that may result in dangerous substances, overpressure, or runaways and establishing the process design and conditions disfavoring these reactions;
- minimizing the quantities of hazardous substances, preferentially by using safer substances, or by *in situ* production and use;
- avoiding external sources that could trigger runaways, for instance by using heat exchange fluids which do not react in a dangerous way with the reaction mixture;
- operating the process well within the region of stability.

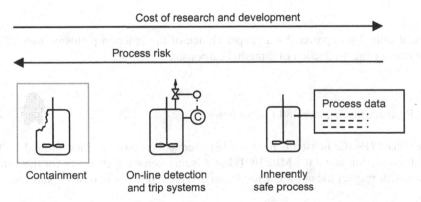

Figure 12.22 *Three routes for protection against chemical process hazards.*

The choice of reactor is a major decision. Batch reactors are only suitable for non-hazardous processes of low or moderate rate. If batch operation is too risky, semi-batch operation is preferred. Continuous reactors are the best solution for fast and highly exothermic reactions.

With respect to safety the "human" factor should not be underestimated. Box 12.10 illustrated that a better process configuration reduces the chance of making mistakes.

Box 12.10   Different Reactors for Different Stages [68]

In a certain process with two reaction stages, originally both reactions were performed in one reactor (Figure B12.10.1a). If by mistake, the first-stage reactants (A and B) were added during the second stage, or if second-stage reactants (C and D) were added during the first stage, a runaway reaction could occur. Originally, interlocks and training were used to prevent such incidents.

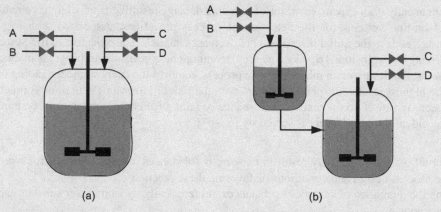

(a)   (b)

Figure B12.10.1   *Two-stage batch reaction: (a) original design; (b) later design.*

An inherently safer method is to perform the two reaction stages in two different reactors, with A and B fed only to the first reactor and C and D fed only to the second (Figure B12.10.1b).

Safety can be drastically improved by a proper choice of the reaction pathway. Box 12.11 describes a phosgene-free route for the production of dimethyl carbonate.

Box 12.11   Production of Dimethyl Carbonate

Dimethyl carbonate (DMC) in itself is hardly a fine chemical (plant capacities: around 10 000 t/a, total current world production about 0.5 Mt/a [69]) but it is an interesting chemical for the manufacture of fine chemicals with respect to both waste reduction and the replacement of toxic substances with safer ones.

The traditional route to dimethyl carbonate (DMC) involves reaction of methanol with phosgene:

$$2\,CH_3OH + Cl_2CO \rightarrow (CH_3O)_2CO + 2\,HCl \qquad \Delta_r H_{298} = -864 \text{ kJ/mol}$$

Apart from the very dangerous effects of phosgene (which was used as a poison gas in World War I), its industrial application leads to problems concerning the removal and disposal of polluting waste (HCl and chlorinated hydrocarbons). A more recently developed process is based on the oxidative carbonylation of methanol:

$$2\,CH_3OH + CO + 0.5\,O_2 \rightarrow (CH_3O)_2CO + H_2O \qquad \Delta_r H_{298} = -489 \text{ kJ/mol}$$

The reaction takes place in a suspension reactor in the presence of copper salts as the catalyst. This reaction carries much less risk than the phosgene route and no HCl is produced. The technology is evolving rapidly because of the great advantage of DMC over phosgene and other toxic compounds as a specialty solvent and as an intermediate in the production of fine chemicals and pharmaceuticals.

QUESTIONS:

Why is DMC an interesting chemical for the manufacture of fine chemicals? Can you think of a potentially even more attractive reaction for the production of DMC?

References

[1] Cybulski, A., Moulijn, J.A., Sharma, M.M., and Sheldon, R.A. (2001) *Fine Chemicals Manufacture – Technology and Engineering*. Elsevier Science, Amsterdam, The Netherlands, p. 551.

[2] Anon. (2012) Atorvastatin, http://en.wikipedia.org/wiki/Atorvastatin (last accessed 20 December 2012).

[3] Kleemann, A. and Engel, J. (2001) *Pharmaceutical Substances*, 4th edn, Thieme-Verlag. Stuttgart, New York, pp. 146–150.

[4] Sheldon, R.A. (1994) Consider the environmental quotient. *CHEMTECH*, **March**, 38–47.

[5] Sheldon, R.A. (2000) Atom efficiency and catalysis in organic synthesis. *Pure and Applied Chemistry*, **72**, 1233–1246.

[6] Sheldon, R.A. (1993) The role of catalysis in waste minimzation, in *Precision Process Technology; Perspectives for Pollution Prevention* (eds M.P.C. Weijnen and A.A.H. Drinkenburg). Kluwer, Dordrecht, The Netherlands.

[7] Trost, B.M. (1991) The atom economy – a search for synthetic efficiency. *Science*, **254**, 1471–1477.

[8] Li, C.J. and Trost, B.M. (2008) Green chemistry for chemical synthesis. *Proceedings of the National Academy of Sciences*, **105**, 13197–13202.

[9] Sheldon, R.A. (2007) The E Factor: fifteen years on. *Green Chemistry*, **9**, 1273–1283.

[10] Dunn, P.J. (2012) The importance of green chemistry in process research and development. *Chemical Society Reviews*, **41**, 1452–1461.

[11] Ishino, M. (2006) Development of new propylene oxide process. 16th Saudi Arabia–Japan Joint Symposium, 1–11.

[12] Rohan, D., Canaff, C., Fromentin, E., and Guisnet, M. (1998) Acetylation of anisole by acetic anhydride over a HBEA zeolite — Origin of deactivation of the catalyst. *Journal of Catalysis*, **177**, 296–305.

[13] Métivier, P. (2001) Friedel–Crafts acylation, in *Fine Chemicals Through Heterogeneous Catalysis* (eds R.A. Sheldon and H. van Bekkum). Wiley-VCH Verlag GmbH, Weinheim, pp. 161–172.

[14] Arends, I., Sheldon, R.A., and Hanefeld, U. (2007) Introduction: green chemistry and catalysis, in *Green Chemistry and Catalysis*. Wiley-VCH Verlag GmbH, Weinheim, pp. 1–47.

[15] Guisnet, M. and Guidotti, M. (2009) Applications in synthesis of commodities and fine chemicals, in *Zeolite Characterization and Catalysis* (eds A.W. Chester and E.G. Derouane). Springer, The Netherlands, pp. 275–347.

[16] Arends, I., Sheldon, R.A., and Hanefeld, U. (2007) *Green Chemistry and Catalysis*. Wiley-VCH Verlag GmbH, Weinheim.

[17] Blaser, H.U., Malan, C., Pugin, B.T., Spindler, F., Steiner, H. and Studer, M. (2003) Selective hydrogenation for fine chemicals: recent trends and new developments. *Advanced Synthesis & Catalysis*, **345**, 103–151.

[18] Blaser, H.U. (2006) A golden boost to an old reaction. *Science*, **313**, 312–313.

[19] Corma, A. and Serna, P. (2006) Chemoselective hydrogenation of nitro compounds with supported gold catalysts. *Science*, **313**, 332–334.

[20] Huisgen, R. (1984) 1,3-Dipolar cycloaddition – Introduction, survey, mechanism, in *1,3-Dipolar Cycloadditional Chemistry* (ed. A. Padwa). John Wiley & Sons, Inc., New York, pp. 1–176.

[21] Sha, C.K. and Mohanakrishnan, A.K. (2002) Azides, in *Synthetic Applications of 1,3-Dipolar Cycloaddition Chemistry Toward Heterocycles and Natural Products*, vol. **59** (eds A. Padwa and W.H. Pearson). John Wiley & Sons, Inc., New York, pp. 623–679.

[22] Rostovtsev, V.V., Green, L.G., Fokin, V.V., and Sharpless, K.B. (2002) A stepwise huisgen cycloaddition process: copper(I)-catalyzed regioselective "Ligation" of azides and terminal alkynes. *Angewandte Chemie International Edition*, **41**, 2596–2599.

[23] Zhang, L., Chen, X., Xue, P., Sun, H.H.Y., Williams, I.D., Sharpless, K.B., Fokin, V.V. and Jia, G. (2005) Ruthenium-Catalyzed Cycloaddition of Alkynes and Organic Azides. *Journal of the American Chemical Society*, **127**, 15998–15999.

[24] Rasmussen, L.K., Boren, B.C., and Fokin, V.V. (2007) Ruthenium-catalyzed cycloaddition of aryl azides and alkynes. *Organic Letters*, **9**, 5337–5339.

[25] Kolb, H.C., Finn, M.G., and Sharpless, K.B. (2001) Click chemistry: diverse chemical function from a few good reactions. *Angewandte Chemie International Edition* **40**, 2004–2021.

[26] Lin, G.Q., Zhang, J.G., and Cheng, J.F. (2011) Overview of chirality and chiral drugs, in *Chiral Drugs*. John Wiley & Sons, Inc. Hoboken, pp. 3–28.

[27] Knowles, W.S. (1983) Asymmetric hydrogenation. *Accounts of Chemical Research*, **16**, 106–112.

[28] Mills, P.L., Ramachandran, P.A., and Chaudhari, R.V. (1992) Multiphase reaction engineering for fine chemicals and pharmaceuticals. *Reviews in Chemical Engineering*, **8**, 5–176.

[29] Parshall, G.W. and Nugent, W.A. (1988) Making pharmaceuticals via homogeneous catalysis, part 1. *CHEMTECH*, **3**, 184–190.

[30] Mills, P.L. and Chaudhari, R.V. (1997) Multiphase catalytic reactor engineering and design for pharmaceuticals and fine chemicals. *Catalysis Today*, **37**, 367–404.

[31] Leffingwell, J. and Leffingwell, D. (2011) Chiral chemistry in flavours & fragrances. *Specialty Chemicals Magazine*, **March**, 30–33.

[32] Hagen, J. (2006) Asymmetric catalysis, in *Industrial Catalysis: A Practical Approach*, 2nd edn. Wiley-VCH Verlag GmbH, Weinheim, pp. 59–82.

[33] Bruggink, A., Roos, E.C., and De Vroom, E. (1998) Penicillin acylase in the industrial production of β-lactam antibiotics. *Organic Process Research & Development*, **2**, 128–133.

[34] Wegman, M.A., Janssen, M.H., Van Rantwijk, F., and Sheldon, R.A. (2001) Towards biocatalytic synthesis of β-lactam antibiotics. *Advanced Synthesis & Catalysis*, **343**, 559–576.

[35] Deaguero, A.L., Blum, J.K., Bommarius, A.S., and Flickinger, M.C. (2009) Biocatalytic synthesis of β-lactam antibiotics, in *Encyclopedia of Industrial Biotechnology*. John Wiley & Sons, Inc., Hoboken.

[36] Oyama, K. (1992) The industrial production of aspartame, in *Chirality in Industry* (eds A.N. Collins, G.N. Sheldrake and J. Crosby). John Wiley & Sons, Inc., New York, pp. 237–247.

[37] Ritter, S.K. (2010) Greening up process chemistry. *Chemical & Engineering News*, **88**, 45–47.

[38] Sheldon, R.A. (2007) Enzyme immobilization: the quest for optimum performance. *Advanced Synthesis & Catalysis*, **349**, 1289–1307.

[39] Jiménez-González, C., Curzons, A., Constable, D., and Cunningham, V. (2004) Cradle-to-gate life cycle inventory and assessment of pharmaceutical compounds. *The International Journal of Life Cycle Assessment*, **9**, 114–121.

[40] Constable, D.J.C., Jimenez-Gonzalez, C., and Henderson, R.K. (2006) Perspective on solvent use in the pharmaceutical industry. *Organic Process Research & Development*, **11**, 133–137.

[41] Alfonsi, K., Colberg, J., Dunn, P.J., Fevig, T., Jennings, S., Johnson, T.A., Kleine, H.P., Knight, C., Nagy, M.A., Perry, D.A. and Stefaniak, M. (2008) Green chemistry tools to influence a medicinal chemistry and research chemistry based organisation. *Green Chemistry*, **10**, 31–36.

[42] Taber, G.P., Pfisterer, D.M., and Colberg, J.C. (2004) A New and Simplified Process for Preparing N-[4-(3,4-Dichlorophenyl)-3,4-dihydro-1(2H)-naphthalenylidene]methanamine and a Telescoped Process for the Synthesis of (1S-cis)-4-(3,4-Dichlorophenol)-1,2,3,4-tetrahydro-N-methyl-1-naphthalenamine Mandelate: Key Intermediates in the Synthesis of Sertraline Hydrochloride. *Organic Process Research & Development*, **8**, 385–388.

[43] Manley, J.B., Anastas, P.T., and Cue, J. (2008) Frontiers in Green Chemistry: meeting the grand challenges for sustainability in R&D and manufacturing. *Journal of Cleaner Production*, **16**, 743–750.

[44] Varma, R.S. (2007) Clean chemical synthesis in water. *Organic Chemistry Highlights*, **February 1**, 1–7.

[45] Leitner, W. (2002) Supercritical carbon dioxide as a green reaction medium for catalysis. *Accounts of Chemical Research*, **35**, 746–756.

[46] Jessop, P.G. and Leitner, W. (1999) *Chemical Synthesis Using Supercritical Fluids* (eds P.G. Jessop and W. Leitner). Wiley-VCH Verlag GmbH, Weinheim, pp. 1–480.

[47] Licence, P., Ke, J., Sokolova, M., Ross, S.K., and Poliakoff, M. (2003) Chemical reactions in supercritical carbon dioxide: from laboratory to commercial plant. *Green Chemistry* **5**, 99–104.

[48] Zhao, D., Wu, M., Kou, Y., and Min, E. (2002) Ionic liquids: applications in catalysis. *Catalysis Today*, **74**, 157–189.

[49] Dupont, J., de Souza, R.F., and Suarez, P.A.Z. (2002) Ionic liquid (molten salt) phase organometallic catalysis. *Chemical Reviews*, **102**, 3667–3692.

[50] Song, C.E. (2004) Enantioselective chemo- and bio-catalysis in ionic liquids. *Chemical Communications*, 1033–1043.

[51] Párvulescu, V.I. and Hardacre, C. (2007) Catalysis in ionic liquids. *Chemical Reviews*, **107**, 2615–2665.

[52] Van Rantwijk, F. and Sheldon, R.A. (2007) Biocatalysis in ionic liquids. *Chemical Reviews*, **107**, 2757–2785.

[53] Moniruzzaman, M., Nakashima, K., Kamiya, N., and Goto, M. (2010) Recent advances of enzymatic reactions in ionic liquids. *Biochemical Engineering Journal*, **48**, 295–314.

[54] Meindersma, G.W., Maase, M., and De Haan, A.B. (2000) Ionic liquids, in *Ullmann's Encyclopedia of Industrial Chemistry*. Wiley-VCH Verlag GmbH, Weinheim. doi: 10.1002/14356007.l14_l01

[55] Stevens, J.G., Gomez, P., Bourne, R.A., Drage, T.C., George, M.W. and Poliakoff, M. (2011) Could the energy cost of using supercritical fluids be mitigated by using CO_2 from carbon capture and storage (CCS)? *Green Chemistry*, **13**, 2727–2733.

[56] Rogers, R.D. and Seddon, K.R. (2003) Ionic liquids – solvents of the future? *Science*, **302**, 792–793.

[57] Seddon, K.R. (2003) Ionic liquids: a taste of the future. *Nature Materials*, **2**, 363–365.

[58] Trambouze, P., Van Landeghem, H., and Wauquier, J.P. (1988) *Chemical Reactors, Design/Engineering/Operation*, Gulf Publishing Company. Houston, TX.

[59] Thomas, M.R. (1997) Salicylic acid and related compounds, in *Kirk-Othmer Encyclopedia of Chemical Technology*, 4th edn, vol. **21** (eds J.I. Kroschwitz and M. Howe-Grant). John Wiley & Sons, Inc., New York, pp. 601–626.

[60] Lowenheim, F.A. and Moran, M.K. (1975) *Faith, Keyes and Clark's Industrial Chemicals*, 4th edn. John Wiley & Sons, Inc., New York, pp. 117–120.

[61] Bruggink, A. (1998) Growth and efficiency in the (fine) chemical industry. *Chimica Oggi (Chemistry Today)*, **16**, 44–47.

[62] Irandoust, S., Cybulski, A., and Moulijn, J. A. (1997) The use of monolithic catalysts for three-phase reactions, in *Structured Catalysts and Reactors* (eds A. Cybulski and J.A. Moulijn). Dekker, New York.

[63] van Dierendonck, L.L., Zahradnik, J., and Linek, V. (1998) Loop venturi reactor – a feasible alternative to stirred tank reactors? *Industrial & Engineering Chemistry Research*, **37**, 734–738.

[64] Nardin, D. (1995) Trends and opportunities with modern Buss loop reactor technology. Chemspec Europe 95 BACS Symposium.

[65] Rasmussen, B. (1987) Unwanted Chemical Reactions in the Chemical Process Industry. Report Risoe-M-2631, The Risø National Laboratory, Roskilde, Denmark.

[66] Regenass, W., Osterwalder, U., and Brogli, F. (1984) Reactor engineering for inherent safety. Eight International Symposium on Chemical Reaction Engineering, Symposium Series No. 87, Institution of Chemical Engineers, Rugby, UK, pp. 369–376.

[67] Regenass, W. (1984) The control of exothermic reactors. Proceedings of the Symposium on Protection of Exothermic Reactions, Chester, April.

[68] Kletz, T.A. and Amyotte, P. (2010) Limitation of effects, in *Process Plants*, 2nd edn. CRC Press, pp. 113–126.

[69] Quadrelli, E.A., Centi, G., Duplan, J.L., and Perathoner, S. (2011) Carbon Dioxide Recycling: Emerging large-scale technologies with industrial potential. *ChemSusChem*, **4**, 1194–1215.

[70] de Lathouder, K.M., Lozano-Castelló, D., Linares-Solano, A., Wallin, S.A., Kapteijn, F. and Moulijn, J.A. (2007). Carbon-ceramic composites for enzyme immobilization, *Microporous and Mesoporous Materials*, **99**, 216–223.

13

Biotechnology

13.1 Introduction

Biotechnology is both an old and a novel discipline. It is old in the sense that it has been applied in traditional processes such as the production of beer and wine since before 6000 BC. However, it is fair to state that it was more an art than a scientific discipline until a few decades ago but this has changed. Biotechnology now encompasses an array of subdisciplines, such as microbiology, biochemistry, cell biology, and genetics. It has become possible to describe life processes now in great depth at cellular and molecular level.

Just like chemical engineering, bioengineering includes kinetics, transport phenomena, reactor design, and unit operations; it is not surprising that chemical engineers contribute significantly to the field. It is frequently stated that biotechnology has a large potential for the future. However, A lot of progress has been made and already today the life sciences affect over 30% of the global economic turnover, mainly in the sectors of environment, healthcare, food and energy, and agriculture and forestry. The economic impact will grow as biotechnology provides new ways of influencing raw material processing. Biotechnology has enabled breakthroughs in the manufacture of new pharmaceuticals and the development of gene therapies for treatment of previously incurable diseases. Biotechnology is also very important in solving environmental problems, such as in wastewater treatment. Although in the production of chemicals, biotechnology does not yet play a large role, its role is increasing, as is illustrated in Chapter 7.

It is interesting to observe that the character of the biotechnological industry has changed from a problem-area-oriented industry to a more generic one, based on enabling technologies such as genetic modification by recombinant DNA and cell fusion techniques. The chemical industry has undergone a similar development. In the past, this industry was organized along product lines, for example, sugar industry, oil refining, and so on. Later it was discovered that generic subdisciplines can be distinguished, namely unit operations, chemical reactors, shaping techniques, and so on. Subsequently, underlying disciplines such as transport phenomena, interface chemistry, and chemical reaction engineering were emphasized. More recently, integration has become more central, as reflected in the acknowledgment of areas such as chemical reaction engineering and process integration. The same holds for biotechnology, where application areas as, for instance, beer production and water purification, have developed independently from each other.

The biotechnological industry is based largely on renewable and recyclable materials, so is able to adapt to the needs of a society in which energy is becoming increasingly expensive and scarce. Biotechnology can play

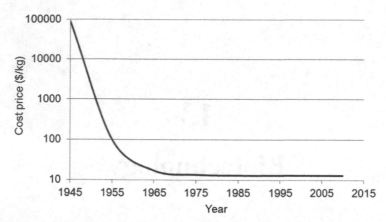

Figure 13.1   *Historical development of the cost price of penicillin.*

a substantial role in the reduction of the greenhouse effect and, in more general terms, in the realization of a sustainable society. Although biotechnology is generally considered clean technology, it shares with chemical engineering the development of processes producing waste streams, which present an environmental burden.

Typical products of biotechnological processes are (see also Chapter 7):

- cell biomass: yeast, single-cell protein (Section 13.3);
- metabolic products of the cells (Sections 13.4 and 13.5):
  o   anaerobic: alcohols, organic acids, hydrogen, carbon dioxide (Sections 13.4.1, 13.4.2, and 13.5);
  o   aerobic: citrate, glutamate, lactate, antibiotics, hydrocarbons, polysaccharides (Section 13.5);
- products of reactions catalyzed by enzymes (Section 13.6); virtually all types of chemical reactions can be catalyzed by enzymes.

In general — and biotechnology is no exception — the cost price of a product decreases with growing demand and, of course, production rate, due to development of the market with time. Figure 13.1 shows the historic development of the cost price of penicillin. It is expected that such a trend will be seen for many products that are still very expensive today.

Figure 13.2 shows an overview of some selected products and their prices. As would be expected, a distinct relationship exists between cost price and production capacity. The prices differ enormously (note the logarithmic scale). For comparison, crude oil and a number of petrochemical products (ethene, propene, and benzene) are included.

QUESTION:

> *Explain why the price of benzene is much lower than that of L-glutamic acid, while their production volumes are approximately the same.*

13.2   Principles of Fermentation Technology

To grow and/or to produce a metabolic product, microorganisms used in fermentation need a source of carbon (the substrate), energy, nitrogen, minerals, trace elements, and, frequently, vitamins. The substrate can consist of pure C-containing species such as polysaccharides, hydrocarbons, alcohols, and carbon dioxide or it can

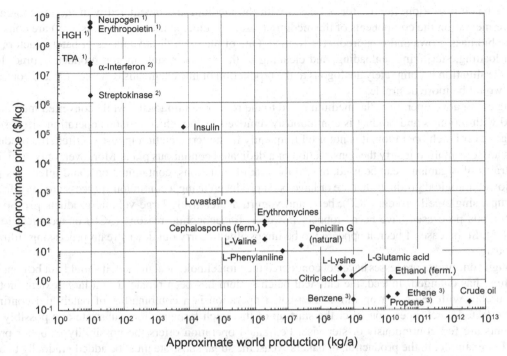

Figure 13.2 *Overview of biotechnological products and their production capacities and prices [A. Bruggink, personal communication.]. Most realistic figures for 2000; [1]1992 data; [2]estimated production costs for 1995, market price may be up to 100 times higher; [3]bulk petrochemicals.*

be a more complex material like molasses, cellulose waste liquors, pharmaceutical media, and so on. Sources of nitrogen are, for instance, ammonia, urea, and amino acids. Minerals are organic salts such as phosphates, sulfates, and chlorides, while the most important trace elements are potassium, sodium, magnesium, calcium, iron, cobalt, and zinc. In addition, aerobic organisms require oxygen for growth.

Similar to the statement that a catalyst particle is a reactor in itself, a microorganism can be viewed as a reactor. Biotransformations can be carried out by whole cells (yeast, plant or animal cells) or by part of the cell, in particular isolated enzymes, which may be referred to as biocatalysts. Although up to now their characteristics were mainly dictated by the microorganism itself, they can be modified to a large extent using recombinant DNA technology. This so-called genetic engineering, in principle, offers unlimited opportunities to create new combinations of genes that do not exist under natural conditions. This field of research is currently booming. The technique also applies to plants and animals but then is referred to as pharming. However, a critical note can be made. If society is moving towards a more sustainable world, as in all new fields of technology limits of the natural resources will be encountered. The availability of land, water, minerals, and so on, and the potential poisoning of groundwater or the disturbance of the ecosystem in rivers and lakes have to be taken into account.

13.2.1 Mode of Operation

Fermenters are often operated in batch or fed-batch (in the chemical industry usually referred to as semi-batch) mode with respect to the substrate, but continuous systems are also used.

In pure batch operation, the reactor is loaded with the medium and inoculated with the microorganism. During fermentation the components of the medium (carbon source, nutrients, vitamins, etc.) are consumed, while the biomass grows and/or a product is formed. One of the main disadvantages of batch reactors is the fact that loading, sterilizing, unloading, and cleaning of the reactor results in non-productive time. In this respect, the situation is completely analogous to the production of fine chemicals (Chapter 12) and continuous reactors would be more desirable.

During continuous operation, the medium is fed to the reactor continuously, while converted medium that is loaded with biomass and product is continuously removed. Although continuous operation offers several advantages over batch operation, it is not used frequently in the fermentation industry. Often, the production volumes are too small to justify the construction of a dedicated continuous plant. Moreover, only sufficiently stable strains of organisms can be used, due to the risk of mutations, contaminations, and infections [1, 2]. Some biotechnological products that are or have been produced using a continuous process are high-fructose corn syrups, single-cell protein (SCP), beer, and yogurt, all relatively large volume products [3]. Some of the continuous processes have been stopped, however, for economic reasons (SCP) or technical reasons (beer) [4, 5]. In the case of beer, it appeared to be impossible to carry out downstream processing (filtration) continuously.

Although continuous processes have not conquered the biotechnological market, it should not be concluded that nothing has changed. Indeed, the classical batch system has been replaced to a large extent, not with continuous but with fed-batch operation. Fed-batch operation is a combination of batch and continuous operation. In fed-batch operation, after the start-up of the batch fermentation, substrate and possibly other components are fed continuously or stepwise. Fed-batch operation offers the possibility of better process control. For example, in the production of bakers' yeast the sugar substrate must be added gradually to assure the continuous growth of biomass and to prevent the formation of alcohol, which occurs in the case of too high sugar concentration (Section 13.3).

The fed-batch system has been developed further over the last two decades. For example, in repeated-fed-batch systems periodic withdrawal of 10–60% of the medium volume is applied. Another more sophisticated fed-batch strategy is the cell-retention system: the cells are retained in the reactor, while liquid that contains products or compounds that are toxic to the cells is continuously removed [3, 5]. This is achieved by recycling part of the reactor contents through a membrane that separates the cells from the liquid. Many other modes of operation can be thought of. In fact, these are fully analogous to those used in the chemical industry.

QUESTIONS:

> *Why is the productivity of a continuous reactor higher than that of a semi- or fed-batch reactor? What are advantages and disadvantages of the operation modes discussed?*
> *Compare semi-batch processing in the production of fine chemicals (Chapter 12) and fed-batch operation in biotechnology. Are there fundamental differences?*
> *Can aerobic fermentation be carried out batchwise?*
> *The concept of cell-retention systems is similar to concepts used in the chemical industry. Give a few examples.*

13.2.2   Reactor Types

As in the chemical industry, several basic reactor types and variations are used in practice. Fermentation reactors are nearly always multiphase systems comprising a gas phase containing oxygen and/or nitrogen, one or more liquid phases, and a solid phase, including the microorganisms. Based on the way of contacting the microorganisms with the substrate (and air in the case of aerobic reactors), a distinction can be made between reactors where the microorganisms are mobile and reactors where they are at a fixed position.

In chemical process technology the first class is referred to as slurry reactors, or, for certain types, fluidized bed reactors. In biotechnology often the term *submerged reactors* is used. The second class includes the fixed bed reactors. In biotechnology, these are usually referred to as *biofilm reactors* or *surface reactors*. In biofilm reactors, either the culture adheres to a solid surface that is continuously supplied with air and substrate, or the culture floats on the substrate as a mycelium (a kind of network of threads).

From a conceptual reaction engineering point of view, reactors for biotechnological and chemical processes are very similar. However, the specific properties of the medium and the quite different industrial tradition justify a separate discussion of bioreactors.

In aerobic processes, the most important design factors are the contacting area between the microorganisms and the surrounding liquid and the rate of oxygen diffusion.

13.2.2.1   Submerged Reactors

Figure 13.3 shows the most common representatives of three classes of submerged reactors (in batch mode). The major difference between the types shown is the way in which mixing is accomplished. For aerobic processes, good mixing is very important in order to guarantee sufficient oxygen transfer from the gas phase to the liquid phase and to avoid anaerobic regions in the reactor. The energy input for mixing is provided as follows:

* mechanical stirring: mechanically-stirred-tank reactor;
* forced convection of the liquid: plunging jet reactor;
* operation using pressurized air: bubble column and air lift reactor.

QUESTION:

> Compare these reactors with those treated in previous chapters. What is specific for biotechnology?

13.2.2.1.1   Mechanically-Stirred-Tank Reactor

The mechanically-stirred-tank reactor is still used almost universally in the fermentation industry, although it is not necessarily the best solution. Figure 13.3 shows a standard mechanically-stirred fermenter. The reactor consists of a cylindrical vessel with a height-to-diameter ratio of 1–3, often equipped with baffles to avoid the formation of vortices. The reactor is usually only filled for two-third in order to leave room for foam formation, which is a major problem in many fermentation processes. A foam breaker can be incorporated at the off-gas outlet, while in addition antifoaming agents can be used. In the case of aerobic fermentation, air or oxygen is sparged into the reactor at the bottom. Mixing is achieved by a mechanical impeller equipped with one or more stirrers. An external jacket or an internal coil provides the surface area required for cooling or heating of the reactor contents.

In some cases, the mechanically-stirred tank is not a suitable reactor. Its maximum size is limited to typically 800–1500 m^3, depending on the application. During scale-up, problems are mainly concerned with heat and mass transfer, in particular the transfer of oxygen. This transfer may even require such a large power input that the use of a simple mechanical stirrer is impossible. These problems become more severe with highly viscous or non-Newtonian reaction mixtures. Moreover, the biomass might suffer from too high shear. Therefore, reactors have been developed that are better adapted for the specific circumstances of a certain fermentation (Sections 13.2.2.1.2 and 13.2.2.1.3).

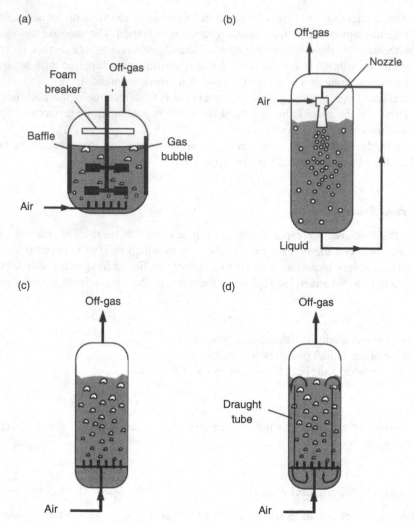

Figure 13.3　*Examples of submerged reactors: (a) mechanically-stirred-tank reactor; (b) plunging jet reactor; (c) bubble column reactor; (d) air lift reactor.*

13.2.2.1.2　Plunging Jet Reactor

In the chemical industry plunging jet reactors reactors are commonly referred to as loop reactors and occasionally as Buss reactors. In these reactors, the gas is dispersed by a liquid stream, the jet, created in a nozzle placed above the liquid surface. In the constriction of this nozzle, the liquid velocity is increased to typically 8–12 m/s. The jet entrains a mantle of gas, impinges on the surface, breaks through it, and penetrates into the liquid volume. The jet exists as long as it is surrounded by the gas mantle. The breakdown of this mantle leads to the formation of a swarm of small bubbles, which move downward and sideward in the liquid.

　　Plunging jet reactors are used primarily for purposes of improving heat and mass transfer. A heat exchanger can easily be incorporated in the external loop, which enables independent control of mass and heat transfer. Other advantages compared to the conventional mechanically-stirred-tank reactor are the lower power requirement per amount of oxygen transferred, the more reliable scale-up, and the possibility of the use of larger reactors.

Although the most important field of application for plunging jet reactors is still in wastewater treatment, they are also used on a large scale for other applications. Examples are the production of yeast in reactors that have been scaled up to 2000 m^3, and the production of single-cell protein (SCP) [6].

QUESTION:

In the plunging jet reactor, liquid is recycled by an external pump. Does the pump have to fulfill specific requirements (and if so, which)?

13.2.2.1.3 Bubble Column Reactor and Air Lift Reactor

The principle of both the bubble column reactor and the air lift reactor is that mixing takes place solely by the dispersion of pressurized air into the reactor. The bubble column is the simplest reactor type and has long been used in the chemical industry, because of its low investment and operating costs, as well as its simple mechanical construction. It is characterized by a large height-to-diameter ratio. Sparging air at the bottom of the reactor in most cases results in sufficient mixing.

Air lift reactors are similar to bubble columns but have additional provisions for control of the bulk liquid flow. The circulation of the liquid is due to the difference in the densities (gas contents) of the gas–liquid mixture in the aerated section and the downcomer regions. Air is sparged at the bottom of the reactor and in its upward movement drags the liquid along. At the top of the column, most bubbles are separated, resulting in a larger apparent density of the mixture in the downcomer, which will flow downward. Advantages of air lift reactors over mechanically-stirred reactors are the simple mechanical construction, the easier scale-up, better bulk mixing, and the possibility of using larger reactors. On the other hand, the investment costs for large-scale reactors are high and the energy costs are higher than in mechanically-stirred reactors because a greater air throughput is necessary at higher pressure. Air lift reactors can be scaled up to over 5000 m^3.

Bubble columns are employed on a large scale in the production of, for example, beer and vinegar. Air lift reactors are employed in, for example, the production of SCP. Air lift reactors with internal separation and recirculation of the biomass (biofilm airlift suspension reactors) are also used in aerobic wastewater treatment [7, 8].

QUESTION:

Compare the four reactors in Figure 13.3 with respect to pressure drop and mass transfer (see also Table 13.1). What would be typical biomass production rates per unit volume of reactor?

Table 13.1 *Important parameters in submerged and surface reactors.*

Reactor type	$k_l a_l$ (s^{-1})	$k_s a_s$ (s^{-1})	solids hold-up (m$_s^3$ m$_r^{-3}$)	specific biofilm area (m^2 biofilm m$_r^{-3}$)
Stirred tank	0.15–0.5	0.1–0.5	0.01–0.1	–
Bubble column	0.05–0.24	~0.25	0.01–0.1	–
Plunging Jet	0.2–1.5	0.1–1	0.01–0.1	–
Three-phase fluidized bed	0.05–0.3	0.1–0.5	0.1–0.5	~2000
Trickle bed	0.01–0.3	0.06	0.55–0.6	~200

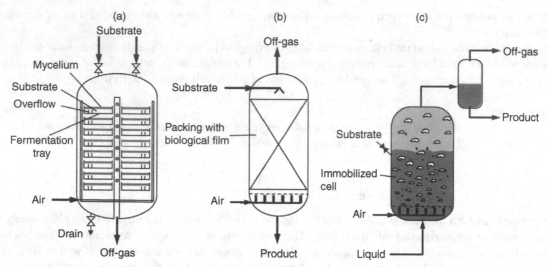

Figure 13.4   *Examples of surface reactors: (a) tray reactor; (b) trickle bed reactor; (c) three-phase fluidized bed reactor.*

13.2.2.2   Surface Reactors

Figure 13.4 shows three types of surface reactors.

13.2.2.2.1   Tray Reactor

The tray reactor is a classical surface reactor, in which the microorganisms float on the substrate as a mycelium. The substrate flows from the top to the bottom via overflow of the liquid from one tray to another. The application of this type of reactor is limited to cell cultures that can form a coating that is sufficiently stable to be reused. An example of the use of a tray reactor is the production of citric acid.

13.2.2.2.2   Trickle bed Reactor

Trickle bed reactors (trickling filters) are frequently used in the oil refining and chemical industries. They are also increasingly used in biotechnology, particularly in wastewater treatment, but also in, for example, the production of vinegar. In trickle bed reactors, the microorganisms are attached to the packing as a "biological film". The nutrient solution is evenly distributed through a feed device and flows downward. In contrast to customary operation in, for instance, hydrodesulfurization of a heavy gas oil fraction (Section 3.4.5), the required air is often fed from the bottom, leading to countercurrent operation. The flow of air is initiated by the fact that it is warmed by the heat of fermentation and rises due to natural convection. The disadvantage of countercurrent operation in fixed bed reactors is that the gas velocity must be kept low to avoid flooding and foaming.

QUESTION:

Why is countercurrent operation common in biotechnology, while it is not often employed in the production of bulk chemicals?

13.2.2.2.3   Three-Phase Fluidized Bed Reactor

The use of fluidized bed reactors in biotechnology has increased considerably in recent years. They are mainly used with cells that are immobilized on solid particles, for example, in wastewater treatment. Fluidized bed reactors in biotechnology are operated with one fluid phase (liquid, anaerobic) or two fluid phases (air and liquid, aerobic). To increase the fluid velocity (in the case of relatively heavy particles) a liquid recirculation loop can be added. In addition to the general advantages of fluidized bed reactors, such as the superior mass and heat transfer characteristics and good mixing, in biotechnological applications a particular advantage of this type of reactor is that it is suitable for cells that are sensitive to shear stresses (animal and plant cells). Another advantage is that plugging of the reactor does not occur as easily as in a fixed bed reactor.

Trickle bed reactors and fluidized bed reactors are also referred to as biofilm reactors.

QUESTION:

> *Why is the fluidized bed reactor considered a surface reactor and not a submerged reactor? In biotechnology the kinetics are often of the so-called Michaelis–Menten type with respect to reactant concentration. This is similar to Langmuir–Hinshelwood kinetics in catalysis. Furthermore, in biotechnology often products are inhibiting. Rank the reactors described in order of decreasing suitability with respect to this type of kinetics.*

13.2.2.3   Oxygen Supply in Fermenters

The oxygen supply to fermentation processes is often limited because of its low solubility (approximately 7–8 g/m_l^3), which depends on the type of substrate, the temperature, the oxygen partial pressure, and so on. Fast-growing microorganisms consume oxygen at a rate of between 2 and 6 $g/m_l^3/s$ [9]. This explains why, even in batch processes, oxygen has to be fed continuously. In most aerobic fermentation processes oxygen transfer from the gas phase to the liquid phase is the limiting step. Oxygen transfer is enhanced by increasing the gas–liquid interfacial area, a_l, and/or the mass transfer coefficient, k_l, which are usually combined in one parameter (Table 13.1). The main operation variables controlling this variable are the intensity of mixing (power input) and the gas velocity.

QUESTIONS:

> *Estimate the value of the gas–liquid mass transfer parameter, $k_l a_l$, for fast-growing microorganisms in a mechanically-stirred-tank reactor and a trickle bed reactor and compare with the values given in Table 13.1. What is your conclusion?*

Most fermentations are carried out in submerged reactors. Surface reactors are used for applications that use slow-growing organisms or for diluted feed (substrate) streams [8]. A typical example is the treatment of large volume wastewater streams with very low substrate concentration. Here, it is important to retain the microorganisms in the reactor for sufficient growth of biomass (maximum specific growth rate up to 0.1 h^{-1}) and, as a consequence, sufficient conversion. In this case the rate of oxygen transfer is less important than in typical "production" fermentations with fast-growing organisms or with concentrated feed streams [8].

Characteristic design parameters for the major reactor types are summarized in Table 13.1. The most important design parameter of aerobic surface reactors is the biofilm area per unit volume of reactor. For high reactor capacities, oxygen transfer may become limiting as a result of insufficient specific biofilm surface area. A typical value for the oxygen flux in surface reactors is $0.11–0.14 \times 10^{-3}$ g $O_2/(m^2_{biofilm} \cdot s)$.

QUESTIONS:

Estimate (calculate using estimated values) the mass transfer parameters and the oxygen transfer rate per unit volume of reactor for the trickle bed reactor and the three-phase fluidized bed reactor. Assume an oxygen concentration of 7 g/m_l^3. Are the values you calculated reasonable?

13.2.3   Sterilization

The presence of contaminating microorganisms and the changes upon infections in a bioprocess may have unfavorable consequences, such as loss of productivity (the medium has to support the growth of both the production organism and the contaminant), product contamination (e.g., single-cell protein, bakers' yeast), product degradation (antibiotic fermentations), and so on [10].

Sterilization practices for biotechnological media must achieve maximum kill of contaminating microorganisms, with minimum temperature damage to the components of the medium. The most convenient method of sterilization is by heating to a sufficiently high temperature to kill living organisms, maintaining that temperature long enough to achieve sterility, and then cooling to culture temperature. For materials liable to damage by heat sterilization, for example, some nutrient media, alternative methods should be used, such as filtration, radiation, or treatment with a chemical sterilization agent (e.g., ethene oxide) [9].

Two alternative heat sterilization procedures are available, namely combined or *in situ* sterilization and separate sterilization. Combined sterilization involves loading of the reactor with part or all of the growth medium and subsequent sterilization. In separate sterilization, the sterilized medium is charged aseptically to the already sterile reactor. For batch processes, both procedures can be used, while continuous processes require separate sterilization.

The advantage of separate sterilization is that the medium can be sterilized in a specifically designed unit that provides sterile medium for several fermenters. The disadvantage is the risk of contamination during transfer from the sterilizing unit to the fermenter.

In the case of combined sterilization in a batch reactor, heating is carried out by passing steam through the bioreactor coils or jacket or by direct sparging of steam into the liquid medium. The latter results in very rapid heating but in many media it leads to excessive foaming. Continuous sterilization of a medium is carried out by passing it through heat exchangers or a venturi steam injection device. Sterilization of empty vessels is commonly carried out by direct sparging with wet steam, which results in a much more rapid sterilization than the use of dry saturated or superheated steam at a given temperature [9].

The rate at which organisms are killed increases rapidly with temperature. For example, in a particular continuous sterilization system with steam injection, the holding time ranged from two minutes at 403 K to three seconds at 423 K [10]. Of course, the sterilization time also depends on the type of microorganism.

For the sterilization of air for aerobic processes, filtration is most commonly used [10]. Filters for the removal of microorganisms may be divided into two groups, so-called "absolute filters", with pores of smaller size than the particles to be removed, and fibrous-type filters, with pores larger than the particles to be removed. The former type of filter (e.g., ceramic or plastic membrane) is claimed to be 100% efficient, explaining the name "absolute filter". Filters of the latter type are made of materials such as cotton, steel wool, and so on, and, in theory, the removal of microorganisms cannot be complete. The advantages of fibrous filters are their robustness, cheapness, and lower pressure drop compared to the absolute filters.

QUESTIONS:

What factors will determine whether to use separate or combined sterilization for batch processes?

Why does sterilization with wet steam result in much more rapid sterilization than sterilization with dry saturated or superheated steam?

13.3 Cell Biomass – Bakers' Yeast Production

A typical industrial example of fermentation technology for biomass production is the production of bakers' yeast. Bakers' yeast is used in the production of bread. It provides the rising power of the dough and, therefore, the airiness of the bread during its baking process. Bakers' yeast is produced by the growth of microorganisms on a substrate consisting of sugars (e.g., glucose, molasses) under aerobic conditions, that is, with excess oxygen. Under anaerobic conditions, that is, with a shortage of oxygen, ethanol will be produced, which is not a desired reaction.

13.3.1 Process Layout

Bakers' yeast is produced in the fed-batch mode. This has the advantage of a higher efficiency, a better control of the dynamics of the overall process and a better quality of the bakers' yeast compared to batch processing. As the fermentation reaction is exothermic and the optimal process temperature is 298–303 K, cooling with cooling water (typically 283–288 K) is limited, which is a common problem in industrial bioprocesses.

Yeast plants are typically run between five and seven days a week. In every production cycle, a series of reactors with increasing size is used (small, medium, large volume) as shown in Figure 13.5. Prior to the production, the reactors are cleaned and sterilized with steam.

Beet and cane molasses serve as the sugar providers. Molasses is a waste product from the sugar industry and the cheapest source of fermentable sugars. Even so, the substrate costs are responsible for 60–70% of

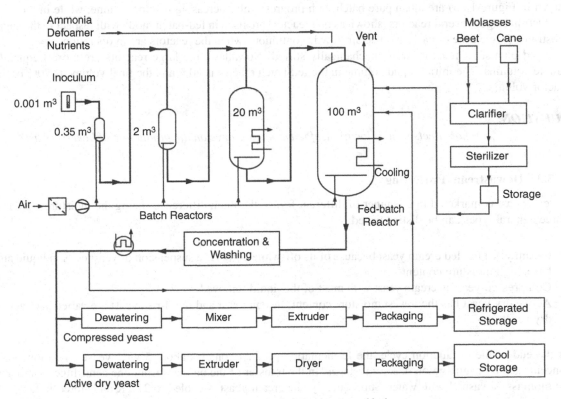

Figure 13.5 *Process steps in the production of bakers' yeast.*

the cost of bakers' yeast. Beet and cane molasses differ in sugar composition, proteins, salts, and vitamins content. Therefore, an additional supply of nutrients (salts, vitamins) is necessary.

The raw molasses is diluted to facilitate pumping and fermentation and treated with acid (to pH = 4) by which precipitation of some organic material occurs. Then the molasses is centrifuged and sterilized ("heat shocked") at 410 K, usually by steam injection for a couple of seconds. The sterilized solution is stored in sterilized vessels. Ammonia is added to supply the necessary nitrogen for yeast growth and to adjust the pH in the fermenter.

The specific growth rate of the biomass may vary between 0.05 and 0.6 h^{-1}. In the beginning, the pH of the substrate solution is around four to ensure optimal growth of the biomass and to prevent contamination as a result of incomplete removal of proteins from the substrate. At the end of the production cycle, the pH is increased to a value of five to prevent strong coloring of the yeast.

QUESTIONS:

The raw molasses is treated with acid causing some precipitation of organic material. Would you have expected this precipitation? Why or why not?
Which sterilization procedure is used? What is a clarifier?

13.3.2   Cultivation Equipment

To start up simply using one large reactor is not optimal. Due to the long residence time for this type of operation, the chances of contamination of the substrate solution are large. The process consists of a number of growth stages, the first of which is the laboratory batch culture (inoculum). The next stages (two are shown in Figure 13.5) are again pure batch with progressively increasing reactor volume, while in the final production stages several reactors (shown as one reactor) are used in fed-batch mode with respect to the sugar substrates (cycle time circa 16 h). The air flow is continuous when the reactors are in operation.

In old plants, all reactors were mechanically stirred. Nowadays, the large reactors are more frequently bubble columns. The initial liquid volume in the fed-batch reactor is 20% and the final volume is 70% of the reactor volume.

QUESTION:

What effects would (local) insufficient mixing have on the production of bakers' yeast?

13.3.3   Downstream Processing

Baker's yeast is marketed in a number of different forms, the main differences being the moisture contents. Three general types can be distinguished:

- Cream yeast (called cream yeast because of its off-white color) is a suspension of yeast cells in liquid and has a high moisture content.
- Compressed yeast is cream yeast with most of the liquid removed.
- Active dry yeast has the lowest moisture content. Instant yeast and rapid-rise yeast are varieties of active dry yeast.

At the end of the production cycle, the fermentation product contains about 5 wt% yeast. The biomass is concentrated by centrifugion in several stages. In the transfer of the biomass from one centrifuge to another, the biomass is washed with water. Subsequently, the cream yeast is cooled to 277 K and stored in tanks. A small part of the cream yeast, after acid treatment, is used as starting material for the next fermentation cycle. The remainder is processed into either compressed or active dry yeast.

13.3.3.1   *Compressed Yeast*

The cream yeast is filtered and the dewatered yeast is continuously cut from the filter surface. It is mixed with emulsifiers and the moisture content adjusted to 70 wt%. The yeast is then extruded in the shape of thick strands, cut, packaged, and stored at low temperature.

13.3.3.2   *Active Dry Yeast*

Although the quality of dried bakers' yeast is less than that of compressed yeast, part of the bakers' yeast is sold as active dry yeast. The reason is that it has a better stability, and hence can be used in (sub)tropical countries. Production and downstream processing of so-called active dry yeast is similar to that of compressed yeast. The yeast is extruded to fine strands (2–3 mm thick) directly after filtration. These strands are chopped to a length of about 7 mm and then dried. Dryers are commonly of the fluidized bed dryer type. The active dry yeast can be stored at higher temperature than compressed yeast.

13.4   Metabolic Products — Biomass as Source of Renewable Energy

The progressive depletion of fossil fuel reserves has led to the consideration of other materials as sources of energy. A possible source is biomass, which can be converted by biotechnology into more useful and valuable fuels. Generally, sugars in biomass are fermented to produce bioethanol and biobutanol (see also Chapter 7). Biogas (a gas mainly consisting of methane) can be produced by anaerobic digestion of organic wastes. Bioethanol and biobutanol are identical to ethanol and butanol produced by chemical routes; the prefix "bio" is used to distinguish between the different production methods.

Thermal cracking of biomass to obtain gasoline-like products, gasification of biomass to produce synthesis gas, and combustion of dried biomass for power generation, are also feasible, at least technically (Chapter 7).

13.4.1   Bioethanol and Biobutanol

As discussed in Section 13.3, yeast cells grow on sugars in the presence of oxygen, while they produce ethanol under anaerobic conditions:

$$\underset{\text{glucose}}{C_6H_{12}O_6} \xrightarrow{\text{yeast}} 2\,\underset{\text{ethanol}}{C_2H_5OH} + 2\,CO_2 \qquad (13.1)$$

Figure 13.6 shows a schematic of the production of ethanol by yeast. The yeast has been cultivated in advance in aerated fermenters as discussed for the production of bakers' yeast (Section 13.3). Table 13.2 shows the operating conditions of the fermenter.

The yeast is added to a mixture of cane molasses and water in large anaerobic fermenters. Hydrochloric or sulfuric acid is added to obtain an acidic medium. Heat is removed by external heat exchangers. In downstream

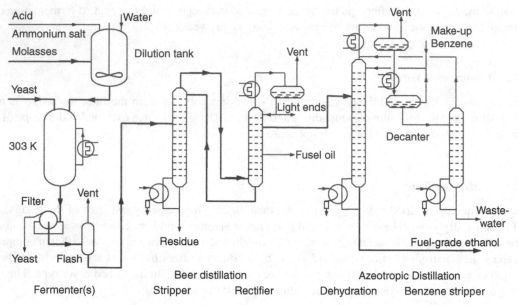

Figure 13.6   *Production of fuel-grade ethanol from biomass in an early plant.*

processing, the yeast is separated by filtration in a rotary drum vacuum filter, and either recycled or used in animal feed after further handling.

QUESTION:

Why is an acidic environment created in the fermenter?

The reactor product, called beer, is first fed to a flash drum and then distilled in a series of four distillation columns. The separation of ethanol and water is very energy intensive. Furthermore, the complete separation of ethanol and water, which is required for fuel-grade ethanol, cannot be attained by normal distillation. Ethanol and water form a homogeneous minimum boiling azeotrope at a temperature of 351 K, where the mixture contains 96 wt% ethanol. Therefore, it is necessary to resort to azeotropic distillation, which is used in many early fuel-grade ethanol plants, or to more advanced separation methods, for example, based on membrane technology or molecular sieve adsorption technology. The latter technology has been adopted by the majority of modern ethanol plants.

Table 13.2   *Characteristics of ethanol production from biomass.*

Characteristic	Value
Temperature (K)	303
pH	4.5
Residence time (h)	40
Final ethanol concentration (vol.%)	8–12

In the process layout of Figure 13.6, the bottom product from the first distillation column (non-fermentable molasses solids) is an additive for animal feed. The overhead vapor contains a mixture of ethanol (approximately 50 vol.%), water, and other volatile components (e.g., acetaldehyde) and is fed to the base of a rectifying column. In the rectifying column light ends and so-called fusel oil (a mixture of higher alcohols) are removed overhead and as a side stream, respectively. The azeotrope is removed as a liquid side stream somewhat below the top of the column and fed to the azeotropic distillation unit to which benzene is also added.

The overhead vapor of the dehydration column is a ternary minimum boiling azeotrope of ethanol, water, and benzene (the azeotropic agent), while anhydrous fuel-grade ethanol is produced at the bottom. The azeotrope is separated by distillation and benzene is recycled to the dehydration column. In this separation sequence, heat integration and energy recovery play a vital role in reducing energy requirements.

To obtain high yields of ethanol on an industrial scale in an economical way, yeast strains must be selected that are resistant to ethanol. As far as the yeast is concerned, ethanol is not only a waste product but it is also harmful. Normally, microorganisms are killed when the alcohol concentration exceeds 12–15 vol.%. New types of yeast have been developed that withstand higher ethanol levels, so that downstream processing is facilitated.

At present, the raw materials for the production of most of the ethanol are food crops, and thus ethanol production may be in competition with the food chain supply. Currently, routes are explored based on lignocellulose biomass (Chapter 7).

QUESTIONS:

What is the composition of the gas leaving the flash drum?

In fact, the higher alcohols are later blended with ethanol and serve to add fuel value. What could be the reason that they are first removed?

Where in Figure 13.6 do you see possibilities for heat integration?

Explain how azeotropic distillation works. What are your thoughts about benzene as the azeotropic agent?

As alternatives for azeotropic distillation, membrane technology (pervaporation) and molecular sieve technology are applied. What are the principles of these technologies, specifically for the dehydration of ethanol?

Butanol was originally produced by the anaerobic conversion of carbohydrates into acetone, *n*-butanol, and ethanol using a specific bacterium. This process is known as the ABE (Acetone, Butanol, Ethanol) process and was commercialized during the first world war. In the 1950s, when the price of petrochemical feedstocks dropped to below that of starch and sugar substrates, butanol shifted to becoming a petrochemically-derived product. Today, virtually all of the butanol is produced by chemical processes but in recent years ABE plants have been built or rebuilt in China [11].

At present, however, modified bacterium strains have been developed allowing a more efficient and selectively biochemical production of butanol:

$$\underset{\text{glucose}}{C_6H_{12}O_6} \xrightarrow{\text{yeast}} \underset{\text{butanol}}{C_4H_9OH} + 2CO_2 \qquad (13.2)$$

This fermentation process is very similar to that for the production of ethanol; there only are minor changes in the distillation section.

QUESTION:

Compare ethanol and butanol as transportation fuels. Give pros and cons.

13.4.2   Biogas

Organic wastes can be converted into methane-containing gas, called biogas, by anaerobic digestion. This has been done in anaerobic purification of wastewater for more than a century (Section 13.5.4). The process has also been used for ages on a small scale in agriculture, based on conversion of animal and vegetable waste. More recently, the process has also gained interest in large-scale production of biogas. Fermentation of the waste, in the absence of air, produces biogas containing mainly methane and carbon dioxide. The solid residue left after fermentation is a good fertilizer.

13.5   Environmental Application – Wastewater Treatment

13.5.1   Introduction

Wastewater treatment is the largest application of biotechnology. Wastewater is water-carried waste arising from domestic and industrial use. The decomposition of organic materials in wastewater can produce foul smelling gases and lead to a reduction in the dissolved oxygen content, thus killing aquatic life. Furthermore, wastewater can contain microorganisms, heavy metals, and other toxic compounds that may be detrimental to both plant and animal life. Removal of these potentially hazardous components from the wastewater is essential. Table 13.3 summarizes the major contaminants of concern. It is clear that wastewater is a very complex mixture.

A major challenge in wastewater treatment is the varying composition and flow rate with time. In particular, the composition of industrial wastewaters can vary considerably, depending on the nature of the chemical process. The flow rate also is an important factor. Daily as well as seasonal variations in flow rate occur.

13.5.2   Process Layout

Wastewater treatment proceeds in three general steps, namely so-called primary, secondary, and tertiary treatment processes; they are also referred to as pretreatment, biological treatment, and advanced treatment (Figure 13.7).

The aim of the pretreatment process is to remove suspended solids (e.g., sand and possibly fats). This is done by means of physical processes such as sedimentation and flotation. Industrial wastes sometimes have

Table 13.3   *Major wastewater contaminants.*

Contaminant	Comments
Suspended solids	lead to sludge deposits
Biodegradable organics	mainly proteins, carbohydrates, and fats, leading to reduced dissolved oxygen concentration
Pathogens	can cause diseases in humans and animals
Nutrients (N and P)	can cause eutrophication of lakes and reservoirs leading to algae blooms
Priority pollutants	may be carcinogenic, mutagenic, teratogenic, or highly toxic (e.g., benzene, and chloro-hydrocarbons)
Refractory organics	include surfactants, phenolics, and pesticides, and are often not removed by conventional wastewater treatment processes
Heavy metals	may arise from industrial processes
Dissolved inorganics	mainly calcium, sodium, and sulfate arising from domestic use

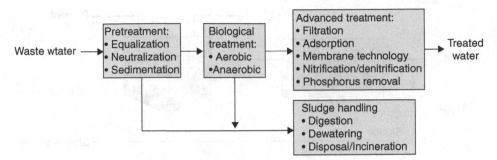

Figure 13.7   *Outline of wastewater treatment.*

pH values far from the pH range required for optimal performance of the subsequent biological processes. In such cases, neutralization is required as pretreatment step. Apart from reducing the load on subsequent processes, primary treatment also has the effect of minimizing variations in the wastewater flow rate.

In secondary treatment processes, the concentration of organics (suspended and soluble) is reduced. Secondary treatment processes are biological processes and can be divided into aerobic and anaerobic processes. The choice of process depends on the concentration of biodegradable organics and the flow rate of the wastewater to be treated. Aerobic treatment is generally used for mildly polluted (relatively low concentration of biodegradable organics) and cooler wastewaters, whereas anaerobic treatment is often employed as a pretreatment step for highly polluted and warmer wastewaters.

QUESTION:
   Explain the logic of the statement concerning aerobic and anaerobic treatment.

The microorganisms responsible for the degradation of the organic compounds are already present in the wastewater. Thus, there is no need for a dedicated production process like in the production of bioethanol.

Tertiary treatment processes aim to further improve the quality of the wastewater. Conventional biological treatment produces an effluent containing typically about 30 g/m^3 suspended solids and 20 g/m^3 BOD[1]. In tertiary treatment processes these values are both reduced to 2 g/m^3. In addition, the total nitrogen and phosphorus concentrations are reduced as well as the amounts of heavy metals and pathogenic microorganisms. Various processes are used, including grass plots (gently sloping grass fields, on which water purification is achieved by the action of soil microorganisms, by the uptake of nutrients by the grass, and by the filtering action of the soil), filtration, adsorption, and membrane technology.

In addition to the treatment of the wastewater, processing and disposal of the resulting sludge is required. This sludge, which is formed during primary and secondary treatment in large amounts, is not very concentrated (about 98% water) and has a bad odor. Stabilization of the sludge takes place by either aerobic or anaerobic digestion similarly to the secondary treatment discussed in Section 13.5.3 (activated sludge process). Reduction of the water content is achieved by various methods, such as sedimentation, air drying, filtering, and centrifuging. The sludge can then be disposed of on land or incinerated. The latter option is increasingly used.

---

[1] BOD = Biochemical Oxygen Demand: The amount of oxygen required for the biochemical oxidation per unit volume of water at a given temperature and for a given time. It is used as a measure of the degree of organic pollution of water. The more organic matter the water contains the more oxygen is used by the microorganisms [10]. (COD = Chemical Oxygen Demand = amount of oxygen required for chemical oxidation).

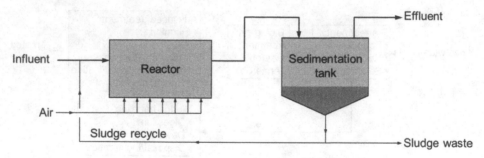

Figure 13.8   *Schematic of activated sludge process.*

13.5.3   Aerobic Treatment Processes

In aerobic treatment of wastewater basically the following reaction takes place:

$$\text{organics} + O_2 + N + P \xrightarrow{\text{Cells}} \text{new cells} + CO_2 + H_2O + \text{soluble microbial products} \qquad (13.3)$$

Roughly one half of the organics removed is oxidized to carbon dioxide and water, the other half is synthesized to biomass. Besides a source of carbon and oxygen, the microorganisms also require nitrogen (as ammonia or nitrate) and phosphorus (as phosphate), sulfur, and trace quantities of many other substances for growth. Domestic sewage normally contains a balanced supply of nutrients. For industrial effluents this is not the case. These are usually mixed with nutrient-rich domestic wastewater.

13.5.3.1   *Activated Sludge Process*

The so-called activated sludge process, developed at the beginning of the twentieth century [12], is still widely used today. Figure 13.8 shows a schematic of this process.

The wastewater is introduced into the reactor, where bacteria convert the organic compounds according to reaction 13.3. The mixture of cells and treated water is then passed into a sedimentation tank. Part of the settled sludge is recycled to the reactor while the remainder is further treated and disposed of.

In the conventional process aeration takes place uniformly over the length of the reactor tank. This is not optimal: the oxygen requirement at the inlet of the aeration tank (where the substrate concentration is largest) is larger than that at the outlet and, as a consequence, an oxygen deficiency might occur at the inlet, while an oxygen surplus can occur at the outlet, which is unnecessarily expensive. Furthermore, peaks in the wastewater supply and poisonous compounds could disturb the purification process.

Solutions are staged introduction of either the air or the wastewater. In the first case, "tapered aeration", aeration is decreased along the reactor tank by reducing the number of aerators per section. In the second case, wastewater is introduced along the length of the tank instead of only at one point.

QUESTIONS:

>   *What are advantages and disadvantages of both configurations? Why is part of the sludge recycled?*

In the activated sludge processes described the microorganisms are suspended in the wastewater (submerged reactor). An alternative is the use of so-called fixed-film processes (surface reactor). In the latter case, the natural ability of organisms to grow on surfaces can be exploited to keep the biomass in the reactor and avoid the need for separation from the treated water. Two examples of such surface reactors have already been discussed in Section 13.2, namely the countercurrent trickle bed reactor (also called "trickling filter") and the three-phase fluidized bed reactor. The most important design parameter is the biofilm area per unit volume of reactor (Section 13.2).

In the case of wastewater treatment, the trickle bed reactor is filled with stone or plastic media on which the microorganisms grow as a film. These microorganisms utilize the soluble organic material in the wastewater but little or no degradation of suspended organics is achieved. Hence, although separation of the microorganism is not required, a sedimentation tank is present for the removal of these suspended organics.

In fluidized bed reactors for wastewater treatment, usually sand is used as support for the microorganisms to grow on.

More advanced particle-based surface reactors with large specific surface areas (up to 3000 $m^2_{biofilm}/m^3_{reactor}$) have been developed. These reactors are capable of processing a higher load of wastewater and are more efficient than trickle bed and fluidized bed reactors [8]. A novel innovative type of surface reactor is the rotating biological contactor (Box 13.1).

QUESTION:

Compare the reactor designs with respect to the surface area of the biofilm. Why can the biofilm area per unit volume of reactor be so much larger in the fluidized bed reactor than in the trickle bed reactor? Suggest some advanced designs resulting in larger biofilm areas.

Box 13.1 Rotating Biological Contactor

A special type of surface reactor is the rotating biological contactor (Figure B.13.1.1). This reactor comprises a series of discs (2–3 m in diameter) arranged in compartments and mounted on a rotating horizontal shaft, which is typically positioned above the liquid level so that the discs are only partially immersed in the wastewater. The biofilm growing on the discs is alternately exposed to the atmosphere, where oxygen is absorbed, and to the liquid phase where soluble organic material is utilized.

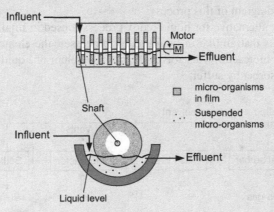

Figure B.13.1.1 *Rotating biological disk contactor; top: reactor; bottom: one disk.*

13.5.3.2   *Nitrogen and Phosphorus Removal*

Biological treatment of wastewater presently aims at the removal not only of organic substances, but also of nitrogen and phosphorus compounds, because these compounds cause eutrophication of surface waters. Nitrogen removal is based on two different processes called nitrification and denitrification. In the latter process, readily biodegradable organic compounds such as methanol or acetic acid have to be present.

Nitrification:

$$NH_4^+ + 2\,O_2 \rightarrow NO_3^- + H_2O + 2\,H^+ \tag{13.4}$$

Denitrification:

$$6\,NO_3^- + 5\,CH_3OH \rightarrow 3\,N_2 + 6\,OH^- + 5\,CO_2 + 7\,H_2O \tag{13.5}$$

Since denitrification requires the presence of organic compounds, it seems logical to carry out the denitrification and the aerobic removal of organics simultaneously. However, for denitrification an oxygen-free environment is required, whereas the degradation of organic compounds and nitrification require oxygen. Hence, a separate denitrification reactor is required [13]. For the biological removal of phosphorus, an additional reactor is required.

Polluted gases [14] and solid materials [15] can also be treated successfully by biological methods. Box 13.2 illustrates the use of biological treatment for polluted gas streams.

Box 13.2   Aerobic Treatment of Gases

An example of gas treatment is the biological oxidation of hydrogen sulfide in gas streams (e.g., natural gas, Claus plant tail gas [14]. Hydrogen sulfide is absorbed in an absorber at high pressure (up to 60 bar). In the bioreactor, the dissolved sulfide is subsequently oxidized into elemental sulfur at atmospheric pressure. The following reactions proceed:

$$\text{Absorption and hydrolysis of } H_2S: \quad H_2S + OH^- \rightarrow HS^- + H_2O \tag{13.6}$$

$$\text{Biological sulfur formation}: \quad HS^- + \tfrac{1}{2}O_2 \rightarrow S + OH^- \tag{13.7}$$

The solvent used for absorption of hydrogen sulfide is regenerated in the bioreactor. Figure B.13.2.1 shows a simplified block diagram of this process.

This process is a viable alternative for liquid redox processes used in small-scale operation (Section 8.3). Its main advantage is that, unlike in liquid redox processes, the elemental sulfur is not formed in the absorber but in the bioreactor. This solves the main problem of liquid redox processes, namely severe clogging of the absorber by sulfur.

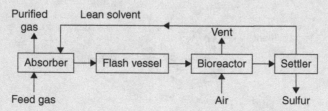

Figure B.13.2.1   *Simplified block diagram of a process for aerobic gas treatment.*

13.5.4 Anaerobic Treatment Processes

Anaerobic treatment of wastewater has important advantages over aerobic treatment, such as lower sludge production, the formation of a gas that can be utilized (biogas), and lower processing costs (aeration requires investment and energy costs). However, anaerobic processes require much more careful control of specific process parameters and the microorganisms grow quite slowly. Furthermore, the degree of removal of biodegradable substance is generally lower than in aerobic processes. Therefore, anaerobic treatment is usually applied as a pretreatment for wastewaters with a high concentration of biodegradable components, while the finishing step is an aerobic process.

The anaerobic decomposition of organic substances in wastewater is accomplished by a complex mixture of microorganisms, which convert the organic material into methane and carbon dioxide [16]. Scheme 13.1 shows the steps involved in this conversion.

1. In the first step, high molecular weight compounds, such as carbohydrates, proteins, and lipids, are hydrolyzed to yield soluble monomer compounds, such as sugars, amino acids, fatty acids, and alcohols.
2. The monomers produced by hydrolysis are fermented to organic acids, carbon dioxide, and hydrogen.
3. Organic acids with three or more carbon atoms (propionic and butyric acid) are converted to acetic acid, carbon dioxide, and hydrogn by so-called acetogenic bacteria.
4. The last, "methanogenic", steps produce methane from acetic acid, carbon dioxide, and hydrogen:

$$CH_3COOH \rightleftarrows CH_4 + CO_2 \qquad \Delta_r H_{298} = -34 \text{ kJ/mol} \tag{13.8}$$

$$CO_2 + 4H_2 \rightleftarrows CH_4 + 2H_2O \qquad \Delta_r H_{298} = -164 \text{ kJ/mol} \tag{13.9}$$

The choice of reactor (Figure 13.9) for the anaerobic fermentation of the organics in the wastewater stream depends on the physical and chemical properties of the substances present, that is, the amount of particles present, the reactivity of the organic components, and so on. Table 13.4 summarizes characteristics and applications of the main types of reactors used for different wastewater streams.

In the stirred-tank reactor, mixing of the contents can take place by mechanical agitation (as shown), a water recirculation pump, or by injection of oxygen-free gas. The residence time of the sludge is increased by the presence of a settling tank from which settled particles (sludge and unconverted substrate in suspension) are recycled. This is necessary because the methane-producing microorganisms grow slowly (the minimum residence time of sludge is about 20 days).

The upflow anaerobic sludge blanket (UASB) reactor usually consists of three zones, namely a dense sludge bed, a sludge "blanket" (with less concentrated sludge), and a separation zone where entrained sludge

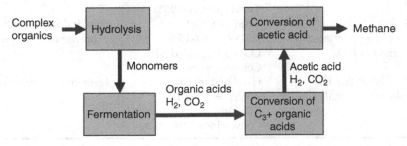

Scheme 13.1   *Anaerobic decomposition of organic matter.*

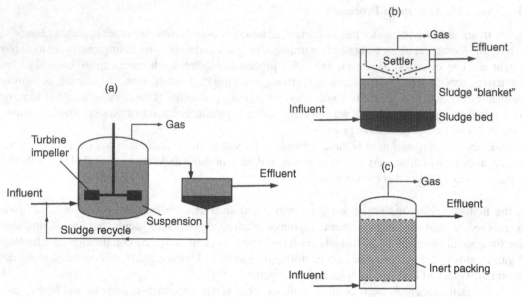

Figure 13.9   *Reactors for anaerobic wastewater treatment: (a) stirred-tank reactor; (b) upflow anaerobic sludge blanket (UASB) reactor; (c) fixed bed reactor.*

particles are separated from the liquid and gas. For this type of reactor, the sludge must have a relatively large density in order to prevent it from being washed out of the reactor with the wastewater flowing upward (although with low velocity). Advantages of the UASB are its simple construction, the high obtainable throughput, and the lack of mixing costs (mixing occurs by the gas produced). Disadvantages are the need for an efficient gas/sludge separator and a good water distributor. Furthermore, at small loads gas or treated water recirculation is required. The reason is that at small loads not enough gas is produced for sufficient mixing.

Table 13.4   *Anaerobic reactors for different types of wastewater [17, 18].*

Water type	Reactor[a]	Examples
Water with 0.5–25 kg/m^3 easily digestible lower carbohydrates, possibly with large particle content	Stirred-tank reactor WHSV = 5–10 kg COD/(m$^3_{reactor}$ · day) Conversion: 80–95%	Manure, distilleries, chemical industry, pulp and paper industry, preserved food plants
Water with 0.5–25 kg/m^3 easily digestible higher carbohydrates, small particle and protein content	UASB reactor[b] WHSV = 10–15 kg COD/(m$^3_{reactor}$ · day) Conversion: ~95%	Food industry, starch industry, breweries
Water with 25–200 kg/m^3 slowly digestible carbohydrates, small particle content	Fixed bed reactor WHSV = 10–20 kg COD/(m$^3_{reactor}$ · day) Conversion: 90–98%	Fermentation industry, dairy industry

[a]COD = chemical oxygen demand; [b]UASB = upflow anaerobic sludge blanket.

The fixed bed reactor is filled with an inert packing material, such as gravel, rock, or plastic. Microorganisms attach themselves to the packing material and thus form a biofilm that remains in the reactor. The wastewater may flow upward or downward through the reactor. This reactor type is similar to an aerobic trickle bed reactor. These reactors are suitable for high throughput, especially for decomposition of soluble, easily convertible substrates. Fixed bed reactors are suited mainly for the treatment of dissolved organic matter. In other uses particles would plug the channels in the bed and thus decrease its efficiency.

QUESTIONS:

> *Is anaerobic water treatment a multiphase process? What phases are present?*
> *Instead of the combination of a stirred tank and a settling tank a stirred-tank reactor without a settling tank could be used. Compare both configurations with respects to (reactor and total) volume and operation parameters (energy requirements for stirring/pumping). Discuss mixing in the UASB reactor.*

13.6 Enzyme Technology – Biocatalysts for Transformations

Enzymes in principle are homogeneous catalysts, although a colloidal solution is a better description. They are complex chains of amino acids with a molecular weight ranging from 20 000 up to 200 000. Enzymes are present in living cells where they act as catalysts. Although in nature enzymes are only formed in living cells, they can also function outside the cell. The use of enzymes can be of two kinds, namely as biocatalysts in a reaction system, or as the final product, for example, as food additives. Here, the focus is on the former application.

13.6.1 General Aspects

Enzymes can operate as part of a living cell but they can also be extracted from living cells while keeping their catalytic activity. The immobilization of enzymes on a fixed support is a relatively recent development that shows promise for the future. Immobilization was developed in the 1960s because the amount of enzymes present in biomass was very small, resulting in diffusion and stability problems. However, current recombinant DNA techniques enable the cheap production of enzymes, which makes them suitable for once-only use. This has the advantage that in many processes the enzymes can be used directly, without the need for immobilization. Enzymes compete with conventional chemo-catalysts, mainly in fine chemicals production (Chapter 12). Table 13.5 compares enzyme technology with fermentation and chemical technology.

An important advantage of enzymes over most conventional catalysts is their stereoselectivity, of which enantioselectivity is a special case (see also Section 12.2). In general, chemical processes lack the ability to discriminate between **R** and **S** enantiomers of asymmetric (chiral) molecules and, as a consequence, produce a racemic mixture. It should be noted that chemo-catalysts can also possess stereoselectivity. A good example is shape-selective catalysis (Section 10.4).

A commercial example of a chemical process route in fine chemistry with a enantioselective homogeneous catalyst is in the production of L-dopa (Section 12.2) [19]. Enantioselectivity allows the production of one enantiomer exclusively. This is often important, especially in pharmaceuticals production. Understandably, the separation of two optical isomers is very difficult using common separation methods, such as distillation, crystallization, and so on, because nearly all physical properties are identical for both enantiomers. Furthermore, even if the undesired enantiomer has no adverse effects, the productivity of a process decreases by its formation. Obviously, it is preferable to produce the desired enantiomer exclusively.

Table 13.5   *Comparison of technologies for chemicals production.*

Parameter	Classical fermentation	Enzymes	Chemo-catalysts
Catalyst	living cells	enzymes	metals, acids, . . .
Catalyst concentration (kg/m^3)	10–200	50–500	50–1000
Specific reactions	sometimes	often	often
Stereoselectivity	often	often	sometimes
Reaction conditions	moderate	moderate	moderate–extreme
Sterility	yes	yes	no
Yield (%)	10–95	70–99	70–99
Important cost item	cooling water	enzyme	varies
Challenges	regulation of microorganisms, selectivity	stability	selectivity, stability

13.6.2   Immobilization of Enzymes

Analogously to the immobilization of homogeneous catalysts, much research is carried out on the subject of immobilizing enzymes. Supports such as polymers or membrane reactors are used. A novel approach is the use of a monolithic stirrer reactor (Figure 13.10). In this reactor, monolithic structures are used as stirrer blades. Enzymes can be immobilized on the channel walls of the monolithic stirrer. [20, 21].

The advantage of immobilized enzymes over their homogeneous analogues is the possibility of recovering the enzymes from a reaction mixture and reusing them repeatedly, like heterogeneous catalysts in the chemical industry. In addition, it has been discovered that in immobilized form enzymes are thermally more stable, while they can also be used in non-aqueous environments. The amount of waste produced is also reduced, leading to lower effluent treatment costs. Finally, the product is not contaminated with the enzyme, which is particularly important in food and pharmaceutical applications [22]. Of course, the additional cost of development time and immobilization must be balanced against these advantages. The production of enzymes and immobilization techniques have been covered elsewhere [20, 22, 23].

Figure 13.10   *Monolithic stirrer reactor design. Reprinted with permission from [21] Copyright (2011) Elsevier Ltd.*

The choice between the use of whole cells and enzymes, and between soluble and immobilized form, depends on many factors, such as the nature of the conversion, stability and reuse, technical improvements, and, eventually, cost. In principle, enzymes contain orders of magnitude more active sites per unit mass than whole cells. On the other hand, the activity per site usually is considerably lower. Technical improvements can result directly from immobilization (e.g., increased product purity and/or yield, reduced waste production) but also indirectly. Immobilization of cells or enzymes enables the use of continuous rather than batch operation, thus simplifying control and reducing labor costs due to the possibility of increased automation of the process. Immobilized enzymes are mainly used in the production of fine chemicals and pharmaceuticals; currently they cannot compete economically with conventional catalysts in the bulk chemical industry. Present commercial applications of immobilized enzymes include the production of L-amino acids, organic acids, and fructose syrup.

13.6.3 Production of L-Amino Acids

The demand for L-amino acids for food and medical applications is growing fast. Both chemical and microbial processes can be used for their production. However, the chemical routes lack stereoselectivity, thus leading to lower productivity. In Japan the immobilized enzyme aminoacylase has been since 1969 used for the production of L-amino acids, of which methionine is the most important. Aminoacylase catalyzes the stereospecific de-acylation of acyl-amino acids; in a mixture of acyl-L- and D-amino acids, only the acyl-L-amino acid is converted into its corresponding amino acid:

acyl-L-amino acid   L-amino acid

$$\tag{13.10}$$

Figure 13.11 shows a flow scheme of the Tanabe Seiyaku process that utilizes immobilized aminoacylase for the continuous production of L-amino acids.

The de-acylation reaction is carried out in a reactor packed with aminoacylase adsorbed on a support. The desired L-amino acid can be separated from the unconverted acyl-D-amino acid by crystallization, as a result of their different solubilities. Subsequently, the acyl-D-amino acid is racemized and the resulting racemic mixture is returned to the enzyme reactor together with the feed mixture of acyl-L- and D-amino acid. Racemization of acyl-amino acids can be accomplished, for example, by heating at 373 K with acetic anhydride in an acetic acid solution [24].

Aminoacylase can also be used in solution, that is, as a homogeneous catalyst. Figure 13.12 shows a comparison of the relative costs involved in the production of amino acids with homogeneous aminoacylase in a batch process and with immobilized aminoacylase in a continuous process.

Clearly, the main differences can be attributed to catalyst (enzyme or enzyme + support) and operating costs. The latter include labor, fuel, and so on. Particularly the labor costs are greatly reduced in the continuous process. The reduction in catalyst costs when using the continuous process is a result of the higher stability and the reusability of immobilized aminoacylase.

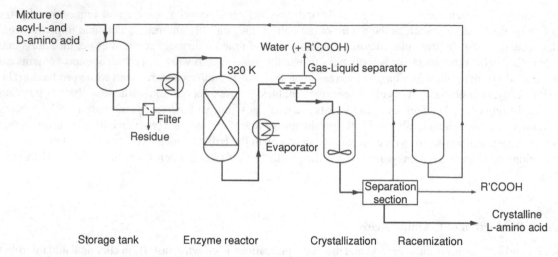

Figure 13.11   *Production of L-amino acids using aminoacylase.*

QUESTION:

Analyze the batch and continuous process in terms of the factors mentioned in Section 13.1. Does Figure 13.12 represent a fair comparison?

13.6.4   Production of Artificial Sweeteners

Probably the most successful application of immobilized enzymes is the production of fructose syrups using immobilized glucose isomerase for the conversion of glucose. Another artificial sweetener, D-Mannitol can also be produced using this enzyme.

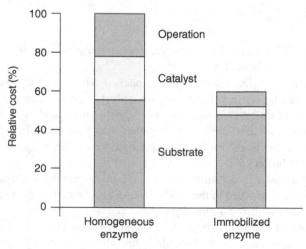

Figure 13.12   *Relative production costs of L-amino acids using homogeneous (batch process) and immobilized (continuous process) aminoacylase.*

Table 13.6 *Relative sweetness of sucrose and alternative sugars.*

Sugar	Relative sweetness (weight basis)
Saccharose (Sucrose)	100
Glucose syrups	40–60
Fructose	114

At the end of the 1970s a shortage of sucrose (natural sugar) occurred, resulting in high prices. It was attempted to replace sucrose by substitutes. Table 13.6 gives the relative sweetness of sucrose and possible alternatives. Glucose has a much lower sweetness than sucrose, whereas fructose has a higher sweetness.

Glucose can be obtained by depolymerization of starch (from corn) by acid hydrolysis and can then be isomerized using an alkaline catalyst into a mixture of glucose and fructose. This product, referred to as high-fructose corn syrup (HFCS), is highly desirable but in the production process large amounts of by-products are formed. A novel process became feasible in which the depolymerization is catalyzed by enzymes. In addition, the enzyme glucose isomerase was discovered, which catalyzes the conversion of glucose into fructose.

Initially (in the mid-1960s), the isomerization of glucose into fructose was carried out in batch reactors with soluble glucose isomerase. In the late 1960s the cost-reducing advantages of using immobilized glucose isomerase were demonstrated; this led to the rapid commercialization of immobilized glucose isomerase technology [25]. The production of HFCS (ca. 10^7 t/y) is the largest commercial application of immobilized enzymes. An alternative to immobilization of the enzyme is the immobilization of the complete cell. This technique has also been used for glucose isomerase [25].

The conversion of glucose into fructose is an equilibrium reaction:

$$(13.11)$$

An approximately 1:1 mixture of glucose and fructose is obtained at temperatures between 300 and 350 K [25] (Figure 13.13).

QUESTION:

In practice, temperatures of 325–335 K are used. Why, despite the better equilibrium, are higher temperatures not used?

Figure 13.14 shows a simplified flow scheme of the continuous production of HFCS starting from glucose. Glucose isomerase requires the presence of metal ions for its catalytic activity. Usually, a magnesium salt is added. In addition, salts are added for adjustment of the pH.

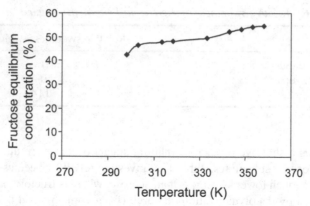

Figure 13.13 *Equilibrium concentration of fructose in a glucose/fructose mixture as a function of temperature* [25,26].

Treatment of the product with carbon and ion exchange resins is required for the removal of the added salts and impurities that result in undesired coloring. Two ion exchangers are used; the first is a strong acid cation exchange resin in the hydrogen form, the second a weak base anion exchange resin [25]. Subsequently, the pH is adjusted for maximum stability and the final step is concentration of the syrup by vacuum evaporation.

Enzyme systems might be compressed in conventional fixed bed operation. In that case, multiple shallow beds, as shown in Figure 13.14, can be used or the enzyme can be supported on a porous support. The additional advantage of such systems is that they reduce the pressure drop [25].

The production of mannitol (Box 13.3) is an example of a process in which bio- and chemo-catalysis are combined.

QUESTION:

> *Processing under vacuum conditions is relatively costly. Why is vacuum evaporation employed in the process shown in Figure 13.13?*

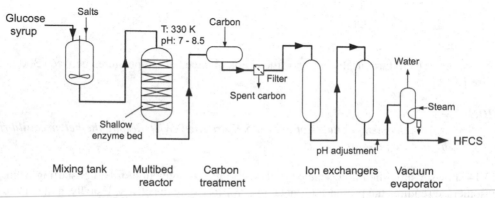

Figure 13.14 *Continuous production of HFCS (high-fructose corn syrup).*

Box 13.3 Production of Mannitol

D-mannitol is a valuable sweetener, because it has a low hygroscopicity and has no tooth-decaying effects. It does not become sticky until the humidity level is 98%. This makes mannitol attractive as a coating for hard candies, dried fruits, chewing gums, and in medicines.

Mannitol can be produced in the "Combi-Process" (Scheme B.13.3.1) by combination of the enzymatic interconversion of glucose and fructose with the simultaneous selective hydrogenation of fructose into mannitol.

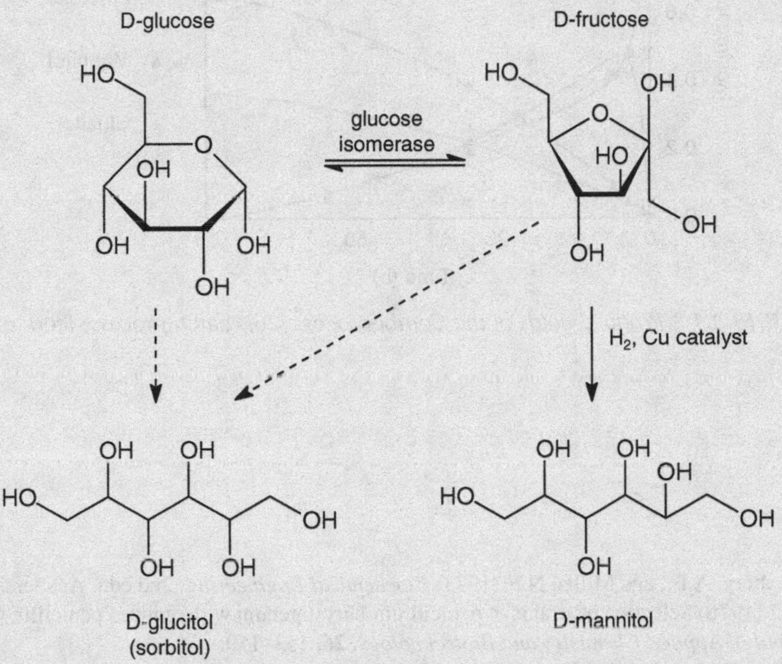

Scheme B.13.3.1 *Combi-process: simultaneous enzymatic isomerization and metal-catalyzed hydrogenation of D-glucose/D-fructose mixtures to produce D-mannitol [26]. The dotted arrows indicate slow reactions.*

In this case, the equilibrium of the isomerization reaction is shifted towards D-fructose by the selective hydrogenation of D-fructose to D-mannitol. For the conversion of glucose into fructose glucose isomerase is used, while the selective hydrogenation is carried out with a solid (chemo-)catalyst.

An optimal mannitol yield requires:

- selective hydrogenation of fructose;
- high selectivity towards mannitol in the hydrogenation of fructose;
- relatively fast conversion of glucose to fructose.

Mannitol yields over 60% are obtained when starting with glucose (Figure B.13.3.1). When a mixture of glucose and fructose is used, essentially the same yield is obtained, confirming that the isomerization reaction is fast and the hydrogenation reaction is relatively slow.

QUESTION:

How would you separate mannitol from a mixture of mannitol and sorbitol?

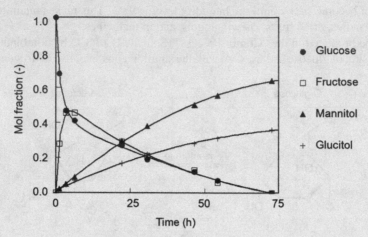

Figure B.13.3.1   *Product yields of the Combi-process [26]; batch process; feed: glucose.*

References

[1] Aiba, S., Humphrey, A.E., and Millis, N.F. (1973) *Biochemical Engineering*, 2nd edn. Academic Press, New York.

[2] Righelato, R.C. (1976) Selection of strains of penicillium-chrysogenum with reduced penicillin yields in continuous cultures. *Journal of Applied Chemistry and Biotechnology*, **26**, 153–159.

[3] Heijnen, J.J., Terwissscha van Scheltinga, A.H., and Straathof, A.J. (1992) Fundamental bottlenecks in the application of continuous bioprocesses. *Journal of Biotechnology*, **22**, 3–20.

[4] Hough, J.S., Keevil, C.W., Maric, V., Philliskirk, G., and Young, T.W. (1975) Continuous Culture in Brewing, in *Continuous culture 6: Applications and new fields* (eds A.C.R. Dean, D.C. Ellwood, C.G.T. Evans, and J. Melling). Ellis Horwood, Chichester, 226–237.

[5] Lee, Y.L. and Chang, H.N. (1990) High cell-density culture of a recombinant *Escherichia-Coli* producing penicillin acylase in a membrane cell recycle fermenter. *Biotechnology and Bioengineering*, **36**, 330–337.

[6] Liers, S. (1997) Bioreactoren, in *Procestechnieken En Eginering*, p. 37540 (in Dutch).

[7] van Benthum, W.A.J., van den Hoogen, J.H.A., van der Lans, R.G.J.M., van Loosdrecht, M.C.M., and Heijnen, J.J. (1999) The biofilm airlift suspension extension reactor. Part I: design and two-phase hydrodynamics. *Chemical Engineering Science*, **54**, 1909–1924.

[8] Nicolella, C., van Loosdrecht, M.C.M. and Heijnen, S.J. (2000) Particle-based biofilm reactor technology. *Trends in Biotechnology*, **18**, 312–320.

[9] Winkler, M.A. (1983) Application of the principles of fermentation engineering to biotechnology, in *Principles of Biotechnology* (ed. A. Wiseman). Surrey University Press, New York, pp. 94–143.

[10] Stanbury, P. and Whitaker, A. (1984) *Principles of Fermentation Technology*. Pergamon Press, Oxford, Chapter 5.

[11] Dong, H., Tao, W., Dai, Z., Yang, L., Gong, F., Zhang, Y., and Li, Y. (2012) *Biobutanol Biotechnology in China III: Biofuels and Bioenergy*, vol. **128** (eds F.W. Bai, C.G. Liu, H. Huang and G.T. Tsao). Springer, Berlin/Heidelberg, pp. 85–100.

[12] Snape, J.B., Dunn, I.J., Ingham, J., and Prenosil, J.E. (1995) *Dynamics of Environmental Bioprocesses, Modelling and Simulation*. VCH, Weinheim, Chapter 2.

[13] Simmler, W. and Mann, T. (1992) Water, in *Ullmann's Encyclopedia of Industrial Chemistry*, 5th edn, vol. **B8** (ed. W. Gerhartz). VCH, Weinheim, pp. 36–37.

[14] Janssen, A.J.H., Dijkman, H., and Janssen, G. (2000) Novel biological processes for the removal of H_2S and SO_2 from gas streams, in *Environmental Technologies to Treat Sulfur Pollution* (eds P.N.L. Lens and L. Hulshoff Pol). IWA Publishing, London, pp. 265–280.

[15] Tichý, R. (2000) Treatment of solid materials containing inorganic sulfur compounds, in *Environmental Technologies to Treat Sulfur Pollution* (eds P.N.L. Lens and L. Hulshoff Pol). IWA Publishing, London, pp. 329–354.

[16] Erickson, L.E. and Fung, D.Y.C. (eds) (1988) *Handbook of Anaerobic Fermentations*. Marcel Dekker, New York, p. 325.

[17] Hosten, L. and Van Vaerenbergh, E. (2009) Biologische eenheidsbewerkingen in *Procestechnieken En Engineering*. p. 34170 (in Dutch).

[18] Barford, J.P. (1988) Start-up, dynamics and control of anaerobic digesters, in *Handbook of Anaerobic Fermentations* (eds L.E. Erickson and D.Y.C. Fung). Marcel Dekker, New York, p. 803.

[19] Parshall, G.W. and Nugent, W.A. (1988) Making pharmaceuticals via homogeneous catalysis, part 1. *CHEMTECH*, **3**, 184–190.

[20] de Lathouder, K.M., Marques Fló, T., Kapteijn, F., and Moulijn, J.A. (2005) A novel structured bioreactor: Development of a monolithic stirrer reactor with immobilized lipase. *Catalysis Today*, **105**, 443–447.

[21] Moulijn, J.A., Kreutzer, M.T., Nijhuis, T.A., and Kapteijn, F. (2011) Monolithic catalysts and reactors: high precision with low energy consumption, in *Advances in Catalysis*, vol. **54** (eds B.C. Gates and H. Knözinger). Academic Press, New York, pp. 249–327.

[22] Messing, R.A. (1975) *Immobilized Enzymes for Industrial Reactors*. Academic Press, New York.

[23] Bohak, Z. and Sharon, N. (1977) *Biotechnological Applications of Proteins and Enzymes*. Academic Press, New York.

[24] Chibata, I. (1974) Optical resolution of DL-amino acids, in *Synthetic Production and Utilization of Amino Acids* (eds T. Kaneko, Y. Izumi, I. Chibata and T. Itoh). John Wiley & Sons, Inc., New York.

[25] Antrim, R.L., Colilla, W. and Schnyder, B.J. (1979) Glucose isomerase production of high-fructose syrups, in *Applied Biochemistry and Bioengineering*, vol. **2** (eds L.B. Wingard Jr., E. Katchalski-Katzir and L. Goldstein). Academic Press, New York, pp. 97–155.

[26] Makkee, M. (1984) Combined Action of Enzyme and Metal Catalyst, Applied to the Preparation of D-Mannitol. PhD Thesis, Delft University of Technology, Delft, The Netherlands.

General Literature

Averill, B.A., Laane, N.W.M., Straathof, A.J.J., and Tramper, J. (1999) Biocatalysis, in *Catalysis: an Integrated Approach to Homogeneous, Heterogeneous and Industrial Catalysis* (eds R.A. van Santen, P.W.N.M. van Leeuwen, J.A. Moulijn and B.A. Averill), 2nd edn. Elsevier, Amsterdam, The Netherlands.

Flickinger, M.C. and Drew, S.W. (eds) (1999) *Encyclopedia of Bioprocess Technology: Fermentation, Biocatalysis, and Bioseparation*. John Wiley and Sons, Inc., New York.

Ho, C.S. and Oldshue, J.Y. (eds) (1987) *Biotechnology Processes, Scale-up and Mixing*. AIChE, New York.

Präve, P., Faust, U., Sittig, W., and Sukatsch, D.A. (eds) (1987) *Basic Biotechnology, a Student's Guide*. VCH, Weinheim.

Wingard, L.B., Katchalski-Katzir, E., and Goldstein, L. (eds) (1979) *Applied Biochemistry and Bioengineering*, vol. **2**. Academic Press, New York.

Zaborsky, O.R. (1973) *Immobilized Enzymes*. CRC Press, Cleveland, OH.

14

Process Intensification

14.1 Introduction

A few decades ago a general feeling developed that the discipline of Chemical Engineering was reaching maturity. This would imply that new breakthrough-type developments were not to be expected anymore. However, this static picture has changed profoundly.

On the one hand, it became felt that the field of chemical engineering was much broader than often realized. In particular, "Product Technology" was seen as a new paradigm. The basis for this was the observation that many products are not pure compounds like methanol, acetic acid, and so on, but complex mixtures sold for their performance instead of their composition. Examples are the replacement of washing powders by washing tablets, slow-release medicine formulations, food products, batteries, and so on [1].

On the other hand, it was believed that even in traditional topics breakthroughs are still possible, leading to a really advanced Process Industry. Often the term Process Intensification (PI) is used for lumping together several breakthrough-type new developments in chemical engineering.

14.1.1 What is Process Intensification

Initially the term Process Intensification was introduced as a concept aimed at reducing the size of plants by orders of magnitude (10–100) [2]. Miniaturization of the plant and integration of reaction and separation have been the objectives of the pioneers in this field and, in fact, still are seen as key objectives. Figure 14.1 shows a classical picture of the intended transformation of a conventional plant into a future plant. In contrast, Box 14.1 shows the similarity between a sixteenth century plant and a contemporary one.

Later the definition was broadened to the development and design of more compact, cleaner, more cost-effective, and safer plants. There is still an on-going discussion about the definition of Process Intensification [3, 4].

Chemical Process Technology, Second Edition. Jacob A. Moulijn, Michiel Makkee, and Annelies E. van Diepen.
© 2013 John Wiley & Sons, Ltd. Published 2013 by John Wiley & Sons, Ltd.

Figure 14.1   *An artist's impression of the drastic change from a conventional plant in 2002 (left) into a plant based on process intensification concepts (right). Reprinted under the terms of the STM agreement from [5] Copyright (2004) CRC Press.*

Box 14.1   Process Technology in the Production of Gold, Anno 1556

It is rewarding to learn from history. In 1556 (!), Agricola published the classic work "De Re Metallica" [6], describing the state-of-the-art of mining, refining, and smelting metals. This work has been of great significance for the development of science and technology. For chemical engineers, it is an eye opener in many respects. A good example from the detailed descriptions in the book is a woodcut describing a continuous process for the production of gold from ore, part of which is shown in Figure B14.1.1 (top). When comparing this with the equipment in modern plants (Figure B14.1.1, bottom), we should become modest: there is a striking similarity between the processing equipment of the sixteenth century plant and plants of today [3]. Even the stirrers do not seem hopelessly old-fashioned. Thus, there should be room for breakthroughs in equipment/operation and design, supporting the idea of introducing PI.

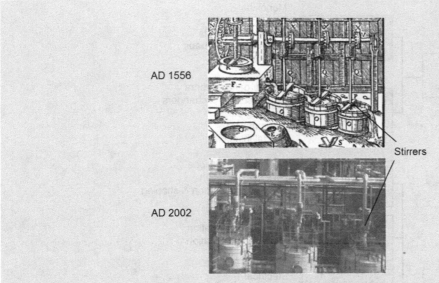

Figure B14.1.1 *Detail from a plant for the production of gold from ore (1556) compared with that of a modern plant (2002) showing a popular reactor configuration that is remarkably similar to the 1556 ore plant. Reprinted under the terms of the STM agreement from [5] Copyright (2004) CRC Press.*

14.1.2 How to Intensify Processes

In the early stages of PI, tools were divided in Intensifying Equipment and Intensifying Methods, in a way analogous to the distinction between hardware and software [3, 5]. Figure 14.2 is based on this division. Figure 14.2 focuses on the meso- and macroscale. However, the microscale and the nanoscale are the scales where the basic phenomena take place and the relevance of this scale for PI is obvious. In the future, probably the real impact will be on these scales. An obvious example is catalysis: a new excellent catalyst may allow a completely new process route that might be a prime example of process intensification. In this book a large number of breakthrough technologies have been discussed based on the discovery of novel catalyst systems. Examples of this type of PI are given, among others, in Chapter 12.

Several attempts have been made to define PI in general terms. A "PI approach" has been summarized by the following guidelines [4]:

- optimize kinetics (maximize the reaction rates of the desired reactions, minimize the rates of undesired reactions);
- give each molecule the same processing experience;
- optimize the driving forces and the associated surface areas;
- maximize synergies by combining subprocesses.

The first guideline refers to the molecular scale. This is the domain of *chemistry and catalysis*. The length scale is in the nanorange. A good example is multifunctional catalysis, where two sets of catalytic sites work in harmony, thereby increasing selectivity. Returning to the remark on new process routes we add:

- select the optimal chemical route.

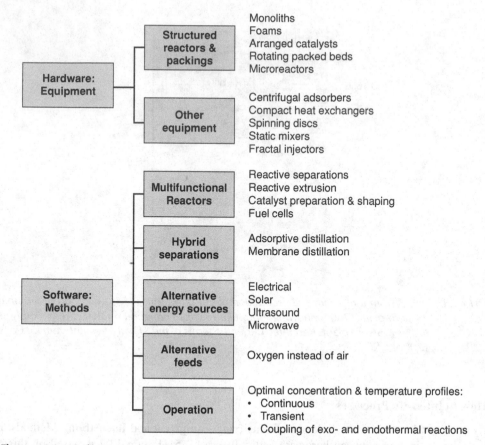

Figure 14.2   *Software and hardware tools for process intensification in chemical processes [3].*

The other guidelines resemble guidelines in *chemical engineering* culture. In basic reactor theory courses, the continuous stirred-tank reactor is compared with the plug-flow reactor. For nearly all kinetic models under isothermal conditions, the selectivity in a plug-flow reactor is higher because of the fact that all molecules are treated exactly the same. Molar flow rates across interfaces are maximal at maximum driving force (dc/dz) and specific interface area ($A_{\text{interface}}/V_{\text{equipment}}$). Equipment design is often based on optimization of these parameters.

Combining reaction and separation or combining two separation techniques are elegant concepts, which are at the basis of most PI successes. A good example is reactive distillation, in which distillation and reaction are combined (Section 4.3.1). In fact, for many chemical engineers this is the first example coming to mind when thinking of PI.

QUESTIONS:

    Compare the guidelines with the tools in Figure 14.3. Give an example of a new chemical route in which PI plays an important role.
    What makes a fuel cell a multifunctional reactor?
    Give an example of a configuration in which an exo- and an endothermic reaction are coupled or can be coupled.
    Give examples of PI-types of breakthroughs in oil refining (Hint: check conversions based on acid catalysis) and in the production of fine chemicals.

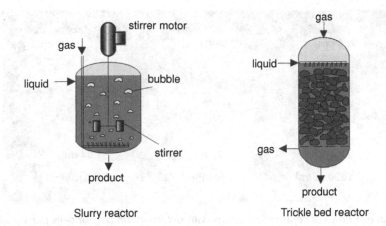

Figure 14.3 *The workhorses in the chemical industry for solid-catalyzed reactions involving gas phase and liquid phase reactants.*

In this chapter it is not attempted to present a full description of PI. The discussion is limited to a hardware/equipment item, "Structured catalytic reactors", and a software/methods item, "Reactive separation".

14.2 Structured Catalytic Reactors

In a reactor, more often than not, by-products are generated along with the desired products. The formation of by-products requires separation equipment and usually also recycle streams. Thus, improvement of the reactor design can have a large impact on the product yields. Advanced reactors are an obvious part of the PI toolbox.

Heterogeneous catalysis with a single fluid phase or multiple fluid phases is present everywhere in chemical plants. The most popular reactors used in industry for multiphase applications are slurry reactors and trickle bed reactors (Figure 14.3).

QUESTION:

> *Often, the selectivity in slurry reactors is larger than in trickle bed reactors. Explain. (Hint: the size of the catalyst particles is generally different in the two reactor designs, Section 10.2.)*

Instead of accepting the random and chaotic behavior of conventional reactors, reactors can be designed and built that are characterized by regular spatial structures. Such reactors are referred to as structured reactors [7]. The distinction between the catalyst particle and the reactor vanishes. In nearly all respects, structured catalysts and reactors outperform random particles and random/chaotic reactors. The use of a structured catalyst and/or reactor permits almost total control over all relevant length scales for mass transfer and catalysis, which is in contrast to the compromises encountered in the design of conventional catalysts/reactors. The desired properties for PI can be translated into maximizing the gradients (dc/dz, dT/dz) and the related specific area ($A_{\text{diffusive transport}}/V_{\text{reactor}}$).

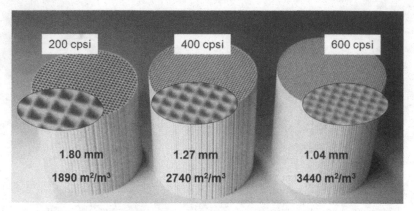

Figure 14.4   *Typical examples of ceramic monoliths with different numbers of cells per square inch (cpsi); the numbers in bold font are the diameters of the channels and the specific geometric surface areas ($m^2_{geometric}/m^3_{reactor}$).*

14.2.1   Types of Structured Catalysts and Reactors

Monoliths, the most popular structured reactors [8], are continuous structures consisting of a large number of parallel channels (Figure 14.4).

QUESTION:
>   *Compare randomly packed beds with monoliths regarding selectivity.*

Figure 14.5 shows a solid foam structure. Foams are three-dimensional cellular materials made of interconnected pores. They can be modeled as the negative image of a packed bed. Foams combine a high porosity with a high specific surface area. They are formed from a wide variety of materials such as metals and ceramics.

Figure 14.6 (left) shows a static mixer. Structures like this are widely used in distillation towers because of their low pressure drop combined with high specific surface area. By application of such packings, the capacities of distillation columns can be increased in a simple way, even by retrofitting existing columns. A related structure, shown in Figure 14.6 (right) is used in catalytic distillation: the catalyst particles are placed in the pockets of a structured wire packing [10]. This configuration might not be optimal, but it has the great advantage of allowing the use of commercially available catalyst particles [11].

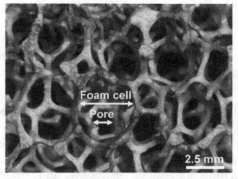

Figure 14.5   *Solid foam (aluminum); note the open, cellular structure. Reprinted with permission from [9] Copyright (2011) Elsevier Ltd.*

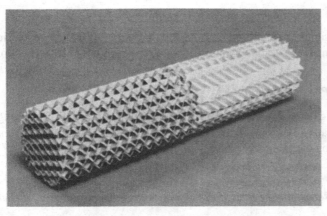

Figure 14.6   *Static mixer (left); structured packed bed of catalyst particles used in catalytic distillation (right). Reprinted with permission from [9] Copyright (2011) Elsevier Ltd.*

QUESTIONS:

Why are structured packings often used in retrofitting of distillation columns? Why in particular in vacuum distillation?
Explain the term "structured packed bed".

Microreactors (Figure 14.7) also belong to the family of structured reactors. Current microfabrication techniques allow fabrication of small structured catalytic reactors [12]. The versatile fabrication possibilities for chip-based reactors have led to the simultaneous development of structured and unstructured reactors, but in the final analysis the structured version is usually favored.

QUESTION:

Earlier it was stated that reactors should be designed such that concentration and temperature gradients and the related surface area-to-volume ratios are as large as possible. Have structured reactors been designed in this way?

Figure 14.7   *Microfabricated packed bed in which pillars take on the function of catalyst particles. Reprinted with permission from [9] Copyright (2011) Elsevier Ltd.*

14.2.2 Monoliths

Monolith reactors are conceptually the simplest structured reactors and they are most wide-spread. Monoliths are continuous structures consisting of narrow parallel channels (typically 1–3 mm diameter) (Section 10.6).

14.2.2.1 *Monolith Reactors for Gas Phase Reactions*

Although monolith reactors are already extensively used for environmental purposes, such as cleaning of flue gas from power plants and nitric acid plants (Section 6.4) and exhaust gas from cars (Section 10.6), examples of their use in the commercial production of chemicals are very limited.

QUESTION:

> *What would be the reason(s) for the limited use of monoliths in the production of chemicals?*

An example is the production of phthalic anhydride [13], an important feedstock for the production of plasticizers and plastics. The production of phthalic anhydride involves the heterogeneously catalyzed selective oxidation of *o*-xylene in air:

$$\Delta_r H = -1117 \text{ kJ/mol} \tag{14.1}$$

The reaction is highly exothermic and, as a consequence, control of the reactor temperature is a major challenge. The process is carried out in externally cooled multitubular reactors. To reduce investment and utility costs, many efforts have been spent in recent year to enhance the *o*-xylene inlet concentration from about 60 g/m^3 to 100 g/m^3 (STP) or more. Due to the high exothermicity of the process, such a concentration rise results in an increased reactor temperature. This results in deactivation of the catalyst, and eventually the *o*-xylene conversion and phthalic anhydride selectivity drop below the process constraints (especially the quality of the purified product).

A solution to this problem is the implementation of catalytic post reactors for the conversion of unreacted *o*-xylene to phthalic anhydride. As a result, an increase of product yield and quality is achieved [14, 15] (Box 14.2).

Box 14.2 Enhanced Phthalic Anhydride Yield with Catalytic Post Reactor

Lurgi, GEA, and Wacker have jointly developed a post reactor system using a honeycomb monolith. The use of monolithic post reactors enables retrofitting a plant using existing cooling facilities. The post reactor is operated adiabatically and is controlled only by regulation of the inlet gas temperature by means of pre-cooling. Figure B14.2.1 shows a simplified flow scheme of the reaction section.

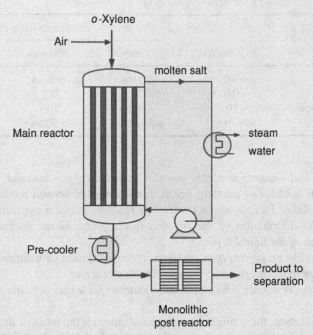

Figure B14.2.1 *Simplified flow scheme of the reaction section in a phthalic anhydride plant retrofitted with a monolithic post reactor.*

The installation of the monolithic post reactor allows for higher conversion and higher selectivity. In addition, longer operation between catalyst replacements is possible, because a significant portion of the *o*-xylene and intermediate underoxidation products are converted to phthalic anhydride in this reactor.

QUESTION:

> *What is the function of the molten salt recirculation? Compare the production of phthalic anhydride with that of terephthalic acid (Section 9.5). Why is no adiabatic post reactor used in the latter case?*

14.2.2.2 *Monolith Reactors for Gas–Liquid Reactions*

Because of the small size of the monolith channels the flow is laminar. In principle, a laminar flow velocity profile is associated with low mass transfer rates and a wide residence time distribution. At first sight this might seem killing for the idea of using monoliths as catalytic reactors for liquid phase processes. This can be understood from Table 14.1, which gives typical values for the time scales of diffusional transport.

The data in Table 14.1 directly give information on the expected reactor performance. For instance, for particle sizes in the millimeter range, typical for most trickle bed reactors, internal diffusion will often be rate limiting. Fortunately, for gases radial transport by diffusion is fast and, as a consequence, the residence time distribution is sharp.

Table 14.1   *Characteristic time scales for diffusion in capillaries and pores [9].*

Characteristic	D_{eff} (m^2/s)	Characteristic dimension			
		1 mm	100 μm	10 μm	1 μm
Gas	~10^{-5}	50 ms	500 μs	5 μs	50 ns
Liquid	~10^{-9}	500 s	5 s	50 ms	0.5 ms
Liquid in typical catalyst pore	~10^{-10}	5000 s	50 s	0.5 s	5 ms
Liquid in zeolite pore	< ~10^{-11}	>50.000 s	>500 s	>5 s	>50 ms

QUESTIONS:

Trickle bed reactors are very popular in practice, for instance in hydrodesulfurization (Section 3.4.5). The particle size is in the range of several millimeters. Considering the data in Table 14.1, how can trickle bed reactors perform satisfactory in practice? (Hint: draw the distribution of the gas and liquid phase in the packed bed and estimate the thickness of the liquid layers.)

Evaluate the properties of monolith reactors in relation with the principles attributed to PI. Compare monolith reactors with packed bed reactors.

Why is the residence time distribution sharper for larger diffusivity?

In operations with liquid streams, there might be some pessimism on the basis of the data presented in Table 14.1. Compared to diffusion in the gas phase, the diffusivity in the liquid phase is four orders of magnitude smaller, suggesting unrealistically long residence times. However, in a reactor configuration based on a monolithic stirrer (Figure 13.10) satisfactory performance was observed in many cases. The reason is the combination of modest rates of reaction and short pieces of monoliths leading to enhanced mass transfer. In addition, most practical liquid phase applications involve mixtures of gas and liquid phases rather than just liquids, and the hydrodynamics of gas–liquid systems has some spectacular characteristics that often lead to high mass transfer rates.

14.2.2.3   Hydrodynamics

In multiphase catalytic reactors hydrodynamics plays an important role in reactor performance. The flow regimes in capillaries have been investigated extensively [9, 16, 17]. Figure 14.8 shows the well-known two-phase flow patterns observed in tubes.

Going from left to right in the Figure 14.8, the flow pattern changes from bubble flow to film flow. In the former, the liquid is a continuous phase and the gas is a discontinuous phase, being present as bubbles; in the latter the liquid flows downward as a film along the walls, while the gas is a continuous phase that can flow upward or downward. In monoliths the two important flow patterns are film flow and Taylor flow, also referred to as "slug flow" or "segmented flow" (Figure 14.9). Taylor flow combines a low pressure drop with very high mass transfer rates and near-plug-flow behavior [18].

Film flow is suited for catalytic distillation. In monolithic structures this regime is feasible, whereas in trickle bed reactors this is usually not the case. The most used flow regime in gas–liquid systems is the Taylor flow regime (Box 14.3).

QUESTION:

Why is countercurrent gas–liquid operation of conventional fixed bed reactors generally not feasible?

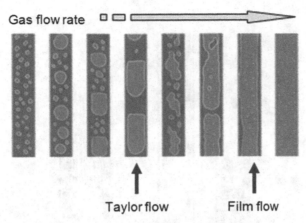

Figure 14.8   *Hydrodynamic regimes regimes reported in gas–liquid cocurrent downflow in capillaries. The liquid flow rate is constant and the gas flow rate increases going from left to right.*

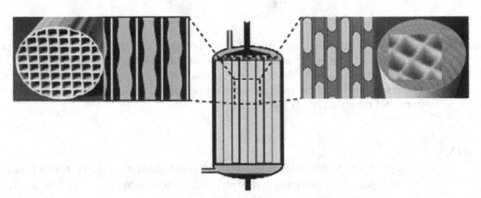

Figure 14.9   *Film flow (left) versus Taylor flow (right). Reprinted with permission from [9] Copyright (2011) Elsevier Ltd.*

Box 14.3   Taylor Flow

In cocurrent gas–liquid flow in small-diameter channels several flow regimes can be observed, of which the preferred one is usually Taylor flow. This type of flow is characterized by gas bubbles and liquid slugs flowing consecutively through the small monolith channels. The gas bubbles occupy (nearly) the whole cross-section of the channels and are elongated. Only a thin liquid film separates the gas bubbles from the catalyst (Figure B14.3.1).

The rate of mass transfer in cocurrent operation under Taylor flow conditions is high for two reasons [16, 19]. Firstly, the liquid layer between bubble and catalyst coating is very thin. Secondly, the liquid in the slugs circulates internally like a cylindrical moving belt. Moreover, the gas bubbles push the liquid slug forward as a piston, resulting in predominantly plug-flow characteristics.

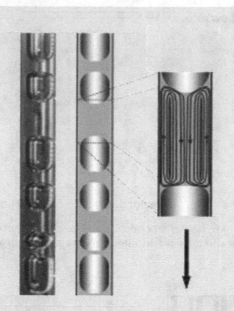

Figure B14.3.1   *Taylor flow in a capillary. Left: image of air–water flow; middle: schematic representation of the gas-liquid slugs; right: velocity pattern in a liquid slug obtained by CFD (computational fluid dynamics) showing the liquid recirculation. The thickness of the liquid film between the gas bubbles and the walls is of the order of 20–100 μm. Reprinted with permission from [9] Copyright (2011) Elsevier Ltd.*

QUESTIONS:

> *Compare Taylor flow with single-phase liquid flow in a capillary. If in a liquid phase process plug-flow behavior is desired, could a monolith reactor still be used?*

14.2.2.4   *Comparison of Monolith Reactors with Conventional Reactors*

For solid-catalyzed gas–liquid reactions monoliths compete with slurry reactors (mainly in fine chemicals) and trickle bed reactors (bulk chemicals and oil refining). Table 14.2 compares these reactors with the three-phase monolith reactor.

QUESTION:

> *The pressure drop in a monolith reactor is very small. It can even be negative! Explain.*

The main advantage of the slurry reactor is that the relatively small particles result in good internal and liquid–solid mass transfer characteristics leading to good catalyst utilization. Easy temperature control is another advantage. Its main disadvantages are catalyst attrition problems, the need for a catalyst separation step, and the chaotic behavior of the reactor, which makes scale-up difficult.

For trickle bed reactors these characteristics are reversed, that is, there are no attrition problems, no catalyst separation is needed, and scale-up is relatively easy, while catalyst utilization is worse for conventional

Table 14.2   *Comparison of slurry reactor, trickle bed reactor, and multiphase monolith reactor for catalytic gas–liquid reactions [9, 20–22].*

Characteristic	Slurry reactor	Trickle-flow reactor	Multiphase monolith
Particle diameter/wash coat thickness (mm)	0.001–0.2	1.5–6	0.01–0.2
Fraction of catalyst (m^3_{cat}/m^3_r)	0.001–0.01	0.55–0.6	0.07–0.15
Liquid hold-up (m^3_l/m^3_r)	0.8–0.9	0.05–0.25	0.1–0.5
Pressure drop (kPa/m)		50	3
Volumetric mass transfer coefficient			
Gas–liquid, $k_l a_l$ (s^{-1})	0.03–0.3	0.01–0.1	>1
Liquid–solid, $k_s a_s$ (s^{-1})	0.1–0.5	0.06	0.03–0.09

catalyst shapes. The choice of the optimal particle diameter is a result of a compromise between good catalyst utilization (small particles) and low pressure drop (large particles). In addition, temperature control is more difficult.

Three-phase monolith reactors combine the advantages of conventional reactors, while eliminating their main disadvantages.

Monolith reactors also have disadvantages. They are more difficult to fabricate than conventional catalyst particles and therefore are more expensive. Catalyst deactivation is an important issue in mololiths, as it is in trickle bed reactors. Catalyst life should be sufficiently long. Although the scale-up of monoliths is straightforward, there are some questions as to how monolith blocks should be stacked and what the effect is on performance. Moreover, radial distribution over the reactor cross-section should be good, because radial dispersion in the reactor does not occur.

QUESTION:

> *Compare monolith reactors with foams and with microreactors. Give similarities and differences.*

An example where the use of monolith reactors could be advantageous is in catalytic hydrogenations. Currently, in most catalytic hydrogenation processes for the production of fine and specialty chemicals slurry reactors are employed. These processes can cause problems such as contamination of the environment with metals, excessive waste production, and safety issues. These problems could be solved by employing monolith reactors (Box 14.4).

Box 14.4   Monolith Loop Reactor for Catalytic Hydrogenation Reactions

Air Products and Chemicals has developed a monolith loop reactor (MLR) in which catalysts developed by Johnson Matthey are used [23]. The MLR can be retrofitted into existing stirred-tank reactor systems to increase reaction rates (by a factor of 10–100) and avoid the use of a slurry catalyst. This system eliminates all slurry catalyst handling, reducing or eliminating the associated environmental and safety problems.

Figure B14.4.1 shows a flow scheme of a catalytic hydrogenation process using an MLR. The contents of the stirred-tank reactor are recirculated through the MLR using an external pump until the reaction is complete. The ejector enhances the gas-to-liquid mass transfer rate.

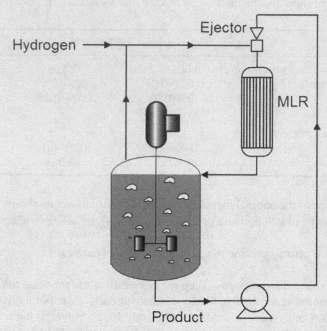

Figure B14.4.1   *Use of a monolith loop reactor for catalytic hydrogenation reactions.*

QUESTIONS:

> *Is the process shown in Figure 14.4 a continuous or a batch process?*
> *What are other options for the use of monoliths with existing stirred-tank reactors?*

14.2.3   Microreactors

Recently, small-scale continuous-flow reactors have received a great deal of attention, firstly as laboratory reactors, then later also for industrial-scale production of fine chemicals, pharmaceuticals, and specialty chemicals, and even bulk chemicals. These so-called microreactors (or microflow reactors) are devices consisting of single or multiple small-diameter channels (typically between 10 and 1000 μm) that allow reactions to be carried out on the submillimeter scale.

Several reactions that are currently carried out in batch or semi-batch reactors could benefit from continuous operation in microreactors. Such reactions are generally fast or very fast and/or involve thermal hazards [24, 25]. Examples of reactions that can be successfully carried out in microreactors are given elsewhere [26–29]. Some examples of the use of microreactors in industrial-scale production are given in Table 14.3.

Some of the technological advantages of microreactors over large batch reactors are:

- easier control – this is particularly relevant for safely conducting highly exothermic reactions such as nitrations and oxidations;
- much better mixing and mass and heat transfer efficiency, resulting in less by-products;

Table 14.3   *Industrial examples of the use of microreactors [30].*

Type of reaction	Intermediate in synthesis of	Company	Remarks
Nitration	naproxcinod	DSM	highly exothermic reaction
Azidation	NK1 antagonist	Eli Lilly	risk of detonation in batch process
Lithiation	H1 antagonist	Neurocrine biosciences	safer with continuous flow
Nitration and bromination of heterocycles	Various	AstraZeneca	safer and less impurities with continuous flow

- extremely small material hold-up – this, in combination with the easier control, leads to inherently safer processes;
- high flexibility owing to their modular design (Figure 14.10);
- large-scale production can be achieved by "numbering up" (i.e., increasing the number of reactors) instead of the traditional "scaling up" (i.e., increasing the reactor size).

QUESTION:

Why is "numbering up" instead of scaling up an advantage of microreactors? What about economies of scale?

Compare monolith reactors with microreactors. Give pros and cons.

Besides reaction kinetics, another factor to take into account is the different phases involved (solid–liquid–gas). In many reactions a solid is present, whether as reactant, catalyst, or product. The microreactors currently available can only handle solids very poorly; solid particles may easily block the microchannels. Thus, the (multipurpose) use of microreactors is still limited to homogeneous reactions and, to some extent, to gas–liquid and liquid–liquid reactions. This reduces the amount of possible reaction candidates for microreactor technology. A further technological development will be needed to develop microreactors capable of handling solids in a flexible way [24]. Moreover, corrosion might be more of an issue for microreactors than for large-scale reactors.

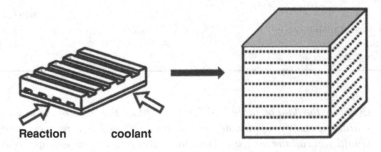

Figure 14.10   *Illustration of the modular design of microreactors.*

QUESTIONS:

> *Assume you are considering microreactor technology for a solid-catalyzed reaction. What general bottlenecks do you expect during the development of the process?*
> *Give examples of reactions that are suited for microreactor technology. For which type of reactions would you consider this technology unsuitable?*

Box 14.5 illustrates the difference in operation and performance characteristics between a large batch reactor and a microreactor.

Box 14.5 Example: Hydrogenation of *o*-cresol

The catalytic hydrogenation of *o*-cresol (Scheme B14.5.1) is a highly exothermic reaction. The reaction rate is limited by the low solubility of hydrogen. A reactor has been developed having superior heat and mass transfer properties. This intensified reactor operates as a slurry reactor at high pressure and is equipped with a very efficient heat exchanger. Data are given in Table B14.5.1. From this table, it can be seen that the characteristics of the intensified reactor differ much from those of a conventional batch reactor.

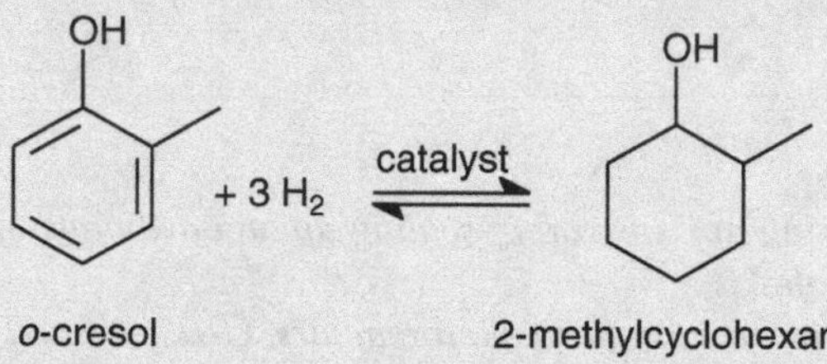

o-cresol   2-methylcyclohexanol

Scheme B14.5.1 *Hydrogenation of o-cresol. Reprinted with permission from [32] Copyright (2009) Tekno Scienze.*

Table B14.5.1 *Comparison of operation and performance of a batch reactor and an intensified reactor for the hydrogenation of o-cresol [31].*

Characteristic	Batch reactor	Intensified reactor
Volume (m^3)	6	0.0007
Surface/volume (m^{-1})	5.3	150
Pressure (bar)	15	200
Temperature (K)	370–430	440–540
Catalyst (%)	4	0.4
Solvent (%)	75	no solvent
Gas–liquid mass transfer, $k_l a_l$ (s^{-1})	0.01–0.2	0.1–0.3
Residence time	4–5 h	<3 min
Conversion (%)	95	>99

QUESTIONS:

> *Evaluate the data of Table B14.5.1 and explain the basis of the intensification. When scaling up, how many intensified reactors are needed to replace the batch reactor? Would you call the intensified reactor of this example an example of a microreactor? The authors refer to it as a "minireactor". Comment on this.*

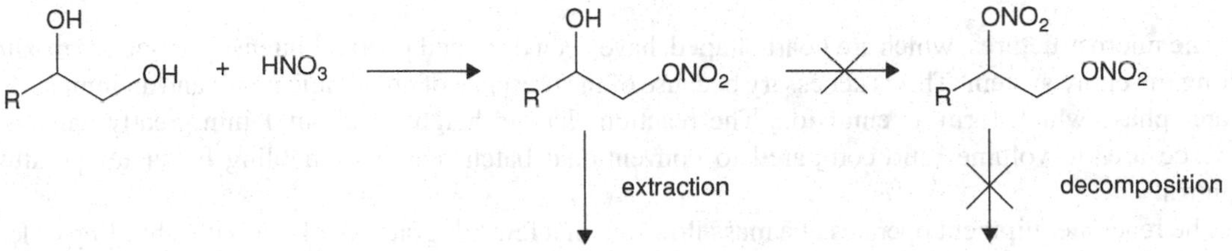

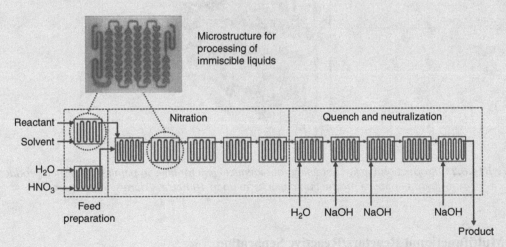

Scheme 14.1   *Typical example of a selective nitration reaction.*

An example of the successful application of microreactors on an industrial scale is in selective nitration reactions. Nitration reactions are difficult to handle due to the exothermic properties of organic nitrates. Although the reaction enthalpy of the nitration itself is moderate, the decomposition of the nitrated product is highly exothermic and violent decomposition can occur by improper handling [32]. Scheme 14.1 shows an example of the selective nitration of only one of two identical functional groups.

Strict control of reaction parameters such as temperature, stoichiometry, and residence time is crucial for both product quality and safety. To achieve high selectivity, conventional nitration processes use a high degree of dilution. In a process developed by DSM using microreactors (Box 14.6) such high-dilution conditions in the reaction zone are not necessary.

Box 14.6 Selective Nitration In a Microreactor System [25, 32]

Figure B14.6.1 shows the functions performed in the microreactor, which has been engineered and manufactured by Corning. Besides the reaction itself, feed preparation, dilution, and neutralized with caustic soda are performed in the microreactor consisting of 12 microstructures.

Figure B14.6.1   *Glass microstructure and schematic of a reactor made from 12 glass microstructures enabling the mass and heat transfer functions required (heat exchangers on each microstructure are not shown) [32]. Reproduced by permission of Teknoscienze. Reprinted wih permission from [25] Copyright (2011) American institute of Chemical Engineers.*

The microstructures, which are heart-shaped, have been designed to effect intense continuous mixing along the entire system. This is necessary because of the presence of an organic phase and an immiscible water phase which form an emulsion. The reaction channel height of about 1 mm greatly enhances surface area-to-volume ratio compared to conventional batch reactors, enabling better temperature control.

The reactor equipment operates at a mass flow rate of a few kilograms per hour with a hold-up of less than 150 ml. Scale-up, or rather numbering up, is easily achieved by assembling a number of reactors in production banks (Figure B14.6.2). The total flow processed by the production unit was approximately 100 kg/h, which corresponds to approximately 800 tons per year based on 8000 hours of continuous operation.

Figure B14.6.2   *Production bank, containing four identical reactor lines in parallel. Two such production banks enable a production capacity of up to 1000 t/a. Courtesy of Corning.*

14.3   Multifunctional Reactors/Reactive Separation

The potential of multifunctional systems for PI is obvious. In multifunctional systems different processes are coupled in one piece of equipment, usually reaction and separation or several separation processes.

The most important benefit of incorporating two different functions in one piece of equipment, of course, is reduction of the capital investment: why not place catalyst particles on the trays of the distillation column?

One process step is eliminated, along with the associated pumps, piping, and instrumentation. Other benefits depend on the specific reaction. In reactive distillation, optimal heat integration is possible. Equilibrium-limited reactions are an obvious class of reactions that would benefit from the continuous removal of one of the products *in situ*.

QUESTIONS:

How can a process be intensified by in situ removal of a component? Give an example.
Why is optimal heat integration possible in reactive distillation?
Give an example of the coupling of two separation methods.

In coupling reaction and separation the term "multifunctional reactor" is commonly used; when emphasizing the separation, "reactive separation" is usually spoken of.

Various separation techniques exist, for example, distillation, adsorption on a solid, absorption in a solvent, and membrane separation. The reaction will usually imply the use of a catalyst and the coupling of reaction and separation in principle allows enhancement of selectivities and yields. Coupling reaction and separation is the basis of most successful practical examples of PI. Here, discussion is limited to reactive distillation, membrane reactors, and reactive adsorption.

Although the combination of reaction and separation in one vessel is an elegant solution, in practice there is a fundamental limitation. The conditions of both processes cannot be optimized separately, so, for example, the temperature and pressure needed for the reaction must be in harmony with the conditions for separation. An important question is always whether there is a satisfactory window of operation.

14.3.1 Reactive Distillation

Reactive distillation involves a combination of reaction and distillation in a single column. If the reaction is catalyzed, the process is referred to as catalytic distillation. Reactive distillation is a two-phase flow process with gas and liquid flowing in countercurrent mode. If a solid catalyst is present, this requires special catalyst morphology.

QUESTION:

Why is a special catalyst morphology required for countercurrent operation?

In the bulk chemical industry, catalytic distillation is already employed in a number of processes, for example, for the production of MTBE (methyl *tert*-butyl ether) and other butyl ethers, the production of cumene and ethylbenzene, and hydrogenation of aromatics [35]. An example of the use of reactive distillation in the production of fine chemicals is the synthesis of acetoacetarylamides [37].

14.3.1.1 Production of MTBE

The formation of MTBE by etherification of isobutene with methanol (Section 6.3.4) is an equilibrium-limited exothermic reaction:

$$CH_3OH + \underset{H_3C}{\overset{H_3C}{>}}C{=}CH_2 \rightleftharpoons H_3C{-}\underset{CH_3}{\overset{CH_3}{\underset{|}{C}}}{-}O{-}CH_3 \qquad \Delta_r H_{298,liq} = -37.5 \text{ kJ/mol} \qquad (14.2)$$

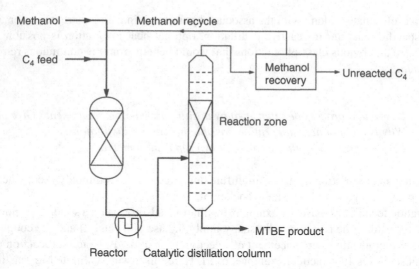

Figure 14.11   *Production of MTBE using catalytic distillation.*

The isobutene conversion can be enhanced by continuous removal of the product. Catalytic distillation offers this possibility. Figure 14.11 shows a schematic of a process for the production of MTBE, employing catalytic distillation.

In practice, the catalytic distillation column is preceded by a conventional fixed bed reactor. In this reactor the largest part of the conversion takes place at relatively high temperature (high reaction rate, low equilibrium conversion), while the remainder takes place in the catalytic distillation column at lower temperature (low reaction rate, high equilibrium conversion). The MTBE formed is continuously removed from the reaction section. Methanol, isobutene, and the inert hydrocarbons flow upward through the catalyst bed and then to a rectification section, which returns liquid MTBE. Isobutene conversions of over 99% can thus be achieved

An additional advantage of catalytic distillation is that the exothermic heat of reaction is directly used for the distillative separation of reactants and products. Hence, perfect heat integration is achieved: the heat of reaction does not increase the temperature resulting in near-isothermal operation in the reaction zone.

QUESTION:

Why is a conventional fixed bed reactor placed upstream of the catalytic distillation column?

14.3.1.2   Production of Ethylbenzene

Ethylbenzene is produced by alkylation of benzene with ethene:

$$\Delta_r H_{298} = -114 \text{ kJ/mol} \qquad (14.3)$$

Catalytic distillation is an alternative for the processes described in Section 10.4.3 for the production of ethylbenzene. The selectivity strongly depends on the ethene/benzene ratio: at low values the selectivity approaches 100%. Figure 14.12 shows a simplified flow scheme. In this configuration a low concentration of ethene in the reactor is maintained because of the recycle of unreacted benzene.

The heat of the highly exothermic alkylation reaction is now used directly in the distillation of reactants and products: benzene evaporates from the liquid mixture in the top of the distillation column. The vapor is then

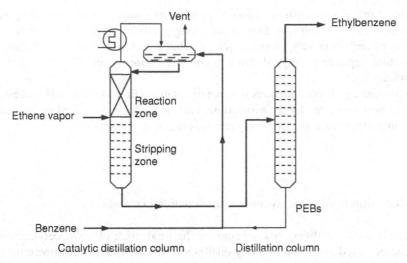

Figure 14.12 *Simplified flow scheme of catalytic distillation process to produce ethylbenzene. PEBs = poly(ethylbenzenes).*

condensed in the overhead condenser and returned to the reactor. It is claimed that this process requires only half the amount of energy compared to other processes, while the investment costs are reduced by 25% [34]. Furthermore, due to the "internal recirculation" of excess benzene, there is no need to (externally) recycle large amounts of benzene, reducing the operating costs of the process.

QUESTIONS:

Why are PEBs recycled?
Find the boiling points of the main components. Is the reactor configuration of Figure 14.12 feasible? What would be the reaction conditions (check Chapter 10). Do you expect that there is a feasible window of operation?
Why is the catalyst bed incorporated in the top section of the distillation column?

14.3.1.3 Production of Acetoacetarylamides

Acetoacetarylamides are important starting materials for the preparation of colored pigments and agrochemicals. They are manufactured in thousands of tons worldwide by reacting diketene with aromatic amines in a variety of solvents [37]. The simplest acetoacetarylamide is acetoacetylaniline (reaction 14.4).

$$\Delta_r H_{298} \ll 0 \text{ kJ/mol} \qquad\qquad (14.4)$$

In the conventional process, downstream processing includes completely batch-wise operated crystalliza-tion, solid–liquid separation, and drying. Due to the strong exothermic reaction, diketene has to be slowly added to a solution of aniline in an appropriate solvent. Batch cycle times of several hours are required to crystallize the product, separate the crystals from the mother liquor by centrifugation, and remove residual solvent by vacuum drying.

An alternative process employs continuous reactive distillation (Box 14.7). In this process, the problem of the large amount of heat generated by the exothermic reaction is elegantly solved by introducing water into the system. The water evaporates and efficiently removes the heat of reaction.

Box 14.7 Production of Acetoacetylaniline by Reactive Distillation

Figure B14.7.1 shows a simplified flow scheme for the production of acetoacetylaniline by reactive distillation. Aniline is fed to the top of a tray distillation column and moves downward reacting with diketene, which is fed a few stages above the bottom and moves upward, together with the water. The high-boiling acetoacetylaniline leaves from the bottom as a single melt phase and is transferred to a cooled steel band where it immediately solidifies to flakes.

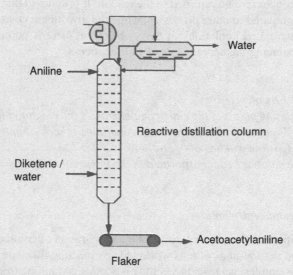

Figure B14.7.1 *Production of acetoacetylaniline by reactive distillation.*

This continuous technology offers a number of advantages over the batch process [37]:

- reduced number of unit operations (4 vs 6) by removing the crystallization and separation steps;
- higher equipment utilization (20 vs 0.3 t/m^3/d) leading to increased capacity and lower investment costs for a new plant;
- use of "green" solvents by performing the reaction in water.

14.3.2   Coupling Reaction and Membrane Separation

In a catalytic membrane reactor the presence of a membrane (see also Section 6.1.6.2), thanks to its permselectivity, affects the course of the reactions, allowing improvements of either the achievable conversion (e.g., equilibrium-limited reactions) or the selectivity towards intermediate products (e.g., consecutive reaction schemes) [38].

Three major application areas exist for catalytic membrane reactors [39]:

- *yield enhancement of equilibrium-limited reactions*: a reaction product is selectively permeating through the membrane, thereby enhancing the per pass conversion compared to conventional fixed bed reactors (e.g., for dehydrogenations);
- *selectivity enhancement*: accomplished by selective permeation or controlled addition of a reactant through the membrane (e.g., partial oxidations or hydrogenations);
- *membrane bioreactors*: coupling membranes and biochemical reactions promoted by (immobilized) enzymes or microorganisms.

QUESTIONS:

> *What does "permselective" mean? What would be the origin of permselectivity in porous ceramic membranes?*

14.3.2.1   *Yield Enhancement of Equilibrium-Limited Reactions*

The first field for which membrane reactors have been conceived is in equilibrium enhancement. A classical bench-scale example of the use of a catalytic membrane reactor, showing its potential, is the dehydrogenation of ethane:

$$H_3C\text{–}CH_3 \underset{}{\overset{Pt}{\rightleftharpoons}} H_2C = CH_2 + H_2 \qquad \Delta_r H_{298} = 137 \text{ kJ/mol} \qquad (14.5)$$

The reactor consists of a porous alumina membrane tube covered with platinum crystallites showing good catalytic activity for dehydrogenation (Figure 14.13). The ceramic tube consists of a multilayered composite porous alumina tube.

For the gases playing a role in the reaction, Figure 14.14 shows the gas flow rate as a function of the pressure drop over the membrane. Clearly, the transport of hydrogen is much faster than that of ethane and ethene. Hence, in principle, it is possible to shift the equilibrium to the side of ethene by removal of hydrogen.

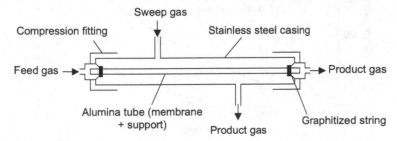

Figure 14.13   *Membrane reactor for the dehydrogenation of ethane [40].*

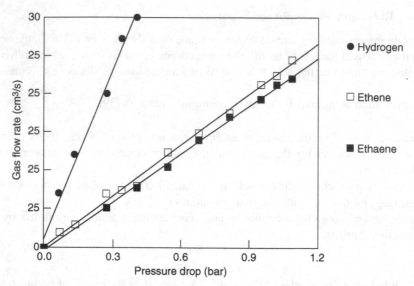

Figure 14.14   *Permeation rate of hydrogen, ethane and ethene as a function of pressure [40].*

Indeed, it has been shown that with this reactor conversions far beyond thermodynamic equilibrium can be realized (Figure 14.15).

QUESTIONS:

Why are two equilibrium conversion curves shown in Figure 14.15?
Evaluate the function of the sweep gas. What are the pros and cons of a high flow rate of the sweep gas?

Although membrane reactors certainly open opportunities, additional challenges have to be faced in the design and production of robust modules allowing high rates of mass (and heat) transfer. For a feasible process the production of hydrogen and the flow rate through the membrane must be of similar magnitude.

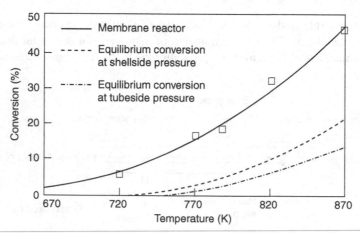

Figure 14.15   *Conversion in a membrane reactor compared to thermodynamic equilibrium [40].*

QUESTIONS:

> *In a membrane reactor design, one degree of freedom is the choice of the location of the membrane, close to the inlet of the reactor or further away. What is your choice and why? In this example a ceramic membrane is used. Give pros and cons of ceramic membranes compared to polymeric membranes.*

14.3.2.2 Selectivity Enhancement of Consecutive Reactions

Another interesting application opportunity of membrane reactors is gradually supplying one of the reactants along the reactor through the membrane. This may be particularly useful in partial oxidation reactions [39], examples of which are oxidative coupling of methane (Section 4.5.5) and oxygen membrane reforming (Section 5.2.4); with a membrane that is permselective towards oxygen and thus prevents the inert nitrogen from reaching the location where the reaction takes place, air can be used as the source of oxygen instead of pure oxygen.

Furthermore, the controlled staged addition of a reactant along the reactor length can have favorable effects on reaction selectivity [38, 39]. Some reactions, such as aforementioned partial oxidation reactions and hydrogenation reactions, can be driven to high selectivity by keeping the oxygen or hydrogen concentrations in the reacting mixture relatively low. This can be accomplished through a membrane, which can be used to dose the reactant in all parts of the reactor at once at the desired rate.

An additional benefit of the membrane reactor concept for partial oxidations is that the separation of the oxidant and organic molecules, except on the catalytic sites, creates reactor conditions less prone to explosions and other undesirable safety effects (Box 14.8).

Box 14.8 Alternative Process for the Production of Ethene Oxide Using a Catalytic Membrane Reactor

In the production of ethene oxide from ethene by partial oxidation (Section 10.5.2, the flammability limits restrict the feed composition, which in conventional reactors leads to the necessity of a large excess of the hydrocarbon. By adopting a multitubular cooled catalytic membrane reactor, oxygen and ethene can be kept separated, avoiding any flammability constraints. Figure B14.8.1 illustrates this.

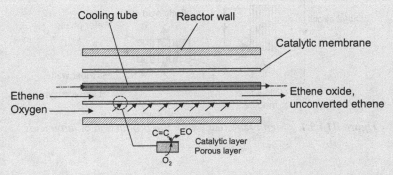

Figure B14.8.1 *Catalytic membrane reactor for the partial oxidation of ethene to ethene oxide.*

Oxygen diffuses through the membrane at a limited rate at various points along the reactor, so its concentration is kept low. Hence, besides avoiding flammability constraints, the selectivity of the reaction is also enhanced.

14.3.2.3 *Membrane Bioreactors*

Another example of the use of a membrane reactor is the production of fatty acids by fat hydrolysis (Scheme 14.2). Usually, this reaction is carried out at high temperatures and sometimes in the presence of a catalyst. A major challenge of the process is the fact that the remaining water phase is heavily polluted, thus making it uneconomical to recover the produced glycerol. Box 14.9 illustrates an alternative process using a membrane bioreactor.

QUESTION:

> *Why is the water phase heavily polluted?*

Box 14.9 Production of Fatty Acids by Fat Hydrolysis in a Membrane Reactor

A development at laboratory scale is the application of an enzyme (lipase) to catalyze the hydrolysis: Water and fat are mixed at low temperature (300 K) in continuous stirred-tank reactors. The water phase contains the enzyme. A much purer glycerol solution is obtained than in the conventional process. The challenge of this type of process is that the equilibrium is not favorable.

An elegant solution has been proposed based on a membrane reactor consisting of a module with hollow cellulose fibers (Figure B14.9.1) [41]. The enzyme is placed at the inner side of the fibers, to which the fat is fed. Water passes at the outside and diffuses through the membrane to react at the fat/lipase interface. The fatty acid formed stays in the oil phase, whereas the glycerol formed is transported through the membrane into the water phase. Laboratory studies show nearly complete conversions.

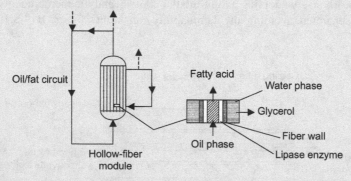

Figure B14.9.1   *Membrane reactor for the production of fatty acids.*

QUESTION:

> *What is the origin of the permselectivity in this case?*

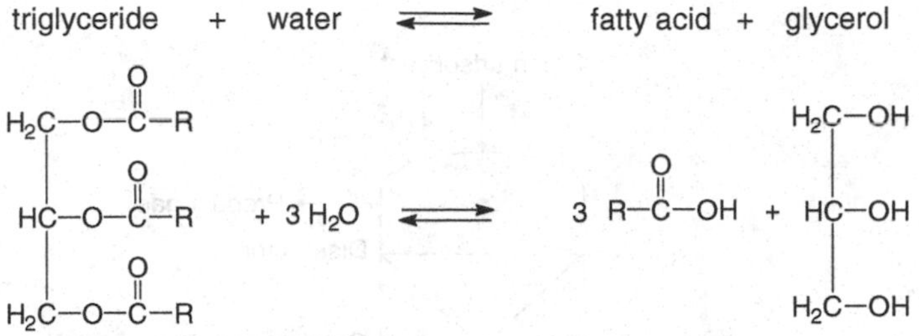

triglyceride + water ⇌ fatty acid + glycerol

$$H_2C-O-\overset{\overset{O}{\parallel}}{C}-R$$

$$HC-O-\overset{\overset{O}{\parallel}}{C}-R \quad + 3\,H_2O \quad \rightleftharpoons \quad 3\ R-\overset{\overset{O}{\parallel}}{C}-OH \quad + \quad HC-OH$$

$$H_2C-O-\overset{\overset{O}{\parallel}}{C}-R$$

$$H_2C-OH$$

$$H_2C-OH$$

Scheme 14.2 *Production of fatty acids.*

14.3.3   Coupling Reaction and Adsorption

An alternative multifunctional reactor combines reaction with adsorption. At a laboratory scale, a combination of reaction and adsorption has been employed for a few reactions, including the classical ammonia and methanol syntheses (Chapter 6).

Although since their discovery ammonia synthesis and methanol synthesis have been optimized to a high degree of sophistication, it is believed that process improvements might still be possible. In particular, energy losses might be decreased further. The major energy losses in these processes are due to:

- condensation (inerts are present in the stream from which ammonia must be condensed, lowering the heat transfer rates);
- recycles (compressor, heating, etc.), due to the limited conversion per pass.

Moreover, there are energy losses having a kinetic origin. This can be explained as follows. Over the length of the catalyst bed the product concentration increases, which causes a decrease of the reaction rate. Hence, longer beds are needed and, as a consequence, the pressure drop over the catalyst bed increases.

It has been suggested to employ a multifunctional reactor by adding an adsorbent, which enables a higher conversion per pass (Box 14.10).

Box 14.10 Coupling Reaction and Adsorption in the Synthesis of Methanol

Figure B14.10.1 shows the principle of the use of an adsorbent for methanol synthesis. The reactor is a fixed bed reactor, with the adsorbent and the gas flowing countercurrently [42, 43]. A porous powder "trickles" over the catalyst pellets. It adsorbs the methanol produced and hence the equilibrium is shifted towards methanol. In this specific example, an FCC catalyst appeared to be a useful, selective adsorbent for methanol. In a laboratory-scale reactor it was confirmed that conversions exceeding equilibrium values could be obtained.

QUESTIONS:

> *Give advantages and disadvantages of this technology. Is the optimal temperature a point of concern?*
> *In practice, the reactor temperature is much higher than that used in the laboratory work cited. Does this have consequences for the expected performance?*

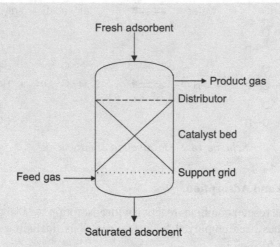

Fresh adsorbent

Product gas

Distributor

Catalyst bed

Support grid

Feed gas

Saturated adsorbent

Figure B14.10.1   *Gas–solid–solid trickle bed reactor for methanol synthesis.*

References

[1] Voncken, R.M., Broekhuis, A.A., Heeres, H.J., and Jonker, G.H. (2004) The many facets of product technology. *Chemical Engineering Research and Design*, **82**, 1411–1424.

[2] Ramshaw, C. (1983) Higee distillation – an example of process intensification. *Chemical Engineering (London)*, **389**, 13–14.

[3] Stankiewicz, A.I. and Moulijn, J.A. (2000) Process intensification: transforming chemical engineering. *Chemical Engineering Progress*, **96**, 22–33.

[4] Van Gerven, T. and Stankiewicz, A. (2009) Structure, energy, synergy, time – the fundamentals of process intensification. *Industrial & Engineering Chemistry Research*, **48**, 2465–2474.

[5] Stankiewicz, A. and Drinkenburg, A.A.H. (2004) Process Intensification: History, Philosophy, Principles, in *Re-Engineering the Chemical Processing Plant; Process Intensification* (eds A. Stankiewicz and J.A. Moulijn). CRC Press, pp. 1–32.

[6] Agricola, G. (1556) *De Re Metallica Libris XII*. J. Froben & N. Episopius, Basel, Switzerland.

[7] Cybulski, A. and Moulijn, J.A. (2005) *Structured Catalysts and Reactors*, 2nd edn. CRC Press, pp. 856.

[8] Kapteijn, F., Heiszwolf, J., Nijhuis, T.A., and Moulijn, J.A. (1999) Monoliths in multiphase catalytic processes – aspects and prospects. *CATTECH*, **3**, 24–40.

[9] Moulijn, J.A., Kreutzer, M.T., Nijhuis, T.A., and Kapteijn, F. (2011) Monolithic catalysts and reactors: high precision with low energy consumption, in *Advances in Catalysis*, vol. **54** (eds B.C. Gates and H. Knözinger). Academic Press, pp. 249–327.

[10] Behrens, M., Olujic, Ž., and Jansens, P.J. (2006) Liquid flow behavior in catalyst-containing pockets of modular catalytic structured packing katapak SP. *Industrial & Engineering Chemistry Research*, **46**, 3884–3890.

[11] Schildhauer, T.J., Kapteijn, F., Heibel, A.K., Yawalkar, A.A., and Moulijn, J.A. (2005) Reactive stripping in structured catalytic reactors: hydrodynamics and reaction performance, in *Integrated Chemical Processes: Synthesis, Operation, Analysis, and Control* (eds K. Sundmacher, A. Kienle and A. Seidel-Morgenstern). Wiley-VCH Verlag GmbH, Weinheim, pp. 233–264.

[12] de Loos, S., van der Schaaf, J., Tiggelaar, R., Nijhuis, T., de Croon, M., and Schouten, J. (2010) Gas–liquid dynamics at low Reynolds numbers in pillared rectangular micro channels. *Microfluidics and Nanofluidics*, **9**, 131–144.

[13] Lorz, P.M., Towae, F.K., Enke, W., Jäckh, R., Bhargava, N., and Hillesheim, W. (2000) Phthalic acid and derivatives, in *Ullmann's Encyclopedia of Industrial Chemistry*. Wiley-VCH Verlag GmbH, Weinheim. doi: 10.1002/14356007.a20_181.pub2

[14] Eberle, H.J., Breimair, J., Domes, H., and Gutermuth, T. (2000) Post reactor technology in phthalic anhydride plants. *Petroleum Technology Quarterly*, **June**, 84.

[15] Groppi, G., Tronconi, E., and Beretta, A. (2005) Monolithic catalysts for gas-phase syntheses of chemicals, in *Structured Catalysts and Reactors*, 2nd edn (eds A. Cybulski and J.A. Moulijn). CRC Press, pp. 243–310.

[16] Heiszwolf, J.J., Kreutzer, M.T., van den Eijnden, M.G., Kapteijn, F., and Moulijn, J.A. (2001) Gas−liquid mass transfer of aqueous Taylor flow in monoliths. *Catalysis Today*, **69**, 51–55.

[17] Kreutzer, M.T., Kapteijn, F., Moulijn, J.A., Kleijn, C.R., and Heiszwolf, J.J. (2005) Inertial and interfacial effects on pressure drop of Taylor flow in capillaries. *AIChE Journal*, **51**, 2428–2440.

[18] Moulijn, J.A., Kreutzer, M.T., Cybulski, A., and Andersson, B. (2005) Two-phase segmented flow in capillaries and monolith reactors, in *Structured Catalysts and Reactors* (eds A. Cybulski and J.A. Moulijn). CRC Press, Boca Raton, USA, pp. 393–434.

[19] Kreutzer, M.T., Du, P., Heiszwolf, J.J., Kapteijn, F., and Moulijn, J.A. (2001) Mass transfer characteristics of three-phase monolith reactors. *Chemical Engineering Science*, **56**, 6015–6023.

[20] Trambouze, P., Van Landeghem, H., and Wauquier, J.P. (1988) *Chemical Reactors, Design/Engineering/Operation*. Gulf Publishing Company, Houston, TX.

[21] Irandoust, S., Cybulski, A., and Moulijn, J.A. (1997) The use of monolithic catalysts for three-phase reactions, in *Structured Catalysts and Reactors* (eds A. Cybulski and J.A. Moulijn). Dekker, New York.

[22] Ramachandran, P.A. and Chaudhari, R.V. (1983) Three phase catalytic reactors, in *Topics in Chemical Engineering*, vol. **3** (ed. Hughes, R.). Gordon and Breach, New York.

[23] Anon. (2003) Air Products and Johnson Matthey debut monolith catalyst technology. *Chemical Engineering Progress*, **January**, 9.

[24] Roberge, D.M., Ducry, L., Bieler, N., Cretton, P., and Zimmermann, B. (2005) Microreactor technology: a revolution for the fine chemical and pharmaceutical industries? *Chemical Engineering Technology*, **28**, 318–323.

[25] Calabrese, G.S. and Pissavini, S. (2011) From batch to continuous flow processing in chemicals manufacturing. *AIChE Journal*, **57**, 828–834.

[26] Mason, B.P., Price, K.E., Steinbacher, J.L., Bogdan, A.R., and McQuade, D.T. (2007) Greener approaches to organic synthesis using microreactor technology. *Chemical Reviews*, **107**, 2300–2318.

[27] Webb, D. and Jamison, T.F. (2010) Continuous flow multi-step organic synthesis. *Chemical Science*, **1**, 675–680.

[28] Watts, P. and Wiles, C. (2007) Recent advances in synthetic micro reaction technology. *Chemical Communications*, 443–467.

[29] Jähnisch, K., Hessel, V., Löwe, H., and Baerns, M. (2004) Chemistry in Microstructured Reactors. *Angewandte Chemie International Edition*, **43**, 406–446.

[30] Wiles, C. and Watts, P. (2010) The scale-up of organic synthesis using micro reactors. *Chemistry Today*, **28**, 3–5.

[31] Li, S., Hu, Y. and Zong, X. (2012) The feasibility research of using an intensified continuous mini-reactor to replace a discontinuous reactor. *Advanced Materials Research*, 391–392, 894–899.

[32] Braune, S., Pöchlauer, P., Reintjens, R., Steinhofer, S., Winter, M., Lobet, O., Guidat, R., Woehl, P., and Guermeur, C. (2009) Selective nitration in a microreactor for pharmaceutical production under cGMP conditions. *Chemistry Today*, **27**, 26–29.

[33] DeGarmo, J.L., Parulekar, V.N., and Pinjala, V. (1992) Consider reactive distillation. *Chemical Engineering Progress*, **88**, 43–50.

[34] Podrebarac, G.G., Ng, F.T.T., and Rempel, G.L. (1997) More uses for catalytic distillation. *CHEMTECH*, **27**, 37–45.

[35] Harmsen, G.J. (2007) Reactive distillation: The front-runner of industrial process intensification: A full review of commercial applications, research, scale-up, design and operation. *Chemical Engineering Progress*, **46**, 774–780.

[36] Järvelin, H. (2004) Commercial production of ethers, in *Handbook of MTBE and Other Gasoline Oxygenates* (eds H. Hamid and M.A. Ali). CRC Press, pp. 194–212.

[37] Heckmann, G., Previdoli, F., Riedel, T., Ruppen, T., Veghini, D., and Zacher, U. (2006) Process development and production concepts for the manufacturing of organic fine chemicals at LONZA. *Chimia*, **60**, 530–533.

[38] Krishna, R. (2002) Reactive separations: more ways to skin a cat. *Chemical Engineering Science*, **57**, 1491–1504.

[39] Specchia, V., Fino, D., and Saracco, G. (2005) Inorganic membrane reactors, in *Structured Catalysts and Reactors* (eds A. Cybulski and J.A. Moulijn). CRC Press, pp. 615–662.

[40] Champagnie, A.M., Tsotsis, T.T., Minet, R.G., and Webster, A.I. (1990) A high temperature catalytic membrane reactor for ethane dehydrogenation. *Chemical Engineering Science*, **45**, 2423–2429.

[41] Van den Broek, J. (1985) Vette subsidie voor membraanreactor. *Chemisch Magazine*, **April**, 182–183 (in Dutch).

[42] Westerterp, K.R. (1992) Multifunctional reactors. *Chemical Engineering Science*, **47**, 2195–2206.

[43] Westerterp, K.R., Bodewes, T.N., Vrijland, M.S.A., and Kuczynski, M. (1988) Two new methanol converters. *Hydrocarbon Process*, **67**, 69–73.

15

Process Development

The strategy for developing new technologies or products is an important concern in research and development (R&D) departments for the chemical industry. The development of a particular process is successful if the desired product is produced in time at the planned rates, the projected manufacturing cost, and the desired quality standards. Included in the term cost are obvious items such as raw materials cost, but also safety and environmental compatibility. Timing is also a critical factor: A technically perfect plant that comes on stream too late and, therefore, has to operate in a changed market, might turn out to be worthless. So, cost and planning have to be carefully monitored throughout the route from laboratory to process plant.

15.1 Dependence of Strategy on Product Type and Raw Materials

Referring to Figure 2.1, the position of a product in the tree structure determines the type of R&D that is conducted. R&D for base chemicals and intermediates (bulk chemicals, also referred to as commodities) is basically different from that for consumer products (specialties), as shown in Table 15.1. R&D related to fine chemicals is somewhere in between.

The development of new technologies for the production of base chemicals and intermediates may also be dictated by the raw materials situation. For instance, acetaldehyde production was based on coal-derived ethyne until the 1960s, but when the oil era started it became much cheaper and safer to use oil-derived ethene as the feedstock. With crude oil reserves decreasing, C_1 chemistry based on natural gas is gaining importance for the production of many base and intermediate chemicals that are currently derived from oil. In addition, in recent years, much R&D focus has been on the conversion of biomass into chemicals (Chapter 7).

Bulk chemicals have a wide range of possible uses and, therefore, have a long life. The cost of process development is justified by the increased profit that will result from the lower costs of a new process. Examples are the introduction of synthetic methanol after World War I and the commercialization of the methanol carbonylation route to acetic acid in the 1960s, which has quickly replaced the old acetaldehyde oxidation process (Section 9.2).

On the other hand, consumer products will often be replaced by better ones eventually. Thus, process development for bulk chemicals usually aims at *process improvement*, while *product improvement* is the common strategy for consumer products.

Chemical Process Technology, Second Edition. Jacob A. Moulijn, Michiel Makkee, and Annelies E. van Diepen.
© 2013 John Wiley & Sons, Ltd. Published 2013 by John Wiley & Sons, Ltd.

Table 15.1   *Differences in process development for bulk and specialty chemicals.*

Feature	Bulk chemicals	Specialty chemicals
Product Life Cycle	Long (>30 years)	Short (<10 years)
Focus of R&D	Driven by cost and environment	Driven by yield
	Process improvement (modified or new technology)	Product improvement
Competing processes	One route usually the best	Competitors may use variety of routes
Route selection		
Number of feasible routes	Few	Many, possibly hundreds
Number of feasible feedstocks	Few	Many
Number of reaction steps	Usually one or two	Several (from 3–20)
Effort as a proportion of total R&D	Low	High
Impact of poor route selection	High	Moderate
Patent protection	On processes or technologies	On chemical structure and chemistry, often to block competitors' use of similar routes
Technology	Usually continuous	Batch and continuous
Scale-up	Pilot plants and simulation tools commonly used	Use of rules of thumb prevails

At present, the technologies for the production of the base chemicals and intermediates are generally well established. Therefore, development activities will usually result in minor process improvements (e.g., energy savings in methanol and ammonia production, environmental measures, and improved catalysts), which however, can have a large impact on total profits due to the large volumes involved. Still, it is possible that novel concepts are introduced, for example, the use of a multifunctional reactor for the production of MTBE (Chapter 14) and the ongoing efforts to replace the chlorohydrin process for the production of propene oxide by the direct oxidation route (Section 10.5.3). A more general drive is towards sustainability and process intensification [1] (Section 14.3.1).

In the specialty chemicals (consumer products) industry the development of modified or new products and their accompanying new processes is much more motivated by market demands concerning product quality and by safety considerations. Examples of product modifications are the enormously changed quality of car tires (rubber), paints, and detergents in the past 30 years. Examples of new products developed to meet current and future demands are advanced materials such as composites (fiber-reinforced resins, e.g., for use in aviation and aerospace) and engineering plastics (high-grade plastics for construction uses, e.g., in cars). Another example is the increasing interest in chiral drugs (Section 12.2). Catalysts and biocatalysts also belong to the class of new and modified product development.

For bulk chemicals, which can be produced from only a limited number of feedstocks by a limited number of reactions, the majority of the R&D activities concerns process design and scale-up. In contrast, in R&D for specialty chemicals a lot of effort is put into the selection of the chemical route. This is the result of the large number of potentially feasible routes; pharmaceuticals and agrochemicals can be manufactured by hundreds of different routes.

In the following sections the focus is on research and development of continuous bulk chemicals processes. A focus on fine chemicals can be found eleswhere [2].

15.2 The Course of Process Development

Process development is a continuous interaction between experimental programs, process design, and economic studies. The starting point is a chemical reaction discovered in the laboratory, often accompanied by a suitable catalyst, and the outcome is the production plant. A large part of the development activities is concerned with scaling up from laboratory to full-scale industrial plant, with the intermediate miniplant and pilot plant stages.

Until about World War II, scale-up basically was an empirical process. The common way to deal with scale-up was enlargement in small steps, as schematically depicted in Figure 15.1 for a chemical reactor (drawn to scale). This procedure is still quite common in the production of fine and specialty chemicals, where most processes are batch operations.

Figure 15.1 also shows the approximate production rates and the time intervals associated with the design/construction and operation. Depending on the nature of the process these time intervals may differ greatly from those given. Large safety margins were incorporated, which often resulted in plants with a much larger capacity than what they were designed for. Clearly, scale-up in small steps is very expensive and insecure as long as no predictive models are used.

The time scale in Figure 15.1 is extremely long according to modern standards. Hardly any process development will be approved by management if the horizon for the start-up is eight years or more. There is a need for a drastic reduction of the time for process development. Fortunately, the discipline of chemical engineering has matured and promising developments are taking place. Progress has been and is still being made in software and hardware improvements. Many process steps at present may be captured in a mathematical model with sufficient predictive value. So, the scale-up factor may be enhanced, making it possible to extrapolate directly from miniplant to production scale, thus saving time and money. Therefore, nowadays the usual practice in scale-up is to design an industrial process based on chemical engineering principles and

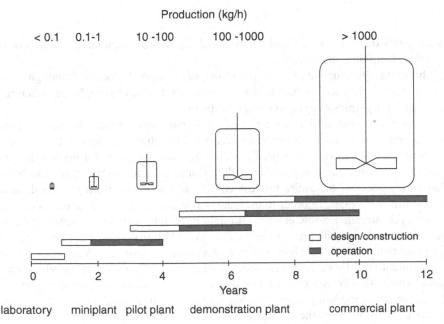

Figure 15.1 *Conventional scale-up procedure (production data apply to bulk chemicals).*

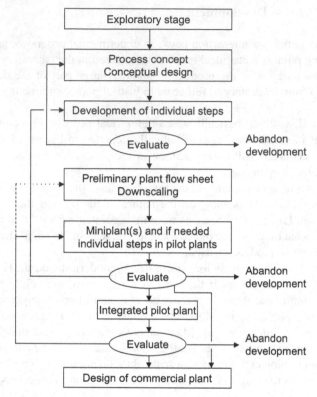

Figure 15.2   *Steps in the development of a process [3].*

laboratory experiments on a very small scale. Next, analysis shows which parts of the process need to be examined in further detail on a small scale.

Figure 15.2 shows the cyclic nature of the development of a chemical process. Although the steps outlined in Figure 15.2 are typical, they are not mandatory. For instance, in fine chemicals manufacture the sequence of steps often is laboratory–miniplant–commercial production.

Figure 15.2 assumes a successful completion of the exploratory phase, that is, the reaction provides satisfactory yields and selectivities in the laboratory reactor. The laboratory (and literature) data are the basis for the process concept. Once this has been produced, the individual steps are further developed and tested on laboratory scale (Is a certain separation possible at all?). Once all the individual steps have been successfully tested, a reliable flow sheet of the entire process can be drawn. A miniplant is then designed, based on scale-down of the full-scale plant, in order to evaluate the performance of the entire process (e.g., build-up of impurities in recycle streams). Some equipment might have to be tested on a larger scale, the pilot stage, in order to be able to scale-up without too large a risk. An additional step might be to design and build a complete pilot plant before the full-scale production plant comes on stream. After each step in the process development an evaluation follows. It is then decided whether to go on with the development, stop or start again at an earlier stage. This decision is based on technical, cost, and market considerations.

It should be stressed that modern process development is done in a parallel rather than a serial way. This calls for close cooperation between the various disciplines involved. One of the advantages is that the time required for development can be greatly reduced.

Also in experimentation the world is changing. Catalyst selection can be done in a combinatorial way and the time-consuming phase of catalyst performance testing benefits from parallel reactor units. Additionally, in other parts of laboratory work more automation and a robotic approach are useful. The latter includes experimental programs in miniplants and pilot plants. Revolutionary changes in experimental procedures can be expected.

QUESTION:

> *Imagine a future process development laboratory for the production of fine chemicals and one for the production of bulk chemicals. Which new developments do you expect?*

15.3   Development of Individual Steps

This section examines the chemical and chemical engineering steps (shown in Figure 15.2) in the development of a new process. Of course, the technical and economic feasibility of a process is not the only important factor in the successful development of laboratory results to a process plant. Other factors determining the success are the market situation, the patenting and licensing situation, environmental impact, and so on. This section does not focus upon retrofitting of existing chemical processes, where the changes are only marginal from a scientific point of view.

15.3.1   Exploratory Phase

Process development typically starts with the discovery and subsequent research of a promising new product, or new chemical synthesis route to an existing product. Research in this phase focuses on chemical reactions, catalysis, and possibly reactor design. It includes obtaining information on which reactions take place, the thermodynamics and kinetics of the reaction(s), dependence of selectivity and conversion on the process parameters, catalyst and catalyst deactivation rate, and so on.

During process development, research in the laboratory continues. For instance, in the early stages kinetic data might not yet be available, but as the development continues these data will have to be collected for proper reactor and process design.

15.3.2   From Process Concept to Preliminary Flow Sheet

After optimization of the laboratory synthesis and assembly of other important information, a process concept is developed and the economic potential is determined. The availability and quality of raw materials must also be determined. In this phase conceptual process designs are made to identify promising options [4]. At this point several different versions of the process may be drawn up (and even different synthesis routes might have come out of the laboratory), from which it is hard to choose the best one.

Many of the alternatives can, of course, be eliminated for obvious reasons (industrial, safety, economic, legal, etc.). Still, a large number usually remains and it is impossible to calculate costs for all of them. Finding the optimum choice requires knowledge, experience, and creativity. Tools that help making the decision are continuously improving and include data banks and simulation programs.

It is important that an economic evaluation is part of the design process, because only then can promising alternatives be identified from the start and hopeless ones discarded as soon as possible. At the early stages, a comparison of the raw material costs with the product value can already give an indication; if even at 100% conversion the product value does not exceed the raw material costs, the process is certainly not viable at that time.

15.3.2.1   *Reactor System*

The more information that becomes available about the process, the more detailed the conceptual design can become. The reactor system (reactor, including fresh feed, product, and gas and liquid recycle streams) must be specified early on because it influences the product distribution and the separation section. The effect of reactor type on product distribution is illustrated by the coal gasification process (Section 5.3). In this process, the product distribution in a moving bed, fluidized bed, and entrained flow reactor are very different (e.g., much more hydrogen and carbon monoxide are formed in an entrained flow reactor, while the content of methane and other hydrocarbons in the product gas from a moving bed reactor is very large).

The definite selection of the reactor is not always possible at the early stages of conceptual design, because a kinetic model has not been developed yet. However, factors such as desired temperature and pressure (determined by the thermodynamic equilibrium for reversible reactions), type of catalyst (heterogeneous with small or large particles, homogeneous, none), and the phases of the reactants and products, will narrow the choice. It has to be noticed that sometimes the particle size can only be decided upon after elucidation of the kinetics. Boxes 15.1 and 15.2 show examples of conceptual reactor designs.

Box 15.1   Ammonia Synthesis

In ammonia synthesis (Section 6.1), a solid-catalyzed gas phase reaction, low temperatures are required to achieve acceptable conversions, while the rate of reaction increases with temperature. (This is a dilemma all reversible exothermic reactions have in common.) Hence, the reaction is carried out in an adiabatic fixed bed reactor with intermediate cooling, thus compromising between high temperatures (small reactor volumes) and high equilibrium conversion. The exploratory laboratory research revealed that high conversions are not possible, even at high pressures. Therefore, a gas recycle loop has to be incorporated. A purge is needed to prevent impurities building up.

Because of the damaging effect of carbon monoxide and carbon dioxide in the synthesis gas on the catalyst activity, purification of the feed is necessary. Figure B15.1.1 shows a simplified conceptual

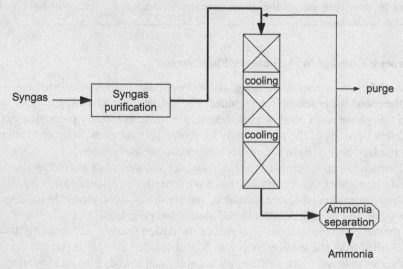

Figure B15.1.1   *Simplified initial conceptual flow sheet of ammonia synthesis.*

flow sheet, which incorporates these results. The source of synthesis gas (coal, natural gas, etc.) has not been specified. The choice of feedstock and process depends on (local) availability, cost, and so on, but nowadays steam reforming of natural gas, followed by partial oxidation in air, is the predominant technology (Section 5.2). Decisions must also be made concerning feed purification steps (shift reactions, CO removal by methanation or some type of physical separation process (Section 5.4)). Further research and development activities determine the ammonia reactor design in more detail (i.e., whether intermediate cooling will be accomplished by cold gas injection or in heat exchangers, etc.). The reactor size can be determined once the kinetic data have become available.

QUESTIONS:

> *What aspects do you expect to be critical for a successful development. (Hint: consider Figure 15.2. When would an evaluation lead to "Abandon development"?)*

Box 15.2   Fluid Catalytic Cracking (FCC)

A process where the size of the catalyst particles plays an important role is catalytic cracking (Section 3.4.2). The earliest commercial catalytic cracking process was based on fixed bed reactors, developed and patented by E. Houdry. Oil companies were very interested in the process, but because of the high licensing costs they decided to develop their own processes. In these processes small catalyst particles were used instead of the larger particles used in a fixed bed and, thus, they evaded the patents. The advantage of the usage of small particles is the high catalyst effectiveness factor. Another reason for using fine particles was that transport between reactor and regenerator would be possible. The primary advantage of catalyst recirculation is the fact that the heat needed for the endothermic cracking reactions can be supplied by the combustion of coke in the regenerator, the catalyst being the heat carrier. In the Houdry process this "heat balanced" operation is not possible.

When using small catalyst particles, the choice of reactors is limited. If the reactants are in the liquid phase, a slurry reactor can be used. However, because of the high temperatures required for the endothermic cracking reactions, the hydrocarbons are present in the gas phase, so this option can be discarded immediately. With gaseous reactants a fluidized bed or entrained flow reactor can be used. Figure B15.2.1 shows a simplified conceptual flow sheet for the FCC process.

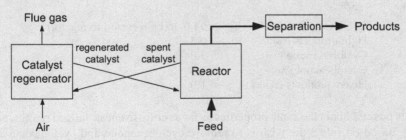

Figure B15.2.1   *Simplified initial conceptual flow sheet of the FCC process.*

The use of small catalyst particles, either in a fluidized bed or entrained flow reactor, implies the need for separation of the gas and the catalyst. The hydrocarbon products are also separated to obtain gasoline and other products. The most difficult part of the process is the transport of the catalyst between the reactor and regenerator and vice versa. Much development effort has to be put in the design of a reliable and safe catalyst recirculation system.

15.3.2.2 Separation System

When the reaction product contains multiple components, it can be difficult to decide which is the best separation system. An example of the determination of a distillation sequence is given in Box 15.3.

Box 15.3 Cyclohexanone Production

In the production of cyclohexanone from cyclohexane, a series of four distillation columns is required to separate the product mixture, which comprises five components. This means that 14 possible sequences can be distinguished. The question: what is the best sequence? In this case it is relatively easy to come up with the right answer, based on common heuristic rules. These rules, which apply to other separation processes also, are [4]:

1. Remove corrosive or hazardous components as soon as possible.
2. Remove reactive components or monomers as soon as possible.
3. Remove products as distillates.
4. Remove recycles as distillates, particularly if they are recycled to a fixed bed reactor.
5. Remove most plentiful first.
6. Remove lightest first.
7. Perform High-recovery separation last.
8. Perform Difficult separation last.

QUESTION:
> *Explain the logic of these rules.*

The product mixture from the cyclohexane oxidation reactor has the following composition (the products are arranged in order of increasing boiling point):

Cyclohexane (%)	94.0 (to be recycled to reactor)
Light products (%)	0.5
Cyclohexanone (%)	3.0
Cyclohexanol (%)	1.5
Heavy products (%)	1.0

All components possess about the same properties as far as corrosiveness, hazard ratings, and reactivity are concerned. Based on rule 3 the wish is to remove cyclohexanone and cyclohexane as distillates. Cyclohexane is both the most plentiful and the lightest component, so it is removed first. The separation of cyclohexanone and cyclohexanol is most difficult and requires the highest recovery, so this separation is saved till last. This only leaves the choice of whether to remove the heavy or the light components first. The heavy components are more plentiful, so they are removed first. The distillation sequence is thus as shown in Figure B15.3.1.

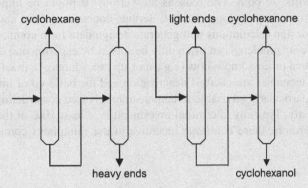

Figure B15.3.1   *Distillation sequence in cyclohexanone production.*

QUESTION:
    Check the sequence of distillation in the flow sheets for, FCC (Section 3.4.2), steam cracking (Chapter 4), methanol production (Section 6.2), and acetic acid production (Section 9.2).

15.3.2.3   Testing of the Individual Steps

When reactor design and separation processes have been decided upon, a conceptual flow sheet for the industrial plant can be drawn. The individual unit operations (reactor, distillation, etc.) are designed based on existing information (approximate size of columns is determined, etc.). The process design at this early stage is, of course, preliminary: not all data are available yet and some pieces of equipment either are not included in the flow sheet or are not sized accurately. Individual process steps have to be examined in more detail and the results fed back to the conceptual flow sheet stage to be able to produce a more reliable preliminary process flow sheet. For instance, the first conceptual flow sheet is often based on the assumption that the (liquid) reaction products can relatively easily be separated by distillation, because no vapor–liquid equilibria have been determined yet. When the individual steps are developed and tested, the occurrence of azeotropes, and so on, might be observed, resulting in the need for more expensive azeotropic distillation or some other separation technique. Temperature limitations may also be discovered, for example, in the distillation of the products from styrene manufacture, distillation under reduced pressure is necessary in order to prevent styrene polymerizing.

The laboratory experiments are usually carried out with a mixture prepared from pure materials. Therefore, accumulation of impurities or by-products in recycle streams is not accounted for in the tests. Furthermore, laboratory experiments often are of short duration and usually initial catalyst activities are measured. Catalysts may appear very active in the laboratory, but they could lose their activity and/or selectivity within days or even hours (e.g., as a result of coke formation by undesired side reactions). This at least calls for special measures concerning catalyst replacement and/or regeneration, but might even lead to the abandonment of the entire process development. Miniplants and pilot plants can reveal such problems.

15.3.3   Pilot Plants/Miniplants

A pilot plant may be defined as an experimental system, at least part of which displays operation that is representative (either identical or transposable) of the part which will correspond to it in the industrial unit [5]. Pilot plants range in size from a laboratory unit (miniplant) to an almost commercial unit (demonstration plant) but are usually intermediate in size. The development of a new process or product usually requires miniplant or pilot plant work, or both. The reasons are various. It might be important to generate larger quantities of a product in order to develop a market. Serious doubts may exist about individual steps. It is often necessary to check design calculations and generate design data for a commercial plant. In particular with novel processes, scale-up problems can often only be solved by experimental demonstration. Often it is rewarding to gain operational process know-how (e.g., start-up procedures). In practice, long-term effects are often revealed. Notorious examples are catalyst deactivation and the build-up of impurities. In the case of a really novel process, it is particularly advisable to demonstrate it before constructing a commercial plant.

Pilot plants are very costly. Typically, the initial investment is at least 10% of the investment costs of the industrial scale plant. Therefore, there is a large incentive to use miniplants combined with mathematical simulation of the process.

15.3.3.1   Miniplants

If the objective of a pilot plant is to evaluate the feasibility of a new process technology or generate design data for a new process, then the miniplant strategy may be most appropriate. The miniplant serves to test the entire process, including all recycles, on a small scale. Its design is based on scaling down of the projected commercial plant. The miniplant technique has the following characteristics:

- The miniplant includes all recycling paths and it can consequently be extrapolated with a high degree of reliability.
- The components used (columns, pumps, etc.) are often the same as those used in the laboratory. In addition, miniplants usually contain standardized equipment that can be reused in other miniplants, resulting in reduced investment costs and high flexibility.
- The miniplant is operated continuously for weeks or months and, therefore, is automated to a large degree.
- The miniplant is used in combination with the mathematical simulation of the industrial scale process.

The size of a miniplant depends on the system used. In addition, there is a trend towards further miniaturization. While earlier often a range of 0.1–1 kg product per hour was realistic, nowadays a range of 0.01–1 kg product per hour is more typical. In many, but not all cases a miniplant can also be regarded as a small pilot plant.

QUESTIONS:

> *What are the advantages of a miniplant compared to a large pilot plant? What would be the minimum size of the miniplant in a development program for improving a HDS process in an oil refinery (Section 3.4.5)? Is a pilot plant required? Answer the same question for the development of a new production process based on a new feedstock (for instance, biomass instead of oil fractions).*

The use of a miniplant together with mathematical simulation often makes it possible to skip the pilot plant stage, though retaining the same scale-up reliability. As a result it is possible to speed up process development, which can have a noticeable effect on the cost effectiveness of a new process.

Miniplants are particularly suitable to test long-term catalyst stability under practical conditions, that is, with real feeds and recycle streams. The incorporation of recycle streams is important to detect the effects of

Table 15.2 Typical production rates and scale-up factors.

	Production rate (kg/h)	Scale-up factor
Industrial plant	1000–10 000	—
Pilot plant	10–100	10–1000
Miniplant	0.01–1	1000–1 000 000

trace impurities or accumulated components that are not observed in laboratory-scale experiments (see also Section 15.4.2.4).

The miniplant technique has certain limitations, however, in that the scale-up of individual process steps, such as extraction, crystallization, and fluidized bed operation, from miniplant to full-scale plant is often too risky in technical terms. Table 15.2 shows typical production rates and scale-up factors (size of industrial plant/size of mini or pilot plant). For some important process steps, Table 15.3 shows typical maximum scale-up values above which reliable scale-up is no longer possible.

As can be seen, fluidized bed reactors are governed by a much smaller scale-up factor than other reactor types. The scale-up problems are mainly caused by the change of flow characteristics when going from a small unit to an industrial reactor. Calculations show that the conversion in fluidized beds may vary from that in plug-flow to well below that in mixed flow, and it is very difficult to reliably estimate the conversion for a new situation. The problem is that in a laboratory-scale reactor the gas bubbles are of the same order of magnitude as the column diameter (about 0.1 m). This bubble diameter is only slightly dependent on the scale of the fluidized bed, which means that in an industrial reactor the bubbles are much smaller than the reactor diameter. Figure 15.3 illustrates this. The result of the different bubble-to-column diameter ratios is that the flow patterns on the small and large scale differ and therefore mixing, mass transfer, and heat transfer also differ.

The disadvantage of the limited scale-up factor can be circumvented by isolating the critical process steps and working on them at an intermediate stage, the pilot stage. If enough representative feed can be supplied to the critical step for a sufficiently long time, then the entire plant can again be extrapolated to full scale without too much risk. In addition, modeling of equipment has become a very powerful tool, leading to lower scale-up costs.

Table 15.3 Typical maximum scale-up values of some important process steps [6].

Process step	Maximum scale-up value
Reactors	
Multitubular and adiabatic fixed bed	>10 000[a]
Homogeneous tube and stirred tank	>10 000
Bubble column	<1000
Gas–solid fluidized bed	50–100
Separation processes	
Distillation and rectification	1000–50 000
Absorption	1000–50 000
Extraction	500–1000
Drying	20–50
Crystallization	20–30

[a]Scale-up factors of over 50 000 have already been achieved.

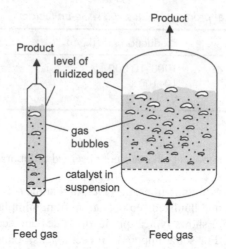

Figure 15.3   *Small- and large-scale fluidized bed reactors.*

QUESTIONS:

> *Why are the hydrodynamics for multiphase processes in a laboratory-size fluidized bed often very different from those in the commercial one?*
> *Modeling advances often reduce the costs of scale-up. Give examples of situations where you expect this to be the case.*

15.3.3.2   *Pilot Plants*

A pilot plant should be designed as a scaled-down version of the industrial plant, not as a larger copy of the existing miniplant. Pilot plants vary greatly in size and complexity. In the case of new technology and for market testing an integrated pilot plant is required. When the feasibility of a process cannot be proven on miniplant scale, a demonstration plant may be built. This is often the case with solid handling processes, for example, coal gasification and coal liquefaction. Its minimum size is determined by the minimum size of a particular unit operation or piece of equipment. In coal liquefaction, for instance, the process chemistry can be proved in miniplant operation. However, the uncertainty in the scale-up of the liquid/solid separation equipment requires pilot plants of larger scale.

Critical steps may be examined in individual pilot units. These pilot units need not be a complete but smaller replica of the envisioned commercial unit, as long as they yield the required data for scale-up. Some examples are given in Section 15.4.

15.3.3.3   *Mock-Ups*

Mock-ups are useful tools in scale-up. Mock-ups or "cold flow models" are usually used to simulate those aspects of the process related to fluid flow. They do not necessarily physically resemble the part of the process that is studied. Compared to pilot units they are relatively cheap. An example of the use of a mock-up at the miniplant stage is found in FCC catalyst testing (Box 15.4).

QUESTION:

> *Could a mock-up play the role of a pilot plant in specific situations?*

Box 15.4   FCC Catalyst Testing in a Miniplant

Figure B15.4.1 shows a schematic of a miniplant for the testing of FCC catalysts [7]. Table B15.4.1 shows scale-up data for an industrial FCC riser reactor, the miniplant, and the miniplant mock-up.

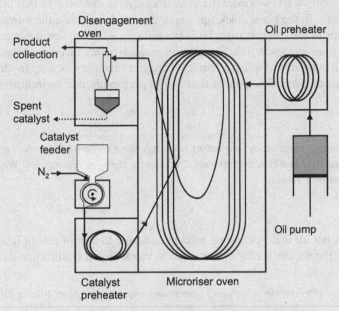

Figure B15.4.1   *Miniplant for testing FCC catalysts.*

Table B15.4.1   *Data for industrial FCC unit, miniplant for FCC catalyst testing, and mock-up.*

	Industrial reactor	Miniplant	Mock-up
Feed	FCC feed + steam	FCC feed + nitrogen	nitrogen
Temperature (K)	790	790	298
Reactor length (m)	20–30	20	4–20
Reactor diameter (m)	1–2	0.004	0.004
Residence time (s)	2–5	2	0.4–2
Reynolds $(\rho u D/\eta)^a$	$\approx 5 \cdot 10^5 - 10^6$	≈ 3000	≈ 3000
Catalyst feed rate (kg/min)	$10^4 - 3 \cdot 10^4$	≈ 0.001	≈ 0.001

$^a \rho$ = fluid density (kg/m^3); u = fluid velocity (m/s); D = reactor diameter (m); η = dynamic viscosity (kg/(m·s)).

Although the miniplant (microriser) and industrial unit (Section 3.4.2) have quite different appearances, the miniplant still resembles the industrial unit, for example, similar length, residence time, oil/catalyst contact time, contacting pattern, and so on. Industrial riser reactors have a length of 20–30 m and a diameter of 1–2 m. In the miniplant, the same length is used but due to spatial restrictions the tube has to be folded. It is important to determine whether folding of the reactor influences the contact between gas and catalyst, and thus conversion, product distribution, and so on. It is conceivable, for instance, that the catalyst particles and the gas are separated in the bends due to centrifugal forces.

The mock-up, a glass model of the folded miniplant riser reactor, is ideal to examine this phenomenon. As the problem is essentially of a hydrodynamic nature, it is important to keep the Reynolds number constant (hydrodynamic similarity, see also Section 15.4 for the use of similarity in scale-up). This is achieved by using a glass reactor with the same diameter as the pilot plant, and nitrogen at room temperature as feed. This way, the kinematic viscosity (η/ρ) is the same as that of the miniplant feed at reaction temperature. In the glass mock-up, which can be shorter than the miniplant reactor, observation of the behavior of the catalyst particles is possible. In addition, this behavior can be quantified with techniques such as residence time distribution measurements of colored catalyst particles. The results obtained from both the miniplant (chemical behavior) and the mock-up (hydrodynamics) can be incorporated into a reactor model to predict the catalyst performance in an industrial riser.

QUESTIONS:

> *During catalyst development research for FCC usually fixed bed and occasionally fluidized bed reactors are used. Compare these reactor types. Which type would you choose? Explain.*

As stated previously, not all unit operations need to be tested in a pilot plant (Table 15.4). For instance, for distillation columns the required information, such as vapor–liquid equilibrium data, and so on, is either

Table 15.4 *Rules of thumb to decide which unit operations require pilot plant testing [8].*

Operation	Pilot plant required?	Comments
Distillation	Usually not. Sometimes needed to determine tray efficiency.	Foaming may be a problem.
Fluid flow	Usually not for single phase. Often for two-phase flow.	Some single-phase polymer systems are also difficult to predict. CFD[a] can be an important tool.
Reactors	Frequently.	Scale-up from laboratory often justified for homogeneous systems and single-phase fixed bed reactors.
Evaporators, reboilers, coolers, condensers, heat exchangers	Usually not unless there is a possibility of fouling.	
Dryers, solids handling, crystallization	Almost always.	Usually done using vendor equipment[b]
Extraction	Almost always.	

[a]CFD = Computational Fluid Dynamics.
[b]Vendor tests are a common form of pilot plant testing for these unit operations. The equipment supplier has miniaturized the equipment to the smallest size where reliable design data can still be obtained. The advantage of vendor testing is that the charge is usually low (the vendor expects to sell a large piece of equipment) and the testing can usually be done at short notice. A limitation is that equipment vendors often cannot carry out measurements at the envisaged reaction conditions and cannot handle toxic or corrosive materials.

available from literature or can easily be determined in the laboratory. The methodology for design of distillation equipment is well known and straightforward. Reactors, on the other hand, often require pilot plant testing.

15.4 Scale-up

This section is not intended to treat the theoretical aspects of scale-up, but rather to give examples of the use of scale-up methods in the development of processes that have been treated in this book. The reactor is the core of the process plant. It also is the part of the plant where unsolved scale-up problems can have a large negative impact. This section concentrates on reactor scale-up. Scale-up makes use of laboratory and/or pilot plant data, complemented by mock-up studies and mathematical modeling, to determine the size and dimensions of the industrial unit. Two reactor categories are dealt with in this section, namely (i) reactors with a single fluid phase, possibly with a homogeneous catalyst, and (ii) fixed bed reactors with one or more fluid phases. Attention is also paid to a typical scale-up problem: the build-up of impurities in recycle streams.

15.4.1 Reactors with a Single Fluid Phase

Reactors involving a single fluid phase present the least difficulty in scaling up. Homogeneous reactions may be carried out using one of the following reactor systems:

- (semi-)batch reactor (fine chemicals, biotechnology);
- continuous tubular reactor (steam cracking);
- continuous stirred-tank reactor/cascade of stirred-tank reactors (homogeneous catalysis).

15.4.1.1 Batch Reactor

Batch and semi-batch reactors are frequently used for the production of fine chemicals and specialties, such as cosmetic products, floor polishes, and so on. Bioreactors are also often batch reactors. Concerns involved with the scale-up of batch reactors have been dealt with in Section 12.6. As discussed there, the greatest problems arise when highly exothermic reactions are involved. The solution to this problem is to separate reaction volume and heat exchange area, for example, by applying external heat exchangers. In this respect it is convenient to work at the boiling point of the reaction mixture, allowing heat removal with a reflux system. Then, the batch reactor may be scaled up from laboratory size to industrial plant simply based on reaction time.

Nowadays a special type of laboratory reactor is often used to determine relationships to describe heat production and transfer. The reactor is a calorimeter in which parameters such as the thermal resistances and the heat evolved by reactions can be measured accurately. From the results, reactor specific constants can be derived, based on which scale-up calculations for reactors of similar geometry can be made.

15.4.1.2 Continuous-Flow Tubular Reactor

An example of a process employing a continuous-flow tubular reactor involving one fluid phase is the steam cracking of ethane and naphtha (Chapter 4). The most important concern in the scale-up of a steam cracking unit is heat transfer and, associated with this, the temperature profile. On a small scale often an electric furnace is used, whereas on an industrial scale the furnace is always heated by burners [9]. To be able to translate the laboratory results into the industrial unit, a kinetic model of the reaction system is necessary because the interaction of temperature and chemical reactions has to be predicted. The problems are that many different molecules are involved in the reactions and that naphtha does not always have the same composition. An accurate kinetic model can be derived from experimentation on bench/miniplant scale and pilot plant scale together with the introduction of the "lumping principle", that is, various molecules are grouped together to form for example, the *n*-alkanes, the iso-alkanes, and so on lumps. The model treats these lumps as if they were one species, where in fact the lumps consist of several species (*n*-alkanes: C_4, C_5, etc.). The model will then be able to predict the effect of the operating parameters on the performance of cracking of different types of naphtha, based on their composition (i.e., % *n*-alkanes, % iso-alkanes, etc.), without extensive additional experimentation.

This "lumping" approach is also commonly used in modeling of, for instance, the kinetics of catalytic cracking (Box 15.5) and catalytic reforming. The problem is to determine to what extent lumping is tolerated, varying from considering every molecule separately to putting all molecules in one lump.

Box 15.5   Lumping Models for Catalytic Cracking [10, 11]

Figure B15.5.1 shows two lumping models that are used in catalytic cracking.

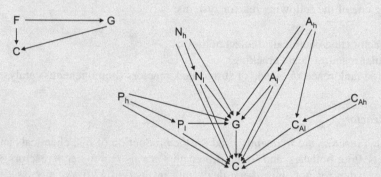

Figure B15.5.1   *Three-lump model and ten-lump model for catalytic cracking (symbols are explained in the text).*

The three-lump model considers a gas oil feed (F), a gasoline (G) and a light gas + coke lump (C). In the ten-lump model the gas oil lump is divided into a heavy and a light gas oil (subscripts h and l), which each contain four lumps, that is, alkanes (P), naphthenes (N), aromatic substituent groups (A) and aromatic rings (C_A). The two remaining lumps again are the gasoline (G) and the light gas + coke lump (C). The arrows represent the conversion of one lump into the other with the corresponding rate coefficients.

The three-lump model is useful to model data obtained from the laboratory but has no predictive value for other feedstocks. One of the reasons for this is the oversimplification. Another, more fundamental reason is that the definition of conversions and selectivities is not straightforward. For instance, the feed F already contains gasoline components, which are partly converted to gas and coke. The ten-lump model is more complex and does have more predictive value. However, determination of the model parameters in this case is more elaborate.

Nowadays, due to increasingly faster computers, lumping models may contain hundreds of different lumps. A good example is naphtha cracking, where models are used based on real elementary reaction steps. In fact, extended de-lumping has taken place. Another example is combustion of methane, in which essentially all reaction steps can be taken into consideration in a model. This leads to several thousands of equations for the component concentrations and the associated reaction rate coefficients. It is not surprising that the best examples are non-catalyzed reactions, but even for FCC and hydrotreating de-lumping is practiced by applying sophisticated models. It goes without saying that for the latter examples de-lumping is much more difficult than for non-catalyzed reactions. It is expected that molecular modeling will give this field a push.

15.4.1.3 Continuous-Flow Stirred-Tank Reactor

The scale-up of this type of reactor usually involves the transposition from a batch laboratory reactor to a continuous stirred tank industrial reactor (or cascade of reactors). If the kinetics are known from small-scale experimentation, this transposition is relatively easy because scaling up is possible on the basis of the kinetic model. However, a pilot plant reactor may still be built. The motivation then is not a reactor scale-up problem but the need to answer other questions, for instance about the influence of impurities on the catalyst activity, the mechanical stability of the catalyst particles, and catalyst removal, particularly in homogeneously catalyzed systems.

15.4.2 Fixed Bed Catalytic Reactors with One or More Fluid Phases

Many applications of heterogeneous catalysis can be found in oil refining and the chemical industry. Some examples that have been treated in this book are steam reforming, catalytic reforming, ammonia synthesis, methanol synthesis, and hydrotreating. Major issues in the design and scale-up of these processes are:

- temperature control;
- pressure drop;
- catalyst deactivation and reactivation.

15.4.2.1 Temperature Control

A distinction is made between endothermic and exothermic reactions. In very endothermic reactions such as steam reforming, the temperature decrease may be so severe that the reactor would have to be excessively long to compensate for the low reaction rate. In this case, the reaction mixture has to be heated rapidly to high temperature in order to achieve satisfactory production. In steam reforming heating is accomplished by mounting tubes containing the catalyst in a furnace in which the heat flux is very high. To achieve uniform heating in general the maximum tube diameter must not exceed 0.1 m.

Exothermic reactions such as ammonia production and hydrodesulfurization are usually carried out in multibed reactors with intermediate cooling. Cooling is either done in external heat exchangers or by injection of cold feed gas.

15.4.2.2 Pressure Drop

The pressure drop in a catalyst bed must be limited, particularly when recycle streams are involved (e.g., in the synthesis of ammonia, methanol, and ethene oxide). To decrease the pressure drop the bed height may be

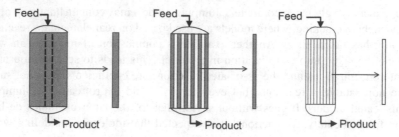

Figure 15.4 *"Show-tube" concept for scaling up a large diameter fixed bed reactor (left), multitubular reactor (middle) and monolith reactor (right).*

reduced or larger catalyst particles may be used. Both solutions have their drawbacks. Another solutions are to use radial flow or a structured reactor.

15.4.2.3   Catalyst Deactivation

In process development, information on the stability of the materials involved is crucial. This applies in particular to the catalysts to be used. A good example is deactivation of the catalyst by coke deposition. To maintain the desired conversion level, usually regeneration by burning off the coke is necessary. This may be achieved by installing a number of parallel reactors, or by using moving bed or fluidized bed operation. Examples are catalytic reforming and catalytic cracking.

In the scale-up of fixed bed reactors the "show-tube" concept (Figure 15.4) is very useful. The show-tube concept is based on similarity. It is highly successful, since most, if not all, parameters are kept the same. The concept is shown for a large diameter fixed bed reactor (used in e.g., hydrotreating), for a multitubular system (used in, e.g., the Lurgi methanol process), and for a monolith reactor.

In the scale-up of a large diameter fixed bed reactor a characteristic slice of the bed is studied; the reactor is scaled up in diameter and not in height. In the case of a multitubular reactor or a monolith reactor, the show tube can be one full-sized tube or one monolith channel, respectively, with the same dimensions as the ones used on the commercial scale. Scale-up is then simply linear, adding more tubes or channels in a bundle. Some points of concern in the use of the show-tube concept are:

- fluid distribution over bed or multiple tubes is not addressed;
- heat distribution over bed or tubes is not addressed;
- inconveniently long pilot reactors are needed.

The problem of fluid distribution is most severe in trickle bed reactors (gas/liquid/catalyst), because situations may prevail in which liquid preferentially flows through a certain part of the bed, while the gas flows through another part (Figure 15.5). In fact, often not only the non-ideality of the bed itself is the cause, but the inlet device also plays a crucial role. This holds even more for monolith reactors: if the distribution at the inlet is right it will be right everywhere in the reactor, while maldistribution at the inlet will persist throughout the monolith. In this case a mock-up is very useful to study the liquid phase distribution [5]. Recently, Computational Fluid Dynamics (CFD) has become a great help in designing inlets and other internals.

Presently, the design criteria for fixed beds are well known. When a fixed bed reactor is used in catalyst testing, downscaling is the objective. The issue now is to find the smallest reactor that can provide accurate

Liquid   Gas

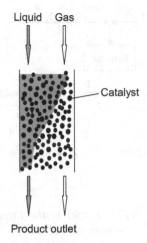

Catalyst

Product outlet

Figure 15.5   *Non-ideal trickle-flow reactor.*

data, for example, in determining the kinetics. The trend is to scale down to microflow level [12–15]. The show-tube principle is no longer used in this case. Table 15.5 shows some typical data for trickle bed reactors of varying size.

The liquid hourly space velocity (LHSV) is kept constant (and thus the average residence time) but, due to the smaller reactor length, the liquid velocity has to be lower in going to a smaller scale. This results in different fluid dynamics in the large and the small reactor.

The particle diameter, d_p, is the main factor determining the minimum dimensions of the small-scale reactor. A particle size of 1.5 mm is commonly used in hydrotreating processes. With this particle size, reactors with bed volumes of the order of 10^{-4} m^3 (single fluid) and 10^{-3} m^3 (two fluid phases) would be required to insure an even flow distribution. These volumes are based on relationships determining the minimum D/d_p and L/d_p ratios required to minimize wall effects and deviations from plug flow, respectively. One way to still be able to use a microreactor is to dilute the catalyst bed with smaller inert particles. Then the hydrodynamics are determined by the smaller particles and the kinetics by the 1.5 mm diameter catalyst particles. The catalyst bed dilution technique has proven to be a convenient tool to allow downscaling of laboratory reactors to microreactor size, both for single-phase and two-phase fluid flow, while generating data that are relevant for commercial operation. The small size has many advantages such as lower costs, low materials consumption, enhanced safety, and so on.

Table 15.5   *Data for commercial, pilot, and micro trickle bed reactors (LHSV = 2 $m^3{}_{liq}/m^3{}_{cat}\cdot h$, $d_p = 1.5$ mm) [13].*

	Commercial	Pilot plant	Microreactor
Catalyst volume (m^3)	100	0.01	8×10^{-6}
Diameter (m)	2.5	0.04	0.01
Length (m)	20	8	0.1
Liquid velocity (cm/s)	1.1	0.4	0.006
Reynolds number, Re_p	55	22	0.3

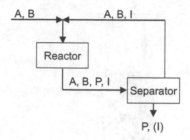

Figure 15.6 *Accumulation of an impurity in a recycle stream.*

QUESTIONS:

> *Why is the length not kept constant in the three reactors in Table 15.5? What is the consequence of the choice to vary the LHSV?*
>
> *In multiphase fixed bed reactors, the fluid distribution over the diameter of the bed can be non-homogeneous. When comparing microreactors used in the laboratory with industrial reactors, in which case do you think the problem is largest? (Hint: in a laboratory reactor the particle size is usually smaller than in an industrial reactor. What is the consequence for the hydrodynamics?)*

15.4.2.4 Catalyst Stability and Accumulation of Impurities

Catalyst stability is a very important aspect of every catalytic process. It is often frustrating that it is close to impossible to obtain experimental data on deactivation. In particular, a problem exists when deactivation is caused by the accumulation of impurities whose presence was not detected on the laboratory scale.

Impurities virtually become lost in small-scale experimentation, for example, due to settling as a result of long-standing feedstocks [16]. Impurities may not be found in laboratory work because their levels are too small to be detected. In the laboratory the reaction is often carried out batch-wise without recycle streams. Furthermore, usually short runs are carried out that do not allow sufficient time for the impurities to build up for their effect to be noticeable. However, in commercial operations with long operating periods, impurities may well reach concentrations that are significant and sometimes very damaging. Examples are deactivation of catalysts due to very small quantities of impurities in feed streams or the build-up of impurities in recycle streams, and lowering of product purity due to enrichment of a solvent with by-products. Figure 15.6 shows an example of the accumulation of impurities in a recycle stream.

Two gaseous reactants A and B react to produce liquid product P. In the separator the gas and liquid phase are separated. The gas phase is recycled to the reactor. However, some gaseous by-product I, which is only slightly soluble in P, is formed during the reaction. In a once-through laboratory process this often remains unnoticed. The danger is that the impurity unexpectedly accumulates in the system. Impurities can be detected in miniplants or integrated pilot plants that include all the recycle streams. Once they have been detected, the appropriate measures can be taken, for example, installing pretreating units, purging some of the recycle stream, and so on.

QUESTIONS:

> *Catalyst stability is generally crucial for the successful development of a catalytic process. The experimental studies described above are performed in a continuous miniplant or pilot plant. Would it be useful to use batch reactors instead? Explain your answer.*

15.5 Safety and Loss Prevention

The chemical process industry has grown rapidly, with corresponding increases in the quantities of hazardous materials being processed, stored, and transported. Plants are becoming larger and are often situated in or close to densely populated areas. Accidents in chemical plants can result in large human and/or economic losses and environmental pollution. Increasing attention must therefore be paid to the control of these hazards and, in particular, to the use of appropriate technologies to identify the hazards of a chemical process plant and to eliminate them before an accident occurs; this *is loss prevention.*

Major hazards in chemical plants are generally associated with the possibility of explosion, fire, or the release and dispersion of toxic substances, with some plants combining some or all of these hazards. As shown in Table 15.6, fires are the most common chemical hazards, while explosions in particular unconfined vapor cloud explosions lead to the largest economic losses. The potential for fatalities is highest in the case of toxic release (the most notorious being the Bhopal disaster in 1984, which killed over 2500 civilians at the time (the death toll had exceeded 8000 in 1990) and seriously injured an estimated 200 000 more [17], Section 15.5.1).

A number of the most severe accidents are described throughout this section but it should be noticed that lesser incidents with smaller consequences occur every day across the globe. Major disasters often lead to major changes in legislation. Unfortunately, minor incidents are seldom documented, often inadequately investigated, and the lessons that could be learnt are quickly forgotten. For instance, prior to the Bhopal disaster, minor incidents had already occurred at the plant site.

Similar accidents as in Bhopal, although improbable, could happen. An analysis that compared chemical incidents in the United States in the early-to-mid-1980s to the Bhopal incident revealed that of the 29 incidents considered, 17 incidents released sufficient volumes of chemicals with such toxicity that the potential consequences could have been more severe than those in Bhopal (depending on weather conditions and plant location) [18].

Figure 15.7 shows the main causes of chemical plant accidents. Most of the losses as a result of accidents can be attributed to human error. For instance, mechanical failures could all be due to human error as a result of improper design, maintenance, inspection or management. Piping system failures represent the largest part of the incidents (for instance, the Flixborough disaster in 1974, Section 15.5.1).

The analysis of accidents has shown that they are a result of a chain of events, often starting with a relatively trivial incident which (either because it goes unnoticed or because of inappropriate response) triggers a chain reaction that can rapidly lead to a catastrophe.

15.5.1 Safety Issues

Most accidents in chemical plants result in spills of flammable, explosive, and toxic materials, for example, from holes and cracks in tanks and pipes, and from leaks in flanges, pumps, and valves.

Table 15.6 Types of chemical plant accidents [17].

Type of accident	Fire	Explosion	Toxic release
Probability of occurrence	high	intermediate	low
Potential for fatalities	low	intermediate	high
Potential for economic loss	intermediate	high	low

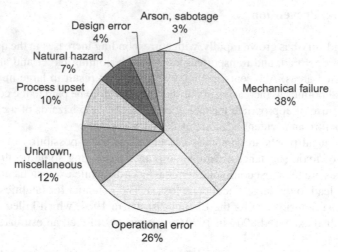

Figure 15.7   *Causes of chemical plant accidents [19].*

15.5.1.1   *Flammability – Fires and Explosions*

The hazard caused by a flammable material depends on a number of factors, that is, its flashpoint, its autoignition temperature, its flammability limits, and the energy released in combustion (Tables 15.7 and 15.8).

The main difference between fires and explosions is the rate of energy release. Fires release energy slowly, while explosions release energy very rapidly (microseconds typically). Whether the presence of a flammable material will result in a fire or explosion depends on several factors. The essential elements for combustion are fuel (e.g., gasoline, wood, propane), oxidizer (e.g., oxygen, chlorine, hydrogen peroxide), and ignition source (e.g., spark, flame, static electricity, heat). When one of these is lacking (or is present in too low a quantity) a fire will not occur. Note that if the autoignition temperature of a substance is relatively low, even hot surfaces can initiate an explosion.

Table 15.7   *Flammability characteristics of liquids and gases.*

Characteristic	Description
Flashpoint	Lowest temperature at which a liquid will ignite from an open flame.
Autoignition temperature	Temperature at which a material will ignite spontaneously in air, without any external source of ignition (flame, spark, etc.).
Flammability limits	Lowest and highest concentrations of a substance in air, at normal pressure and temperature, at which a flame will propagate through the mixture.
Lower flammability limit (LFL)	Below LFL mixture is too lean to burn (not enough fuel).
Upper flammability limit (UFL)	Above UFL mixture is too rich to burn (not enough oxygen).

Table 15.8 Toxicity and flammability characteristics of selected liquids and gases [17, 20, 21].

Compound	TLV[a] (ppm)	Flash point (K)	LFL (vol% in air)	UFL (vol% in air)	Autoignition temperature (K)	Heat of combustion (MJ/kg)
Acetone	750	253	2.5	13	738	28.6
Ethyne	2500[b]	Gas	2.5	100	578	48.2
Benzene	10[c]	262	1.3	7.9	771	40.2
Butane	800	213	1.6	8.4	678	45.8
Cyclohexane	300	255	1.3	8	518	43.5
Ethanol	1000	286	3.3	19	636	26.8
Ethene	2700[b]	Gas	2.7	36.0	763	47.3
Ethene oxide	1[c]	244	3.0	100	700	27.7
Hydrogen	4000[b]	Gas	4.0	75	773	120.0
Methane	5000[b]	85	5.0	15	811	50.2
Toluene	100 (skin)	278	1.2	7.1	809	31.3

[a]TLV = threshold limit value.
[b]Simple asphyxiant; value shown is 10% of LFL.
[c]Suspected carcinogen; exposures should be carefully controlled to levels as low as reasonably achievable below TLV.

With the kind of information presented in Table 15.8, process design can eliminate or reduce the existence of flammable mixtures in the process during start-up, steady-state operation, and shut-down.

Box 15.6 Basel, Switzerland

In November 1986, the Sandoz chemical factory in Basel, Switzerland, had a warehouse fire. While the firemen were extinguishing the flames they sprayed water over drums of chemicals that were exploding due to the heat released. The water/chemicals mixture was washed into the Rhine, dumping approximately 30 metric tons of pesticides and other toxic chemicals into the river. As a result the river life died up to 150 kilometers downstream. Things could have been worse though. A nearby building contained sodium, a metal that reacts violently with water. If the fire hoses had been sprayed on the stored sodium, the explosion could have destroyed a group of storage tanks holding the nerve gas phosgene. An investigation of the spill revealed that there were no adequate catch basins at the site for collecting the runoff firewater with the chemicals.

After the fire had been put out, the German government checked the water as it passed through Germany. It discovered twelve major pollution incidents (not related to Sandoz) within one month.

QUESTION:

Into which category does this accident fit (Figure 15.7)?

Many accidents have occurred as a result of plant modifications which had unforeseen side effects. Modifications may have to be made for start-up or during maintenance; temporary modifications may be necessary if a piece of equipment fails, or process modifications may be required as a result of changed feedstock, and so on. A well-known example of an accident involving a temporary modification is the Flixborough disaster (Box 15.7).

Box 15.7 Flixborough, UK

The accident at the Nypro caprolactam plant outside of Flixborough, UK, (June, 1974) is a typical example of a vapor cloud explosion. An intermediate in the production of caprolactam is cyclohexanol, which is produced by partial oxidation of cyclohexane (Figure B15.7.1). The conversion must be kept low (usually under 10%) to avoid complete oxidation of the feed and product [22].

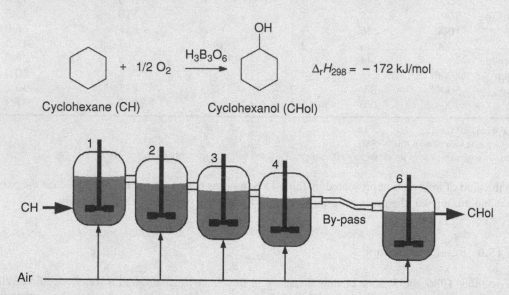

$$\Delta_r H_{298} = -172 \text{ kJ/mol}$$

Cyclohexane (CH)   Cyclohexanol (CHol)

Figure B15.7.1 *Oxidation reaction and arrangement of reactors and temporary pipe for by-passing reactor no. 5.*

A sudden failure of a temporary cyclohexane line bypassing a reactor that had been removed for repair (Figure B15.7.1) led to the vaporization of an estimated 30 metric tons of cyclohexane. The vapor cloud dispersed throughout the plant and was ignited by an unknown source. The entire plant site was leveled and 28 people were killed.

The failure of the by-pass pipe was attributed to inadequate support of the pipe and a poor design (consisting only of a full-scale sketch in chalk on the workshop floor [23]) that failed to account for the movement of the pipe due to the pressure inside it. When the pressure rose a little above the normal level, the bellows at each end of the pipe began to rotate and failed.

QUESTIONS:

What could have been done to prevent this incident, or, at least, limit its consequences?
What is meant by "bellows"?

15.5.1.2   Toxicity – Toxic Releases

Nearly all of the substances used in the production of chemicals are poisonous to some extent. The potential hazard will depend on the inherent toxicity of the substance and on the frequency and duration of exposure.

Table 15.9 *LD$_{50}$ values of some substances (oral, rat) [20].*

Substance	LD$_{50}$ value (mg/kg)	Category (EEC guideline)[a]
Dioxine (TCDD)	0.001	Very toxic
Potassium cyanide	10	Very toxic
Tetraethyl lead	35	Toxic
Lead	100	Toxic
DDT	150	Toxic
Aspirin	1000	Harmful
Sodium chloride	4000	—

[a] <25 very toxic, 25–200 toxic, 200–2000 harmful.

Toxic effects on humans can be acute (short-term effects) or chronic (long-term effects). The inherent toxicity of a substance is measured by tests on animals. The acute effects of substances are expressed by the LD$_{50}$ value, the lethal dose at which 50% of the test animals are killed. Table 15.9 gives LD$_{50}$ values for a selection of substances.

The LD$_{50}$ value only gives a crude indication on possible chronic effects. The most commonly used guide for controlling long-term exposure of workers to contaminated air is the "Threshold Limit Value" (TLV). The TLV is defined as the concentration to which it is believed the average worker could be exposed, for eight hours a day, day by day, five days a week, without suffering harm. Table 15.8 shows TLV values for some substances. Chronic effects are much less easy to identify, since years may pass before these effects become visible. Furthermore, the cause then might be difficult to trace, for example, the effect of asbestos and solvents in the paints industries. The most notorious example of the hazards involved when dealing with highly toxic materials is the Bhopal, India accident (3 December 1984) (Box 15.8).

Box 15.8 Bhopal, India

The Bhopal plant (partially owned by Union Carbide and partially owned locally) produced pesticides, amongst others carbaryl. An intermediate chemical in this process is methyl isocyanate (MIC), which is extremely dangerous. It is reactive, toxic, volatile, and flammable. The TLV for MIC is 0.02 ppm.

The disaster started with the contamination of a large MIC storage tank with a large amount of water (how is still unclear), leading to an exothermic chemical reaction. Insufficient cooling of the storage tank led to a runaway reaction with an accompanying temperature increase past the boiling point of MIC (312 K). The MIC vapors traveled through a pressure relief system and into a scrubber and flare system installed to consume the MIC in case of a release. Unfortunately, the scrubbing and flare systems were not operating. An estimated 40 metric tons of MIC vapor was released, the toxic cloud spread to the adjacent town (vapors stayed close to the ground due to the density of about twice that of air), killing over 2500 civilians and injuring an estimated 200 000 more.

The exact cause for the contamination of the MIC is not known. However, several design measures could have prevented this disaster [17]:

- a well-executed safety review/HAZOP study (Section 15.5.3.1 and references) would possibly have identified the problem;
- adequate refrigeration of the storage tank could have prevented the runaway reaction;

- the scrubber and flare system should have been fully operational to prevent the release;
- the inventory of MIC should have been minimized, for example, by redesigning the process to produce and consume MIC in a highly localized area (inventory <10 kg [17]); although storage of MIC was convenient, it was not necessary;
- an alternative reaction route, not involving MIC, could have been adopted (B15.8.1).

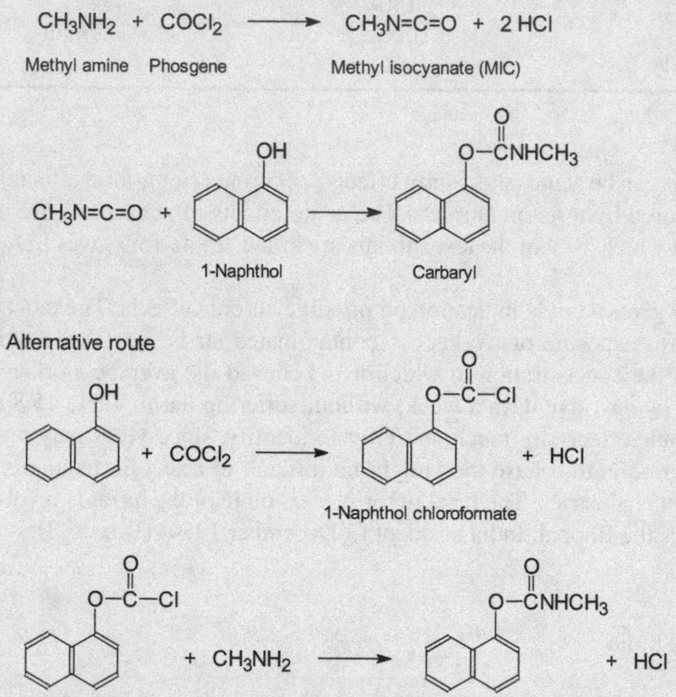

Scheme B15.8.1   *Alternative routes for the production of the insecticide carbaryl.*

The Bhopal event and other major chemical plant incidents have changed the chemical processing industry profoundly. The result has been new legislation with better enforcement, enhancement in process safety, development of inherently safer plants, harsher court judgements, management willing to invest in safety-related equipment and training, and so on.

QUESTION:

The high toxicity of MIC had dramatic consequences. What about the toxicity of the other chemicals used?

Which route would you prefer? Explain.

Some points to consider in the minimization of the risks involved when using hazardous substances are substitution (using less hazardous substances), containment (avoid leaks), ventilation, disposal provisions (vent stacks, vent scrubbers), and good operational practice (written instructions, training of personnel, protective clothing, monitoring of the plant environment to check exposure levels, etc.).

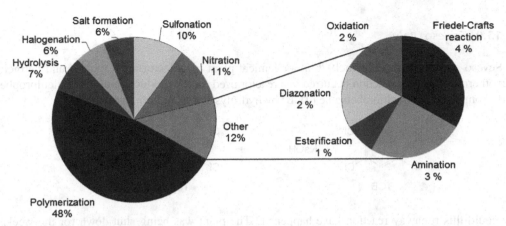

Figure 15.8   *Contribution of processes to chemical plant accidents involving runaway reactions (% of incidents) [19].*

15.5.2   Reactivity Hazards

Every year there are a significant number of exothermic runaway reactions, particularly in batch and semi-batch processes, that can halt production and result in serious injuries, or even death, to plant operators. At the top of the critical list are reactions such as polymerization, nitration, sulfonation, and hydrolysis (Figure 15.8).

QUESTIONS:
> *What do all of the processes in Figure 15.8 have in common?*
> *The number of accidents in batch processes is much larger than that in continuous processes. Give at least three reasons for this.*

The consequences of exothermic runaway reaction can be as severe as those from the ignition and explosion of a fuel/air mixture (Box 15.9). In addition, release of toxic substances can occur. Chemical hazards principally arise from:

* thermal instability of reactants, intermediates, and/or products;
* rapid exothermic reactions, which can raise the temperature to the decomposition temperature of the substances involved or cause violent boiling of these substances;
* rapid gas evolution, which can pressurize and possibly rupture the plant.

These chemical hazards must be considered in assessing whether a particular chemical process can be operated safely on an industrial scale. Runaway reactions frequently result from poor understanding of the chemistry involved, leading to a badly designed plant. Knowledge of the reaction enthalpies associated with the desired reaction and possible undesired reactions, and information on the thermal stability, that is, the temperature at which any decomposition reaction may occur, are essential for evaluating the hazards. The consequences of possible process maloperation must also be considered, for example, agitation or cooling water failure, overcharging or omitting one of the reactants, or charging the wrong reactant. Adequate control and safety backup systems should be provided, as well as adequate operational procedures, including training.

Box 15.9   Seveso, Italy

The Seveso accident happened in 1976 at a chemical plant manufacturing pesticides and herbicides. An exothermic runaway reaction occurred in a reactor used for the production of 2,4,5-trichlorophenol (TCP) from 1,2,4,5 tetrachlorobenzene (TCB) by hydrolysis with sodium hydroxide:

How could this runaway reaction have happened? The plant was being shut down for the weekend, leaving the reactor half full of unreacted material at elevated temperature. The runaway reaction in the unstirred mixture was triggered by the slight, unexpected, heat input from the hot, dry wall of the reactor to the upper layer of the reaction mixture. The temperature of this upper layer reached a level at which slow and weak exothermic reactions started. After seven hours (!) these reactions resulted in the start of other, more rapid, exothermic reactions [24]. The rupture disk in the safety valve burst as a result of excessive pressure, and an aerosol cloud containing 2,3,7,8-tetrachlorodibenzo-*p*-dioxin (TCDD) was released into the air. TCDD was a by-product of the uncontrolled exothermic reaction:

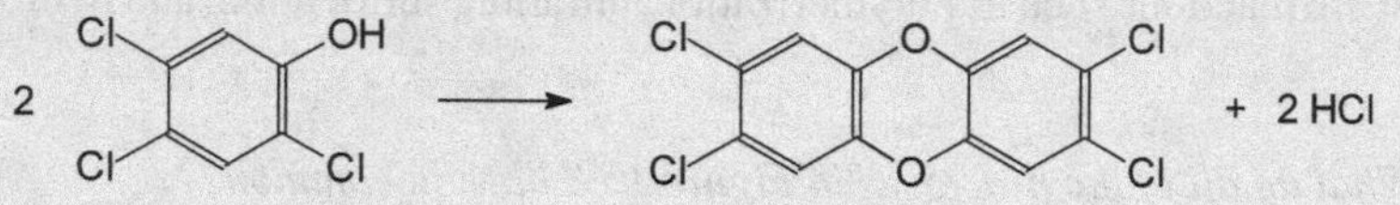

The reactor had no automatic cooling system and only maintenance personnel were there at the time of the accident, so there was no operator to start cooling manually and suppress the reaction. Fortunately, a worker noticed the cloud and stopped the release after only 20 minutes. TCDD is commonly known as dioxin, which is widely believed to be one of the most toxic man-made chemicals. Although no immediate fatalities were reported, between a few 100 grams and a few kilograms of dioxin (lethal to man even in microgram doses, Table 15.9) were widely dispersed, which resulted in an immediate contamination of some 30 square kilometers of land and vegetation. More than 600 people had to be evacuated and 2000 were treated for dioxin poisoning.

The best-known consequence of the Seveso disaster and previous serious chemical accidents (for example, Flixborough) was the impulse that it gave to the creation of the European Community's Seveso Directive in 1982, a new system of industrial regulation for ensuring the safety of hazardous operations.

The occurrence of runaway reactions is not limited to chemical reactors; these reactions may also start in storage tanks (Section 12.7 and Box 15.8).

It is necessary to be aware of the hazards associated with the use of catalysts. Many metal-based catalysts are pyrophoric in their reduced state and must be kept away from flammable substances. When not in use, they must be blanketed with an inert gas, such as nitrogen.

In, for example, FCC or hydrocracking operation, when unloading a coked catalyst there is a danger of iron sulfide fires. Iron sulfide will ignite spontaneously when exposed to air and, therefore, must be wetted with water to prevent it from igniting vapors. Coked catalyst may be either cooled below 320 K before being dumped from the reactor, or dumped into containers that have been purged and made inert with nitrogen and then cooled before further handling.

Catalysts might form dangerous compounds when exposed to gases containing carbon monoxide. For instance, catalysts containing metallic nickel might form $Ni(CO)_4$ at temperatures below 430 K. $Ni(CO)_4$ is an extremely toxic, almost odorless gas that is stable at low temperatures. It might be formed, for instance, in methanation reactors (Section 5.4) upon cooling down, unless the system has been thoroughly purged with nitrogen.

15.5.3   Design Approaches to Safety

Many of the chemicals used in industry can present safety, health, and environmental problems, particularly if their inherent hazards or those arising from specific operations have not been identified and evaluated, and a basis for safe operation of the process has not been developed and implemented [25]. Furthermore, possible (probable) deviations from the design operating conditions (e.g., temperature and pressure in a reactor) that could result in hazardous situations must be foreseen and proper action taken to prevent these deviations from occurring or to limit their effect. Processes that cause no or negligible danger under all foreseeable circumstances (all possible deviations from the design operating conditions) are inherently safe, and naturally are preferred if possible. However, most chemical processes are inherently unsafe to a greater or lesser extent. There are opportunities for hazard reduction at every level of the design process. In the following sections the steps to be taken in the design of a safe chemical plant are summarized.

15.5.3.1   Identification and Assessment of Hazards

For the safe design of a chemical process it is important that hazardous chemical and physical properties of all reactants and products involved (including undesired by-products) are known. These properties are often not inherent characteristics of the substance, but depend on the plant situation (e.g., operating temperature and pressure, presence of other chemicals, etc.). Important properties from a safety viewpoint include vapor pressure (related to overpressure), flash points and flammability limits, and toxicity (Tables 15.8 and 15.9). Furthermore, all reactions (desired and undesired) should be identified if possible, especially those that may result in overpressure or the production of dangerous chemicals (e.g., explosive or toxic). Data are required on reaction rates and enthalpies for exothermic reactions and unstable chemicals, on temperature limits beyond which explosive decompositions or other undesirable behavior can occur, and on rates of gas or vapor generation (overpressure). The strength and corrosion rates of materials of construction should also be evaluated.

The operating conditions under which hazardous situations could occur (during normal operation and as a result of failures) should be identified. Normally, the design team prepares a HAZOP (hazard and operability) study, in which all of the possible paths to an accident are identified. Then, a fault tree is created and the probability of the occurrence of each potential accident computed. A brief outline of a HAZOP study and references to more detailed accounts are available elsewhere [21].

15.5.3.2   Control of Hazards

Control of material hazards can be achieved, for example, by the replacement of hazardous chemicals by less dangerous chemicals, if possible, and by minimizing the inventory of hazardous substances present. Similarly, replacing a hazardous process route with an alternative route or minimizing the number of hazardous operations leads to increased safety.

One method of preventing fires and explosions is "inerting", that is, the addition of an inert gas to reduce the oxygen concentration so that the mixture is below its LFL (Lower Flammability Limit). This method is used in, for example, the production of ethene oxide, where methane is used for inerting (Section 10.5.2.1). Another method involves avoiding the build-up of static electricity and its release in a spark, for example, by the installation of grounding devices and the use of antistatic additives. In addition, explosion-proof equipment and instruments are often installed, for example, explosion-proof housings that absorb the explosion shock and prevent the combustion from spreading beyond the enclosure. Another approach is good ventilation or construction in the open air to reduce the possibility of creating flammable mixtures that could ignite.

In the design of the plant, containment of flammable and toxic materials and using less severe operating conditions of pressure and temperature will result in a less hazardous process. Furthermore, inventories of hazardous chemicals should be minimized.

15.5.3.3   *Control of the Process*

The safety of a process that is inherently unsafe depends on proper design of the process, provision of automatic control systems, alarms, interlocks, trips, and so on, together with good operating practices and management.

15.5.3.4   *Loss Limitation*

If an accident occurs, the damage and injury caused should be minimized. Some measures are the installation of sprinkler systems, provision of fire-fighting equipment, and a safe plant layout (separate people from processes, etc.). In processes where pressures can build up rapidly (e.g., ethene polymerization, Chapter 11) it is crucial that pressure relief devices have been installed for venting to the atmosphere or to scrubbers, flares, and condensers. The devices include relief and safety valves, knock-out drums, rupture disks, and so on.

15.6   Process Evaluation

After each development stage the status of the process should be evaluated. The basis for this evaluation is the documentation of the knowledge of the process gained so far. The important questions to be answered are:

- Is the production process technically feasible in principle?
- What is the economic attractiveness of the process?
- How big is the risk in economic and technical terms?

The technical feasibility of a process is proven by research in the laboratory, miniplant, and pilot plant. If certain process steps present difficulties, these problems should be solved either by improving these steps or by selecting a different procedure. The technical evaluation of a process aims to steer the process research and development in the right direction. Factors increasing the technical risk of a process are:

- exceeding of limits (e.g., if the dimensions of the largest extraction column previously operated by a company are considerably exceeded);
- unfamiliarity of the company with the particular technology (e.g., continuous processes, high pressure, per-acids, fluidized beds);

- use of units that are difficult to scale up (e.g., for solids processing);
- use of technically non-established equipment or unit operations (e.g., monolith reactors in multiphase application).

The technical risk of a process can be reduced in a number of ways. The first is to increase the expenditure on research and development of the weak points, for example, pilot plant testing of the problematic unit, find well-established alternative equipment or other ways to perform the unit operation. A second option is to develop failure scenarios, that is, determine what can be done if problems should occur (e.g., backup units, additional instrumentation). The decision basically is an economic question; the costs of increased research and development expenditure have to be weighed against the cost of eliminating the risk when the plant is started up or in operation. It should be noted that safety should have an even higher weighting than economy, that is, processes that are unsafe should not be brought to commercialization.

An important factor in the decision to pursue or abandon the development of a process is the economic perspective of a process. As the purpose of investing money in a chemical plant is to earn money, some means of comparing the economic performance or profitability is needed. To be able to assess the profitability of a process, estimates of the investment and operating costs are required. For small projects and for simple choices between alternative processes or equipment, usually a comparison of the investment and operating costs is sufficient. More sophisticated techniques are required if a choice has to be made between large, complex processes with different scope, time scale, and type of product.

The economic evaluation of processes is based on criteria set by the company, such as payback time (*PBT*), net present value (*NPV*), and so on. A detailed description of the economic criteria and their use can be found elsewhere [21, 26 and numerous other texts on the subject]. These kinds of evaluations are usually done by a specialist group within the company, but chemical engineers and industrial chemists should have a basic knowledge in order to be able to communicate.

QUESTION:

> *In principle, a plant will not be built in case of a negative outcome of the economic analysis. Are there situations in which this does not apply?*

15.6.1 Capital Cost Estimation

Several methods are used to estimate the capital cost required for a chemical plant. These range from preliminary estimates (or study estimates) with an accuracy of 20–30% to detailed estimates with an accuracy of within 5%. The latter are usually only obtained in the final development stage of a process, when the engineering drawings have been completed and all equipment has been specified. Here, the focus is on the preliminary estimates, which are adequate for comparing processes in the early stages of development, and for left-or-right and go-or-no go decisions (Figure 15.2).

Preliminary estimates usually start with the use of cost charts ([26–28] and Section 15.6.1.2) for estimating the purchase cost, C_P, of major equipment items, the other costs being estimated as factors of the equipment cost. Computing systems using equations based on cost charts are already available (e.g., ASPEN PLUS and HYSYS), and more advanced systems (adding more details of equipment design, materials, and so on, and related costs such as site preparation, service facilities, etc.) leading to more accurate estimates are becoming available.

Step-counting methods are also frequently used for order-of-magnitude estimates [29–33]. These techniques are based on the presumption that the capital cost is determined by a number of major processing steps (e.g., a distillation column with its reboiler and condenser) in the overall process. The main weakness

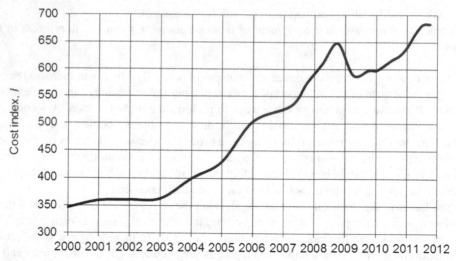

Figure 15.9   *Chemical engineering plant cost index (I = 400 in January 2004) [34].*

of step-counting methods is the ambiguity in defining a "major processing step". Clearly, these methods can only give order-of-magnitude estimates of the plant cost. However, due to their speed and simplicity, they are very useful in the conceptual stage of process design, when comparisons between alternative process routes are being made.

In practice a company developing a new process often has data on part(s) of the flow scheme under consideration. In that case these data allow a solid basis for a quick economic evaluation of these parts of the process design.

15.6.1.1   Cost Indices

The cost of equipment has increased over the years as a result of inflation. The method usually used to update historical cost data makes use of published costs, which relates present costs to past costs:

$$\text{Cost in year Y} = \text{Cost in year X} \times \frac{\text{Index year Y}}{\text{Index year X}} \tag{15.1}$$

An example is the capital cost index published by IHS on a frequent basis [34] and shown graphically in Figure 15.9.

All cost indices should be used with caution and judgment. The longer the period over which the correlation is made the more unreliable the estimate.

15.6.1.2   Estimation of Equipment Costs

Cost charts (and equations based on them) represent purchase costs of equipment items as a function of key equipment sizes, for example, the heat transfer area for heat exchangers, as shown in Figure 15.10. In this

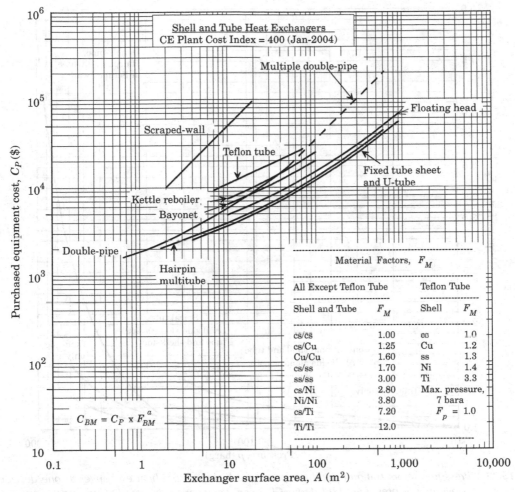

Figure 15.10 *Purchased equipment costs for shell and tube and double-pipe heat exchangers [26]. Bare module factors F_{BM}^a are obtained from Figure 15.12 using material factors given here and pressure factors F_P from Figure 15.11. From Chemical Engineering Process Design and Economics, 2nd edition, by Ulrich and Vasudevan, Process Publishing [2004], page 383, by permission.*

figure, the logarithm of the purchased cost, C_P, is plotted as a function of the logarithm of the heat exchange area A for a variety of shell and tube and double-pipe heat exchangers.

The equipment costs in Figure 15.10 are for carbon steel construction and operation at pressures up to 10 barg[1] . Corrections are needed when more expensive materials (e.g., corrosion resistant or able to resist elevated pressure) are used. Material factors, F_M, are provided in Figure 15.10 for different combinations of shell and tube materials. Figure 15.11 shows the pressure factor, F_P, as a function of pressure.

---

[1] barg is the pressure of a system which one would see displayed on a normal pressure gauge; it is the pressure of the system, over and above atmospheric pressure; bara denotes absolute pressure.

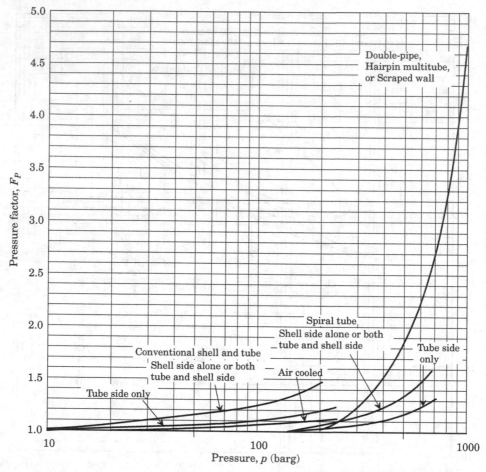

Figure 15.11   *Pressure factors (ratio of purchase price of a high pressure heat exchanger to one designed for conventional pressures) [26]. From Chemical Engineering Process Design and Economics, 2nd edition, by Ulrich and Vasudevan, Process Publishing [2004], page 384, by permission.*

The product $F_P \times F_M$ is used in Figure 15.12 to determine the bare module factor, F_{BM}, which can then be used to obtain the purchased equipment cost of heat exchangers of materials of construction other than carbon steel and/or operating at higher pressure (the so-called bare module cost, C_{BM}):

$$C_{BM} = C_P \times F_{BM} \tag{15.2}$$

It can be seen from Figure 15.10 that for a floating head heat exchanger with $10 \leq A \leq 900$ m^2 the following equation is valid (the slope of the line equals 0.6):

$$\frac{C_{P_1}}{C_{P_2}} = \left(\frac{A_1}{A_2}\right)^{0.6} \tag{15.3}$$

The fact that the exponent is less than unity expresses the *economies of scale*, that is, one large heat exchanger is less costly than several smaller ones having the same overall heat exchange area (and being of identical

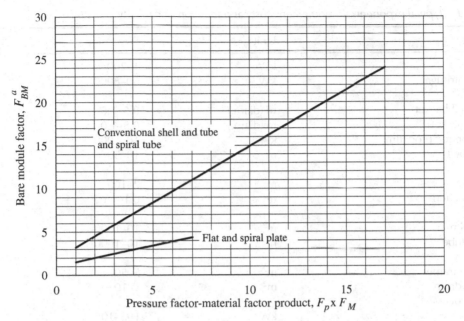

Figure 15.12 *Bare module factors as a function of material and pressure factors for heat exchangers [26]. From Chemical Engineering Process Design and Economics, 2nd edition, by Ulrich and Vasudevan, Process Publishing [2004], page 384, by permission.*

type and material of construction, and at the same moment in time!). Economies of scale apply to many types of equipment, as shown in Table 15.10.

QUESTIONS:

Compare the bare module costs (in 2004) of a double-pipe and a floating head shell and tube heat exchanger made of carbon steel (cs) for a heat exchange surface of 10 m² at a pressure of 10 barg.

Estimate the bare module cost of a floating head shell and tube heat exchanger made of carbon steel (cs) for a heat exchange surface of 20 m² at a pressure of 10 barg in 2004 and 2010.

Why have some equipment items exponents of 1 or even >1? What unit, not present in the table also has an exponent of about 1? (Hint: In which separation technology is the driving force for separation a difference in diffusivity, combined with a flow rate across a surface separating two compartments?) Why are size limits given in the table? (Hint: consider what happens for very small equipment and extremely large equipment.)

Economies of scale also apply to entire plants (Table 15.11), provided that the materials of construction are the same for each capacity figure. Of course, the technologies should also be the same.

Blowers are used for pressure increases from 0.03 to 5 bar, while compressors are used for higher discharge pressures. The cost of compressors may represent a large part of the equipment costs. Compressors are very expensive because they require large volumes and thick walls and involve large moving parts. When comparing Figure 15.13, showing the cost chart for blowers and gas compressors, with Figure 15.10 for

Table 15.10 *Typical exponents for equipment costs as a function of capacity [21, 26].*

Equipment	Unit	Size range	Exponent
Agitators	kW	1–200	0.7
Blower, centrifugal	kW	50–8000	0.9
Centrifuge			
Horizontal basket	m (diam.)	0.5–1.0	1.3
Vertical basket	m (diam.)	0.5–1.0	1.0
Compressor, reciprocating	kW	10–2000	1.0
Conveyor belt	m	10–50	0.8
Dryer			
Drum	m^2	2–100	0.6
Rotary vacuum	m^2	0.1–18	0.4
Evaporator			
Falling film	m^2	10–1000	0.52
Vertical tube	m^2	10–1000	0.53
Filter			
Plate and frame	m^2	5–50	0.6
Vacuum drum	m^2	1–10	0.6
Furnace (process)			
Box	kW	103–105	0.77
Cylindrical	kW	103–104	0.77
Heat exchanger, floating head	m^2	10–900	0.6
Reactor, jacketed			
Agitated	m^3	3–30	0.45
Jacketed	m^3	3–30	0.4
Refrigeration unit	kJ/s	5–10 000	0.6
Stack, carbon steel	m	10–180	1.3
Tank			
Floating roof	m^3	200–70 000	0.6
Spherical, 0–5 barg	m^3	50–5000	0.7
Vessel, process			
Horizontal	m^3	1–50	0.7
Vertical	m^3	2–50	0.9

Table 15.11 *Typical exponents for plants as a function of capacity [35, 36].*

Compound	Process	Exponent
Acetic acid	Methanol conversion	0.59
Ammonia	Steam reforming of natural gas	0.66
Methanol	Steam reforming of natural gas	0.71
Ethene	Steam cracking of ethane	0.60
Ethene oxide	Selective oxidation of ethene	0.78
Polyethene	Low-pressure polymerization process	0.70
Sulfuric acid	Contact process	0.67

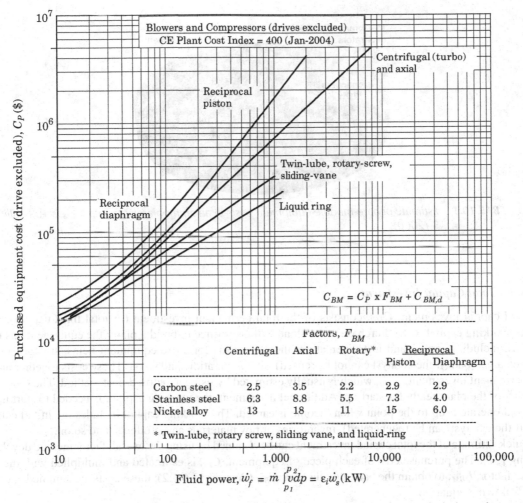

Figure 15.13 *Purchase costs of blowers and compressors [26]. Costs of drives are excluded. To determine the bare module cost of drives $C_{BM,d}$ refer to Figures 5.20 and 5.21 in Ref. [26]. From Chemical Engineering Process Design and Economics, 2nd edition, by Ulrich and Vasudevan, Process Publishing [2004], page 380, by permission.*

heat exchangers, it is immediately evident that for large power input compressors are very costly. Box 15.10 illustrates this for an ammonia plant.

Box 15.10 Compressor Cost in Ammonia Production from Hydrogen and Nitrogen

Ammonia synthesis requires high pressures (Section 6.1). Figure B15.10.1 illustrates the high costs for gas compression.

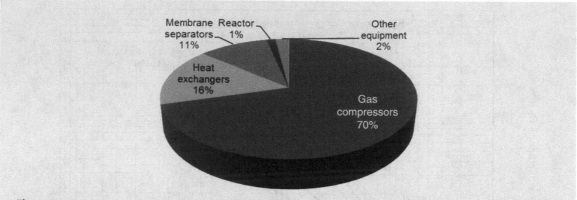

Figure B15.10.1 *Estimate of bare-module equipment costs in an ammonia plant; Total C_{BM} is 109 million $ (2000).*

15.6.1.3 *Total Capital Investment*

The total capital investment, C_{TCI}, consists of the total permanent investment (or total fixed capital), C_{TPI}, and the working capital, C_{WC}. Purchase costs alone will not suffice to build a plant; the equipment has to be installed, including piping and instrumentation, there are costs for spare equipment items, storage vessels, initial catalyst load (if the catalyst cannot be reused), site preparation, and so on. These costs together are the total permanent investment, C_{TPI}, which is usually estimated by using a factored-cost method. The C_{TPI} is the total cost of the plant ready for start-up. Additional investment, the working capital, is needed to start up the plant and operate it up to the point when income is earned. The working capital includes the initial catalyst load (if the catalyst can be regenerated), inventories of raw materials and products, and so on.

A quick estimate of the total capital investment can be obtained by using the overall factor method developed by Lang [37]. The purchase cost of each piece of equipment, C_{Pi}, is estimated and multiplied with the bare module factor, F_{BM}, to obtain the bare module cost, C_{BM} (Section 15.6.1.2); these costs are summed and used in the following equation:

$$C_{TPI} = 1.05 \cdot f_L \cdot \sum (C_{P_i} \cdot F_{BM_i}) = 1.05 \cdot f_L \cdot \sum (C_{BM_i}) \tag{15.4}$$

in which the factor 1.05 covers the delivery cost to the plant site and f_L is the Lang factor, which accounts for costs associated with the installation of the equipment and any additional investment costs. Typical values are given in Table 15.12.

Table 15.12 *Lang factors for use in various plant types [21].*

Plant type	Lang factor f_L
Solids processing	3.1
Mixed solids and fluids processing	3.6
Fluids processing	4.7

Table 15.13   *Summary of operating costs [21].*

Cost item	Typical values
Fixed costs (A)	
1. Maintenance	5–10% of C_{TPI}
2. Operating labor	from manning estimates and labor costs
3. Laboratory costs	20–23% of (2)
4. Supervision	20% of (2)
5. Plant overheads	50% of (2)
6. Depreciation	10% of C_{TPI}
7. Insurance	1% of C_{TPI}
8. Local taxes	2% of C_{TPI}
9. Royalties	1% of C_{TPI}
Variable costs (B)	
10. Raw materials	from flow sheets and unit cost
11. Miscellaneous materials	10% of (10)
12. Utilities	from flow sheets and unit costs
13. Shipping and packaging	usually negligible
Indirect costs (C)	20–30% of (A + B)
Annual operating costs ($)	A + B + C
Operating costs ($/kg)	(A + B + C) / annual production rate

More detailed factor estimates can be obtained by considering each of the items included in the Lang factor separately [21, 27, 28, 38, 39]. Many companies also use a factor to account for the different costs of construction at different plant locations around the world [27].

15.6.2 Operating Costs and Earnings

Apart from the total capital investment, the annual operating costs and annual earnings (pre-tax or after tax) must be estimated in order to obtain an approximate measure of the profitability of the plant.

Operating costs are divided into fixed costs and variable costs. The fixed costs do not vary with production rate and have to be paid whatever the quantity produced. The variable costs, on the other hand, depend on the quantity produced. Table 15.13 shows the main components of the operating costs and how they can be estimated. The variable costs can be calculated from material and energy balances and material prices.

Fixed costs are partly labor related and partly capital related. The latter are estimated as a percentage of fixed capital investment. The indirect operating costs are the company overheads, which include administration, general research and development, and sales expenses.

The most reliable sources of cost data are company records, especially with regard to labor, utilities, overhead, and other general costs. When these are not available, costs are estimated according to generally accepted principles [21, 26–28, 31, 40].

Raw material costs usually constitute a major part (40–60%) of the operating costs, especially for base chemicals. The cost of raw materials follows from raw material consumption (e.g., t/t product) and the unit cost of the particular raw material (e.g., $/t). For bulk chemicals the unit costs can be obtained from *ICIS Pricing* (www.icis.com) and from specialized journals, such as *European Chemical News*, and for refinery hydrocarbons from *The Oil and Gas Journal* for use in preliminary estimates. Other sources of price data, although not on a systematic basis, include *Chemical Engineering Progress, Chemical Engineering News,* and *Chemical Week*. For pharmaceuticals, and so on, prices can be obtained from their manufacturers. See Table A.4 in Appendix A for prices of some bulk chemicals.

Utilities include steam, electricity, process water, cooling water, and so on. Waste disposal is also often included in the utilities. Accurate consumption figures can only be obtained from material and energy balances over major equipment items. Utility prices depend on location and proximity of their source, as well as on whether they are purchased or generated on site. Unit costs are best obtained from company records.

Operating labor costs (*OLC*) are based on an estimation of the number of operators for the process and the cost to the company of one operator. The number of operators depends on the type and arrangement of the equipment, the number of equipment items, the amount of instrumentation and control for the process, and company policy in establishing labor requirements. Supervision and other related costs are then estimated as a fraction of *OLC* [21, 27].

Depreciation is an income tax credit to replace depleted facilities and applies to fixed assets only. Depreciation is calculated as a fraction of the total permanent investment (C_{TPI}). For calculating operating costs a straight line depreciation is common, usually over a 10-year period with zero scrap value at the end, that is, 10% of C_{TPI} per year. Sometimes 15% is used, while for small investments 33% is used.

The total sales revenues can be computed for the products and by-products, based on the quantities produced and the selling prices.

15.6.3 Profitability Measures

One of the simplest methods for estimating process profitability is the *return on investment* (*ROI*), which relates the (estimated) annual profit to the total capital investment (C_{TCI}):

$$ROI = \frac{\text{annual profit (pre-tax or after-tax)}}{\text{total capital investment}} \cdot 100(\%) \tag{15.5}$$

The total capital investment includes working capital. The usual practice is to calculate the *ROI* after-tax based, that is, using the net annual profit. *ROI*s can also be calculated on a cash flow basis or using total permanent investment only. Generally, acceptable *ROI*s for larger projects are in the 20% range; less than 5% is usually a "No" for a new project. When the selling price of a certain product is unavailable, it is common to select a desired *ROI* and calculate the selling price required.

Another simple profitability measure is the *payback period* (*PBP*), the time required to recover the C_{TPI}:

$$PBP = \frac{\text{total permanent investment}}{\text{annual cash flow}} \text{ (year)} \tag{15.6}$$

in which cash flow is:

$$CF = (1\text{-tax rate}) \cdot (\text{annual revenues} - \text{anual operating costs}) + \text{annual depreciation} \tag{15.7}$$

Typical *PBP*s for profitable investments are three to four years. Shorter *PBP*s are common for relatively small replacement or modification investments in an existing plant (e.g., the installation of a heat exchanger to recover waste heat).

Box 15.11 gives an example of the calculation of ROI and PBP values.

Box 15.11 Savings in an Ammonia Plant

In an ammonia plant with a capacity of 357 t/d, the axial ammonia converters have been replaced by radial converters. This has led to a reduction of the required loop pressure from 300 to 250 bar, while the pressure drop was reduced from 5 to 3 bar. As a result, energy savings of 0.85 GJ/t ammonia were achieved. The *operating factor* of the plant is only 0.7675 (i.e., the plant does not operate at full capacity and/or not for 365 days per year). The investment amounted to 1.1 $ million, while the energy costs were 5.35 $/GJ. Calculate the *ROI* and *PBP* for this investment (neglect tax).

Plant ammonia production: $357 \cdot 365 \cdot 0.7675 = 10^5$ t/a
Energy savings: $0.85 \cdot 10^5 \cdot 5.35 = 0.455 = 10^6$ $/a
$ROI = 41\%$
$PBP = 2.4$ year

Estimates of *ROI* and *PBP* are often based on the profit or cash flow in the third year of operation or on an average value. The *ROI* and *PBP* are useful as indicators when many different alternatives are screened but they are unsuitable as profitability instruments because changes in future cash flows and the "time-value of money" are not taken into consideration.

The time-value of money represents the fact that one dollar received now is worth more than the same dollar received after some time, the difference being the interest rate (or discount rate), r, that could have been recovered during that time span:

$$NFV = NPV \cdot (1 + r)^n \tag{15.8}$$

in which *NPV* is the *net present value* (the value now) and *NFV* the *net future value* (what the value will be worth after n years). For example, if $1000 is put into the bank now, at an interest rate of 4%, after ten years it will be worth $ $1000 \cdot (1+0.04)^{10} = \$1480$.

When applying this principle to cash flows received from an investment project such as a chemical plant, this means that money earned in the early years of operation is more valuable than money earned in later years. This is expressed as follows:

$$NPV_{\text{cash flow in year } n} = \frac{\text{estimated cash flow in year } n \ (NFV)}{(1 + r)^n} \tag{15.9}$$

and demonstrated in the example in Box 15.12.

Box 15.12 Choosing between Two Alternatives Differing in the Time-Value of the Money Earned

Assume an investment of $ 10 million in a plant at the beginning of year 1 (end of year 0). During operation, at the end of year 1, year 2, ... year 10 a cash flow of $ 2 million per year is generated. Alternatively, the money could be invested in a plant generating $ 3 million at the end of years 1 to 5 and

$ 1 million at the end of years 6 to 10. $r = 15\%$. In both cases a cash flow of $ 20 million is generated in total in the period of 10 years considered, but the distribution over the years is very different. Which is the most profitable plant?

1.
$$NPV = \sum_{0}^{n} \frac{NFW}{(1+r)^n} =$$

$$-C_{TCI} + 2 \cdot \left\{ \frac{1}{(1+r)} + \frac{1}{(1+r)^2} + \cdots + \frac{1}{(1+r)^{10}} \right\} =$$

$$-C_{TCI} + 2 \cdot \frac{1}{r} \cdot \left\{ 1 - \frac{1}{(1+r)^{10}} \right\} = -10 + 10.04 \approx \$0 \text{ million}$$

2.
$$NPV = \sum_{0}^{n} \frac{NFW}{(1+r)^n} =$$

$$-C_{TCI} + \left\{ \frac{3}{(1+r)} + \cdots + \frac{3}{(1+r)^5} + \frac{1}{(1+r)^6} + \cdots + \frac{1}{(1+r)^{10}} \right\} =$$

$$-C_{TCI} + \frac{1}{r} \cdot \left\{ 1 - \frac{1}{(1+r)^{10}} \right\} + \frac{2}{r} \cdot \left\{ 1 - \frac{1}{(1+r)^5} \right\} = \$1.72 \text{ million.}$$

Hence, option 2 is the best investment based on a net present value analysis. Apparently, at a discount rate of 15%, for situation 1 the investment has only been earned back. This is the so-called *discounted cashflow rate of return* (*DCFRR*). The higher the *DCFRR*, the higher the profitability of the project.

QUESTIONS:

Derive the equation in option 1.
Why is a discount rate of 15% used in this example, and by many companies, instead of the interest rate used by banks? Comment on the value of 15%.
Calculate the DCFRR for situation 2 (trial and error).
Give an example of a plant that might show a cash flow pattern such as given in option 2. (Hint: in some cases governments provide a tax advantage for a specified period to stimulate market penetration of a certain product.)

Of course, this procedure can be extended to different investments at different times (e.g., investment for design, construction, working capital) and variable cash flows (e.g., starting with lower production capacity, decreasing selling price as a result of operation in a different market, expiration of patents, etc.), and other methods of compounding interest (e.g., per month, per day, or continuous). More elaborate information can be found in the literature [27].

15.7   Current and Future Trends

Process research and development, and scale-up in particular, have come a long way since the early days. The original practice of extrapolating from laboratory scale through several intermediate stages to the industrial-scale plant has for a large part been replaced by the miniplant approach. This approach saves both time and money, which are the primary factors in the profitability and competitiveness of a process. Current trends in research and development are:

- Increased use of flow reactor systems instead of batch reactors ("flow chemistry"). This applies to process development and industrial plants.
- Increased use of the miniplant technique combined with mathematical simulation, skipping the pilot stage, resulting in increased scale-up factors.
- Miniaturization of research equipment, especially in catalyst testing. This can extend to "lab-on-a-chip". Application of high-throughput experimental procedures, heavily based on robotics and automation. Robotized synthesis programs have proven to be successful in the development of drugs. They are referred to as combinatorial methods. More recently, it has been realized that in the synthesis of chemicals (e.g., ligands, fine chemicals) combinatorial methods are also powerful. In process development they have large potential as well. One can think of catalyst development, but also of optimization studies. Many companies have started this type of research, for example, HTE, Germany, and Avantium, The Netherlands.
- Development and use of expert (knowledge-based) systems[2]. These are already available for simple systems, such as the optimization of unit operations. Other systems for which expert systems might be used in the (far) future are catalyst development and optimization of entire plants.
- In the case of short-lived consumer products (dyes, detergents, pharmaceuticals, etc.), which are mostly produced in batch processes, Multiproduct or Multipurpose Plants (MPPs) (Section 12.4) are used. However, there is a trend towards dedicated plants using continuous processes, due to the increased use of catalysis and higher quality demands. In addition, continuous-flow microreactor technology is increasingly considered, also for use in MPPs.
- Gaining increased insight into the relation between variability of process conditions and variability in product quality. For instance, polymers need to be extremely pure; a required purity of 99.9999% is not uncommon. Small fluctuations in process conditions may well cause the purity to fall below this value.
- Finding novel routes in bulk chemicals production. For example, steam cracking for the production of light alkenes (Chapter 4) in fact is a crude technology associated with low selectivity and large consumption of exergy. It is expected that selective catalytic processes will replace steam cracking, but currently large investments are done to erect/revamp these "classical units".
- Process Intensification (Chapter 14) will be an important goal in new plant designs. Multifunctional reactors will be part of that.
- Another development that might become important is the trend towards pipe-less plants [2] in the production of fine chemicals and pharmaceuticals (Chapter 12). Conventionally, reactants and products flow from one piece of equipment to the following through pipelines. Cleaning these pipelines between campaigns is time consuming and requires large amounts of solvents. It is conceivable that in many cases it is more efficient to load reactant or product mixtures in vessels, which can serve as mixers or reactors and can be transported between specialized stations. These stations are designed for the various unit operations such as mixing, reaction, distillation, filtration, and so on. One of the advantages of pipe-less plants is their high flexibility. In addition, they are particularly suitable for processing solid materials and slurries, which are difficult to transport through pipelines.

---

[2] Expert systems: The difference between simulation programs and expert systems is that in the former modeling is rigidly embodied by an algorithm, whereas in the latter knowledge of the model is stored in the knowledge base. This knowledge not only consists of facts and relations between facts but also of a heuristic method for storage and acquisition of this information. When a human expert is faced with choosing from a large number of alternatives, (s)he eliminates those which seem unfruitful to find the optimal solution, using heuristic knowledge (rules of thumb), gained by experience in other similar situations. An expert system mimics the human expert.

These trends all promote the saving of time and money. Other future developments will be driven by environmental legislation, safety, and the development of the raw materials market. The latter includes C_1 chemistry, renewable raw materials (biomass), the development of technology based on photons (from the sun), and the usage of waste streams for the production of useful products. An explicit drive for R&D in the chemical industry will be the achievement of a sustainable society.

These trends will deeply influence process development and will stimulate the development of real breakthroughs in the chemical industry. It is the right time to be a chemical engineer or industrial chemist!

References

[1] Stankiewicz, A.I. and Moulijn, J.A. (2000) Process intensification: transforming chemical engineering. *Chemical Engineering Progress*, **96**, 22–33.

[2] Cybulski, A., Moulijn, J.A., Sharma, M.M., and Sheldon, R.A. (2001) *Fine Chemicals Manufacture – Technology and Engineering*. Elsevier Science, Amsterdam, The Netherlands, pp. 551.

[3] Vogel, H. (1992) Process development, in *Ullmann's Encyclopedia of Industrial Chemistry*, 5th edn, vol. **B4** (W. Gerhartz). VCH, Weinheim, pp. 438–475.

[4] Douglas, J.M. (1988) *Conceptual Design of Chemical Processes*. McGraw-Hill, New York.

[5] Trambouze, P., Van Landeghem, H., and Wauquier, J.P. (1988) *Chemical Reactors, Design/Engineering/Operation*. Gulf Publishing Company, Houston, TX.

[6] Krekel, J. and Siekmann, G. (1985) Die Rolle des Experiments in der Verfahrensentwicklung. *Chemie Ingenieur Technik*, **57**, 511–519.

[7] Den Hollander, M.A., Makkee, M., and Moulijn, J.A. (1998) Coke formation during fluid catalytic cracking studied with the microriser. *Catalysis Today*, **46**, 27–35.

[8] Bisio, A. and Kabel, R.L. (1985) *Scale-Up of Chemical Processes*. John Wiley & Sons, Inc., New York.

[9] Van Damme, P.S., Froment, G.F., and Balthasar, W.B. (1981) Scaling up of naphtha cracking coils. *Industrial and Engineering Chemistry Process Design and Development*, **20**, 366–376.

[10] Weekman, V.W. and Nace, D.M. (1970) Kinetics of catalytic cracking selectivity in fixed, moving, and fluid bed reactors. *AIChE Journal*, **16**, 397.

[11] Jacob, S.M., Gross, B., Voltz, S.E., and Weekman, V.W. (1976) Lumping and reaction scheme for catalytic cracking. *AIChE Journal*, **22**, 701–713.

[12] Sie, S.T. and Krishna, R. (1998) Process development and scale up: I. Process development strategy and methodology. *Reviews in Chemical Engineering*, **14**, 47–88.

[13] Sie, S.T. and Krishna, R. (1998) Process development and scale up: III. Scale-up and scale-down of trickle bed processes. *Reviews in Chemical Engineering*, **14**, 203–252.

[14] van Herk, D., Kreutzer, M.T., Makkee, M., and Moulijn, J.A. (2005) Scaling down trickle bed reactors. *Catalysis Today*, **106**, 227–232.

[15] van Herk, D., Castaño, P., Makkee, M., Moulijn, J.A., and Kreutzer, M.T. (2009) Catalyst testing in a multiple-parallel, gas-liquid, powder-packed bed microreactor. *Applied Catalysis A*, **365**, 199–206.

[16] Fleming, R. (1958) *Scale-Up in Practice*. Reinhold Publishing Corporation, New York.

[17] Crowl, D.A. and Louvar, J.F. (1990) *Chemical Process Safety: Fundamentals With Applications*. Prentice Hall, Englewood Cliffs, NY.

[18] Bryce, A. (1999) Bhopal Disaster Spurs U.S. Industry, Legislative Action. United States Chemical Safety and Hazard Investigation Board.

[19] Marrs, G.P., Lees, F.P., Barton, J., and Scilly, N. (1989) Overpressure protection of batch chemical reactors. *Chemical Engineering Research & Design*, **67**, 381–406.

[20] Prugh, R.W. (1996) Plant safety, in *Kirk-Othmer Encyclopedia of Chemical Technology*, 4th edn, vol. **19** (eds J.I. Kroschwitz and M. Howe-Grant). John Wiley & Sons, Inc., New York, pp. 190–225.

[21] Sinnot, R.K. (1996) *Coulson and Richardson's Chemical Engineering*, 2nd edn. Butterworth-Heinemann, Oxford.

[22] Musser, M.T. (1987) Cyclohexanol and cyclohexanone, in *Ullmann's Encyclopedia of Industrial Chemistry*, 5th edn, vol. **A8** (ed. W. Gerhartz). VCH, Weinheim, pp. 217–226.

[23] Kletz, T.A. (1985) *What Went Wrong? Case Histories of Process Plant Disasters*. Gulf Publishing Company, Houston, TX.

[24] Cardillo, P., Girelli, A., and Ferraiolo, G. (1984) The Seveso case and the safety problem in the production of 2,4,5-trichlorophenol. *Journal of Hazardous Materials*, **9**, 221–234.

[25] Sharrat, P.N. (1997) *Handbook of Batch Process Design*. Blackie, London.

[26] Ulrich, G.D. and Vasudevan, P.T. (2004) *A Guide to Chemical Engineering Process Design and Economics*, 2nd edn. Process Publishing.

[27] Seider, W.D., Seader, J.D., and Lewin, D.R. (1999) *Process Design Principles. Synthesis, Analysis, and Evaluation*. John Wiley & Sons, Inc., New York.

[28] Peters, M.S. and Timmerhaus, K.D. (1991) *Plant Design and Economics for Chemical Engineers*, 4th edn. McGraw-Hill, New York.

[29] Viola, J.L. Jr. (1981) Estimate capital costs via a new, shortcut method. *Chemical Engineering*, **74**, 80–86.

[30] Allen, D.H. and Page, R.C. (1975) Revised technique for predesign cost estimating. *Chemical Engineering*, **68**, 142–150.

[31] Cevidalli, G. and Zaidman, B. (1980) Evaluate research projects rapidly. *Chemical Engineering*, **73**, 145–152.

[32] Ward, Th.J. (1984) Predesign estimating of plant capital costs. *Chemical Engineering*, **77**, 121–124.

[33] Zevnik, F.C. and Buchanan, R.L. (1963) Generalized correlation of process investment. *Chemical Engineering Progress*, **59**, 70–77.

[34] IHS (2012) IHS CERA: Capital Costs. http://www.ihs.com/info/cera/ihsindexes/index.aspx (last accessed 22 December 2012).

[35] Couper, J.R. and Rader, W.H. (1986) *Applied Finance and Economic Analysis for Scientists and Engineers*. Van Nostrand-Reinhold, New York.

[36] El-Halwagi, M.M. (2012) Overview of process economics, in *Sustainable Design Through Process Integration*. Butterworth-Heinemann, Oxford, pp. 15–61.

[37] Lang, H.J. (1948) Simplified approach to preliminary cost estimates. *Chemical Engineering*, **55**, 112–113.

[38] Guthrie, K.M. (2012) *Process Plant Estimating, Evaluation, and Control*. Craftsman, Solano Beach, CA.

[39] Guthrie, K.M. (1969) Data and techniques for preliminary capital cost estimating. *Chemical Engineering*, **76**, 114–142.

[40] Wessel, H.E. (1952) New graph correlates operating labor data for chemical processes. *Chemical Engineering*, **59**, 209–210.

General Literature

McKetta, J.J. and Cunningham, W.A. (eds) (1976) *Encyclopedia of Chemical Processing and Design*. Marcel Dekker, New York.

Magazines

Chemical Engineering Magazine
European Chemical News
The Oil and Gas Journal
Chemical Engineering Progress
Chemical & Engineering News
Chemical Week

Appendix A

Chemical Industry – Figures

Table A.1 *Chemical sales (only) of the chemical industry top 20 in the United States (10^9 US $).*

Company	2011	2010
Dow Chemical	60.0	52.9
ExxonMobil	41.9	34.3
DuPont	34.8	29.4
Chevron Phillips	13.9	10.5
PPG Industries (glass)	13.8	12.3
Praxair	11.3	10.0
Huntsman Corp.	11.2	8.8
Mosaic (a.o. phosphates)	9.9	5.2
Air Products	9.7	8.4
Momentive (specialties)	7.8	7.3
Eastman Chemical	7.2	5.6
Celanese	6.7	5.7
Dow Corning	6.4	5.9
Lubrizol	6.1	5.3
CP Industries (paint)	6.1	2.8
Styron	6.0	4.9
Honeywell	5.7	4.6
Occidental Petroleum (exploration)	4.8	3.8
Ecolab (food)	4.6	4.2
Ashland (oil)	4.5	3.9

Source: *Chemical & Engineering News*, 14 May 2012.

Chemical Process Technology, Second Edition. Jacob A. Moulijn, Michiel Makkee, and Annelies E. van Diepen.
© 2013 John Wiley & Sons, Ltd. Published 2013 by John Wiley & Sons, Ltd.

Table A.2 Chemical sales (only) of the chemical industry top 20 in the world (10^9 US \$).

Company	2010	2009
BASF	70.4	45.8
Dow Chemical (Germany)	53.7	43.2
Sinopec	47.4	23.6
ExxonMobil	35.5	24.1
Royal Dutch Shell	35.2	19.9
Formosa Plastics	34.7	24.3
SABIC	33.7	18.2
DuPont	31.3	24.9
LyondellBasell Industries	27.7	17.0
Mitsubishi Chemical	26.0	14.1
Ineos Group Holding	24.8	16.8
Total	23.2	18.8
Bayer	22.5	17.8
AkzoNobel	19.4	18.4
Mitsui Chemicals	18.5	12.1
Sumitomo Chemicals	17.4	13.2
Evonik Industries	17.2	12.3
LG Chem	17.1	12.8
Air Liquide	16.9	14.2
Reliance Industries	14.8	12.7

Source: *Chemical & Engineering News*, 25 July 2011.

Table A.3 Production of some of the largest volume chemicals in the United States (10^3 t)[a].

Chemical[b]	2010	1998	1988
Sulfuric Acid (I)	32 511	48 340	43 264
Ethene (O)	23 975	23 614	16 875
Lime (M)	18 000	22 852	17 326
Propene (O)	14 085	12 979	9627
Sodium carbonate (M)	10 000	11 538	9034
Chlorine (I)	9732	13 050	11 438
Polyethene, LD and LLD (P)	9312	6715	4716
Ethene dichloride (O)	8810	11140[c]	5909
Ammonia (I)	8300	20 070	17 091
Polypropene (P)	7826	6271	3299
Polyethene, HD (P)	7660	5864	3810
Sodium hydroxide (I)	7520	11 681	10 702
Nitric acid (I)	7480	9524	8119
Ammonium nitrate (I)	6878	8758	7625
Polyvinylchloride (P)	6358	6578	3787
Benzene (O)	6015	1011	729
Ethylbenzene (O)	4220	5743[c]	4504
Styrene (O)	4102	5166	4075
Phosphoric acid (I)	3556	14 626	11 846
Cumene (O)	3478	2045	2021
Ethene oxide (O)	2662	3692	2700
Urea (O)	2320	7984	7179
Polystyrene[d] (P)	2293	4284	3517
Butadiene (O)	1580	1844	1437
Polyesters (P)	1088	2006	749

[a]Sources: *Chemical & Engineering News*, 28 June 1999; 4 July 2011; http://minerals.usgs.gov/minerals/pubs/mcs2011/mcs2011.pdf.

[b]I = inorganics; M = minerals; O = organics; P = plastics.

[c]reporting method changed in 1996.

[d]Includes styrene–acrylonitrile, acrylonitrile–butadiene–styrene, and other styrene polymers.

Table A.4 *Indication of prices of some bulk chemicals in Europe and the United States (US $/t), June 2012.*

Chemical	Europe, spot	Europe, contract	US, spot	US, contract
Ethene	845–855	1205	925	1015
Propene	830–840	1105	1145–1210	1145
Butadiene	2000–2050	1850	1760–1805	1985–2160
Benzene	1190–1220	1020	1255–1270	1225
Toluene	1090–1110	na	1125–1160	na
p-Xylene	1310–1330	1090	1330–1340	1695
o-Xylene	1390–1410	1160	1390–1430	1410
Styrene	1300–1330	1347	1390–1410	1455
Methanol	310–330	340	365–370	440
MTBE	1110–1130	na	na	na
Ammonia	600–690	na	690	na

na = not applicable/not available.
Sources: *Chemical Week* Price Watch; ICIS Pricing.

Appendix B
Main Symbols Used in Flow Schemes

B.1   Reactors and Other Vessels

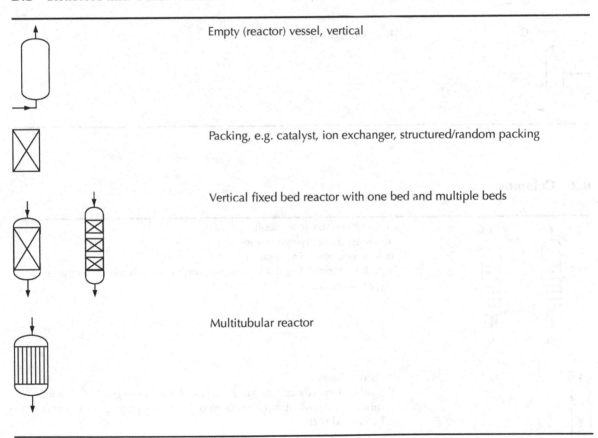

Empty (reactor) vessel, vertical

Packing, e.g. catalyst, ion exchanger, structured/random packing

Vertical fixed bed reactor with one bed and multiple beds

Multitubular reactor

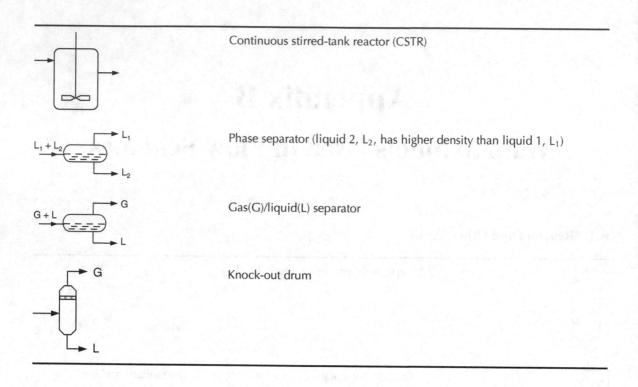

Continuous stirred-tank reactor (CSTR)

Phase separator (liquid 2, L_2, has higher density than liquid 1, L_1)

Gas(G)/liquid(L) separator

Knock-out drum

B.2   Columns

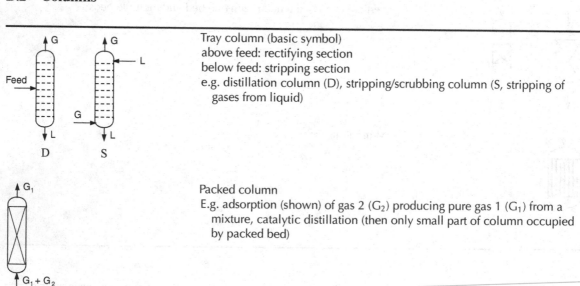

Tray column (basic symbol)
above feed: rectifying section
below feed: stripping section
e.g. distillation column (D), stripping/scrubbing column (S, stripping of
   gases from liquid)

Packed column
E.g. adsorption (shown) of gas 2 (G_2) producing pure gas 1 (G_1) from a
   mixture, catalytic distillation (then only small part of column occupied
   by packed bed)

B.3   Heat Transfer Equipment

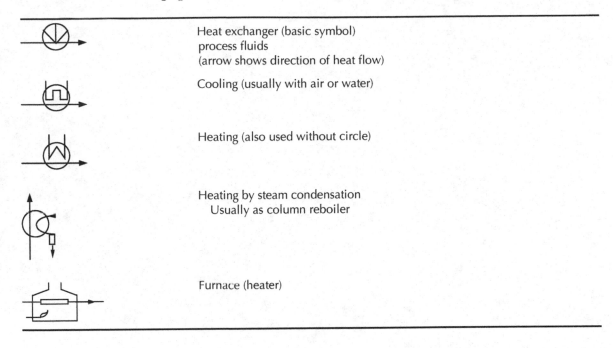

Heat exchanger (basic symbol)
process fluids
(arrow shows direction of heat flow)

Cooling (usually with air or water)

Heating (also used without circle)

Heating by steam condensation
 Usually as column reboiler

Furnace (heater)

B.3.1   Miscellaneous

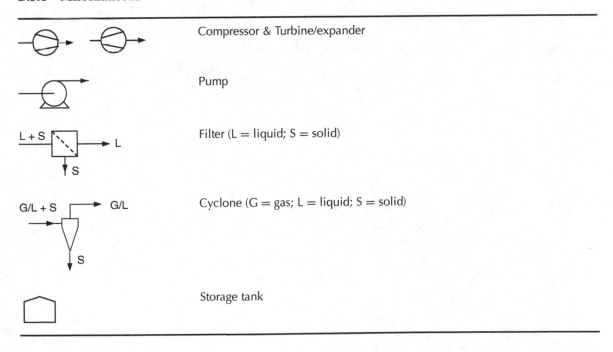

Compressor & Turbine/expander

Pump

Filter (L = liquid; S = solid)

Cyclone (G = gas; L = liquid; S = solid)

Storage tank

Index